Analysis and Control of Complex Nonlinear Processes

in Physics, Chemistry and Biology

WORLD SCIENTIFIC LECTURE NOTES IN COMPLEX SYSTEMS

AIMS AND SCOPE

The aim of this new interdisciplinary series is to promote the exchange of information between scientists working in different fields, who are involved in the study of complex systems, and to foster education and training of young scientists entering this rapidly developing research area.

The scope of the series is broad and will include: Statistical physics of large nonequilibrium systems; problems of nonlinear pattern formation in chemistry; complex organization of intracellular processes and biochemical networks of a living cell; various aspects of cell-to-cell communication; behaviour of bacterial colonies; neural networks; functioning and organization of animal populations and large ecological systems; modeling complex social phenomena; applications of statistical mechanics to studies of economics and financial markets; multi-agent robotics and collective intelligence; the emergence and evolution of large-scale communication networks; general mathematical studies of complex cooperative behaviour in large systems.

Published

Vol. 1 Nonlinear Dynamics: From Lasers to Butterflies
Vol. 2 Emergence of Dynamical Order: Synchronization Phenomena in Complex Systems
Vol. 3 Networks of Interacting Machines
Vol. 4 Lecture Notes on Turbulence and Coherent Structures in Fluids, Plasmas and Nonlinear Media

World Scientific Lecture Notes
in Complex Systems – Vol. 5

editors

L. Schimansky-Geier
Humboldt University, Berlin, Germany

B. Fiedler
Free University Berlin, Germany

J. Kurths
University of Potsdam, Germany

E. Schöll
Technical University Berlin, Germany

Analysis and Control of Complex Nonlinear Processes

in Physics, Chemistry and Biology

World Scientific

NEW JERSEY • LONDON • SINGAPORE • BEIJING • SHANGHAI • HONG KONG • TAIPEI • CHENNAI

Published by

World Scientific Publishing Co. Pte. Ltd.
5 Toh Tuck Link, Singapore 596224
USA office: 27 Warren Street, Suite 401-402, Hackensack, NJ 07601
UK office: 57 Shelton Street, Covent Garden, London WC2H 9HE

British Library Cataloguing-in-Publication Data
A catalogue record for this book is available from the British Library.

World Scientific Lecture Notes in Complex Systems — Vol. 5
ANALYSIS AND CONTROL OF COMPLEX NONLINEAR PROCESSES IN PHYSICS, CHEMISTRY AND BIOLOGY

ISBN-13 978-981-270-583-9
ISBN-10 981-270-583-X

Printed by FuIsland Offset Printing (S) Pte Ltd, Singapore

Dedicated to Werner Ebeling and Gerhard Ertl

Preface

Nonlinear dynamics of complex processes is an active research field with large numbers of publications in basic research and broad applications from diverse fields of science. Nonlinear dynamics as manifested by deterministic and stochastic evolution models of complex behaviour has entered statistical physics, physical chemistry, biophysics, geophysics, astrophysics, theoretical ecology, semiconductor physics and -optics etc. This research has induced a new terminology in science connected with new questions, problems, solutions and methods. New scenarios have emerged for spatio-temporal structures in dynamical systems far from equilibrium. Their analysis and possible control are intriguing and challenging aspects of the current research.

The duality of fundamental and applied research is a focal point of its main attractivity and fascination. Basic topics and foundations are always linked to concrete and precise examples. Models and measurements of complex nonlinear processes evoke and provoke new fundamental questions and diversify and broaden the mathematical concepts and tools. In return, new mathematical approaches to modeling and analysis enlarge the scope and efficiency of applied research.

Fundamental research on complex physical systems, as well as applied interdisciplinary investigations of nonlinear dynamics, are valued highly in research programs and education of universities and research institutes in Berlin and Potsdam (Germany). Initiated by Werner Ebeling (Humboldt-University at Berlin) and Gerhard Ertl (Fritz Haber Institute Berlin) the collaborative research has led to the foundation of the collaborative research center 555 of the German research society (DFG), 9 years ago. This center has stimulated cooperations between the different research groups in the region.

Recent progress on nonlinear complex processes in Berlin and Potsdam is presented in this volume. The topics of our research encompass

- spatio-temporal pattern formation in semiconductors, geophysical, astrophysical, chemical and electrochemical, biological and ecological system,

- the fundamental understanding of elementary structures and their interactions in deterministic and stochastic dynamics in two and three spatial dimensions,

- the role of global and nonlocal feedback and external forces in control of spatio-temporal pattern formation, with applications, e.g., to nonlinear chemical systems and semiconductor nanostructures,

- stochastic and coherence resonance and noise induced behaviour as manifestation of the constructive role of noise in supporting the existence of spatial and temporal structures with examples from semiconductors, chemical and neuronal systems and climatic dynamics,

-the microscopic stochastic modeling of dynamic cellular and intracellular processes as the propagation of Ca-waves above membranes and the action by neurons,

- the understanding of chaos synchronization and of stochastic synchronization,

- the manifold role and the many applications of delayed feedback in chaos control,

- the evidence of synchronization phenomena in population-, climatic and oceanic dynamics as well as their importance in optoelectronic devices.

All chapters were written in the spirit to convince students and outsiders of the attractivity of studying complex nonlinear processes as well as to provide researchers of the field with new details and results. Thus all contributions contain longer introductions of the topic followed by new findings of our research.

The editors are much indebted to David A. Strehober for his helpful technical assistance in the preparation of the book. Further on we acknowledge the proof reading of several parts, as well as collecting of the subject index by Simon Fugmann, Felix Müller, Michael Rading and Tilo Schwalger. The German research foundation (DFG-Sfb 555 "Complex Nonlinear Processes") has generously supported our research during the last 9 years, and the edition of this book. Special thanks are going to Udo Erdmann, the secretary of Sfb 555. We are also grateful to the editor of the World Scientific Lecture Notes in Complex Systems series, Alexander S. Mikhailov, and to Senior Editor Lakshmi Narayan (Ms) for their help and congenial processing of the edition.

Berlin and Potsdam, December 7, 2006

L. Schimansky-Geier, B. Fiedler, J. Kurths and E. Schöll

Contents

Chapter 1

Noise-induced effects in excitable systems with local and global coupling

Xaver R. Sailer[1], Valentina Beato[2], Lutz Schimansky-Geier[1,*] and Harald Engel[2,+]

[1]*Institut für Physik, Humboldt Universität zu Berlin, Newtonstrasse 15, D-12489 Berlin, Germany*

[2]*Institut für Theoretische Physik, Technische Universität Berlin, Hardenbergstraße 36, D–10623 Berlin, Germany*

[*]*alsg@physik.hu-berlin.de,* [+]*h.engel@physik.tu-berlin.de*

Contents

1.1. Introduction

Excitability is a well established phenomenon with applications in various fields of science. Systems that, looked at superficially, have little in common, like neurons [28], chemical reactions [4], wildfires [3], lasers [52], and many others share this property. The variety and diversity of the many examples demonstrates the importance of the concept of excitability.

Fluctuations are ubiquitous in nature. In excitable systems they are considered to be of uttermost importance since they are able to "activate" these systems. Especially if the magnitude of the fluctuations becomes comparable to activation barrier of the system their influence will be determinant. This is true for example in neurons where fluctuating input and fluctuating membrane conductances are commonly agreed upon to functionally impact on their dynamics.

We investigate the influence of fluctuations on excitable dynamics experimentally and theoretically. This study is organized as follows. In section 1.2 we give the general concept of excitability. The most important model for our work, the FitzHugh-Nagumo model, is introduced and its key properties are given. In section 1.3 we treat the analytical techniques that we used. Section 1.4 centers around single individual elements. An experiment on a chemical excitable medium, the light-sensitive Belousov-Zhabotinsky reaction is introduced. We show the setup and explain the experimental implementation. The results from an experiment with fluctuating excitability parameter are given. Results from investigations of coupled systems are given in section 1.5. Globally as well as locally coupled systems are discussed.

1.2. Excitability: What is it and how can we model it?

1.2.1. *General concept*

Consider a dynamical system with a stable fixed point. Since the fixed point is stable a small perturbation (or stimulus) from the fixed point will decay. If a perturbation exceeding a certain threshold is applied, an excitable system responds not with a decay of the perturbation but with a large excursion in phase space until finally re-approaching the stable fixed point. The form and duration of the excitation is scarcely affected by the exact form of the perturbation. Until the system is close to the fixed point again, a new perturbation does hardly affect the system. The corresponding time is called the refractory time. A graphical representation of the scheme is given in Fig. 1.1. In a neuron, for example, the stable fixed point

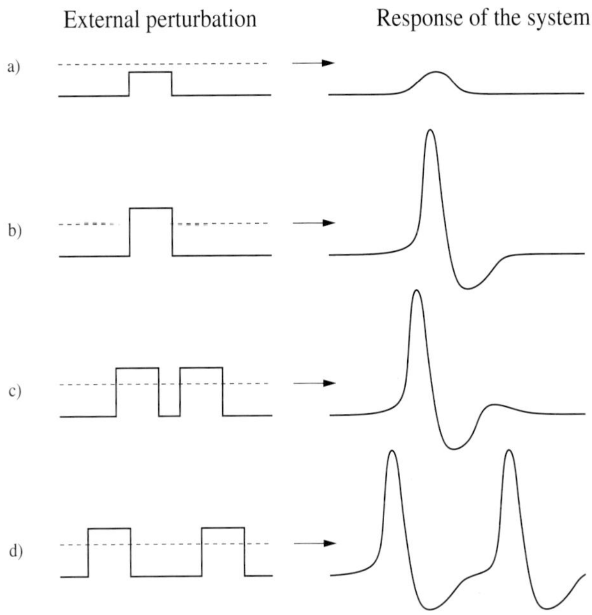

Fig. 1.1. Response (right) of an excitable system to different stimuli (left). The system has a stable fixed point. Panel a): A sub-threshold perturbation generates a small system response. Panel b): A super-threshold perturbation initiates an excitation loop. Panel c): The system can not be excited by a super-threshold perturbation applied during the refractory state. Panel d): Two successive super-threshold perturbations generate excitations only if both are applied to the system in the rest state.

could represent a steady potential drop over the cell membrane. A super-threshold perturbation (for example via input from other neurons) leads to a large electric response of the neuron, a so-called action potential or spike. After some time of insensitivity to new signals the potential drop returns to its steady value.

The precise time from which on a new perturbation can excite the system depends on the stimulus. One therefore usually talks about the relative refractory time.

An excitable system is sometimes modeled to posses three distinct states: The state in which the system is in the vicinity of the fixed point is called the rest state, the state just after excitation is called the firing state, and the state just before the system is close to the fixed point again is called the refractory state. The firing and the refractory state differ by a high (firing) and a low (refractory) value of the activation variable which has a major influence on pattern formation in coupled excitable systems (For further details see project *A*4.).

1.2.2. *A simple model - the FitzHugh–Nagumo system*

In this part we introduce the FitzHugh-Nagumo (FHN) system. Originally derived from the Hodgkin-Huxley model for the giant nerve fiber of a squid, it has extended its application beyond neuron dynamics to all kinds of excitable systems and has by now become an archetype model for systems exhibiting excitability. [1, 14, 36]. First, we neglect coupling and fluctuations and use the following form:

$$\frac{dx}{dt} = \frac{1}{\epsilon}\left(x - \alpha x^3 - y\right) \tag{1.1}$$
$$\frac{dy}{dt} = \gamma x - y + b.$$

Here ϵ, α, and b are real, positive parameters. ϵ is chosen small in order to guarantee a clear timescale separation between the the fast x-variable (activator) and the slow y-variable (inhibitor). The variables α and b determine the position of the so-called nullclines, the two functions $y(x)$ that are determined by setting time derivatives $dx/dt = 0$ and $dy/dt = 0$. Depending on the parameters the FHN system has different dynamical regimes. Fig. 1.2 shows phase space portraits together with the nullclines and timeseries for three qualitatively different cases.

In the upper row we see the excitable regime. The solid lines represent the nullclines of the system, the dashed line a typical trajectory. Each dash represents a fixed time interval, i.e. where the system moves faster through phase space the dashes become longer. The system possesses one fixed point (intersection of the nullclines) which is stable. Small perturbations decay. A super-threshold perturbation leads to a large response (spike) after which the system returns to the fixed point. After that a new perturbation is possible if the system from outside is brought again over the threshold.

In the middle row the oscillatory regime is illustrated. The systems exhibits continuous oscillations. Perturbations have at this stage little influence on the dynamics.

In the lower row we find the bistable parameter regime. The system possesses two stable fixed points. A first overcritical external perturbation

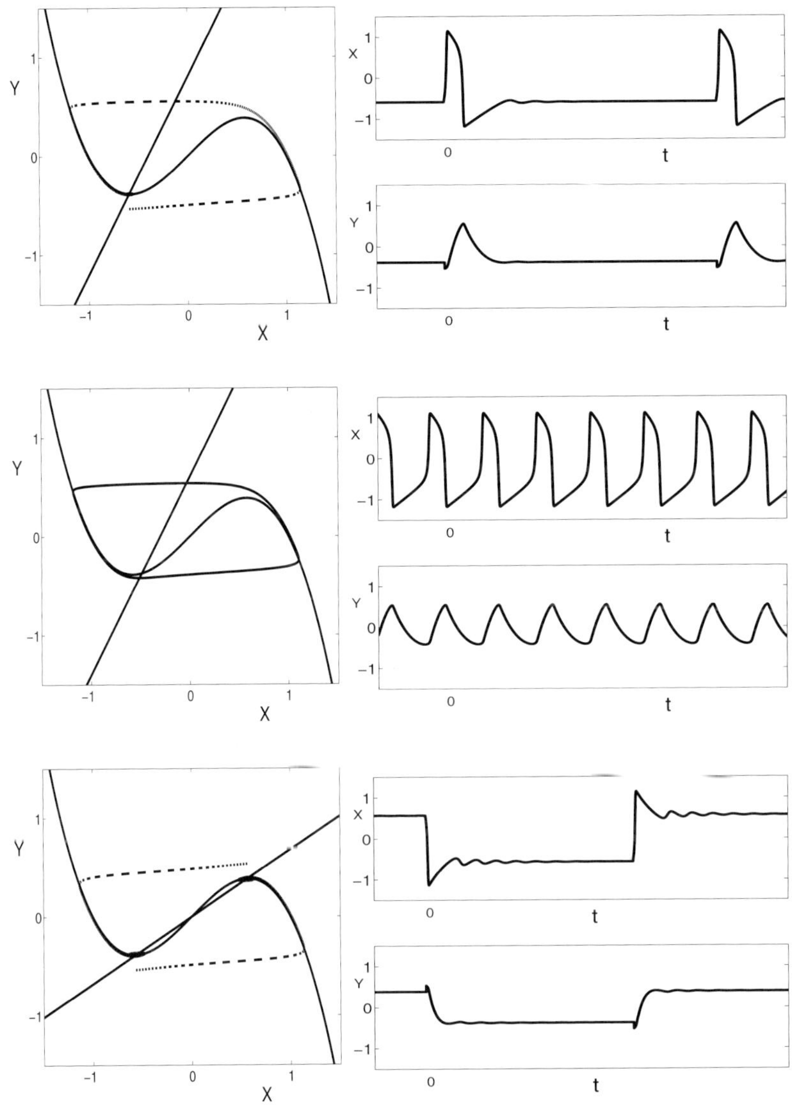

Fig. 1.2. Depending on the parameters the FHN can either exhibit excitable (upper row), oscillatory (middle row) or bistable (lower row) kinetics. $\epsilon = 0.05$.

from one of them leads the system into the basin of attraction of the other one. The basin of attraction of a fixed point is that part of phase space from which systems without further perturbation evolve towards this point.

Once the system is in the vicinity of this second fixed point a new external perturbation leads to a new excursion which ends at the initial point.

The transition from the oscillatory to the excitable parameter regime can, for example, be achieved via an increase of the parameter b. The system then undergoes a Hopf bifurcation and the sole fixed point looses stability. The location of the Hopf bifurcation in parameter space also depends on the parameter ϵ, which governs the separation of the timescales. An illustration for different ϵ-values is shown in Fig. 1.3. With increasing ϵ

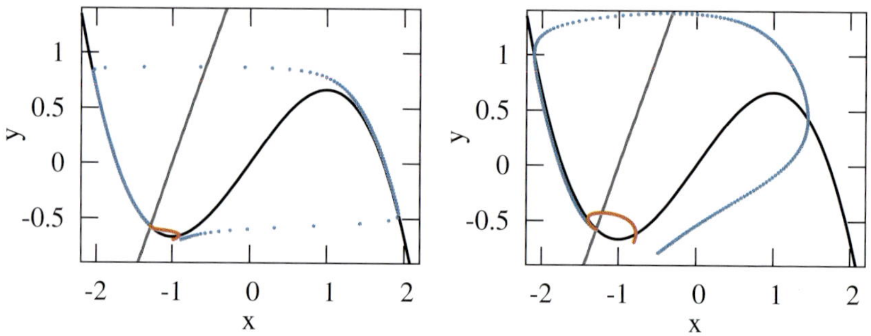

Fig. 1.3. Trajectories of the FitzHugh-Nagumo model for sub- and super-threshold perturbations (red and blue curves respectively). Each point plotted at constant time intervals $t_i = t_{i-1} + \Delta t$ with $\Delta t = 0.005$. Kinetic parameters $b = 1.4$, $\alpha = 1/3$ and $\gamma = 2$. Left: high time scale separation for $\epsilon = 0.01$. Right: low time scale separation for $\epsilon = 0.1$.

the separation of the timescales of the fast activator and the slow inhibitor weakens. In Fig. 1.3 we show typical trajectories for sub- and super-threshold perturbations (respectively red and blue curves) for two different values of the parameter ϵ. In the left panel (high timescale separation) a small perturbation $(\delta x, \delta x)$ to the stable point (x_0, y_0) triggers an excitation loop. It consists in a quick increase of the activator until the right branch of the x-nullcline is reached, followed by a slow inhibitor production: the excited state. After this, the activator x decreases quickly to the left branch of the x-nullcline, and again moves slowly along it towards the fixed point (x_0, y_0): The system is in the refractory state. On the right panel we show a case with low time scale separation between the two system variables. The same perturbation $(\delta x, \delta y)$ now fails to excite the system. The excitation loop in this case does not follow the x-nullcline. As a consequence, the excited and refractory states are not clearly defined anymore and the system responds only to strong perturbations.

1.2.3. *The Oregonator model for the light-sensitive Belousov-Zhabotinsky reaction*

The model that we take into account has been firstly proposed by H. J. Krug and coworkers in 1990 [30], to properly account for the photochemically-induced production of inhibitor bromide in the Belousov-Zhabotinsky reaction (BZ) catalyzed with the ruthenium complex $Ru(bpy)_3^{2+}$ [31, 32]. The Oregonator model was proposed in 1974 [16] on the basis Tyson-Fife reduction of the more complicated Field-Köros-Noyes mechanism [15] for the BZ reaction, the following modified model has been derived:

$$\begin{aligned}
\frac{\partial u}{\partial t} &= \frac{1}{\epsilon}\left[u - u^2 - w\cdot(u-q)\right] \\
\frac{\partial v}{\partial t} &= u - v \\
\frac{\partial w}{\partial t} &= \frac{1}{\epsilon'}\left[fv - w\cdot(u+q) + \phi\right].
\end{aligned} \tag{1.2}$$

The variable u describes the local concentration of bromous acid $HBrO_2$, the variable v the oxidized form of the catalyst $Ru(bpy)_3^{3+}$, and w describes the bromide concentration. Here the parameter ϕ represents the light intensity, and the photochemically-induced production of bromide is assumed to be linearly dependent on it, $d[Br^-]/dt \propto \phi$ [32]. ϵ, ϵ' and q are scaling parameters, and f is a stoichiometric constant [47]. This model can be reduced to the two-component one by adiabatic elimination of the fast variable w (in the limit $\epsilon' << \epsilon$) [47]. In this case one gets the following two-component version of the Oregonator kinetics

$$\begin{aligned}
\frac{du}{dt} &= \frac{1}{\epsilon}\left[u - u^2 - (fv+\phi)\cdot\frac{u-q}{u+q}\right] \\
\frac{dv}{dt} &= u - v.
\end{aligned} \tag{1.3}$$

From the recipe concentrations we obtain $\epsilon = 0.0766$ and $q = 0.002$, f is chosen equal to 1.4.

When ϵ, f, and q are kept fixed, ϕ controls the kinetics in the same way as b does in the FitzHugh-Nagumo model: For small ϕ the kinetics is oscillatory and for $\phi > \phi_{HB} = 4.43 \cdot 10^{-3}$, it becomes excitable via a super-critical Hopf bifurcation.

1.3. Stochastic methods

Excitable systems as considered here are many particle systems far from equilibrium. Hence variables as voltage drop (neurons), light intensity (lasers) or densities (chemical reactions) are always subject to noise and fluctations. Their sources might be of quite different origin, first the thermal motion of the molecules, the discreteness of chemical events and the quantum uncertainness create some unavoidable internal fluctuations. But in excitable systems, more importantly, the crucial role is played by external sources of fluctuations which act always in nonequilibrium and are not counterbalanced by dissipative forces. Hence their intensity and correlation times and lengths can be considered as independent variables and, subsequently, as new control parameters of the nonlinear dynamics.

Normally they can be controlled from outside as, via a random light illumination (chemical reactions) [26] or the pump light (lasers) [49].

The inclusion of fluctuations in the description of nonlinear systems is done by two approaches [50]. On the one hand side one adds fluctuating sources in the nonlinear dynamics, transforming thus the differential equations into stochastic differential equations. The second way is the consideration of probability densities for the considered variables and the formulation of their evolution laws. Both concepts are introduced shortly in the next two subsections.

We underline that the usage of stochastic methods in many particle physics was initiated by Albert Einstein in 1905 working on heavy particles immersed in liquids and which are thus permanently agitated by the molecules of the surrounding liquid. Whereas Einstein formulated an evolution law for the probability $P(r,t)$ to find the particle in a certain position r at time t Paul Langevin formulated a stochastic equation of motion, i.e. a stochastic differential equation for the time dependent position $r(t)$ itself.

1.3.1. *Langevin equation*

Let $x(t)$ be the temporal variable of the excitable system which is also subject to external noise. A corresponding differential equation for the time evolution of $x(t)$ is called a Langevin-equation and includes random parts. We specify to the situation where randomness is added linearly to modify the time derivative of $x(t)$, i.e.

$$\dot{x}(t) = f(x) + g(x)\xi(t)\,. \tag{1.4}$$

Therein the function $f(x)$ is called the deterministic part and stands for the dynamics as introduced previously when considering the FitzHugh Nagumo system or the Oregonator. The new term describes the influence of the randomness or noise and $\xi(t)$ is a stochastic process which requires further definition. It enters linearly and is weighted by the function $g(x)$. Hence, the increment dx during dt at time t gets a second, stochastic contribution being independent of the deterministic one. In case $g(x)$ is a constant the noise acts additively, otherwise it is called multiplicative (parametric) noise and the influence of the noise depends on the actual state $x(t)$ of the system.

The solution of eq. 1.4 depends on the sample of $\xi(t)$. Formally one can interprete the latter as a time dependent parameter and the variable $x(t)$ is found by integration over time. We note that integrals over $\xi(t)$ need a stochastic definition and are defined via the existence of the moments [50]. For this purpose the moments of $\xi(t)$ have to be given.

In practical applications one uses Gaussian sources $\xi(t)$ or the so called Markovian random telegraph process. For both the formulation of the mean and the correlation function is sufficient to define the stochastic process. Later on we will define the value support of $\xi(t)$ (Gaussian or dichotomic) and will give the mean and the correlation function, i.e.

$$\langle \xi(t) \rangle = 0 \tag{1.5}$$

and

$$\langle \xi(t)\xi(t+\tau) \rangle = K(\tau). \tag{1.6}$$

Here we reduced to stationary noise sources. Without loss of generality the mean is set to zero. For the later on considered types of noise this formulation is sufficient to obtain general answers for ensembles and their averages of the stochastic excitable system. Thus we can formulate evolution laws for the probability densities and the other moments. We note that the generalization to cases with more than one noise sources is straightforward and crosscorrelations between the noise source have to be defined.

1.3.2. *Stochastic processes: White and colored noises*

Next we consider several noise sources. They are Gaussian sources if their support of values is due to a Gaussian distribution. Contrary the dichotomic telegraph process assumes two values , i.e. $\xi_1 = \Delta$ and $\xi_2 = \Delta'$. We will always assume $\Delta' = -\Delta$.

The second classification of the noise classifies its temporal correlations. In case of white noise the noise is uncorrelated in time which corresponds to $K(\tau) = 2D\delta(\tau)$ being Dirac's δ-function. Here D scales the intensity of the noise. The power-spectrum of the noise is the Fourier transform of the correlation function. In case of the δ-function it is a constant and independent of the frequency what was the reason to call it white noise. All other noise sources with frequency-dependent power spectrum are thus colored.

The integral over the Gaussian white noise gives the Wiener process which stands for the trajectory of a Brownian particle. The integral during dt

$$dW(t) = \int_t^{t+dt} ds\,\xi(s) \tag{1.7}$$

is the increment in the position of a Brownian particle in this interval. It yields the basis of mathematical proofs. It is Gaussian distributed, the mean vanishes and increments at different times are independent. The variance increases linearly with dt and scales with the intensity $2D$.

$$\langle dW(t)^2 \rangle = 2Ddt\,. \tag{1.8}$$

In case of a Brownian particle D is the spatial diffusion coefficient.

The study of a Brownian particle suspended in a fluid lead also to the introduction of the exponentially correlated Ornstein-Uhlenbeck process [48], the only Markovian Gaussian non-white stochastic process [19, 22]. We present here the Langevin approach to this problem, hence we analyze the forces that act on a single Brownian particle. We suppose the particle having a mass m equal to unity, and we assume the force due to the hits with thermal activated molecules of the fluid to be a stochastic variable. Moreover, due to the viscosity of the fluid, a friction force proportional to the velocity of the particle has to be considered. All this yields the following equation

$$\frac{dv(t)}{dt} = -\gamma v(t) + F(t), \tag{1.9}$$

where γ is the friction constant. The random force $F(t)$ is supposed to be independent of the velocity $v(t)$ of the Brownian particle and to have zero mean. Moreover, the random force is supposed to be extremely rapidly varying compared to $v(t)$. Hence we assume that $F(t) = \xi$ and ξ is Gaussian white noise.

Integrating with respect to the time we get:

$$v(t) = v(0)e^{-\gamma t} + e^{-\gamma t} \int_0^t e^{\gamma s} F(s) ds. \tag{1.10}$$

We suppose the random force to be Gaussian, and a linear operator does not change this property. Thus the velocity $v(t)$ is Gaussian as well if the initial condition $v(0)$ is a random Gaussian variable independent of the random force. The mean value of the velocity reads

$$\langle v(t)\rangle = \langle v(0)\rangle e^{-\gamma t} + e^{-\gamma t}\int_0^t e^{\gamma s}\langle F(s)\rangle ds = \langle v(0)\rangle e^{-\gamma t}. \tag{1.11}$$

For the calculation of the correlation function of the velocity we exploit the assumption that

$$\langle F(t)F(s)\rangle = 2D\delta(t-s). \tag{1.12}$$

The correlation function of the velocity, considering $\Delta t > 0$, is given by

$$\begin{aligned}\langle v(t)v(t+\Delta t)\rangle &= e^{-\gamma(2t+\Delta t)}\langle v(0)^2\rangle \\ &\quad + \int_0^t\int_0^{t+\Delta t} e^{\gamma(t+s)}e^{\gamma(s'-t-\Delta t)}\langle F(s)F(s')\rangle ds ds' \\ &= e^{-\gamma(2t+\Delta t)}\langle v(0)^2\rangle + \frac{D}{\gamma}(e^{2\gamma t}-1)e^{-\gamma(2t+\Delta t)}. \end{aligned}\tag{1.13}$$

Hence the stationary correlation function, obtained for the limit $t \to +\infty$, is

$$\langle v(t)v(t+\Delta t)\rangle \to C(\Delta t) = \frac{D}{\gamma}e^{-\gamma\Delta t}. \tag{1.14}$$

The correlation time of the process is given by

$$\tau = \frac{1}{C(0)}\int_0^{+\infty} C(s)ds = \gamma^{-1}. \tag{1.15}$$

Another stochastic process which yields a non-vanishing correlation time is the dichotomous random telegraph process [20, 22]. It can be described by a phase according to

$$\eta_\pm(t) = \mathsf{sign}\left[\cos(\gamma_\pm(t))\right]\cdot\Delta, \tag{1.16}$$

where the phase increases at each random time t_i by an angle π. Thus the process has zero mean and variance Δ^2. Introducing the Heaviside step function θ we can write the phase as

$$\gamma_\pm(t) = \pi\sum_i \theta(t-t_i), \qquad t_i = t_{i-1} + t'. \tag{1.17}$$

We assume the random variable t' to be exponentially distributed, with distribution

$$p(t') = \frac{1}{\tau}\exp\left(-\frac{t'}{\tau}\right). \tag{1.18}$$

With this assumption the process expressed in Eq. 1.16 has the correlation function

$$\langle \eta_\pm(t)\eta_\pm(s)\rangle = \Delta^2 \cdot \exp\left(-\frac{|t-s|}{\tau}\right), \tag{1.19}$$

where τ is the correlation time of the process $\eta_\pm$ [20, 22]. Thus the random telegraph process, as well as the Ornstein-Uhlenbeck process shown before, presents an exponentially decaying correlation function.

1.3.3. *The Fokker-Planck equation*

In case of Gaussian white noise the probability density obeys a diffusion equation with a drift. In particular, the probability density is the conditioned average [50]

$$p(x,t|x_0,t_0) = \langle \delta(x - x(t))\rangle \tag{1.20}$$

that the sampled trajectories are started in x_0 at time t_0. Then starting from the Chapman-Kolmogorov equation for Markovian processes one finds

$$\frac{\partial P(x,t|x_0,t_0)}{\partial t} = -\frac{\partial}{\partial x}K_1(x.t)P(x,t|x_0,t_0) + \frac{\partial^2}{\partial x^2}K_2(x,t)P(x,t|x_0,t_0). \tag{1.21}$$

Both the drift term $K_1(x)$ and the diffusion coefficient $K_2(x)$ are nonlinear functions of the state variable x. They are defined as the moments of the conditioned increments per unit time, i.e.

$$\begin{aligned} K_n(x,t) &= \frac{1}{n!}\lim_{dt\to 0}\frac{\langle (x(t+dt)-x(t))^n\rangle}{dt} \\ &= \frac{1}{n!}\lim_{dt\to 0}\frac{1}{dt}\int dx'(x')^n P(x',t+dt|x,t) \end{aligned} \tag{1.22}$$

with $n = 1, 2$. These moments can be calculated from the Langevin equation (1.4). For additive noise $g(x) = 1$ it results in

$$K_1(x) = f(x)\,, \qquad K_2(x) = D\,. \tag{1.23}$$

For multiplicative noise the determination of these moments requires a more detailed consideration of the stochastic integral since white noise is too irregular for Riemann integrals to be applied. Application of Stieltjes integration yields a dependence of the moments on how the limit to white noise is taken. If $\xi(t)$ is the limit of the Ornstein-Uhlenbeck -process with $\tau \to 0$ (Stratonovich sense) the coefficients read [50]

$$K_1(x) = f(x) + Dg'(x)g(x), \qquad K_2(x) = Dg^2(x)\,. \tag{1.24}$$

Excitable systems have more than one dynamic variables. Hence the probability density will depend on these variables, i.e. $P(x, y, .., t)$. In case of Gaussian white noise the evolution operator remains a diffusion equation with drift, but in a higher dimensional phase space. We will deal with additive noise in the FitzHugh-Nagumo model in chapter 1.4.2.

For colored noise sources the derivation of evolution equations for the probability densities is more difficult. In a Markovian embedding. i.e. if the Ornstein-Uhlenbeck process is defined via white noise (cf. chapter 1.3.2) and $v(t)$ is part of the phase space one again gets a Fokker-Planck equation for the density $P(x, y, ...v, t)$. Similarly, one finds in case of the telegraph process balance equations for $P(x, y, ..\Delta, t)$ and $P(x, y, ..., \Delta', t)$ which are the densities of the two possible values of the noise Δ and Δ'. They yield a drift term from the deterministic part and a jump part which describes the hopping between the two noise values.

1.3.4. *Moment dynamics*

In this part we will explain a method that we extensively used to describe stochastic dynamical systems. It is based on the dynamics of the moments of a distribution. We applied it successfully to a variety of globally coupled systems. Advantages of the method are simple applicability and quick numerical investigations. Let us consider a globally coupled stochastic system that is described by the following set of Langevin equations:

$$\dot{x}_i = f(x_i, \langle x \rangle) + g(x_i)\xi_i(t) \qquad i = 1..N \tag{1.25}$$

Here, $\langle x \rangle = \sum_{i=1}^{N} x_i$ is the mean value of the system. Special focus will be on systems where the function f has the form

$$f(x_i, \bar{x}) = \hat{f}(x_i) + K(\langle x \rangle - x_j) \tag{1.26}$$

with coupling constant K. $\xi_i(t)$ represents Gaussian white noise determined by

$$\langle \xi_i(t) \rangle = 0 \qquad \langle \xi_i(t_1)\xi_j(t_2) \rangle = 2T\delta_{i,j}\delta(t_2 - t_1) \tag{1.27}$$

We interpret Eq. 1.25 in the Stratonovich sense.

We are especially interested in the mean of the ensemble $\langle x \rangle$. We therefore average over Eq. 1.25 and make a Taylor's expansion of the right hand side around $\langle x \rangle$. We obtain:

$$\dot{\langle x \rangle} = \sum_{n=0}^{\infty} \frac{\mu_n}{n!} \left[f^{(n)}(x, \langle x \rangle)|_{x=\langle x \rangle} + T[g^{(1)}(x)g(x)]^{(n)}|_{x=\langle x \rangle} \right] \tag{1.28}$$

the superscript in $f^{(n)}$ denotes the n-th derivative of f with respect to it's argument. In equation 1.28 we have made use of the central moments $\mu_n = \langle (x - \langle x \rangle)^n \rangle$. They are also time-dependent. For a closed description of the ensemble we need their dynamics, too. It is given by:

$$\dot{\mu}_n = \sum_{m=0}^{\infty} n \frac{\mu_{n-1+m}}{m!} \left[f^{(m)}(x)|_{x=\langle x \rangle} + T(g(x)g^{(1)}(x))^m|_{x=\langle x \rangle} \right]$$
$$+ \sum_{m=0}^{\infty} n(n-1) \frac{\mu_{n-2+m}}{m!} T(g^2(x))^{(m)}|_{x=\langle x \rangle} \tag{1.29}$$

Eq. 1.25 is not the most general form we can treat with the method of the moment dynamics. Especially models with more than one dynamical variable, like the FHN system, are important to us. In this case (Let us call the second variable y) we have to introduce the mixed central moments $\mu_{n,m} = \langle (x - \langle x \rangle)^n (y - \langle y \rangle)^m \rangle$ (and equivalently for more variables).

If we look closely at Eqs. 1.28 and 1.29 we notice that in general they incorporate infinite sums. It is only for polynomials f and g that the sums break off at some final value. Even if we deal with polynomials and the sums break off we notice that the dynamics of the n-th central moment generally depends on other, higher moments. The system of equations 1.29 forms an infinite set of coupled ordinary differential equations. It is only for linear functions f and g that the system decouples.

For more complicated functions we need to apply an approximation to the system of equations 1.28 and 1.29. There are two main ways to do this. One is to neglect the central moments from a certain order on. The other one is to neglect cumulants from a certain order on, instead. The infinite set of equations then reduces to a finite one. Most of the approximation methods are incompatible with a normalizable, nonnegative probability distribution. It is only the trivial method to neglect all moments or equivalently all cumulants above zeroth order and therefore neglect all fluctuations and go with the deterministic description and the *Gaussian approximation* that avoid this problem. For a complete description of any other probability distributions infinitely many cumulants or moments have to be taken into account. The Gaussian approximation consists of neglecting all cumulants above second order:

$$\kappa_n = 0 \qquad n \geq 3 \tag{1.30}$$

The dynamical description then reduces to the dynamics for the mean and the variance (for more dynamical variables to the means, the variances and the covariances).

1.4. Stochastic excitable elements

1.4.1. *The Langevin approach: Phase portraits under fluctuations*

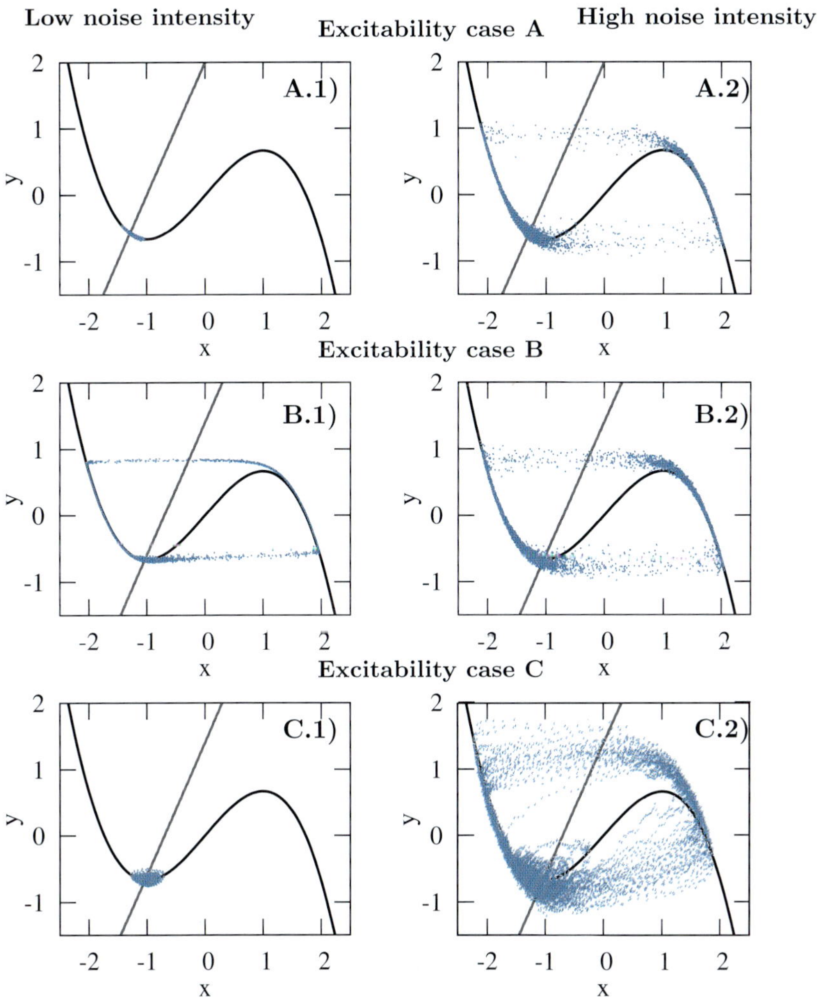

Fig. 1.4. Stochastic trajectories for the FitzHugh-Nagumo model at low noise intensity $\sigma^2 = 0.05$ (left column), and high noise intensity $\sigma^2 = 0.45$ (right column). Panels A): $\epsilon = 0.01$, $b_0 = 2$. Panels B): $\epsilon = 0.01$, $b_0 = 1.4$. Panels C): $\epsilon = 0.1$, $b_0 = 1.4$. The parameter γ is kept fixed equal to 2, the parameter $\alpha = 1/3$.

When the parameter that controls the excitation threshold of an excitable element fluctuates, then we end up with a system of coupled equations of Langevin type. In the case of the FitzHugh-Nagumo system this situation is modeled by the following Eqs.:

$$\frac{dx}{dt} = \frac{1}{\epsilon}\left(x - \alpha x^3 - y\right) \qquad (1.31)$$
$$\frac{dy}{dt} = \gamma x - y + b_0 + \xi(t).$$

Due to fluctuations the stable fixed point can be destabilized and the system is by chance brought out of the rest state. Here $\xi(t)$ is an arbitrary zero mean stochastic process that describes fluctuations in the excitability parameter $b \to b_\xi(t) = b_0 + \xi(t)$ around a mean value b_0. In Fig. 1.4 we show different realizations for the FitzHugh-Nagumo Eqs. 1.31, that permit us to describe its essential properties.

- Small fluctuations result in sub-threshold perturbations, consequently the system explores only a small portion of the phase space near the rest state $(x_0, y_0)_{b_0}$. This is shown both in panel A.1) and C.1) of Fig. 1.4. In panel A.1) the system simply relaxes to the instantaneous fixed points $(x_0, y_0)_{b_\xi(t)}$. In panel C.1), due to a high excitability, small stable limit cycles are induced by noise [a]. If the intensity of the fluctuations is increased the system can occasionally escape the vicinity of the fixed point and performs excitation loops, compare panel A.1) with A.2) or C.1) with C.2).
- For fixed noise intensity, a process, which at low excitability fails to bring the Eqs. 1.31 out of the fixed point vicinity, suffices to induce excitation loops at higher excitability, compare panels A.1), B.1) and C.1).
- The number of states visited by the system during excitations increases with the noise strength, panels B.1) and B.2) for example. Moreover fluctuations of high intensity significantly affect the trajectory of the excitation loops, see panel C.2).

1.4.2. *The Fokker-Planck approach: Numerical solutions*

In the previous section the stochastic FitzHugh-Nagumo system has been treated using the Langevin eqs. 1.31. Alternatively it can be described by the Fokker-Planck equation (FPE) (cf. subsec. 1.3.3). In the case of the

[a]This is because the value b_0 is near the Hopf bifurcation value b_{HB}.

FHN system an analytic solution of this equation cannot be given. Here we therefore use a numeric approach. The equation under study reads:

$$\begin{aligned}\frac{\partial}{\partial t}P(x,y,t) = & -\frac{1}{\epsilon}\frac{\partial}{\partial x}\left((x-x^3-y)P(x,y,t)\right) \\ & -\frac{\partial}{\partial y}\left((\gamma x - y + b)P(x,y,t)\right) \\ & + T_x\frac{\partial^2}{\partial x^2}P(x,y,t) + T_y\frac{\partial^2}{\partial y^2}P(x,y,t)\end{aligned} \tag{1.32}$$

where we have applied uncorrelated additive Gaussian white noise to the activator- and to the inhibitor dynamics. Their intensities are given by T_x and T_y, respectively. In this section we are interested in the stationary solution of eq. 1.32

$$\frac{\partial}{\partial t}P_0(x,y) = 0 \tag{1.33}$$

Eq. 1.32 is a linear FPE. The corresponding system is ergodic and we are faced with a unique, globally attracting solution. Fig. 1.5 shows this solution for varying intensities of the inhibitor noise T_y ($T_x = 0$) [29]. We find qualitative changes in the combination of the extrema and saddle points of the stationary probability density. For low noise we see a single maximum centered near the fixed point of the deterministic system (cf. the intersection of the nullclines in fig. 1.4). For increasing noise the probability density is mainly located close to two elongated maxima. These maxima represent the outer branches of the cubic nullcline. The probability distribution looks crater-like. Systems with the corresponding parameters spend most of their time along the phase space trajectory of a deterministic excited system. Once they enter the vicinity of the fixed point they are quickly reexcited. The corresponding timeseries are characterized by a large coherence. For further increasing noise the minimum of the probability density along with one saddle point vanishes. The corresponding system is not so closely bound to the deterministic trajectory any more.

We want to mention that the probability density further off the maxima becomes extremely small for small noise intensities so that numerical errors will eventually dominate the obtained results. In particular we cannot exclude a second maximum for the low noise case in Fig. 1.5. However we have also performed simulations with varying ϵ (separation of the timescales). For high ϵ (small separation) we find states with clearly one maximum only.

We thus find, depending on the noise intensity and the separation of the timescale three qualitatively different regimes. In these regimes differ-

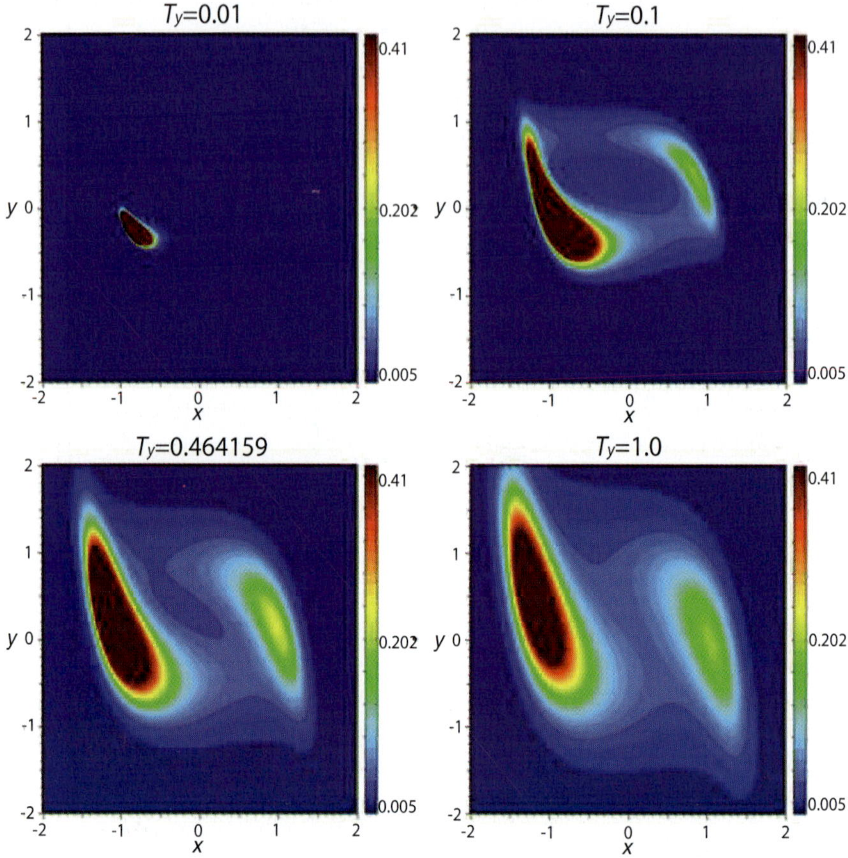

Fig. 1.5. Stationary probability distribution of the FitzHugh–Nagumo System. Inhibitor noise intensity is varied (given above the panels). Other parameters: $\epsilon = 0.1$, $\gamma = 2.$, $b = 1.4$. [29]

ent combinations of maxima, minima, and saddle points in the stationary probability distribution can be observed.

We have also performed simulations with noise in the activator variable. The obtained results differ only quantitatively from the ones presented for inhibitor noise.

1.4.3. *The phenomenon of coherence resonance*

First encountered in oscillatory systems as "*stochastic resonance without periodic forcing*" [17], and later as "*internal stochastic resonance*" [23], it is with the work of A. Pikovsky and J. Kurths on the FitzHugh-Nagumo model [40] that this phenomenon got its present name *coherence resonance* (CR) and that it was associated with excitable systems. In this work the authors showed that the regularity at which an excitable element fires under white noise driving, has a non-monotonous resonant dependence on the noise intensity, and that there exists an optimal noise intensity at which a sequence of noise-induced excitations is most regular. This phenomenon has been studied since more than one decade and has been observed in a huge variety of systems of quite different nature as, for example, anti-CR in excitable systems with feedback [33], CR in coupled chaotic oscillators [54], internal CR in variable size patches of a cell membrane [44, 45], system size CR in globally coupled FitzHugh-Nagumo elements [46], array-enhanced CR in a model for Ca^{2+} release [12], CR at the onset of a saddle-node-bifurcation of limit cycles [34] and of period-doubling bifurcations period-doubling bifurcation [38], and spatial CR in a spatially extended system near a pattern-forming instability [11] (for a comprehensive review see [35]).

To characterize the level of coherence of noise-induced excitations we analyze the time evolution of the activator concentration x in the FitzHugh-Nagumo model, see Fig. 1.6. In this representation the excitation loops shown previously in Fig. 1.4 become spikes spaced out by intervals during which the system performs noisy relaxation oscillations around its stable state. The phenomenon of coherence resonance manifests itself in the three realizations of $x(t)$ for different noise intensities given in Fig. 1.6. For very low noise intensity (upper panel) an excitation is a rare event which happens at random times. In the panel at the bottom, for high noise intensity, the systems fires more easily but still rather randomly. In the panel in the center instead, at an optimal noise intensity, the system fires almost periodically.

The typical oscillation period for the system is given by the mean interspike time interval (ISI) $\langle t_p \rangle$ between two successive noise-induced excitations over many realizations, see enlargements in Fig. 1.6. To it we associate as error the standard deviation. If the system fires regularly, say for simplicity periodically, then the error associated to t_p is zero and consequently the ratio of the standard deviation $srd(t_p)$ to its mean value $\langle t_p \rangle$, i.e. the normalized fluctuations

$$R_p = \frac{\sqrt{\langle (t_p - \langle t_p \rangle)^2 \rangle}}{\langle t_p \rangle}, \tag{1.34}$$

is equal to zero. On the other hand, if the firing is incoherent and takes

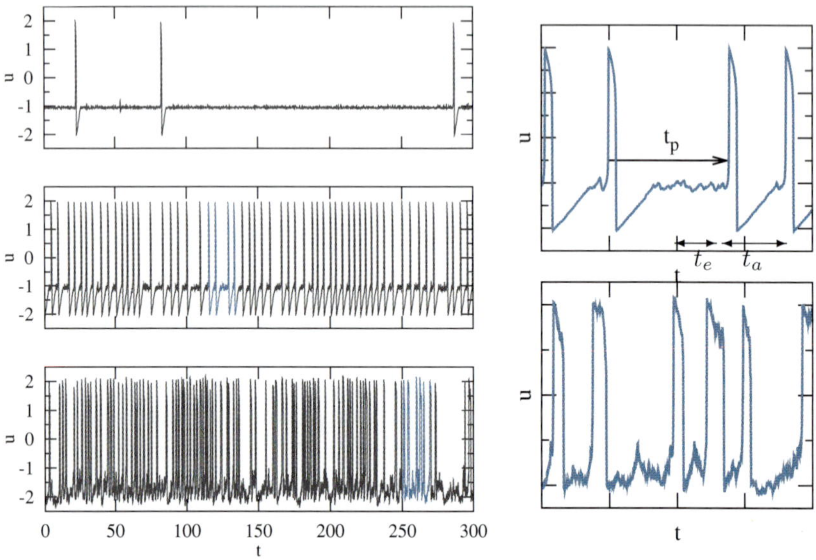

Fig. 1.6. Noise-induced excitations in the stochastic FitzHugh-Nagumo model under inhibitory white noise driving. Parameters for all plots: $b_0 = 1.05$, $\epsilon = 0.01$, $\gamma = 1.4$, $\alpha = 1/3$. Left top: $\sigma^2 = 0.001$. Left center: $\sigma^2 = 0.009$. Left bottom: $\sigma^2 = 0.064$. In the right panels are shown enlargements of the trajectories plotted on the left side. Time scales t_p, t_e and t_a are discussed in the next two pages.

place at random times, then the error associated to t_p is of the same order as the mean ISI, so that $R_p \simeq 1$. Thus R_p is a measure of the coherence of the system response and the minimum of $R_p(\sigma^2)$ characterizes the optimal noise intensity. We plot this dependence for the FitzHugh-Nagumo model in Fig. 1.7.

The phenomenon of CR is due to the presence of two different characteristic time scales in the system which are affected by noise in different manner. One is the time during which the system just fluctuates around the stationary state, which is needed to *activate* an excitation. We call this the activation time t_a. The second time scale, the *excursion* time t_e, is the typical duration of an excitation loop, compare Fig. 1.6. Noise of low intensity does not affect t_e, compare panels B.1 and B.2 of Fig. 1.4. Consequently, assuming $std(t_e)$ small and $t_a >> t_e$ (excitations are rare events), we can write that $R_p \simeq srd(t_a)/\langle t_a \rangle$. In this regime, the spikes are completely random events, so that $srd(t_a)/\langle t_a \rangle = 1$, and, as noise in-

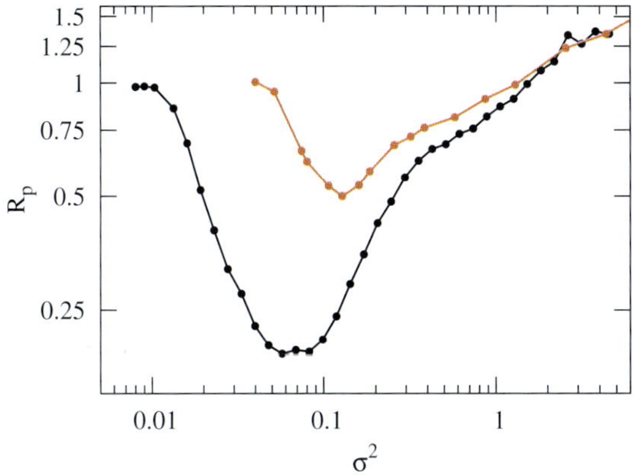

Fig. 1.7. Normalized fluctuations of the inter-spike interval versus the noise intensity for the FitzHugh-Nagumo model. Black curve reproduces the result shown in [40]. Parameters $b_0 = 1.05$, $\epsilon = 0.01$, $\gamma = 1.4$, $\alpha = 1/3$ in Eqs. 1.31. The fixed point is a stable focus. Same results with $b_0 = 1.2$ shown by the red curve, where the fixed point is a stable node.

creases, the coherence of t_a increases too, see Fig. 1.7. Only at moderate noise intensities the excursion time t_e starts to get affected by noise. In this regime the excitation loops, that at low noise intensities possess well defined trajectories, loose regularity with increasing noise intensity, compare panels at the bottom of Fig. 1.6. This phenomenon is also shown in panels B.1 and B.2 of Fig. 1.4. There the trajectories of the noise-induced excitations spread out in the phase space for increasing noise intensity. The transition between this two regimes takes place where the quantity R_p displays its minimum. Then the coherence of the system is highest.

1.4.4. *Coherence resonance with respect to the correlation time*

Most of the research on CR was focused on the case of white noise. White noise is a good approximation as long as the intrinsic time scales of the deterministic system are much larger than the correlation time of the external fluctuations. This is the case for example in neuronal dynamics where a neuron can be externally forced by another randomly bursting neuron. In general, when deterministic and stochastic time scales are not well separated from each other, not only the amplitude but also the temporal correlation is expected to influence noise-induced phenomena as CR. In this subsection we show that also the correlation time of an external noise signal is a control parameter of the coherence of the system. We show this first experimentally employing the light-sensitive Belousov-Zhabotinsky reaction and then we confirm this result numerically performing calculations with the two-component Oregonator model.

In Fig. 1.8 we show schematically the set-up in use in our experiment, for details please see [7]. The set-up adopted for our experiments has as central element an *open gel-reactor*, which allows to maintain constant non-equilibrium conditions during the measurements, see Fig. 1.8. With it, and by means of computer-based spectrophotometry [37], we analyze wave activity in the BZ medium with sufficiently high spatial and temporal resolution.

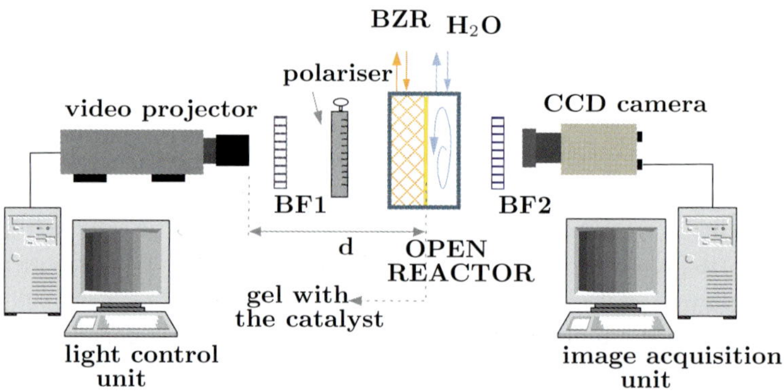

Fig. 1.8. Top: Sketch of the experimental setup. Below: Photograph of the reactor and, behind it, the polarizer, the blue filter and the video projector.

Through a noise signal, precisely a random telegraph signal, we induce nucleations of target patterns in the BZ medium. Under constant illumination I_0 no wave nucleation occurs, but the medium can support excitation waves. At light intensity $I_0 - \Delta I$ the medium is oscillatory and phase waves are induced, which become trigger waves as they propagate towards the surrounding medium maintained at high light intensity. The system is excitable at light intensity $I_0 + \Delta I$ and supports traveling patters. In case of dichotomous fluctuating light, nucleations occur randomly. The regularity of these phenomena is measured recording the activity at a given point of the gel. In this way we get a series of noise-induced nucleations at random time intervals, see data reported in Fig. 1.9.

Already at a rough glance on the noise-induced spikes reported in Fig. 1.9 it is possible to recognize that the coherence is dependent on τ. The spikes plotted in the middle panel show a better regularity than the other two above and below it.

To get a deeper understanding of this phenomenon we perform experiments for correlation times τ in the range between 2 and 60 seconds and we estimate the time t_p between two successive nucleations, and calculate its mean and the standard deviation. In Fig. 1.10 the coherence of the nucleation events is quantified through the normalized fluctuations of t_p,

$$R_p = \frac{\sqrt{\langle (t_p - \langle t_p \rangle)^2 \rangle}}{\langle t_p \rangle}. \tag{1.35}$$

R_p presents a minimum at $\tau \simeq 20$ s, fingerprint of highest coherence. Hence our experimental data give a clear evidence of the existence in the light-sensitive BZ reaction of an optimal correlation time of the fluctuating light driving at which the highest coherence is induced [7].

To validate numerically the above reported results, we perform calculations with the two-component Oregonator model introduced in subsection 1.2.3, see Eqs. 1.3. Here also we are interested in how fluctuations with non-vanishing correlation time in the excitability parameter affect the coherence of the system response. Thus we assume the parameter ϕ, proportional to the light-intensity, to be an exponentially correlated stochastic variable expressed as

$$\phi \to \phi_\eta(t) = \phi_0 \cdot [\eta_\pm(t) + 1], \tag{1.36}$$

being the stochastic process $\eta_\pm(t)$ a random telegraph signal. We choose ϕ_0 in the excitable regime and analyze how the coherence of the Oregonator system changes (Eqs. 1.3) as the correlation time of the process $\eta_\pm$

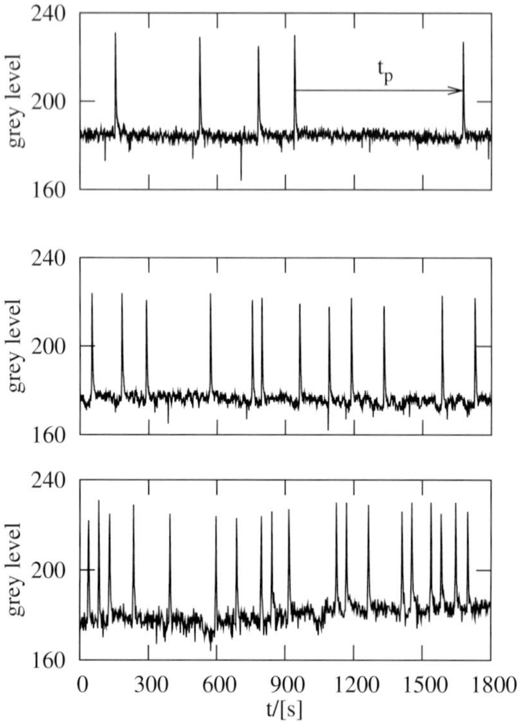

Fig. 1.9. Noise-induced spikes for different correlation time of the fluctuating light. Sequences are obtained tracking the activity at a fixed point of the medium and decoded in gray levels (point chosen in the constant background). From top to bottom: $\tau = 5$ s, $\tau = 15$ s, $\tau = 40$ s. [7]

varies. We choose the parameters ϕ_0 and Δ such that $\phi_- = \phi_0(1 - \Delta)$ to the oscillatory regime. Then for $\phi(t) = \phi_+ = \phi_0(1 + \Delta)$ the system is excitable, therefore from any initial condition it reaches the stationary state and remains there forever.

The results for $R_p(\tau)$ obtained for different values of Δ, see Fig. 1.11, demonstrate that under a random telegraph signal the coherence of noise-induced excitation is enhanced by an optimal choice of the correlation time. Here, the optimal correlation time τ_{opt} decreases as the noise amplitude Δ increases. Further simulations not shown here, confirm that this phenomenon holds for a wide range of the bifurcation parameter ϕ_0, covering almost the whole excitable regime. We emphasize that for not well separated time scales, noise-induced excitations are possible even if both ϕ_-

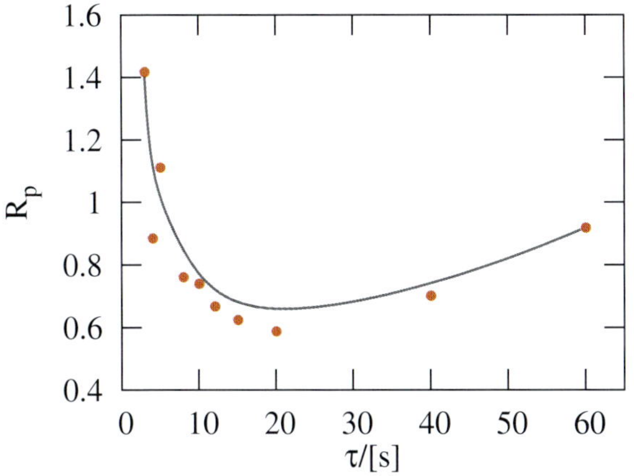

Fig. 1.10. Coherence resonance with respect to the correlation time in the light-sensitive BZ reaction. The normalized fluctuations of the inter-spike times t_p are reported versus the correlation time of the random telegraph signal. [7]

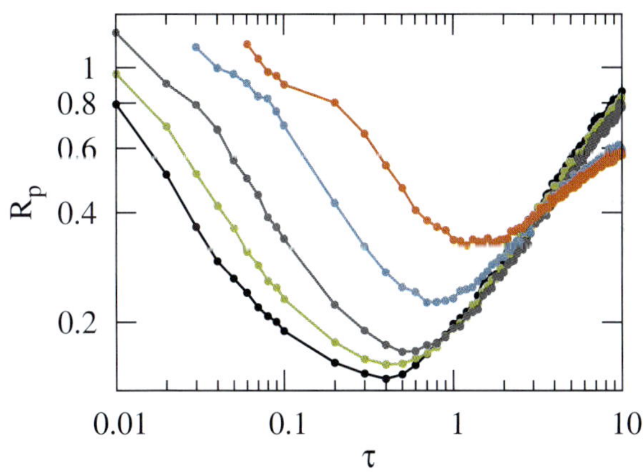

Fig. 1.11. Normalized fluctuations of the inter-spike time versus the correlation time of the noise for the Oregonator model, Eqs. 1.3 with $\phi_0 = 0.0075$. Black curve $\Delta = 0.1$, green $\Delta = 0.8$, gray $\Delta = 0.6$, blue $\Delta = 0.4$, red $\Delta = 0.3$. Each point is an average over $5 \cdot 10^3$ inter-spike intervals. [7]

and ϕ_+ belong to the excitable regime, compare red curve in Fig. 1.11.

1.5. Excitable elements with coupling

1.5.1. *Local coupling: Noise induced nucleations*

Local coupling among excitable elements is realized through diffusive coupling, which introduces spatial degree of freedom in the system. Here we consider the case of an one-dimensional Oregonator system in its three-component version. Thus its local dynamic introduced in subsection 1.2.3, see Eqs, 1.2 has to be supplemented by diffusion terms. As in the experiment with the BZ reaction the catalyst is immobilized in thin gel layer, there is no diffusion in the v variable. This gives

$$\begin{aligned} \frac{\partial u}{\partial t} &= \frac{1}{\epsilon}\left[u - u^2 - w\cdot(u-q)\right] + D_u \frac{\partial^2 u}{\partial x^2} \\ \frac{\partial v}{\partial t} &= u - v \\ \frac{\partial w}{\partial t} &= \frac{1}{\epsilon'}\left[fv - w\cdot(u+q) + \phi\right] + D_w \frac{\partial^2 w}{\partial x^2}. \end{aligned} \tag{1.37}$$

We integrate numerically Eqs. 1.37 in a one dimensional spatial domain of size L with periodic boundary conditions.

We are interested in the effects that fluctuating excitability induces in the emergence of patterns in the spatially extended system. This issue has been already tackled in previous works, most of them concentrating on the non-excitable regime, where no structures can propagate under purely deterministic conditions. In such a situation, an optimal amount of noise has been seen to support patterns as spiral waves [24, 25], coherent structures [51], traveling pulses [27], and pulsating spots [21]. In the excitable regime, noise-induced spiral dynamics [18], spatio-temporal stochastic resonance [25], and noise-enhanced phase synchronization [39] have been reported. Moreover also globally coupled excitable media [41, 53]. Here we focus our attention on how both spatial and temporal correlations influence the process of pattern formation. Thus the excitability parameter ϕ is not a constant value anymore but it is given by

$$\phi \to \phi_\eta(x,t) = \phi_0[\eta(x,t) + 1]. \tag{1.38}$$

Here $\eta(x,t)$ is a spatio-temporally correlated noise process with zero mean, characterized by noise intensity σ^2, correlation time τ and correlation length λ. For more detail on the numerical details see [5].

Suppose the medium is prepared in the homogeneous steady state. Eventually, fluctuations of the light intensity induce spontaneous nucleation

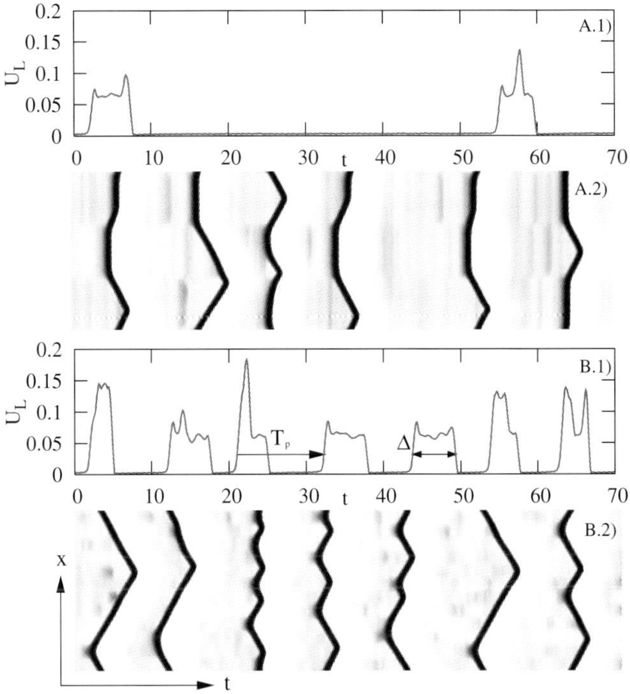

Fig. 1.12. Space averaged activator concentration U_L, panels A.1) and B.1), and corresponding realizations of the activator concentration $u(x,t)$, panels A.2) and B.2). Excitability $\phi_0 = 0.01$ and wavelength $L = 45$. Noise parameters in panels A.1) and A.2): $\sigma^2 = 0.25$, $\tau = 0.75$ and $\lambda = 1.125$; in panels B.1) and B.2): $\sigma^2 = 0.25$, $\tau = 2$ and $\lambda = 2.25$. [5]

of pulse pairs. The pulses propagate in opposite direction and annihilate each other in collision (periodic boundary conditions adopted). For the nucleation of a pulse pair from a local perturbation of the homogeneous steady state a critical nucleus has to be reached. Therefore, besides the correlation time, we expect the correlation length of the stochastic field $\phi(x,t)$ to be an important control parameter.

To characterize the coherence of noise-induced patterns in the distributed excitable medium we calculate the relative fluctuations of the spatial average $U_L(t)$ of the activator field $u(x,t)$

$$U_L(t) = \frac{1}{L} \int_0^L u(s,t)ds. \tag{1.39}$$

Typical space-time plots of $u(x,t)$ and the corresponding behavior of $U_L(t)$

are shown in Fig. 1.12.

The activator spatial average $U_L(t)$ exhibits pronounced spikes, analogously to what was found for the activator $u(t)$ of a single excitable element. Although the variable U_L has no spatial dependence, from its analysis we get information about temporal coherence of the noise-induced patterns, as well as about the spatial coherence. Fig. 1.12 shows indeed that for different combinations of the noise correlations τ and λ the emerging patterns display drastically different characteristics.

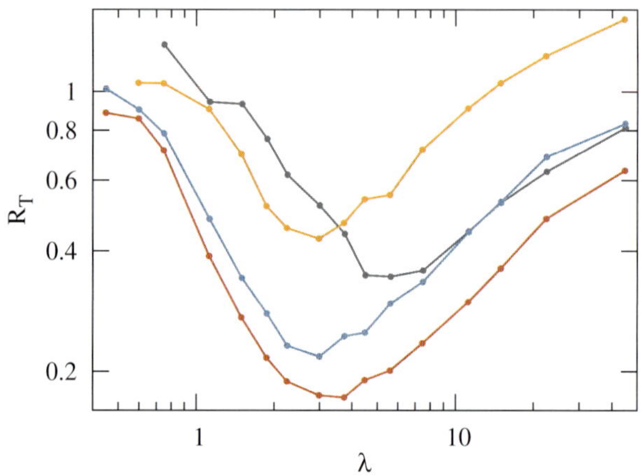

Fig. 1.13. Coherence resonance in the Oregonator model (Eqs. 1.37) with respect to the correlation length for different values of the correlation time τ: gray curve $\tau = 0.3$, red curve $\tau = 2$, blue curve $\tau = 5$, yellow curve $\tau = 20$. Excitability parameter $\phi_0 = 0.01$, system size $L = 45$, noise intensity $\sigma^2 = 0.25$ [5].

In order to quantify the temporal coherence of patterns arising at different correlation lengths, we introduce the quantity

$$R_T = \frac{\sqrt{\langle (T_p - \langle T_p \rangle)^2 \rangle}}{\langle T_p \rangle}, \tag{1.40}$$

which represents the normalized fluctuations of the inter-spike time T_p of the variable $\overline{u}_L(t)$. We keep the values σ^2 and ϕ_0 constant and we look at the dependency $R_T(\lambda)$. Fig. 1.13 shows the influence of the correlation length on the coherence of the noise-induced patterns. In the spatially extended system, for all the considered values of the correlation time τ, we find coherence resonance with respect to the correlation length of the noise driving, in other words there exists an optimal value λ_{opt} of the correlation length at which best coherence is reached. We find that the optimal value λ_{opt} is of the same order of magnitude as the width of the pulse front.

1.5.2. *Propagation of trigger waves in the presence of noise*

We study in this subsection the effect of noise on pulses propagating in the two-component Oregonator model. Also here the excitability parameter ϕ fluctuates, thus the noise enters multiplicatively in the system. The two-component Oregonator model supplied with local diffusive coupling reads

$$\frac{\partial u}{\partial t} = \frac{1}{\epsilon}\left[u - u^2 - [fv + \phi_0 \cdot (\eta(x,t) + 1)] \cdot \frac{u-q}{u+q}\right] + D_u \nabla^2 u$$
$$\frac{\partial v}{\partial t} = u - v, \tag{1.41}$$

where ∇^2 denotes the Laplace operator and $\eta(x,t)$ is a correlated stochastic process with zero mean. Concentration profiles of a typical pulse solution

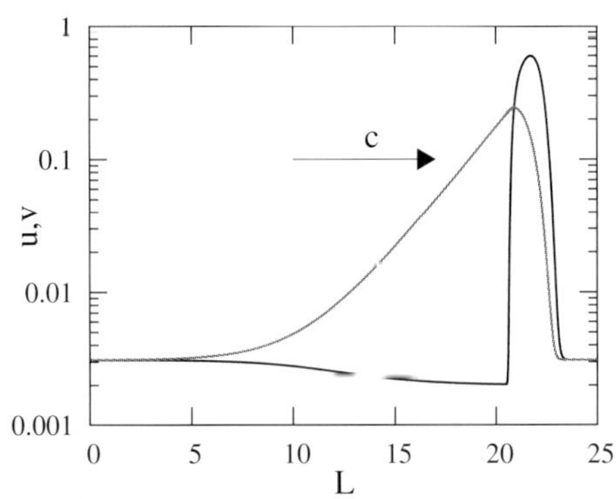

Fig. 1.14. Activator (black curve) and inhibitor (gray line) profile of typical pulse solution of the two-component Oregonator model (Eqs. 1.41) in the excitable kinetic regime ($\phi_0 = 0.01$) and with diffusion coefficient $D_u = 1$. Calculations were performed in a one-dimensional domain of size $L = 50$ applying periodic boundary conditions. The propagation speed is $c = 4.648$. [6]

obtained in the excitable regime are shown in Fig. 1.14. In the case of periodic pulse trains, the dispersion relation $c(L)$ connects the spatial extension of the medium and the velocity of a pulse propagating in it. The slope of the dispersion curve defines whether the interaction between pulses is attractive (positive slope) or repulsive (negative slope). In excitable systems it has been shown that different regimes of dispersion are reached upon variation of the excitability parameter ϕ [8, 9], compare Fig. 1.15.

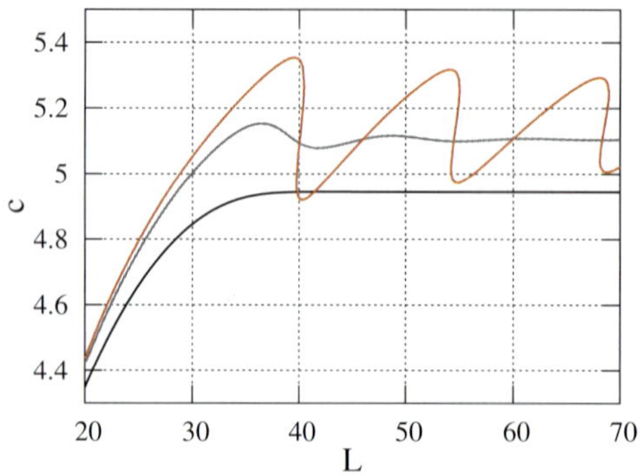

Fig. 1.15. Dispersion curve for periodic pulse trains at different levels of excitability in absence of noise. Decreasing ϕ_0 from relatively large values, normal dispersion is replaced first by oscillatory and finally bistable dispersion (black curve for $\phi_0 = 0.0065$, gray curve for $\phi_0 = 0.0048$ and red curve for $\phi_0 = 0.00433$). [6]

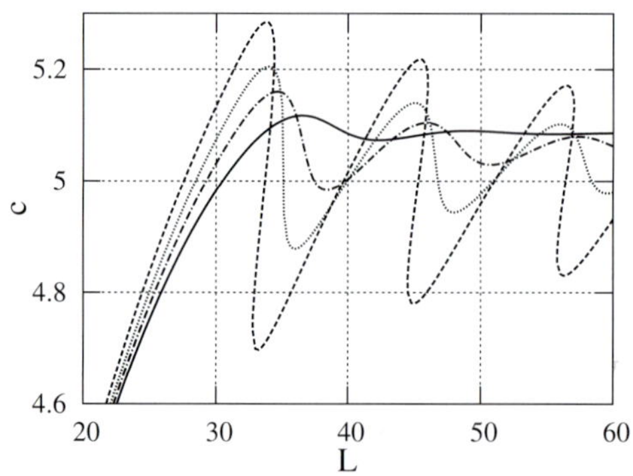

Fig. 1.16. Emergence of bistability in the dispersion relation for increasing noise intensity, black solid curve for the noiseless case with $\phi_0 = 0.005$. At the same excitability, dot-dashed curve for $\sigma^2 = 0.005$, dotted curve $\sigma^2 = 0.01$, dashed curve $\sigma^2 = 0.02$, value at which the dispersion relation becomes bistable. [6]

The same transition from normal to oscillating and bistable dispersion can take place in the case of an excitable medium under fluctuations. In Fig. 1.16 we show the emergence of bistability as the noise intensity of the stochastic signal $\eta(x,t)$ increases. These are results obtained in the framework of a small noise expansion put forward by M. A. Santos, J. M. Sancho and S. Alonso [2, 43]. The numerical calculations have been done employing the continuation software AUTO2000 [13]. For more detail see [6].

1.5.3. *Pattern formation in dichotomously driven, locally coupled FitzHugh-Nagumo systems*

In this subsection we study a spatially extended version of the FitzHugh-Nagumo system:

$$\begin{aligned}\dot{x}_i &= x_i - x_i^3 - y_i + D_x \Delta x_i \\ \dot{y}_i &= \epsilon\,(x_i - a y_i - I_i(t)) + D_y \Delta y_i,\end{aligned} \tag{1.42}$$

where Δ is the discrete version of the Laplace operator. $I(t)$ constitutes a random telegraph process with correlation function (cf. also eq. 1.19)

$$\langle I_i(t_1) I_j(t_2)\rangle = d^2 e^{-2\gamma\tau} K(i,j). \tag{1.43}$$

Here, $K(i,j)$ describes the spatial correlation. We study the influence of the switching rate γ on pattern formation in our system. If not explicitely mentioned otherwise the remaining parameters were kept constant at $\epsilon = 0.05$, $a = 1.475$, $D_x = 0.02$, $D_y = 5.0$, and $d = 0.2$. In order to provide inhomogeneities that might grow we additionally apply tiny additive Gaussian white noise (intensity 10^{-6}) to each side of the system. For these parameters the effect of the random telegraph process is the following: For each state of the telegraph process ($+d$ or $-d$) an individual system ($D_x = D_y = 0$) possesses a single stable fixed point and exhibits excitability. The two fixed points are symmetrically placed around zero, each on a different outer branch of the cubic nullcline (cf. Fig. 1.17). For small switching rates γ a homogeneous extended state approaches one fixed point until a switching occurs and is then attracted by the other fixed point and approaches that one. A random motion of the homogeneous state between the two fixed points is observed. For faster switching rates a new phenomenon occurs. It is schematically shown in Fig. 1.17. Consider a system that for $I = -d$ has reached the right fixed point. Due to the tiny additive noise the systems form a "cloud" around the fixed point. Once a new switching occurs the cloud moves up the right outer branch of the cubic nullcline to its maximum and then quickly towards the left outer branch. If the next switching occurs just at that moment when the cloud crosses the excitation threshold (there is no clear excitation threshold since whether a system is excited or not depends not only on its state but also on that of its neighbors) some of the systems are excited and some return to the nearby fixed point. Due to the strong inhibitory coupling (vertical direction) the excited states do not return to the first fixed point but get trapped on the left outer branch. A spatial structure is formed.

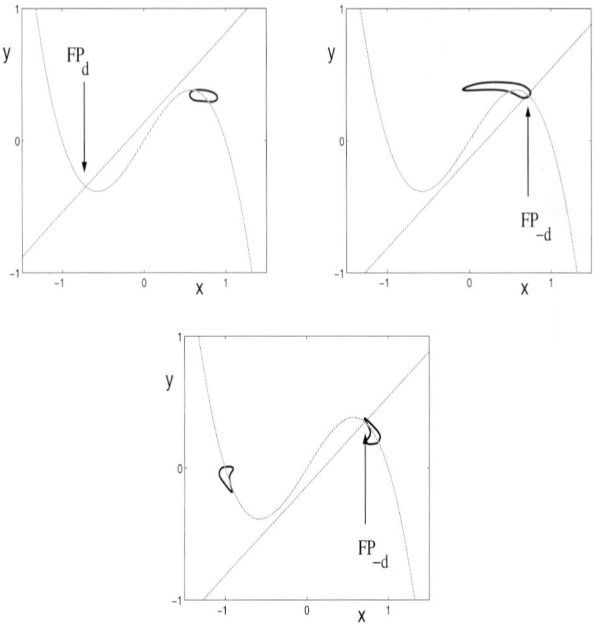

Fig. 1.17. Development of stationary patterns for medium switching rates - phase space. The system with small inhomogeneities (indicated by a cloud) is initially distributed around one of the fixed point. Once switching occurs the cloud starts to move to the second fixed point. If a second switching occurs just as the cloud is quickly passing from the left to the right it can be split into two. The second cloud then does not return to the fixed point of the deterministic system but moves to the other side where it stays. A spatial structure is formed.

In the limit of high switching rates a new phenomenon occurs. Here the switching becomes too fast for the systems to follow. As predicted by Buceta et al. [10] we find an effective dynamics that is obtained by substituting the random telegraph process by its average value $(I(t) = \langle I(t)\rangle = 0$. In this case the system 1.42 has two stable and one unstable fixed point. Linear stability analysis of the extended state shows that the homogeneous state becomes unstable via the Turing mechanism.

The results for the three different switching rate regimes (slow - intermediate - fast) are shown in Fig. 1.18 where we have in the low rate regime artificially provided an inhomogeneity. The mechanism that creates the first inhomogeneity occurs for arbitrary low rates but it may take extremely long. For low switching rates we see that the border separating two

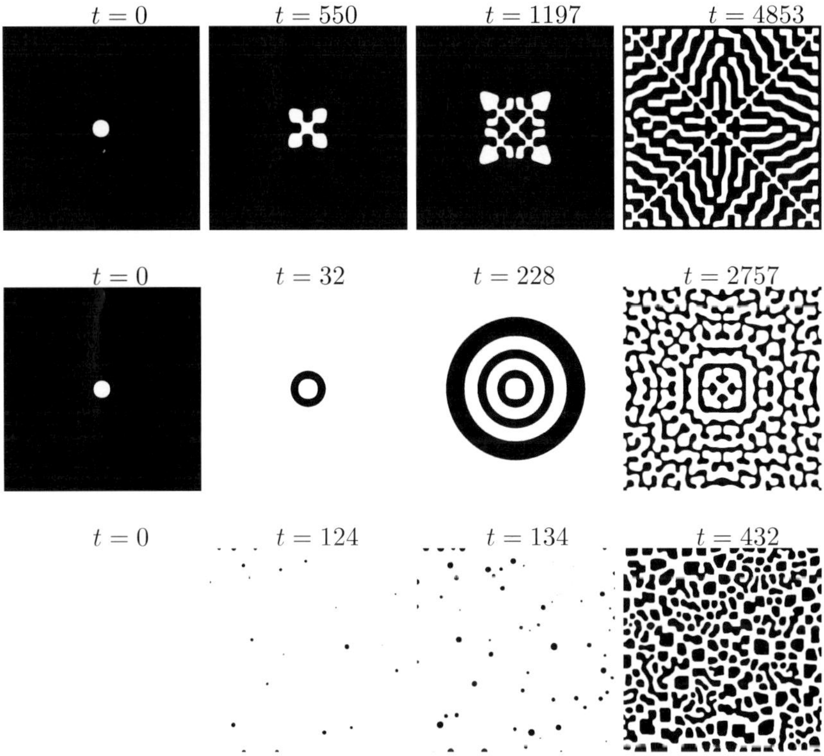

Fig. 1.18. Timeseries with different switching rates. The time is given above each plot. Parameters: $\gamma = 0$ (upper row), 5.0 (middle row), 0.01 (lower row); $\epsilon = 0.05$, $a = 1.475$, $D_x = 0.02$, $D_y = 5.0$, $d = 0.2$. [42]

homogeneous regions is unstable. Inhomogeneities in the border grow until they fill the pattern. This process is very slow (time is given above the pictures). For intermediate switching we see that a new layer is added to the side of the inhomogeneity for each switching of the telegraph process. The pattern evolves much faster into the homogeneous part. For high switching rates the whole homogeneous state becomes unstable. The pattern grows all over the region simultaneously.

We have also applied frozen dichotomous disorder to the system 1.42 and studied the effect of the correlation length of the disorder on pattern formation. The limit of large correlation length corresponds to the limit of slow switching. Here, too, inhomogeneities grow from the border between two large homogeneous regions. The limit of small correlation length is

Fig. 1.19. Coarse fluctuations support pattern formation, fine ones do not. In this parameter regime the system with $I_i(t) = 0$ does not support Turing pattern formation. $D_y = 0.5$, $\Delta r = 0.02$. 2000×2000 points are shown. [42]

equivalent to the fast switching limit. Here we again find an effective bistability. The dynamics can be predicted by using the deterministic dynamics with ($I_i = 0$). For intermediate correlations we again find nontrivial effects. Fig. 1.19 shows noise realizations with resulting patterns for small and intermediate correlation length. For the simulations that lead to these results we have decreased the inhibitor diffusion constant to $D_y = 0.5$. The deterministic dynamics with $I_i = 0$ does not exhibit instability of the homogeneous state. This fits well to the results shown in the upper line. For higher correlation length the homogeneous state is destabilized and a labyrinth pattern emerges. We have thereby significantly increased the parameter regime in which structure patterns emerge.

1.5.4. *Global coupling*

In this subsection we study the following globally coupled system of equations:

$$\begin{aligned} \epsilon\dot{x}_i &= x_i - \frac{x_i^3}{3} - y_i + K\left(\langle x\rangle - x_i\right), \\ \dot{y}_i &= x_i + a + \xi_i(t), \qquad i = 1, \ldots, N. \end{aligned} \tag{1.44}$$

$\xi_i(t)$ is Gaussian, white noise with intensity T and zero average, uncorrelated between the individual subsystems. We fixed the parameters $\epsilon = 0.01$ and $a = 1.05$ to ensure excitability of the individual deterministic ($K = 0$, $T = 0$) system. As order parameter we consider the mean of the N systems $\langle x\rangle = \sum_i x_i$. Fig. 1.20 shows results from direct simulation of eqs. 1.44 for varying noise intensity. For low noise the fixed point of $\langle x\rangle$ is stable. Individual systems spike very rarely and incoherently. For increasing noise these spikes become more frequent until the systems start to move coherently. The spikes of the mean are separated by several small amplitude oscillations. The number of these intermediate oscillations decreases and uninterrupted spiking occurs. For even higher noise intensities the coherence is destroyed again. The individual systems still spike frequently but the mean does not.

In order to have a closer look at the observed phenomena we derive the moment dynamics for the system 1.44 in Gaussian approximation:

$$\begin{aligned} \epsilon\frac{d}{dt}m_x &= m_x - \frac{m_x^3}{3} - m_y - m_x D_x \\ \frac{d}{dt}m_y &= m_x + a \\ \epsilon\frac{d}{dt}D_x &= 2D_x(1 - D_x - m_x^2 - K) - 2D_{xy} \\ \frac{d}{dt}D_y &= 2(D_{xy} + T_{loc}) \\ \epsilon\frac{d}{dt}D_{xy} &= D_{xy}(1 - D_x - m_x^2 - K) - D_y + \epsilon D_x \end{aligned} \tag{1.45}$$

Here, m_x and m_y are the mean values, D_x and D_y the variances, and D_{xy} the covariance of the distribution. Results from simulations of these equations for small noise intensities are shown in Fig. 1.21. For small noise intensities there exists a stable fixed point (not shown). This fixed point becomes unstable for larger noise and a small amplitude oscillation occurs. For further increase of the noise we find a period doubling cascade until chaos emerges. The size of the chaotic attractor is still comparable

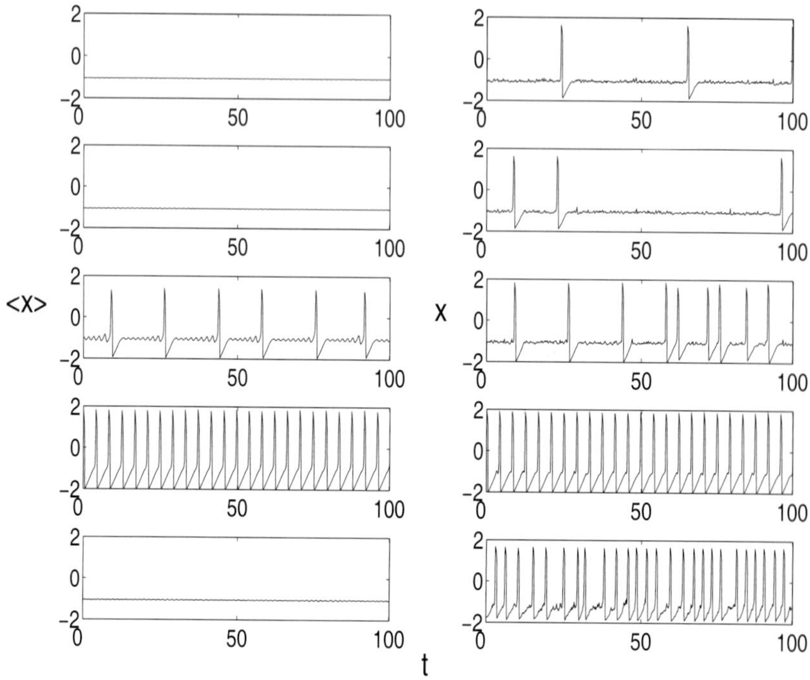

Fig. 1.20. Timeseries for different noise intensities. The left column shows the ensemble average the right column a randomly chosen individual unit. For increasing noise intensity the spike rate of the single element increases quickly and then remains on a high level. The spike rate of $\langle x \rangle$ increases at first but then the synchronization between the individual elements is lost and the spike rate decreases again. Note the small range of noise intensities (from top to bottom): $T_{loc} = 2.7 \times 10^{-4}$, 2.8×10^{-4}, 2.9×10^{-4}, 3.0×10^{-4}, 3.1×10^{-4}. Other parameters: $K = 0.1$. [41]

to the diameter of the small amplitude oscillations. For further increasing noise the chaotic attractor quickly but continuously increases in size. This phenomenon resembles the well-known Canard explosion. Here though, it is not a single limit cycle whose size increases quickly but the chaotic attractor as a whole.

The moment dynamics method allows us to quickly discover the bifurcation diagram. It is given in Fig. 1.22. We see that the transition from stable fixed point of the mean to spiking and back to a stable fixed point is exceptionally complicated. We see a period-doubling cascade followed by a Canard explosion of the chaotic attractor, then a reverse period adding sequence towards a state of uninterrupted spiking. At higher noise inten-

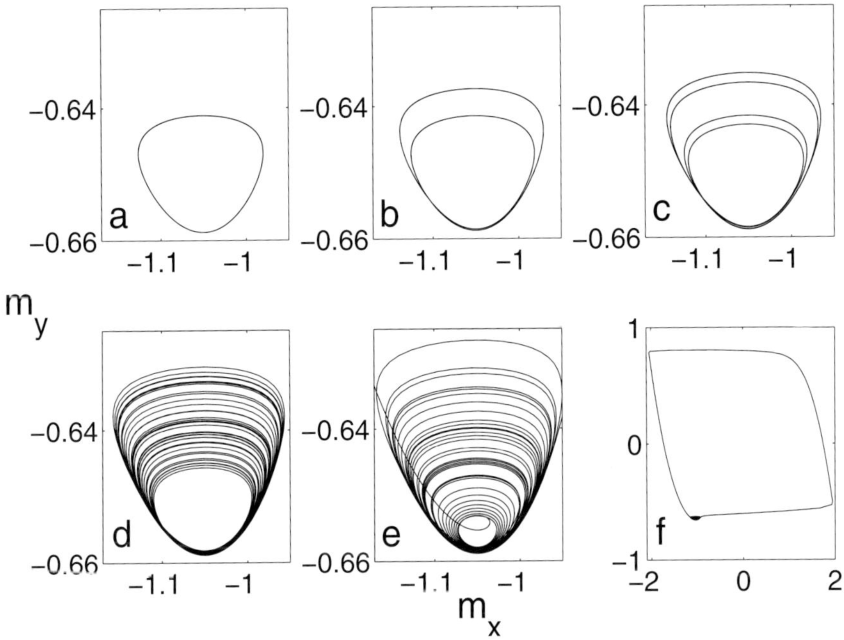

Fig. 1.21. Phase space trajectories for different noise intensities. We see successive period doubling of the small amplitude oscillations until chaos emerges. Soon after, spiking starts. Note the small parameter range (from a to e): T_{loc} = 0.00157, 0.00158, 0.0015826, 0.001585, 0.001586; $K = 0.1$. Subplot e is an enlargement of f. [41]

sities when the spiking decays again there is a new period adding and an inverse period doubling cascade towards the steady fixed point.

These results are supported by simulations of the Langevin equations 1.44. Here the fluctuations in the order parameter are too large to discover the transition in all its detail. Thus, the moment dynamics has has proved a convenient way to increase the resolution of our investigations.

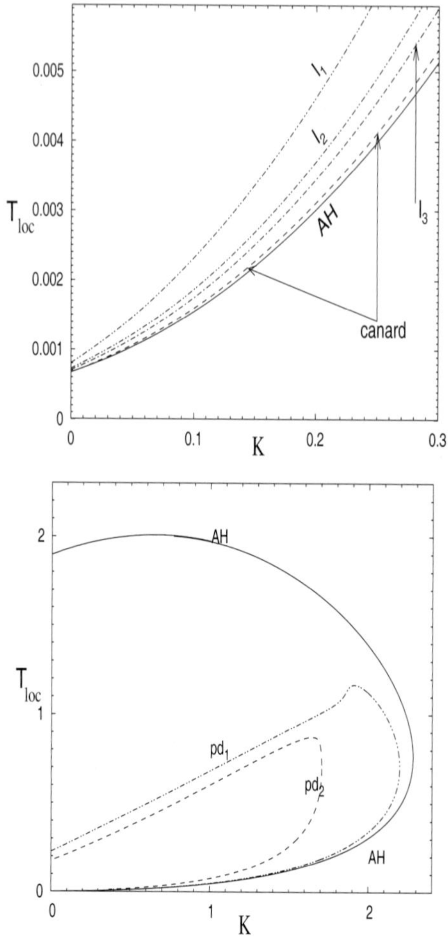

Fig. 1.22. Period doubling and Hopf bifurcations. Top: enlarged part at low coupling strength K and noise intensity T_{loc}. Bottom: Global view of selected bifurcations. AH: Andronov-Hopf bifurcation; l_n: onset of regime with n subthreshold oscillations between two spikes (l_1: onset of non-intermittent spiking). Parameters: $\epsilon = 0.01$, $a = 1.05$. [41]

References

[1] R. R. Aliev and A. V. Panfilov. A simple two-variable model of cardiac excitation. *Chaos, Solitons and Fractals*, 7(3):293–301, 1995.

[2] S. Alonso, F. Sagués, and J. M. Sancho. Excitability transitions and wave dynamics under spatio-temporal structured noise. *Phys. Rev. E*, 65:066107, 2002.

[3] P. Bak, K. Chen, and C. Tang. A forest-fire model and some thoughts on turbulence. *Physics Letters A*, 147, 1990.
[4] M. Bär, M. Eiswirth, H. H. Rotermund, and G. Ertl. Solitary wave phenomena in an excitable surface reaction. *Physical Review Letters*, 69:945–948, 1992.
[5] V. Beato and H. Engel. Coherence resonance phenomena in an excitable system driven by colored noise. *Fluct. Noise Lett.*, 6:L85–L94, 2006.
[6] V. Beato, H. Engel, and L. Schimansky-Geier. Pulse trains propagating through excitable media subjected to external noise. 2006. submitted to PRE.
[7] V. Beato, I. Sendiña-Nadal, I. Gerdes, and H. Engel. Noise-induced wave nucleations in an excitable chemical reaction. *Phys. Rev. E*, 71:035204(R), 2005.
[8] G. Bordiougov and H. Engel. From trigger to phase waves and back again. *Physica D*, 2006. in press.
[9] G. Bordiugov and H. Engel. Oscillatory dispersion and coexisting stable pulse trains in an excitable medium. *Phys. Rev. Lett*, 90:148302, 2003.
[10] J. Buceta, M. Iba nez, J. M. Sancho, and K. Lindenberg. Noise-driven mechanism for pattern formation. *Physical Review E*, 67:021113, 2003.
[11] O. Carrillo, M. A. Santos, J. Garcia-Ojalvo, and J. M. Sancho. Spatial coherence resonance near pattern-forming instabilities. *Europhys. Lett.*, 65:452, 2004.
[12] S. Coombes and Y. Timofeeva. Sparks and waves in a stochastic fire-diffuse-fire model of Ca^{2+} release. *Phys. Rev. E*, 68:021915, 2003.
[13] E. J. Doedel, A. R. Champneys, T. F. Fairgrieve, Y. A. Kuznetsov, B. Sandstede, and X.-J. Wang. *AUTO97: Continuation and bifurcation software for ordinary differential equations*, 1998. available by FTP from ftp.cs.concordia.ca in the directory pub/doedel/auto.
[14] C. Elphick, A. Hagberg, and E. Meron. Dynamic front transitions and spiral-vortex nucleation. *Physical review E*, 51(4):3052, 1995.
[15] R. J. Field, E. Körös, and R. M. Noyes. Oscillations in chemical systems II. Through analysis of temporal oscillations in the bromate-cerium-malonic acid system. *J. Amer. Chem. Soc.*, 94:8649–8664, 1972.
[16] R. J. Field and R. M. Noyes. Oscillations in chemical systems IV. Limit cycle behavior in a model of a real chemical reaction. *J. Amer. Chem. Soc.*, 60:1877–1884, 1974.
[17] H. Gang, T. Ditzinger, C. Z. Ning, and H. Haken. Stochastic resonance without external periodic force. *Phys. Rev. Lett.*, 71:807, 1993.
[18] J. García-Ojalvo and L. Schimansky-Geier. Noise-induced spiral dynamics in excitable media. *Europhys. Lett.*, 47:298–303, 1999.
[19] C. W. Gardiner. *Handbook of stochastic methods*. Springer, Berlin, second edition, 1985.
[20] P. Hänggi and P. Jung. Colored noise in dynamical systems. *Adv. Chem. Phys.*, 89:239, 1995.
[21] H. Hempel, L. Schimansky-Geier, and J. Garcia-Ojalvo. Noise-sustained pulsating patterns and global oscillations in subexcitable media. *Phys. Rev. Lett*,

82(18):3713–3716, 1999.

[22] W. Horsthemke and R. Lefever. *Noise-induced transitions*. Springer, Berlin, 1984.

[23] Z. Hou and H. Xin. Noise-induced oscillations and stochastic resonance in an autonomous chemical reaction system. *Phys. Rev. E*, 60:6329, 1999.

[24] P. Jung, A. Cornell-Bell, K. S. Madden, and F. Moss. Noise-induced spiral waves in astrocyte syncytia show evidence of self-organized criticality. *J. Neurophysiol.*, 79:1098, 1998.

[25] P. Jung and G. Mayer-Kress. Spatiotemporal stochastic resonance in excitable media. *Phys. Rev. Lett.*, 82:855, 1999.

[26] S. Kadar, J. Wang, and K. Showalter. Noise-supported traveling waves in sub-excitable media. *Nature*, 391:770, 1998.

[27] S. Kàdàr, J. Wang, and K. Showalter. Noise-supported traveling waves in sub-excitable media. *Nature (London)*, 391:770–772, 1998.

[28] C. Koch. *Biophysics of Computation: Information Processing in Single Neurons*. Oxford University Press, New York, 1999.

[29] M. Kostur X. Sailer and L. Schimansky-Geier. Stationary probability distributions for FitzHugh-Nagumo systems. *Fluctuation and Noise Letters*, 3:155, 2003.

[30] H. J. Krug, L. Pohlmann, and L. Kuhnert. Analysis of the modified complete Oregonator accounting for oxygen sensitivity and photosensivity of Belousov-Zhabotinsky systems. *J. Phys. Chem.*, 94:4862–4866, 1990.

[31] L. Kuhnert. A new optical photochemical memory device in a light-sensitive chemical active medium. *Nature (London)*, 337:393–394, 1986.

[32] L. Kuhnert, L. Pohlmann, and H. J. Krug. Repressing of chemical waves by photochemical inhibitor releasing. In W. Ebeling and H. Ulbricht, editors, *Selforganization by Nonlinear Irreversible Processes*, pages 128–132. Springer Verlag, Berlin, 1986.

[33] A. M. Lacasta, F. Sagués, and J. M. Sancho. Coherence and anticoherence resonance tuned by noise. *Phys. Rev. E*, 66:045105, 2002.

[34] S. G. Lee, A. Niemann, and S. Kim. Coherence resonance in a Hodgkin-Huxley neuron. *Phys. Rev. E*, 57:3292–3297, 1998.

[35] B. Lindner, J. Garcia-Ojalvo, A. Neiman, and L. Schimansky-Geier. Effects of noise in excitable systems. *Phys. Rep.*, 392(6):321–424, 2003.

[36] K. Martinez, A. L. Lin, R. Kharrazian, X. Sailer, and H. L. Swinney. Resonance in periodically inhibited reaction-diffusion systems. *Physica D*, 168:1–9, 2002.

[37] S. C. Müller, T. Plesser, and B. Hess. Two-dimensional spectrophotometry of spiral wave propagation in the Belousov-Zhabotinskii reaction : I. Experiments and digital data representation. *Physica D*, 24:71, 1987.

[38] A. Neiman, P. I. Saparin, and L. Sone. Coherence resonance at noisy precursors of bifurcations in non linear dynamical systems. *Phys. Rev. E*, 56:270, 1997.

[39] A. Neiman, L. Schimansky-Geier, A. Cornell-Bell, and F. Moss. Noise-enhanced phase synchronization in excitable media. *Phys. Rev. Lett.*, 83:4896–4899, 1999.

[40] A. S. Pikovsky and J. Kurths. Coherence resonance in a noise-driven excitable system. *Phys. Rev. Lett.*, 78:775, 1997.

[41] X. Sailer, M. Zaks, and L. Schimansky-Geier. Collective dynamics in an ensemble of globally coupled fhn systems. *Fluct. Noise Lett.*, 5:299–304, 2005.

[42] X. Sailer, D. Hennig, V. Beato, H. Engel, and L. Schimansky-Geier. Regular patterns in dichotomically driven activator-inhibitor dynamics. *Phys. Rev. E* 73:056209, 2006

[43] M. A. Santos and J. M. Sancho. Front dynamics in presence of spatiotemporal structured noises. *Phys. Rev. E*, 64:016129, 2001.

[44] G. Schmid, I. Goychuk, and P. Hänggi. Stochastic resonance as a collective property of ion channel assemblies. *Europhys. Lett.*, 56:22, 2001.

[45] G. Schmid, I. Goychuk, and P. Hänggi. Channel noise and synchronization in excitable membranes. *Physica A*, 325:165, 2003.

[46] R. Toral, C. Mirasso, and J. D. Gunton. System size coherence resonance in coupled FitzHugh-Nagumo models. *Europhys. Lett.*, 61:162, 2003.

[47] J. J. Tyson and P. C. Fife. Target patterns in a realistic model of the Belousov-Zhabotinsky reaction. *J. Chem. Phys.*, 73:2224, 1980.

[48] G. E. Uhlenbeck and L. S. Ornstein. On the theory of Brownian motion. *Phis. Rev.*, 36:823, 1930.

[49] O. V. Ushakov, H.-J. Wnsche, F. Henneberger, I. A. Khovanov, L. Schimansky-Geier, and M. A. Zaks. Coherence resonance near a hopf bifurcation. *Physical Review Letters*, 95:123903, 2005.

[50] A. Astakhov T. Vadiavasova V. Anishchenko, A. Neiman and L. Schimansky-Geier. *Chaotic and Stochastic Processes in Dynamic Systems.* Springer-Series on Synergetics. Springer Verlag, New York, 2002.

[51] J. Wang, S. Kàdàr, P. Jung, and K. Showalter. Noise driven avalanche behavior in subexcitable media. *Phys. Rev. Lett.*, 74:2130, 1995.

[52] H. J. Wünsche, O. Brox, M. Radziunas, and F. Henneberger. Excitability of a semiconductor laser by a two-mode homoclinic bifurcation. *Physical Review Letters*, 88:023901, 2001.

[53] M. Zaks, X. Sailer, L. Schimansky-Geier, and A. Neiman. Noise induced complexity: From subthreshold oscillations to spiking in coupled excitable systems. *CHAOS*, 15:026117, 2005.

[54] M. Zhang, G. W. Wei, C.-H. Lai, Y.-C. Lai, and Z. Liu. Coherence resonance near the Hopf bifurcation in coupled chaotic oscillators. *Phys. Rev. E*, 66:036201, 2002.

Chapter 2

Synchronization in periodically driven discrete systems

Tobias Prager and Lutz Schimansky-Geier*

Institut für Physik, Humboldt-Universität zu Berlin,
Newtonstr. 15, 12489 Berlin, Germany,
** alsg@physik.hu-berlin.de*

Contents

2.1. Introduction

The impact of periodic signals on dynamical systems has widespread relevance in different fields, ranging from technically generated periodic signals in electronic circuits to the periodic modulation of the solar irradiation on earth. Often the considered systems are intrinsically noisy or external noise influences the dynamics together with the periodic signal. There are different possibilities to quantify the response of such a stochastic system to a periodic signal. Most prominent are spectral based measure like the spectral power amplification or the signal to noise ratio [1, 5]. These measures describe the quality of the system's response to the periodic signal, based on the amplitude of the periodic component in the system's mean output. In many systems, like bistable or excitable dynamics, this periodic component is maximally expressed at some non-vanishing noise level, a counterintuitive effect which is termed *stochastic resonance.*

Another possibility to quantify the response of a stochastic system to periodic signals is to generalize the notion of *synchronization*, which is known from deterministic nonlinear oscillators. We will pursue this idea in what follows. To this end we review in section 2.2 the notion of effective synchronization in stochastic systems. The mean number of synchronized system cycles turns out to be an appropriate quantity to characterize the synchronization properties of the system to the periodic signal. However the task remains to calculate this quantity. This calculation will be based on discrete renewal models for bistable and excitable dynamics. These discrete models are introduced in section 2.3. We first recapitulate the well known two state model for the stochastic dynamics of an overdamped particle in a doublewell system [10] and afterwards introduce a phenomenological discrete model for excitable dynamics. In section 2.4 a theory to calculate the mean frequency and effective diffusion coefficient in periodically driven renewal processes is presented. These two quantities allow to calculate the mean number of synchronized cycles. Finally in section 2.5 we apply this theory to investigate synchronization in bistable and excitable systems.

2.2. Stochastic synchronization in periodically driven systems

2.2.1. *Synchronization in deterministic systems*

Synchronization is a fundamental phenomenon found in nonlinear oscillatory systems [12]. The most prominent example, known since long time ago (Huygens, 1665), is the adjustment to a common frequency of two pendulum clocks with slightly different frequencies, coupled via a common support. This type of synchronization between two coupled systems is called mutual synchronization. In the following we are interested in the synchronization of a system to a periodic driving, called forced synchronization. In a synchronized state the systems dynamics is entrained to the signal, i.e. the system inherits the very same frequency of the signal (1 : 1 synchronization) or the frequencies are locked with some rational $n : m$ relation. This corresponds to the motion of the system's coordinates together with the signal phase on a resonant torus. Depending on the signal amplitude and the difference between system frequency and signal frequency one observes different $m : n$ synchronization regions. These synchronization regions, exhibiting a tongue like shape, are called Arnold tongues. The concept of synchronization has to be extended and revised in order to cope with stochastic systems as will be done in the following subsection.

2.2.2. *Effective synchronization in stochastic systems*

The influence of noise on a dynamical system may have two counteracting effects. On the one hand if the underlying deterministic systems is already oscillatory, like a limit cycle oscillator or a chaotic oscillator, one expects these oscillations to become less regular due to the influence of the noise. On the other hand oscillatory behavior can also be generated by the noise in systems which deterministically do not show any oscillations. A prominent example are excitable systems but also the noise induced hopping between the attractors in a bistable system can be considered as oscillations [1].

Let us consider a stochastic system, for which it is possible to define a cycle, i.e. some behavior which repeatedly happens. This can be for example one turn of a limit cycle oscillator, the hopping from one attractor to the other and back again in a bistable system or an excitation from the rest state to the excited state, followed by a relaxation back to the rest state in an excitable system. As in the case of deterministic systems, forced synchronization of stochastic systems is also considered as an adaption of the cyclic motion to the periodic driving. However due to the stochasticity one can never expect perfect synchronization. Instead there is always a finite probability that an additional or missing cycle of the system with respect to the signal happens. The rarer these *phase slips* occur, the better is the synchronization. Thus synchronization in periodically driven stochastic systems is not an all or none notion but gradually varies from no synchronization to synchronization.

One possibility to quantify the quality of the synchronization is to consider the mean number N_{lock} of synchronized cycles of the system. The larger this number, i.e. the rarer the phase slips happen, the better is the synchronization. As in the case of deterministic synchronization one has to take different frequency locking modes into account. $m : n$ synchronization means that m cycles of the system occur within n periods of the signal.

In order to obtain N_{lock} we introduce the random number $N_{t_0}(t)$ of cycles performed by the system within the time interval (t_0, t). Denoting the period of the signal by $\mathcal{T}$ and its frequency by $\nu := \frac{1}{\mathcal{T}}$, the mean time $T_{\rm lock}$ between two subsequent phase slips, defined as the time after which the difference between system cycles and signal periods has grown to 1, can then be determined by [4]

$$\left\langle \left[N_{t_0}(t_0 + T_{\rm lock}) - \frac{n}{m}\frac{T_{\rm lock}}{\mathcal{T}} \right]^2 \right\rangle = 1. \tag{2.1}$$

In this definition different $m : n$ synchronization modes are already accounted for. Eq.(2.1) can be rewritten as

$$\left\langle N_{t_0}(t_0 + T_{\rm lock})^2 \right\rangle - \left\langle N_{t_0}(t_0 + T_{\rm lock}) \right\rangle^2 + \left[\left\langle N_{t_0}(t_0 + T_{\rm lock}) \right\rangle - \frac{n}{m}\mathcal{T} \right]^2 = 1.$$

Defining the mean frequency of the system as

$$\bar{v} = \lim_{t \to \infty} \frac{\left\langle N_{t_0}(t) \right\rangle}{t} \tag{2.2a}$$

and its effective diffusion coefficient by

$$\bar{D}_{\rm eff} = \lim_{t \to \infty} \frac{\left\langle N_{t_0}^2(t) \right\rangle - \left\langle N_{t_0}(t) \right\rangle^2}{2t} \tag{2.2b}$$

and assuming that $T_{\rm lock}$ is large compared with the signal period $\mathcal{T}$ one obtains [4]

$$2\bar{D}_{\rm eff} T_{\rm lock} + \Delta\nu^2 T_{\rm lock}^2 = 1 \tag{2.3}$$

where $\Delta\nu := \bar{v} - \nu\frac{n}{m}$ denotes the deviation between mean system frequency and signal frequency. Eq. 2.3 reveals, that the mean time between phase slips is determined by two effects, namely a drift part, caused by different frequencies and a diffusion part which persists even if the mean frequency of the system agrees with the signal frequency. This diffusive part is due to the stochastic nature of the system. Solving the quadratic equation for $T_{\rm lock}$ one obtains

$$T_{\rm lock} = \frac{\bar{D}_{\rm eff}}{\Delta\nu^2}\left[\sqrt{1 + \frac{\Delta\nu^2}{\bar{D}_{\rm eff}^2}} - 1 \right].$$

Assuming the mean frequency of the system to be entrained by the signal, i.e. $\Delta\nu \to 0$ the mean locking episodes are solely determined by the effective diffusion coefficient

$$T_{\rm lock} \approx \frac{1}{2\bar{D}_{\rm eff}}.$$

On the other hand if the effective diffusion coefficient vanishes, as is the case for deterministic systems, we have

$$T_{\text{lock}} \approx \frac{1}{\Delta\nu}.$$

Stochastic $m : n$ synchronization can thus be characterized as an adjustment of frequencies, $\bar{v} \approx \frac{n}{m}\nu$ and a reduction in the diffusivity of the process, i.e. a minimal effective diffusion coefficient $\bar{D}_{\text{eff}}$. The mean locking time T_{lock} still depends on the absolute time scale of the process. In order to avoid this dependence, it is reasonable to consider the mean number of locked cycles

$$N_{\text{lock}} = T_{\text{lock}}\bar{v}$$

which gives the average number of system cycles until the difference between system cycles and signal periods has grown to 1. In terms of the Péclet number $\text{Pe} = \bar{v}/\bar{D}_{\text{eff}}$, N_{lock} can be expressed as

$$N_{\text{lock}} = \text{Pe}\left(1 + \sqrt{1 + \text{Pe}^2\frac{\Delta\nu^2}{\bar{v}^2}}\right)^{-1} \tag{2.4}$$

If the frequency difference vanishes, $\Delta\nu \to 0$ the mean number of locked cycles equals half the Péclet number. Eq. (2.4) expresses N_{lock} in terms of two quantities of the periodically driven system, namely the mean frequency $\bar{v}$ and its effective diffusion coefficient $\bar{D}_{\text{eff}}$. As an example we numerically evaluated the mean frequency $\bar{v}$ and the mean number of synchronized system cycles N_{lock} of an excitable FitzHugh-Nagumo system (FHN-system)

$$\dot{x} = x - x^3 - y + \sqrt{2D}\xi(t) \quad \text{and} \quad \dot{y} = \epsilon(x + a_0 - a_1 y - s(t)) \tag{2.5}$$

subjected to a dichotomic periodic driving (n is some integer)

$$s(t) = \begin{cases} A & \text{if } t \in [n\mathcal{T}, (n+\frac{1}{2})\mathcal{T}) \\ -A & \text{if } t \in [(n+\frac{1}{2})\mathcal{T}, (n+1)\mathcal{T}) \end{cases}$$

Tuning the signal frequency we observe different regions of frequency locking, accompanied by a high number N_{lock} of synchronized system cycles, indicating stochastic synchronization (Fig. 2.1).

The following sections are devoted to the calculation of the mean frequency and effective diffusion coefficient in periodically driven bistable and excitable systems. Our approach is based on a modeling of these dynamics as periodic renewal processes, which is presented in the following section.

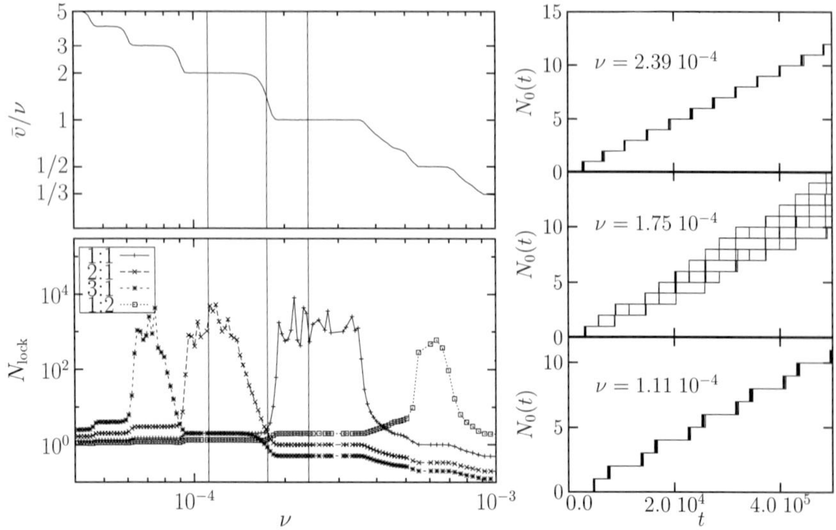

Fig. 2.1. Synchronization in the excitable FHN system. The left figure shows the relation between mean frequency $\bar{v}$ of the system and driving frequency $\nu = \frac{1}{T}$, the effective diffusion coefficient $D_{\rm eff}$ and the mean number of locked cycles $N_{\rm lock}$ according to eq. (2.4) as a function of the driving frequency ν.
Right: For three different driving frequencies, corresponding to 1 : 1 (bottom), 2 : 1, (top) and no synchronization, we have plotted 20 realizations of the corresponding number of cycles $N_0(t)$ which is chosen to increase by 1 at each excitation time. Parameters of the FHN system eqs. (2.5): $a_0 = 0.405$, $a_1 = 0.5$, $\epsilon = 0.001$, $D = 1.0 \times 10^{-5}$, and a dichotomic driving $s = \pm 0.015$, resp. $A = 0.015$. [15]

2.3. Discrete models of continuous stochastic dynamics

One possibility to simplify a continuous stochastic system is the reduction to a description in terms of a few discrete states. The system's behavior is then specified by the transition times between these discrete states. For example when investigating a neurons behavior, the important aspect are often only the times when a spike is emitted and not the complex evolution of the membrane potential [16]. In a doublewell potential system, depending on the questions asked, it may be sufficient to know in which of the two wells the system is located, neglecting the fluctuations in the wells as well as the actual dynamics when crossing from one well to the other. In these cases a reduction to a discrete description can be considered as an appropriate simplification. We first review the two state description of bistable systems [10] and then introduce a phenomenological discrete model for excitable dynamics.

2.3.1. *The doublewell system – a discrete Markovian description*

Let us consider a particle in a symmetric doublewell potential driven by white noise and a signal $s(t)$, Fig. 2.2. The dynamics of this system is

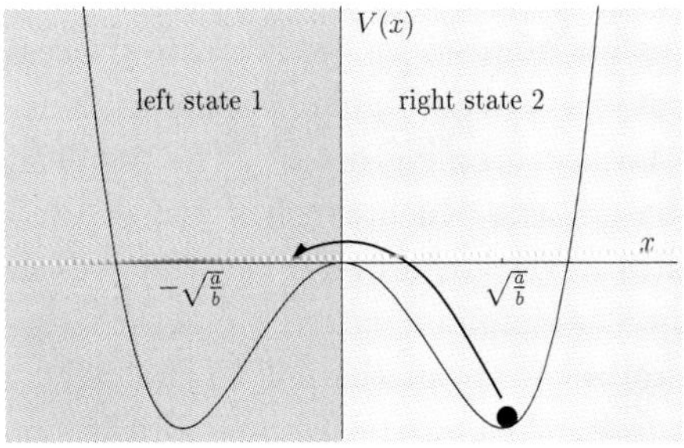

Fig. 2.2. A bistable system as described by eq. (2.6). The reduced two state description is illustrated by the two differently colored half planes.

described by the Langevin equation

$$\dot{x} = -U'(x) + s(t) + \sqrt{2D}\xi(t), \quad U(x) = -\frac{a}{2}x^2 + \frac{b}{4}x^4 \tag{2.6}$$

where $\xi(t)$ is white noise with $\langle\xi(t)\rangle = 0$ and $\langle\xi(t)\xi(t+\tau)\rangle = \delta(\tau)$. If we consider the output (position) $x(t)$ in the course of time (see Fig. 2.2) we notice that the particle is jiggling most of the time in one of the two wells, only occasionally jumping from one well to the other.

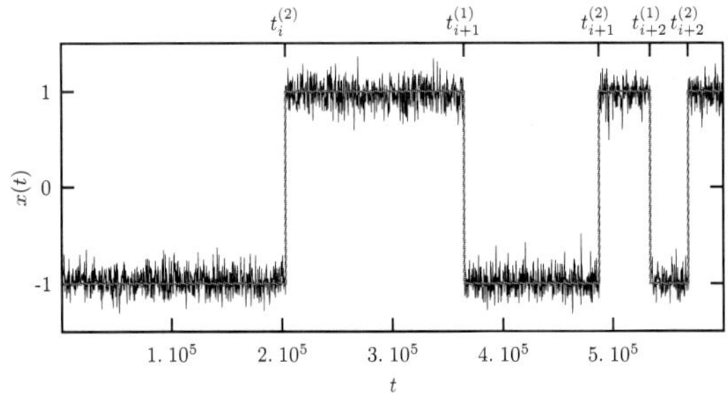

Fig. 2.3. The output $x(t)$ of a bistable system as described by eq. (2.6) with $a = b = 1$ and $D = 0.025$. The reduced two state description is illustrated by the gray line: $s(t) \equiv 0$.

In order to simplify the description of this system one neglects the fast dynamics in the potential wells and considers only the transitions from one well to the other which happen on a much slower time scale. Under the assumption that the potential barrier ΔU between the two wells is large compared to the noise strength D and the relaxation in the wells is fast compared to the time scale of the jumps between the wells, the transitions can be considered as a rate process. Such a rate process has a probability per unit time to cross the barrier, which is independent on the time which has elapsed since the last crossing event. The resulting dynamics in the reduced discrete phase space which consists just of two discrete states *left* and *right* is thus still a Markovian one, i.e. the present state determines the future evolution to a maximal extent.

The transition rates $\gamma^{(1)}$ from the left to the right well and $\gamma^{(2)}$ from the right to the left well, which are both equal for the symmetric doublewell potential, are Kramers rates for the excitation over a potential barrier due to white noise. They can be calculated as [7, 8]

$$\gamma^{(1)} = \frac{\omega_0 \omega^{(1)}}{2\pi\eta} \exp(-\frac{\Delta U^{(1)}}{D}) \quad \text{and} \quad \gamma^{(2)} = \frac{\omega_0 \omega^{(2)}}{2\pi\eta} \exp(-\frac{\Delta U^{(2)}}{D}) \tag{2.7}$$

where ω_0, $\omega^{(1)}$ and $\omega^{(2)}$ are the frequencies (square roots of the modulus of the second derivatives of the potential) at the potential barrier and the left and right minimum respectively, while $\Delta U^{(1)}$ and $\Delta U^{(2)}$ are the potential differences between the potential maximum and the left and right minimum respectively. η is the friction coefficient which is considered to be large (overdamped case). If the potential is modulated by a signal, which varies on a time scale which is slow compared to the relaxation times in the wells and which is sufficiently small such that $\Delta U^{(i)}(t) \gg D$ is satisfied at any time, the two state description still constitutes a rate process, however with time dependent rates

$$\gamma^{(i)}(t) = \frac{\omega_0(t)\omega^{(i)}(t)}{2\pi\eta} \exp(-\frac{\Delta U^{(i)}(t)}{D}). \tag{2.8}$$

Let us consider the waiting time distribution density $w^{(i)}(\tau, t)$ to stay the time τ in state i, if it was entered at time t. For a driven rate process they are known to be

$$w^{(i)}(\tau, t) = \gamma^{(i)}(t+\tau) \exp(-\int_t^{t+\tau} dt' \gamma^{(i)}(t')), \tag{2.9}$$

which in case of a time independent rate reduces to the well known exponential distributions

$$w^{(i)}(\tau) = \gamma^{(i)} \exp(-\gamma^{(i)}\tau).$$

For a periodic signal $s(t) = s(t+\mathcal{T})$ the waiting time distribution densities evidently are also periodic with respect to the entrance time, $w^{(i)}(\tau, t) = w^{(i)}(\tau, t + \mathcal{T})$. Eventually, the density of the times needed for a full cycle $1 \to 2 \to 1$ of the driven system is given by the generalized convolution of the waiting time distribution densities in state 1 and 2

$$w(\tau, t) = \int_0^\tau d\tau' w^{(2)}(\tau - \tau', t + \tau') w^{(1)}(\tau', t). \tag{2.10}$$

This time dependent will be our starting point for the calculation of the synchronization properties in the bistable system in sections 2.4, 2.5.1.

2.3.2. *Excitable dynamics – a phenomenological discrete model*

In contrast to bistable systems an excitable system is monostable, i.e. its dynamics has a single stable fixed point called rest state. Small perturbations of this rest state are damped. However sufficiently large perturbations, which may occur due to noise or an external signal, lead to a strong change in the system's state before it relaxes again to the rest state. Modeling excitable dynamics on a continuous plane phase space, the resulting system has to be at least a two dimensional non potential systems, which already points out some difficulties. However, excitable dynamics can also be modeled as a dynamics on a circle [19] or as a one dimensional dynamics with a superimposed threshold and reset condition like in integrate-and-fire models [17]. In this section we propose a different approach, modeling excitable dynamics as a discrete state system. A prototypical system exhibiting excitable dynamics is the two dimensional FitzHugh-Nagumo system (FHN) which was independently proposed by FitzHugh [3] and Nagumo, Arimoto and Yoshizawa [11] to describe the membrane potential of a neuron. The FitzHugh-Nagumo dynamics is governed by the Langevin equations

$$\dot{x} = x - x^3 - y + s_x(t) + \sqrt{2D}\xi(t) \tag{2.11a}$$

$$\dot{y} = \epsilon(x + a_0 - a_1 y + s_y(t)) \tag{2.11b}$$

where $\xi(t)$ is white noise with $\langle\xi(t)\rangle = 0$ and $\langle\xi(t)\xi(t+\tau)\rangle = \delta(\tau)$ and $s_x(t)$ and $s_y(t)$ denote external signals entering either into the x or y dynamics. The nullclines $\dot{x} = 0$ and $\dot{y} = 0$ together with a typical trajectory are shown in Fig. 2.4. Due to the small parameter ϵ in eqs. (2.11) the x-dynamics is

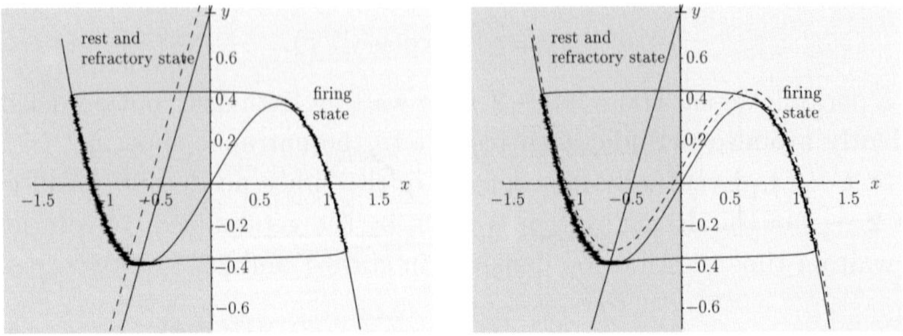

Fig. 2.4. An excitable FHN system subjected to an external driving $s_y(t)$ acting on the slow recovery variable y (left) or $s_x(t)$ acting on the voltage variable x (right) as described by eq. (2.11). The signal applied to y moves the y-nullcline upwards and downwards, thus moving the stable fixed point towards the excitation barrier and back again. The signal applied to x moves the x-nullcline upwards and downwards, thus having a similar effect on the the excitation barrier.

much faster than the y-dynamics.

Let us look in more detail onto the behavior of the FHN system: From the stable fixed point at the intersection of both nullclines, which is also called *rest* state in the neuronal context, the system is excited by a sufficiently large perturbation, leading to a fast transition onto the right branch of the cubic x-nullcline. On the right branch the system takes a high x-value, which at the neuronal level represents a high membrane potential. This state is called *firing* state. After having moved along the right branch the system returns back to the left branch. There, the output (x variable) approaches again a low value like in the rest state, however the system cannot be directly re-excited, it first has to relax back to the rest state. Therefore this state is called *refractory*. A typical output $x(t)$ of the noisy FHN system showing the spiking behavior is presented in Fig. 2.5.

Fig. 2.6 shows the interspike interval distribution density for different constant values of the signal acting additively on either the x or y dynamics. We observe that the interspike intervals are composed of two parts, namely an approximately fixed, and in particular signal independent time, which can be assigned to the motion along the excitation loop, i.e. the stable branches of the cubic nullcline. The second component is an exponentially distributed time, which accounts for the noise induced excitation from the stable fixed point. The corresponding rate strongly depends on the external signal. As for the doublewell system, we assume that for a sufficiently slowly

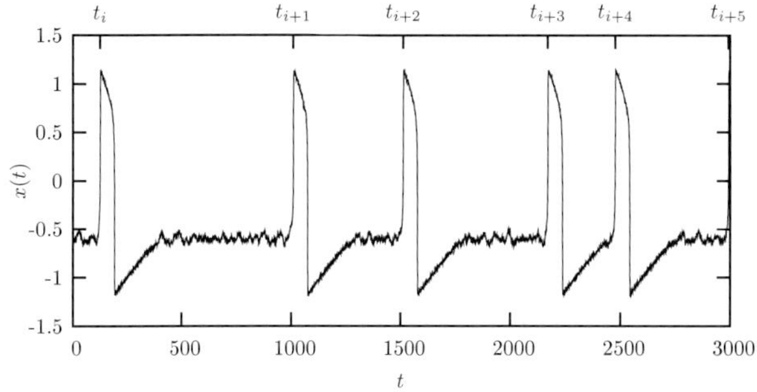

Fig. 2.5. Output $x(t)$ of the FHN system eqs. (2.11) without signal $s_x(t) = s_y(t) = 0$. Other parameters: $a_0 = 0.41$, $a_1 = 0.5$, $D = 0.0001$ and $\epsilon = 0.01$.

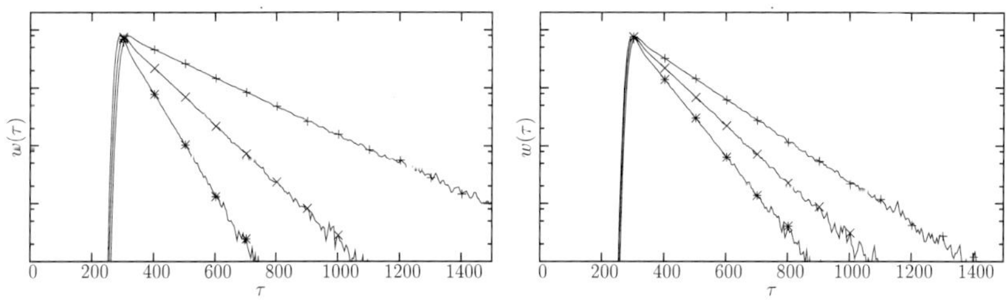

Fig. 2.6. Logarithmic plot of the interspike interval distribution densities of the FHN system eq. (2.11) for different constant values of the external signal $s_y(t)$ (left) or $s_x(t)$ (right) and two different noise levels. $s_{x/y}(t) = 0.01$ ($*$), $s_{x/y}(t) = 0.0$ ($\times$) and $s_{x/y}(t) = -0.01$ ($+$). Other parameters $a_0 = 0.41$, $a_1 = 0.5$, $D = 0.0002$ and $\epsilon = 0.01$. [13]

varying signal, this process remains a rate process, however with a time dependent rate. Assuming the time needed to move along the excitation loop to be distributed according to some waiting time distribution density $w_{\text{loop}}(\tau)$, and denoting the periodic excitation rate by $\gamma(t)$ we obtain the cycle time distribution density, also called interspike interval distribution density,

$$w(\tau, t) = \int_0^\tau d\tau' \gamma(t+\tau) \exp\big[- \int_{t+\tau'}^{t+\tau} dt' \gamma(t') \big] w_{\text{loop}}(\tau') \qquad (2.12)$$

We mention that the interspike interval distribution densities for the FHN system in the different dynamical regimes, including excitable behavior, were already presented in [9]. The calculations were performed under the assumption of a perfect time scale separation and linearization of the nullclines, leading to complex expressions for the Laplace transforms of the waiting time distribution densities in terms of the parameters of the original FHN model eqs. (2.11). However our phenomenological model will be more amenable to further analysis of stochastic synchronization.

In the above considerations we silently assumed subsequent intervals to be correlated only by the external signal and not due to the system dynamics itself. In the undriven case this implies that the spiking times constitute a renewal process, thus rendering subsequent interspike intervals independent. This assumption is justified for a sufficiently strong time scale separation $\epsilon \ll 1$. In this case each excitation happens approximately along the same trajectory in phase space, and thus the information about the duration of the previous interspike interval is lost upon excitation. Therefore subsequent interspike intervals cannot be correlated. If the system is periodically driven, subsequent interspike intervals are certainly correlated, as it is this correlation which leads to the observed synchronization. However the correlation is generated only by the periodic driving. To render this statement more precise, let τ_i and τ_{i+1} denote two subsequent interspike intervals and t_i the time when the first ends and the second starts. Then

$$\langle \tau_i \tau_{i+1} \rangle - \langle \tau_i \rangle \langle \tau_{i+1} \rangle \neq 0$$

in general, i.e. subsequent intervals are correlated. However subsequent intervals, conditioned on the time t_i in between are independent,

$$\langle \tau_i \tau_{i+1} | t_i \rangle - \langle \tau_i | t_i \rangle \langle \tau_{i+1} | t_i \rangle = 0.$$

The remaining part of this paper is devoted to the calculation of the mean frequency $\bar{v}$ and the effective diffusion coefficient $\bar{D}_{\text{eff}}$ as defined in eqs. (2.2) for the discrete model of bistable and excitable systems. To this end we introduce a general theory to calculate these quantities for periodic renewal processes in section 2.4. This theory is then applied to the driven renewal model eqs. (2.9), (2.10) for the doublewell system and eq. (2.12) for the excitable system.

2.4. The effective diffusion coefficient and mean frequency in periodically driven renewal processes

In order to calculate the mean frequency $\bar{v}$ of system cycles and the effective diffusion coefficient $\bar{D}_{\text{eff}}$ defined by eqs. (2.2) we first consider the corresponding instantaneous quantities

$$v(t) = \lim_{t_0 \to -\infty} \frac{d}{dt} \langle N_{t_0}(t) \rangle \tag{2.13}$$

and

$$D_{\text{eff}}(t) = \lim_{t_0 \to -\infty} \frac{d}{dt} \frac{\langle N_{t_0}^2(t) \rangle - \langle N_{t_0}(t) \rangle^2}{2} \tag{2.14}$$

For periodically driven renewal processes these quantities can be shown to be periodic with the period of the driving. $\bar{v}$ and $\bar{D}_{\text{eff}}$ can then be obtained by taking the period average,

$$\bar{v} = \frac{1}{\mathcal{T}} \int_0^{\mathcal{T}} dt v(t) \qquad \text{and} \qquad \bar{D}_{\text{eff}} = \frac{1}{\mathcal{T}} \int_0^{\mathcal{T}} dt D_{\text{eff}}(t) \tag{2.15}$$

In the following we present a method, which relates $v(t)$ and $D_{\text{eff}}(t)$ to the time dependent cycle time distribution density $w(\tau, t)$, which governs the dynamics of the system.

Let us consider the probabilities $p_k(t)$ to have had k cycles up to time t. Furthermore let $j_k(t)$ be the probability flux from state k to state $k+1$, i.e. the probability per time that the $k+1^{\text{st}}$ cycle starts at time t. Taking as initial condition that the first cycle started at time t_0, i.e.

$$j_0(t) = \delta(t - t_0), \tag{2.16}$$

the relation between the probability fluxes $j_k(t)$ of the renewal process is given by

$$j_k(t) = \int_{t_0}^{t} dt' j_{k-1}(t') w(t - t', t'), \quad k \geq 1. \tag{2.17a}$$

Eventually the probabilities $p_k(t)$ are related to the probability fluxes $j_k(t)$ by

$$p_k(t) = \int_{t_0}^{t} dt' j_{k-1}(t') z(t - t', t'), \quad k \geq 1 \tag{2.17b}$$

where the survival probability

$$z(\tau, t) = 1 - \int_0^{\tau} d\tau' w(\tau', t)$$

denotes the probability that the next cycle takes more time than τ, if the last cycle was finished at time t. Based on the probabilities $p_k(t)$ we can define the moments of the number of cycles as

$$M_{t_0}^{[l]}(t) := \sum_{k=0}^{\infty} k^l p_k(t). \tag{2.18}$$

These moments define the corresponding cumulants $K_{t_0}^{[l]}(t)$ from which finally the asymptotic periodic growth coefficients of the cumulants

$$\kappa^{[l]}(t) := \lim_{t_0 \to \infty} \frac{d}{dt} K_{t_0}^{[l]}(t) \tag{2.19}$$

are obtained. The first coefficient $\kappa^{[1]}(t)$ is the instantaneous mean frequency $v(t)$ while $\kappa^{[2]}(t)$ is twice the instantaneous effective diffusion coefficient $D_{\text{eff}}(t)$. In principle they can be calculated from the solutions of eqs. (2.17) according to eqs. (2.18) and (2.19) . However, in practice this is not feasible, as one has to calculate an infinite sum over the $p_k(t)$ where each p_k, according to eqs. (2.17b) and (2.17a), is a k-fold integral involving the waiting time distribution densities $w(\tau, t)$ and the corresponding survival probabilities $z(\tau, t)$.

To find a simpler relation between the periodic coefficients $\kappa^{[n]}(t)$ and the time dependent waiting time $w(\tau, t)$, which governs the microscopic dynamics, we again construct a continuous embedding in the asymptotic limit $t_0 \to -\infty$. Consider the continuous probability distribution density $\mathcal{P}(x, t)$ as an envelope of the discrete probabilities $p_k(t)$. Thus, respecting the normalization, we adopt the relation (cf. Fig. 2.7)

$$p_k(t) = \int_{k-\frac{1}{2}}^{k+\frac{1}{2}} dx \mathcal{P}(x, t). \tag{2.20}$$

We assume that asymptotically, i.e. when the $p_k(t)$ and $\mathcal{P}(x, t)$ approach

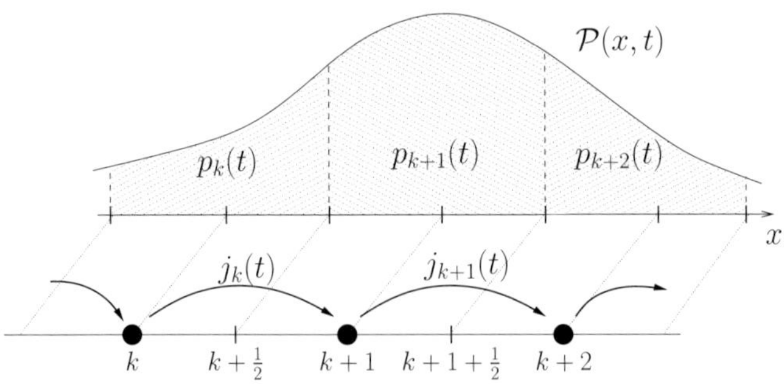

Fig. 2.7. Schematic view of the discrete event dynamics $p_k(t)$ and the continuous description in terms of x. [14]

more and more a uniform density, this equation remains valid in the course of time if the cumulants in both, the discrete and continuous setting grow in the same way. To have the desired growth behavior of the cumulants in the continuous description we assign a Kramers Moyal equation to the evolution of the continuous probability distribution density,

$$\frac{\partial}{\partial t}\mathcal{P}(x,t) = \sum_{n=1}^{\infty} \frac{(-1)^n}{n!}\kappa^{[n]}(t)\frac{\partial^n}{\partial x^n}\mathcal{P}(x,t) \tag{2.21}$$

whose Kramers Moyal coefficients coincide with the periodic cumulant growth coefficients $\kappa^{[n]}(t)$ of the discrete description [18].

Having related the discrete and continuous probability we can likewise relate the probability current $j_k(t)$ of the discrete system to the probability current $\mathcal{J}(x,t)$ of the continuous envelope description. According to the relation between the discrete and continuous probability eq. (2.20), the discrete probability current $j_k(t)$ from k to $k+1$ is equal to the continuous probability current $\mathcal{J}(x,t)$ at $x = k + \frac{1}{2}$ (see Fig. 2.7),

$$j_k(t) = \mathcal{J}(k + \frac{1}{2}, t). \tag{2.22}$$

The continuous probability current $\mathcal{J}(x,t)$ is related to the probability distribution density $\mathcal{P}(x,t)$ by the continuity equation

$$\frac{\partial}{\partial t}\mathcal{P}(x,t) = -\frac{\partial}{\partial x}\mathcal{J}(x,t). \tag{2.23}$$

and therefore according to eq. (2.21)

$$\mathcal{J}(x,t) = -\sum_{n=1}^{\infty} \frac{(-1)^n}{n!}\kappa^{[n]}(t)\frac{\partial^{n-1}}{\partial x^{n-1}}\mathcal{P}(x,t)\,. \tag{2.24}$$

Thus from eq. (2.22) we deduce

$$j_k(t) = -\sum_{n=1}^{\infty} \frac{(-1)^n}{n!}\kappa^{[n]}(t)\frac{\partial^{n-1}}{\partial x^{n-1}}\mathcal{P}(x,t)\bigg|_{x=k+\frac{1}{2}} \tag{2.25}$$

Having fixed the relation between the probabilities and probability fluxes of the discrete renewal process and the continuous embedding, it is now possible to relate the coefficients $\kappa^{[n]}(t)$ appearing in the continuous description (2.21) to the waiting time distribution density $w(\tau,t)$ of the renewal process, involved in the microscopic dynamics (2.17a) and (2.17b). As we are considering the asymptotic behavior we have to pass to the asymptotic limit in eqs. (2.17a) and (2.17b) by shifting the initial time $t_0 \to -\infty$. This results in

$$p_k(t) = \int_0^{\infty} d\tau j_{k-1}(t-\tau)z(\tau, t-\tau). \tag{2.26}$$

and

$$j_k(t) = \int_0^\infty d\tau j_{k-1}(t-\tau) w(\tau, t-\tau). \tag{2.27}$$

Inserting eqs. (2.20) and (2.25) into the above eq. (2.26) we end up with

$$\int_{-\frac{1}{2}}^{\frac{1}{2}} d\Delta x \mathcal{P}(x-\Delta x, t) = -\int_0^\infty d\tau z(\tau, t-\tau) \tag{2.28}$$
$$\sum_{n=1}^{\infty} \frac{(-1)^n}{n!} \kappa^{[n]}(t-\tau) \frac{\partial^{n-1}}{\partial x^{n-1}} \mathcal{P}(x-\frac{1}{2}, t-\tau)$$

with $x = k$. The probability $\mathcal{P}(x-\Delta x, t-\tau)$ can be expressed in terms of the probability $\mathcal{P}(x,t)$ and its derivatives $\frac{\partial^m}{\partial x^m}\mathcal{P}(x,t)$ by performing a Taylor expansion of $\mathcal{P}(x-\Delta x, t-\tau)$ around x, t and converting the time derivatives to derivatives with respect to the state x using the Kramers-Moyal equation (2.21). This results in [14]

$$\mathcal{P}(x-\Delta x, t-\tau) =$$
$$\mathcal{P}(x,t) + c_t^{[1]}(\tau, \Delta x)\frac{\partial}{\partial x}\mathcal{P}(x,t) + c_t^{[2]}(\tau, \Delta x)\frac{\partial^2}{\partial x^2}\mathcal{P}(x,t) + O(3) \tag{2.29}$$

where $O(3)$ denotes terms proportional to $\frac{\partial^m}{\partial x^m}\mathcal{P}(x,t)$ with $m \geq 3$. The functions $c_t^{[1]}(\tau, \Delta x)$ and $c_t^{[2]}(\tau, \Delta x)$ are given by

$$c_t^{[1]}(\tau, \Delta x) = \int_0^\tau d\tau' \kappa^{[1]}(t-\tau') - \Delta x \, .$$

and

$$c_t^{[2]}(\tau, \Delta x) = \frac{\Delta x^2}{2} - \Delta x \int_0^\tau d\tau' \kappa^{[1]}(t-\tau')$$
$$-\frac{1}{2}\int_0^\tau d\tau' \kappa^{[2]}(t-\tau') + \int_0^\tau d\tau' \kappa^{[1]}(t-\tau') \int_0^{\tau'} d\tau'' \kappa^{[1]}(t-\tau'')$$

Equating the coefficients of $\mathcal{P}(x,t)$ and $\frac{\partial}{\partial x}\mathcal{P}(x,t)$ on both sides of eq. (2.28), we end up with ($v(t) \equiv \kappa^{[1]}(t)$ and $D_{\text{eff}}(t) \equiv 2\kappa^{[2]}(t)$)

$$\int_0^\infty d\tau v(t-\tau) z(\tau, t-\tau) = 1 \tag{2.30a}$$

$$\int_0^\infty d\tau D_{\text{eff}}(t-\tau) z(\tau, t-\tau) = \int_0^\infty d\tau v(t-\tau) \int_0^\tau d\tau' v(t-\tau') z(\tau, t-\tau) - \frac{1}{2} \tag{2.30b}$$

These two expressions relate the asymptotic drift and diffusion properties of a periodically driven renewal process as expressed by $v(t)$ and $D_{\text{eff}}(t)$ to its microscopic properties defined by the time dependent survival property $z(\tau, t) = 1 - \int_0^\tau d\tau' w(\tau', t)$. From the periodic solution of these equations one finally obtains the mean frequency $\bar{v}$ and effective diffusion coefficient $\bar{D}_{\text{eff}}$ by taking the period average according to eqs. (2.15).

Equations which govern the higher cumulant growth coefficients $\kappa^{[n]}(t)$, $n \geq 3$, can also be derived using this method [14] by evaluating the coefficients of higher order derivatives of $\mathcal{P}(x, t)$.

2.4.1. *System cycles involving driven rate steps*

Consider one cycle of the system to consist of two independent steps. Examples of such systems include the discrete model for excitable systems, whose cycles consist of an excitation step and the motion along the excitation loop, as well as the bistable two state system, where a full cycle is composed of a transition from left to right and a transition back from right to left again. Denoting the density of the times needed for the first and second step by $w^{(1)}(\tau, t)$ and $w^{(2)}(\tau, t)$ respectively the density of the cycle times is given by the generalized convolution

$$w(\tau, t) = \int_0^\tau d\tau' w^{(2)}(\tau - \tau', t + \tau') w^{(1)}(\tau', t). \tag{2.31}$$

If we assume the second step to be a rate process with time dependent rate $\gamma^{(2)}(t)$, i.e

$$w^{(2)}(\tau, t) = \gamma^{(2)}(t + \tau) \exp\big[- \int_t^{t+\tau} dt' \gamma^{(2)}(t') \big] \tag{2.32}$$

the general eqs. (2.30) can be further simplified. To this end consider a time dependent waiting time distribution density $w(\tau, t)$ and corresponding survival probability $z(\tau, t) := 1 - \int_0^\tau d\tau' w(\tau', t)$. Then

$$\frac{d}{dt} \int_0^\infty d\tau f(t - \tau) z(\tau, t - \tau) = f(t) - \int_0^\infty d\tau f(t - \tau) w(\tau, t - \tau) \tag{2.33}$$

and

$$\frac{d}{dt} \int_0^\infty d\tau f(t - \tau) \int_0^\tau d\tau' f(t - \tau') z(\tau, t - \tau) \tag{2.34}$$
$$= f(t) \int_0^\infty d\tau f(t - \tau) z(\tau, t - \tau) - \int_0^\infty d\tau f(t - \tau) \int_0^\tau d\tau' f(t - \tau') w(\tau, t - \tau).$$

For the special case of the waiting time distribution density defined by eqs. (2.31) and (2.32) we additionally have

$$w(\tau, t - \tau) = \gamma^{(2)}(t) z(\tau, t - \tau) - \gamma^{(2)}(t) z^{(1)}(\tau, t - \tau) \tag{2.35}$$

where $z^{(1)}(\tau,t) := 1 - \int_0^\tau d\tau' w^{(1)}(\tau',t)$ is the survival probability of step 1. Making the Ansatz

$$v(t) = \gamma^{(2)}(t) q_0(t) \qquad \text{and} \qquad D_{\text{eff}}(t) = \gamma^{(2)}(t) q_1(t) \tag{2.36}$$

one can show from eqs. (2.30), using eqs. (2.33), (2.34) and (2.35) that the auxiliary variables $q_0(t)$ and $q_1(t)$ obey

$$q_0(t) + \int_0^\infty d\tau \gamma^{(2)}(t-\tau) q_0(t-\tau) z^{(1)}(\tau, t-\tau) = 1 \tag{2.37a}$$

$$\begin{aligned} q_1(t) + \int_0^\infty d\tau \gamma^{(2)}(t-\tau) q_1(t)(t-\tau) z^{(1)}(\tau, t-\tau) &= -\frac{1}{2} + q_0(t) \qquad (2.37\text{b}) \\ + \int_0^\infty d\tau \gamma^{(2)}(t-\tau) q_0(t-\tau) \int_0^\tau d\tau' \gamma^{(2)}&(t-\tau') q_0(t-\tau') z^{(1)}(\tau, t-\tau). \end{aligned}$$

If in addition the first step is also a rate process with a time dependent rate $\gamma^{(1)}(t)$ i.e.

$$w^{(1)}(\tau,t) = \gamma^{(1)}(t+\tau) \exp\big[-\int_t^{t+\tau} dt' \gamma^{(1)}(t')\big]$$

eqs. (2.37) can be further simplified, finally leading to

$$\frac{d}{dt} q_0(t) = -\big[\gamma^{(1)}(t) + \gamma^{(2)}(t)\big] q_0(t) + \gamma^{(1)}(t) \tag{2.38a}$$

$$\frac{d}{dt} q_1(t) = -\big[\gamma^{(1)}(t) + \gamma^{(2)}(t)\big] q_1(t) - \frac{\gamma^{(1)}(t)}{2} - \gamma^{(2)}(t) q_0^2(t). \tag{2.38b}$$

In the following section eqs. (2.37) and (2.38) will be used to investigate synchronization in excitable and bistable systems, respectively.

2.5. Applications

2.5.1. *Synchronization in a doublewell system*

Consider the discrete model of the periodically driven doublewell potential system. According to subsection 2.3.1 a cycle of this system consists of two driven rate processes. Thus we can employ eqs. (2.36) and (2.38) in order to evaluate the mean frequency and the effective diffusion coefficient. To be more specific, we consider a symmetric doublewell potential with dichotomic driving,

$$\gamma^{(1)}(t) = \begin{cases} r_0 \text{ if } t \in [n\mathcal{T}, (n+\frac{1}{2})\mathcal{T}) \\ r_1 \text{ if } t \in [(n+\frac{1}{2})\mathcal{T}, (n+1)\mathcal{T}) \end{cases} \tag{2.39}$$

and vice versa for $\gamma^{(2)}(t)$. In this special case, eqs. (2.38) can be solved analytically leading after some cumbersome algebra to [4]

$$\bar{v} = \frac{1}{\mathcal{T}} \int_0^{\mathcal{T}} dt v(t) = v_{\rm st} + \alpha\nu \tanh R \tag{2.40}$$

and [2, 13]

$$\begin{aligned} \bar{D}_{\rm eff} &= \frac{1}{\mathcal{T}} \int_0^{\mathcal{T}} dt D_{\rm eff}(t) \\ &= \frac{v_{\rm st}}{2}\Big[\frac{1}{2} + \alpha(\frac{1}{2} + \cosh^{-2} R)\Big] + \frac{\alpha\nu}{2}\Big[\alpha(\frac{1}{2}\cosh^{-2} R + 1) - 1\Big] \tanh R \end{aligned} \tag{2.41}$$

where we have introduced the mean frequency without driving $v_{\rm st} := 1/(\frac{1}{r_0} + \frac{1}{r_1})$, a quantifier for the driving strength $\alpha = \frac{(r_0 - r_1)^2}{(r_0 + r_1)^2}$ and some ratio between inner time scale and driving frequency $R = \frac{(r_0 + r_1)}{4\nu}$. These results agree with the results found independently in [2] using a different approach. Having calculated the effective diffusion coefficient and the mean frequency we can evaluate the mean number of locked cycles $N_{\rm lock}$ according to eq. (2.4).

The transition rates r_0 and r_1 can be calculated according to eqs. (2.7) for the bistable Langevin dynamics eq. (2.6) with a dichotomic driving $s(t) \in \{\pm A\}$. In Fig. 2.8 we have plotted $\bar{v}$ and $N_{\rm lock}$ as a function of the noise strength D. We observe a $1:1$ synchronization region, where the frequency of the system is locked to the driving and the number of locked system cycles $N_{\rm lock}$ becomes very high. The maximum of $N_{\rm lock}$ is very pronounced leading to a more sensitive measure of optimal synchronization than just considering the plateau in the mean system frequency. The theoretical results according to eqs. (2.40) and (2.41) agree perfectly with simulations of the discrete Markovian model and also the simulations of the underlying continuous Langevin dynamics reasonably agree within the range of validity of the rate approximation. Finally we consider $N_{\rm lock}$ as a function of driving frequency. Synchronization shows a maximum as a function of driving frequency, in contrast to spectral based measures of stochastic resonance, like the spectral power amplification (Fig. 2.9). The optimal driving frequency is located between the frequency $1/(\frac{1}{r_0} + \frac{1}{r_1})$ of the adiabatically slowly driven system and the frequency $\frac{1}{2}(r_0 + r_1)$ of the infinitely fast driven system.

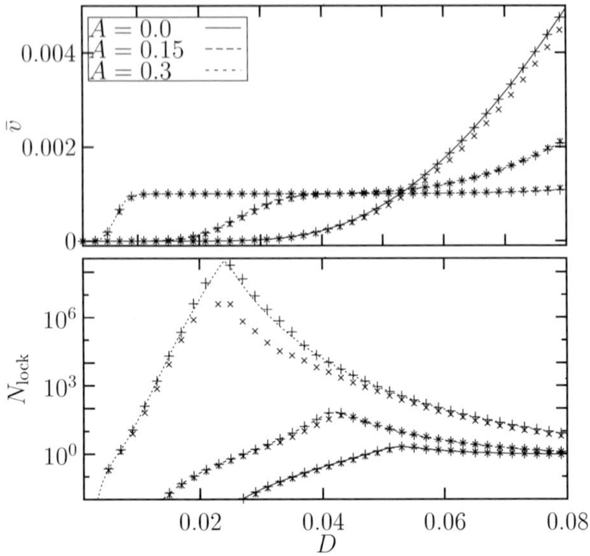

Fig. 2.8. Mean frequency $\bar{v}$ (top) and number of locked cycles $N_{\rm lock}$ (bottom) of the dichotomically periodically driven bistable system eq. (2.6) with $a = b = 1$ ($\times$ symbols) for different values of the driving amplitude A and a driving frequency $\nu = 0.001$. The solid lines show the dependence according to eqs. (2.40) and (2.41). Kramers rates r_0 and r_1 following eq. (2.7) has been inserted. The + symbols present results from simulations of the discrete model. [13]

2.5.2. *Synchronization in the FHN model*

One cycle of the excitable system was argued in subsection 2.3.2 to consist of a signal independent time needed to move along the excitation loop, followed by a rate excitation with signal dependent rate $\gamma(t)$. Thus eqs. (2.37) apply, which however can no longer be solved analytically. Instead we have to resort to numerically methods in order to obtain the periodic solutions $q_0(t)$ and $q_1(t)$ of eqs. (2.37). This is most advantageously done in Fourier space, leading to a infinite set of linear equations for the Fourier coefficients. Truncating the system and solving the remaining finite set of linear equations numerically gives the desired results. Again we consider a dichotomical driving, leading to an excitation rate

$$\gamma(t) = \begin{cases} r_0 \text{ if } t \in [n\mathcal{T}, (n+\frac{1}{2})\mathcal{T}) \\ r_1 \text{ if } t \in [(n+\frac{1}{2})\mathcal{T}, (n+1)\mathcal{T}) \end{cases} \tag{2.42}$$

We further assume a deterministic waiting time T needed to move along the excitation loop (cf. Fig. 2.6), i.e. $w(\tau) = \delta(\tau - T)$. In Fig. 2.10 we have

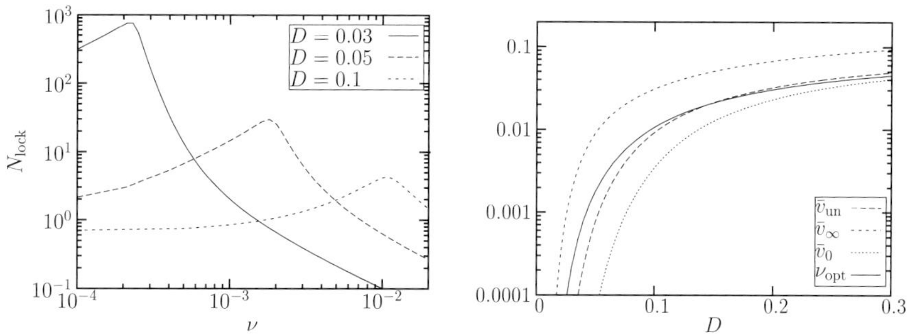

Fig. 2.9. Left: Locked cycles N_{lock} of the Markovian model as a function of driving frequency ν for different noise values. The maximum as a function of driving frequency indicates a "bona fide" resonance. $A = 0.15$, other parameters as in Fig. 2.8. Right: Driving frequency (solid line) at which the number of locked cycles attains its maximum and the mean frequency $v_\infty = \frac{1}{2}(r_0 + r_1)$ of the infinitely fast driven system, $v_0 = 1/(\frac{1}{r_0} + \frac{1}{r_1})$ of the adiabatically slowly driven system v_{un} of the undriven system $A = 0$ as a function of noise strength.

plotted $\bar{v}/\nu$ and N_{lock} as a function of driving frequency. We observe various $m : n$ frequency locking regimes. The corresponding driving frequencies very much depend on the excitation loop time T. Namely $n : 1$ locking is observed if the period of the driving is approximately between $2(n - \frac{1}{2})T$ and $2nT$ while $1 : n$ locking is observed if the period of the driving is approximately between $\frac{1}{n}T$ and $\frac{1}{n-1/2}T$ [13].

Compared with simulations of the two state model we find perfect agreement. Fitting the time T on the excitation loop as well as the two excitation rates r_0 and r_1 to the FHN system, also leads to a reasonable agreement between theory and Langevin simulations of the FHN system.

2.5.3. *Controlling molecular motors*

Finally we apply our methods to investigate the control of molecular motors by a periodic modulation of the fuel concentration. To this end consider a simple two phase model of a molecular motor Fig. 2.11 [15]. The first phase consists of the binding of some fuel molecule X, normally ATP, to the motor protein. This step is a rate process with a rate $\gamma = c[X]$ proportional to the concentration $[X]$ of the fuel molecules. After binding of a fuel molecule a stroke is triggered. Such a stroke is due to a conformational change of the motor protein, which leads to the advancement on the track by one step-size. The time needed for the stroke is assumed to be distributed according to some waiting time distribution density $w_{\text{stroke}}(\tau)$. As a quantifier for the

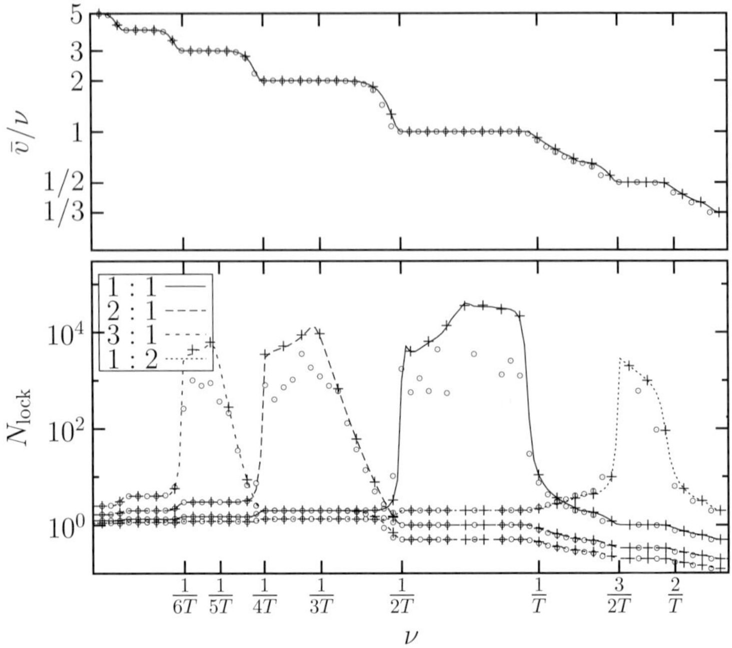

Fig. 2.10. Relation between mean system frequency $\bar{v}$ and driving frequency ν (top) and number of locked cycles N_{lock} (bottom) of the excitable model as a function of driving frequency ν. Lines show dependence as solutions of eqs. (2.36) and (2.37). The + symbols present results from simulations of the discrete system and circles correspond to data from simulations of the FHN system. Parameters for the FHN system eq. (2.11): $a_0 = 0.405$, $a_1 = 0.5$, $\epsilon = 0.001$, $D = 10^{-5}$, $s_x(t) \equiv 0$, $s_y(t) = \pm A$ with $A = 0.015$. Parameters for the two state model (theory and simulations) from simulations of the interspike interval distribution density (cf. Fig. 2.6): $T \approx 2620$, $r_0 \approx 0.0087$ and $r_1 \approx 8.3 \cdot 10^{-8}$. [15]

regularity of the motion of the motor protein we choose the Péclet number

$$\mathrm{Pe} = \frac{\bar{v}}{\bar{D}_{\text{eff}}}.$$

It gives the number of steps after which the variance of the number of steps has grown to one. The higher the Péclet number the more regular is the motion of the motor protein. In order to increase the regularity we consider a periodic modulation of the binding rate $\gamma = c[X]$. This can be obtained by periodically varying the concentration $[X]$ of fuel molecules or likewise, if possible, the reactivity c of the protein.

The resulting system has the very same dynamics as the discrete model

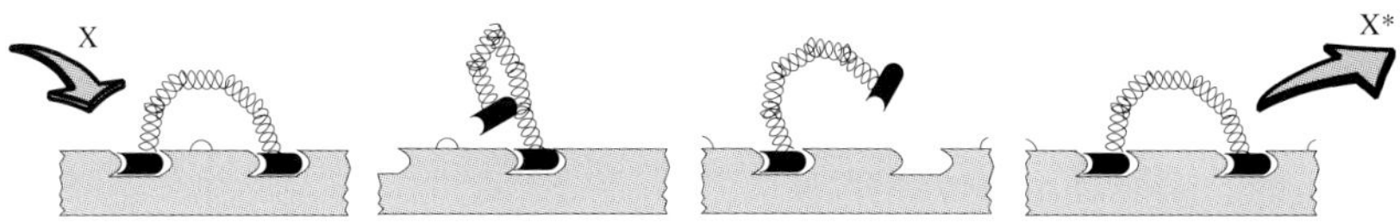

Fig. 2.11. One step of the Brownian stepper. The step is induced by the binding of a fuel molecule X according to a rate process with a rate $\gamma(t)$ which is proportional to the concentration $[X](t)$ of the fuel molecules. Afterwards the motor molecule undergoes conformational changes, thereby releasing the used fuel X^* and advancing by one step Length ℓ. [15]

for excitable dynamics investigated in section 2.5.2, namely the triggering of a stroke which is a rate process with periodically modulated rate followed by the stroke whose time to perform is distributed according to some waiting time distribution density $w_{\text{stroke}}(\tau)$. We can therefore use eqs. (2.36) and (2.37) to numerically evaluate $\bar{v}$ and $\bar{D}_{\text{eff}}$ in order to finally obtain $\text{Pe} = \frac{\bar{v}}{\bar{D}_{\text{eff}}}$.

In the following we consider Γ-distributed stroke times with mean T and variance T^2/n, i.e.

$$w_{\text{stroke}}(\tau) = \frac{1}{\Gamma(n)} \left(\frac{\tau n}{T}\right)^n \frac{\exp(-\frac{\tau n}{T})}{\tau}. \tag{2.43}$$

We further assume a harmonic modulation of the fuel concentration [X],

$$[\text{X}(t)] = \frac{[X]_{\text{max}} + [X]_{\text{min}}}{2} + \frac{[X]_{\text{max}} - [X]_{\text{min}}}{2} \cos 2\pi\nu t.$$

The Péclet number as a function of the modulation frequency is shown in Fig. 2.12. We observe that for optimal modulation frequencies, the regularity of the motor can be strongly enhanced. The resulting Péclet numbers exceed the Péclet number obtained by just applying a constantly high fuel concentration, thereby reducing the stochasticity of each step.

2.6. Conclusion

We have presented a method to calculate the mean frequency and effective diffusion coefficient of the numbers of cycles(events) in periodically driven renewal processes. Based on these two quantities one can evaluate the number of locked cycles in order to quantify stochastic synchronization. Applied to a discrete model of bistable dynamics the theory can be evaluated analytically. The system shows only 1 : 1 synchronization, however in contrast to spectral based stochastic resonance measures the mean number of locked cycles has a maximum at an optimal driving frequency, i.e. the system shows *bona fide* resonance [6]. For the discrete model of

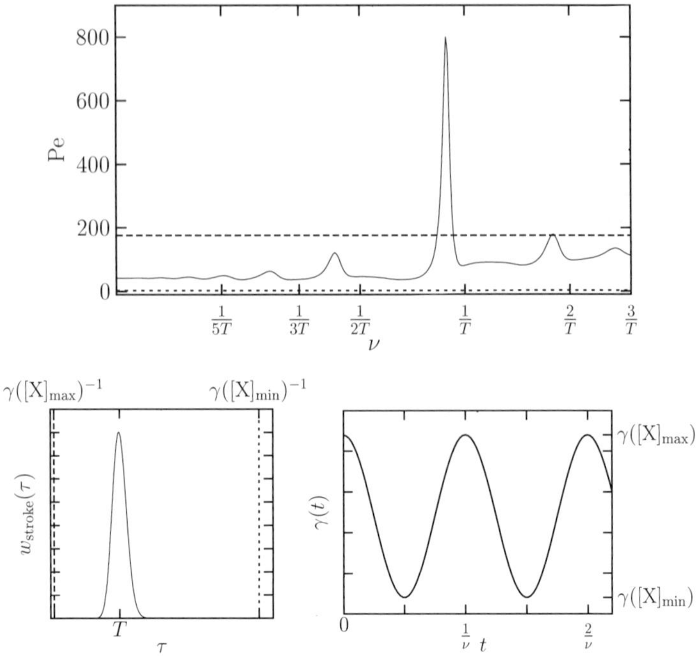

Fig. 2.12. Péclet number as a function of driving frequency (top, solid line) for the molecular motor with a harmonically modulated fuel concentration (bottom right). The stroke time was chosen to be distributed according to a Γ-density eq. (2.43) with $n = 100$. The fastest excitation time $\gamma([X]_{\text{max}})^{-1}$ is 20 times smaller than the mean stroke time T while the slowest excitation time $\gamma([X]_{\text{min}})^{-1}$ is 5 times larger than the mean stroke time T. The dashed and dotted lines in the top plot show the Péclet number for a constant excitation rate $\gamma([X]_{\text{max}})$ and $\gamma([X]_{\text{min}})$ respectively. [15]

excitable dynamics, the theory can be evaluated only numerically. These systems show different $n : m$ synchronization regions, whose corresponding driving frequencies sensitively depend on the time needed to move along the excitation loop.

References

[1] V. S. Anishenko, V. V. Astakhov, A. B. Neiman, T. E. Vadivasova, and L. Schimansky-Geier. *Nonlinear Dynamics of Chaotic and Stochastic Systems.* Springer Verlag, 2002.

[2] J. Casado-Pascual, J. Gómez-Ordóñez, M. Morillo, J. Lehmann an I. Goychuk, and P. Hänggi. Theory of frequency and phase synchronization in a rocked bistable stochastic system. *Physical Review E*, 71:011101, 1 2005.

[3] R. FitzHugh. Impulses and physiological states in theoretical models of nerve membrane. *Biophysical Journal*, 1(6):445–465, 1961.

[4] J. Freund, A.B. Neimann, and L.Schimansky-Geier. Analytic description of noise-induced phase synchronization. *Europhysics Letters*, 50(1):8–14, 2000.

[5] L. Gammaitoni, P. Hänggi, P. Jung, and F. Marchesoni. Stochastic resonance. *Review of Modern Physics*, 70(1):223, 1 1998.

[6] L. Gammaitoni, F. Marchesoni, and S. Santucci. Stochastic resonance as a *Bona Fide* resonance. *Physical Review Letters*, 74(7):1052, 2 1995.

[7] P. Hänggi, P. Talkner, and M. Borkovec. Reaction-rate theory: fifty years after kramers. *Review of Modern Physics*, 62(2):251, 4 1998.

[8] H. A. Kramers. Brownian motion in a field of force and the diffusion model of chemical reactions. *Physica(Utrecht)*, 7:284, 1940.

[9] B. Lindner and L. Schimansky-Geier. Coherence and stochastic resonance in a two-state system. *Physical Review E*, 61:6103, 2000.

[10] B. McNamara and K. Wiesenfeld. Theory of stochastic resonance. *Physical Review A*, 39(9):4854–4869, 5 1989.

[11] J. Nagumo, S. Arimoto, and S. Yoshizawa. An active pulse transmission line simulating nerve axon. *Proc. IRE*, 50(6):2061, 1962.

[12] A. Pikovsky, M. Eosenblum, and J. Kurths. *Synchronization – A universal concept in nonlinear sciences*. Cambridge University Press, 2001.

[13] T. Prager and L. Schimansky-Geier. Phase velocity and phase diffusion in periodically driven discrete-state systems. *Physical Review E*, 71(3):031112, 3 2005.

[14] T. Prager and L. Schimansky-Geier. Phase velocity and phase diffusion in periodically driven renewal processes. *Journal of Statistical Physics*, Vol. 123:391, 2006.

[15] T. Prager, L. Schimansky-Geier, and I. M. Sokolov. Periodic driving controls random motion of brownian steppers. *Journal of Physics Condensed Matter*, 17:S3661–S3672, 2005.

[16] F. Rieke, D. Warland, R. de Ruyter van Stevenick, and W. Bialek. *Spikes: Exploring the neural code*. Cambridge University press, 1996.

[17] R. B. Stein. Some models of neuronal variability. *Biophysical Journal*, 7:37–68, 1967.

[18] R.L. Stratonovich. *Topics in the Theory of Random Noise*. Gordon and Breach, 1969.

[19] K. Wiesenfeld, D. Pierson, E Pantazelou, C. Dames, and F. Moss. Stochastic resonance on a circle. *Physical Review Letters*, 72(14):2125, 4 1994.

Chapter 3

Spiral wave dynamics: Reaction and diffusion versus kinematics

Bernold Fiedler*, Marc Georgi+ and Nihar Jangle!

Institut für Mathematik I, Freie Universität Berlin, Arnimallee 2-6, 14195 Berlin

*fiedler@math.fu-berlin.de, +georgi@math.fu-berlin.de, !jangle@math.fu-berlin.de

Contents

3.1. Phenomena

In nature, spatio-temporal patterns in excitable media occur in seemingly unlimited variety. As early as 1946 Wiener and Rosenblueth [81] introduced the concept of excitable media to explain the propagation of electrical excitation fronts in the heart. Waves of electrical activity in the heart muscle assist its rhythmic contractions. The presence of spiral waves can indicate dangerous fibrillation. This is one of the motivations why the dynamics and control of spiral waves are studied. Furthermore spiral waves are typical, almost ubiquitous, patterns in excitable media; see [88] in this volume.

Mathematically, self-organized spiral patterns are a striking phenomenon of reaction-diffusion systems, in its own right, motivated by a large variety of application areas.

Slime mold aggregation is another example. As long as food in form of bacteria is present the slime mold cells live independently in the soil. As food becomes rare they form a multicellular "organism". This "organism" moves in order to find appropriate conditions for production and dispersal of spores. During the early phase of aggregation, chemotactic movement can proceed in form of spiral waves [17].

Spiral waves also arise in the oxidation of carbon-monoxide on platinum surfaces [10]. In 1972 they have been discovered by Winfree [79] in the photosensitive Belousov-Zhabotinsky (BZ) reaction, see for recent investigations for example [83, 84, 87]. Both reactions are studied in the SFB 555. The classical BZ reaction is a catalytic oxidation of malonic acid, using bromate in an acidic environment. Experimentally it exhibits well reproducible drift, meander and "chaotic" motions of the spiral wave and its tip.

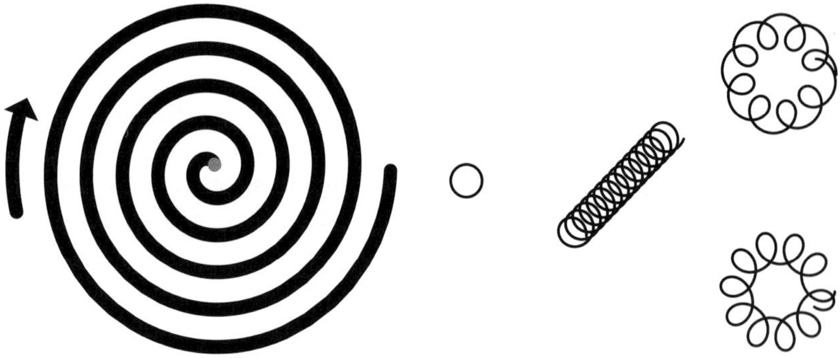

Fig. 3.1. Outward moving spiral wave (left) with different associated tip motions (right: rigid rotation, drift and meander).

In several experiments and numerical simulations, transitions from rigidly rotating spiral waves to other more complicated waves have been observed. The dynamics near rigidly rotating waves and their transition to meandering and drifting spirals has been studied extensively; see, for instance [6, 69, 70]. The transition from planar meandering spirals to seemingly three frequencies has been studied numerically in [63] by calculating the Fourier spectrum of the waves. Analyzing the parameter space of the underlying partial differential equation, different frequency branches have been identified. These frequencies determine the coarse structure and geometry of the pattern. The effect of periodic forcing on rigidly rotating spiral wave solutions has been investigated experimentally in [54] for catalysis on platinum surfaces. The periodic forcing has been realized via an externally modulated temperature. Periodic forcing of a photosensitive variant of the BZ reaction has been studied experimentally in [9, 90]. The forcing is achieved by periodically changing light intensity. In both experimental settings, meanders and resonant drifts occur.

Varying the light intensity enables effective control of the motion of spiral waves, see [7, 37, 75, 88]. Controlling the motion of spiral waves is an important challenge, for example, for the defibrillation of cardiac tissue. The elimination of spiral waves by multiple shocks of external current has been investigated numerically in [64].

Although the examples presented above are of very diverse nature they generate astonishingly similar phenomena. We have chosen the photosensitive BZ reaction as the main experimentally accessible paradigm in the SFB 555; see [88] in this volume. Therefore our paper aims at the mathematical analysis of the dynamics of complex spatio-temporal patterns in nonlinear excitable, spatially or temporally heterogeneous media, with the BZ reaction in mind.

Depending on the space dimension different but related resulting dynamics can be observed. One example are the emission centers of pulse chains in one space dimension. In two space dimensions there are emissions of target patterns and the phenomena of rotating, meandering and drifting spirals. In three space dimensions, formally, these correspond to oscillating and drifting, possibly twisted, filaments of scroll waves.

There are two alternative methods commonly used to model the dynamics of spiral waves. The first one is a description by parabolic partial differential equations (PDEs), specifically *reaction-diffusion systems*. This has been one of the settings used by Wiener, originally, in [81]. The other method is a description of the wave fronts by reduced (mean-)curvature flows of curves and surfaces. This approach, known as *kinematic theory*, also dates back as far as Wiener's original paper [81]. The special case of an *eikonal theory* assumes the normal propagation velocity of the wave front to be independent of curvature, as stated in Huygens' principle and in geometric optics. For more detailed mathematical investigations, and partial justifications in a singular perturbation setting, see also [39], [36], [52], [43], the survey [48], and the references there.

Either approach has advantages and disadvantages when tackling rotating, meandering and drifting spiral wave phenomena.

Reaction-diffusion systems encounter difficulties even for the seemingly simple question of mere existence of rigidly rotating spiral waves. With the technically demanding tool of spatial dynamics in the (logarithmic) radial direction, this difficulty has been overcome for small amplitude waves in a celebrated paper by Scheel [66]. Subsequently, interesting consequences have been derived; most notably a first classification of possible instabilities of defect dynamics, relating to core and far-field break-up [71].

Moreover, under the assumption of pure point critical spectrum, a center manifold description of the associated meandering, drifting, and resonance phenomena has been achieved; see [22], [73], and the survey [21]. We give a brief exposition in section 3.2.1. Specific calculations, however, require disproportionate computational effort to account for the two-dimensional time dependent problem in large (theoretically unbounded) domains.

The unbounded domain causes difficulties, on the one hand, via continuous spectrum which may - and does - interfere with the pure point spectrum required for our bifurcation analysis. The unbounded domain is necessary, on the other hand, to correctly incorporate the Euclidean symmetry of the problem under translations and rotations. In section 3.2 below we will specifically address competing and coexisting pinning and drifting phenomena. Such phenomena are predicted for the light-sensitive BZ system, when full Euclidean symmetry is broken towards a mere translational lattice symmetry, by choosing a spatially periodic lighting.

Kinematic theory, the alternative approach sketched in section 3.3, has the advantage of working with partial differential equations (PDEs) in a single space dimension: a half line which extends from the tip, at $s = 0$, to the Archimedean far-field $s \to +\infty$. Here arc length s parametrizes the wave front. See (3.40) for a PDE for curvature $\kappa = \kappa(t, s)$ as a function of time t and arc length s. A description by curvature $\kappa(t, s)$, rather than position $(x(t, s), y(t, s))$ of the wave front, has the advantage of incorporating Euclidean equivariance into the description, a priori. Indeed, translated and rotated wave fronts are described by identical curvature functions $\kappa(t, s)$. It is therefore surprising, perhaps, that a consistent purely kinematic theory which includes meandering and drifting phenomena has remained elusive, so far. In section 3.3 we first derive existence and multiplicity results for rigidly rotating Archimedean spirals, fairly directly, from a center manifold argument in the associated singular second order ordinary differential equation (ODE) of the kinematic theory; see [15, 16]. For periodic forcing of the wave speed, we then overcome the meandering/drifting barrier of the curvature-independent eikonal theory by modification of the tip dynamics, only, following the presentation in [43]. Following [34] we also derive superspiral patterns, together with a strong stability property, in the periodically forced eikonal case. In addition, we present super-superspirals of higher order, caused by nearly resonant quasi-periodic forcing functions. These eikonal results are based on constant velocity wave propagation, alone, independently of the curvature of the wave front. Such an approach has been suggested in [81], in the guise of wave fronts and Huygens' principle, even though the meandering/drifting effects which we now address were unknown, then.

3.2. Reaction-diffusion spirals

3.2.1. *Center manifold reductions*

An important feature of dynamics and bifurcations of spiral waves is the Euclidean symmetry $SE(2)$ of the plane. The special Euclidean symmetry group $SE(2)$ consists of all planar translations and rotations. Barkley [5] was the first to notice the relevance of this group for meandering spiral wave dynamics. Indeed, let $u(t, x)$, $x \in \mathbb{R}^2$, be any solution of a spatially homogeneous reaction-diffusion system. Let R denote any fixed rotation matrix, and $S \in \mathbb{R}^2$ any fixed translation vector. Then $u(t, R^{-1}(x - S))$ is a solution of the same reaction-diffusion system. Barkley therefore proposed, but could not justify, that the dynamics of planar rigidly rotating or meandering spirals is governed by an $SE(2)$-equivariant vector field. Phenomenologically, he could then interpret the transition to meandering or drifting spiral waves as a Hopf bifurcation. Indeed, he numerically verified the crossing of a pair of simple eigenvalues of the linearization of rotating waves through the imaginary axis. Similarly, Mantel and Barkley [42] described periodic forcing of meandering spirals by periodically forced equivariant equations on the group $SE(2)$ itself. The mathematical tool to justify such reduced descriptions by low-dimensional systems of ODEs, which was still missing at the time, is an $SE(2)$-equivariant version of center manifold theory.

Center manifold theory has become an indispensable tool for the study of ODEs. An equilibrium is called hyperbolic if the linearization of the vector field at that equilibrium does not possess spectrum on the imaginary axis. The local dynamics of ODEs near a hyperbolic equilibrium is determined by that linearization. In particular there exist stable and unstable manifolds. Their tangent spaces at the equilibrium are given by the stable and unstable eigenspaces of the linearization, which are the generalized eigenspaces to eigenvalues with (strictly) negative and positive real parts, respectively. These invariant manifolds contain all solutions which locally approach the equilibrium exponentially fast in forward time and backward time, respectively. The dynamics becomes much more interesting if the equilibrium is non-hyperbolic, i.e., if the linearization possesses purely imaginary eigenvalues. Generically, conserved quantities, time reversibility, or parameters in the system may lead to purely imaginary eigenvalues. These eigenvalues give rise to a local center manifold. The tangent space of the center manifold at the equilibrium is given by the center eigenspace, which is the generalized eigenspace to the purely imaginary eigenvalues. Solutions on the center manifold have only sub-exponential growth or decay. Furthermore all solutions which stay bounded and small for all time are contained in the center manifold [77]. Local center manifolds are not unique, in contrast to stable and unstable manifolds. But center manifold theory simplifies the system. It reduces the dimension and preserves the interesting solutions. One application of center manifold theory is the Hopf bifurcation theorem. Under generic assumptions the Hopf bifurcation theorem states that an equilibrium with a simple pair of purely imaginary eigenvalues, at a specific parameter value, possesses periodic solutions on a center manifold, for nearby parameter values.

Center manifold theory extends to many infinite-dimensional systems, like certain partial differential equations (PDEs). Center manifold reductions can be obtained locally or globally. For local center manifolds of parabolic PDEs see Vanderbauwhede and Iooss [78]. Dimension reductions via global center manifolds for spatially inhomogeneous planar media have been achieved by Jangle [27, 33]; more details will be presented below.

The central question, which we will pursue in the present section, asks for the consequences of symmetry breaking from full $SE(2)$-equivariance to a mere translational lattice symmetry. Experimentally this question is motivated by the photosensitive BZ reaction; see [38, 83–85, 87]. In these experiments, the excitability of the BZ medium in a gel reactor depends sensitively on the intensity of light exposure. This intensity can be varied, both spatially and temporally. For time independent light intensity, for example, with slight spatial variation in form of a checker board light-dark pattern, the spatial homogeneity of the underlying reaction-diffusion system is broken to a mere lattice periodicity, spatially. Mathematically, $SE(2)$-equivariance is thus perturbed to a mere lattice symmetry. This motivates our mathematical study of the possible consequences of *equivariance breaking perturbations* for the dynamics of rotating wave patterns.

But what are possible dynamic consequences of such an equivariance breaking perturbation to a lattice group? As many authors before us, we describe such consequences in terms of the tip motion of meandering spirals. We would like to keep in mind, however, that the prominently visible phenomenon of the tip motion only amounts to a visualization of the translational component of the $SE(2)$ coordinates, which we will derive in section 3.2.2 via an associated center manifold reduction. The additional investment towards a clean, mathematically well-founded framework is amply rewarded. All ambiguities as to various ad-hoc definitions of the spiral tip, for example, are going to disappear in our setting – along with the mystery of their phenomenological equivalence.

Drifting versus pinning are the most prominent spiral tip motions in geometries with lattice symmetry. By *pinning* we denote tip motions which come to rest, converging to a fixed spatial location, as time $t \to +\infty$. *Drifting*, in contrast, indicates motions which are unbounded in the underlying lattice geometry.

In practice, of course, drifting motions will simply sweep the spiral tip to the domain boundary where the spiral will annihilate. Such boundary annihilations - a highly desirable effect in the context of defibrillation - can be understood as a cancellation with the anti-rotating mirror image spiral after local reflection through the Neumann boundary condition. Below, we do not pursue such modifications due to bounded domains.

Instead, we show below how *pinning and drifting motions may coexist*, within the same center manifold of the same underlying system and at the same parameter values. In fact an ever so slight variation of initial conditions may kick the solution from drifting to pinning mode, or back. The respective initial conditions for these behaviors will be interwoven in a Cantor-like structure.

More generally, we will address the question of realizing any flow on the 2-torus $T^2 = \mathbb{R}^2/\mathbb{Z}^2$, which can and should be viewed as the space of tip positions in the plane $\mathbb{R}^2$ modulo the lattice periodicity $\mathbb{Z}^2$. The coexistence of drifting and pinning motions then becomes a corollary to the realization of Cherry flows on 2-tori, see [57], by spiral tip dynamics in lattice symmetries.

3.2.2. *Center manifolds in unbounded domains*

Chemical systems are traditionally modeled by reaction-diffusion systems on suitable domains. As was explained above, our main modeling assumption is that the domain is actually unbounded, that is, we consider governing partial differential equations on the entire plane. This assumption may seem unrealistic: neither experiments nor numerical simulations can be performed on unbounded domains. In our particular context of spiral waves in the BZ reaction, however, experiments indicate that spiral waves behave much as if there were no boundaries. Therefore until boundary annihilation sets in – typically within only one to two wavelengths from the boundary itself – we consider reaction-diffusion systems:

$$\partial_t u(t,x) = D\Delta_x u(t,x) + F(u(t,x)), \qquad u : [0,\infty) \times \mathbb{R}^2 \longrightarrow \mathbb{R}^N, \quad (3.1)$$

where $x \in \mathbb{R}^2$ denotes space, $t \geq 0$ is time, and $u = u(t,x) \in \mathbb{R}^N$ is the concentration vector. The nonlinearity $F : \mathbb{R}^N \longrightarrow \mathbb{R}^N$ is a C^{k+2}-function, $k \geq 1$, and D is a constant diagonal matrix with strictly positive entries.

Under appropriate growth conditions on F there exists an $\alpha \in [0,1)$ such that $F \in C^{k+2}(X^\alpha, X)$, where $X := L^2(\mathbb{R}^2, \mathbb{R}^N)$. The fractional power space X^α is an interpolation space which is more regular than $L^2(\mathbb{R}^2, \mathbb{R}^N)$ but less regular than the Sobolev space $H^2(\mathbb{R}^2, \mathbb{R}^N)$ of functions $u(t,\cdot)$ with square integrable second spatial derivatives. By Henry [29], equation (3.1) generates a C^{k+2} semiflow ϕ_t on X^α. Thus the solution to equation (3.1) at time $t \geq 0$ with initial condition u_0 is given by a function $u = u(t,x)$ such that $u(t,\cdot) = \phi_t(u_0)$.

Equation (3.1) exhibits the symmetry of the homogeneous BZ reaction, namely equivariance under the Euclidean group $SE(2)$ of all translations S and rotations R in the plane. The group multiplication of $(R_i, S_i) \in SE(2)$ is given by

$$(R_1, S_1)(R_2, S_2) = (R_1 R_2, S_1 + R_1 S_2). \quad (3.2)$$

The group action $\rho : SE(2) \longrightarrow GL(X^\alpha)$ of the group $SE(2)$ on the function space of spatial profiles $u(t,\cdot) \in X^\alpha$ is defined by

$$(\rho_{(R,S)} u)(x) := u(R^{-1}(x - S)). \quad (3.3)$$

Thus, an element $g = (R,S)$ of $SE(2)$ simply rotates and translates the profile u. The seemingly strange inverses in (3.3) ensure, together with (3.2), that $\rho_{g_1 g_2} u = \rho_{g_1}\rho_{g_2} u$. Given this bounded linear group action, equivariance of the vector field in equation (3.1) can be stated as:

$$D\Delta(\rho_g u) + F(\rho_g u) = \rho_g(D\Delta u + F(u)) \quad \text{for all } g \in SE(2), u \in H^2. \quad (3.4)$$

The equivariance property (3.4) simply means: whenever $u(t) \in X^\alpha$ solves reaction-diffusion system (3.1), then $\rho_g u(t) \in X^\alpha$ is also a solution, for any fixed choice of $g = (R, S) \in SE(2)$. In terms of the semiflow ϕ_t this property can be expressed as

$$\phi_t(\rho_g u) = \rho_g \phi_t(u) \quad \text{for all } g \in SE(2). \tag{3.5}$$

Substantial mathematical difficulties arise because the group action ρ is only strongly continuous. Moreover the group $SE(2)$ is non-compact due to its translational component. We suppress these technicalities in the following.

Supported by experimental and numerical evidence, we assume there exists a rigidly rotating solution of (3.1). In other words, there exists $u_* \in X^\alpha$ such that the time evolution of this initial condition is a pure rotation with non-zero frequency ω_*:

$$\phi_t(u_*) = \rho_{(R_{\omega_* t},0)} u_*. \tag{3.6}$$

This assumption implies that $\phi_t(u_*)$ is a time-periodic solution of (not necessarily minimal) period $2\pi/\omega_*$. An element u_* of our phase space X^α is called a *relative equilibrium* if the time evolution $\phi_t(u_*)$ lies inside the group orbit $\rho_g(u_*)$. Hence the periodic solution of u_* consists of relative equilibria.

Spectral analysis of the linearized semiflow along a periodic solution is called Floquet theory. The eigenvalues μ of the linearized period map are called *Floquet multipliers*. A *Floquet exponent* is a complex β such that $\exp(\beta\tau)$ is a Floquet multiplier of the system, where τ denotes the minimal period. A periodic solution is hyperbolic if, and only if, it possesses only the trivial Floquet multiplier $\mu = 1$ on the unit circle, and this multiplier has algebraic multiplicity one. Otherwise it is called non-hyperbolic. In ODEs hyperbolic periodic solutions possess stable and unstable manifolds, similarly to the case of hyperbolic equilibria. Non-hyperbolic periodic solutions possess center manifolds.

We now return to our setting (3.1) and assume that $\mu = 1$ is a Floquet multiplier of the periodic solution $\phi_t(u_*)$, with three-dimensional generalized eigenspace. The remaining spectrum we assume to lie strictly in the interior of the complex unit circle. Due to Euclidean symmetry there exist at least three eigenvectors of $D\phi_{2\pi/\omega_*}(u_*)$ corresponding to the eigenvalue 1, see Figure 3.2. The eigenvectors are the three partial derivatives $\partial_t u$, $\partial_{x_1} u$ and $\partial_{x_2} u$, of the solution $\phi_t(u_*)$, at $t = 0$.

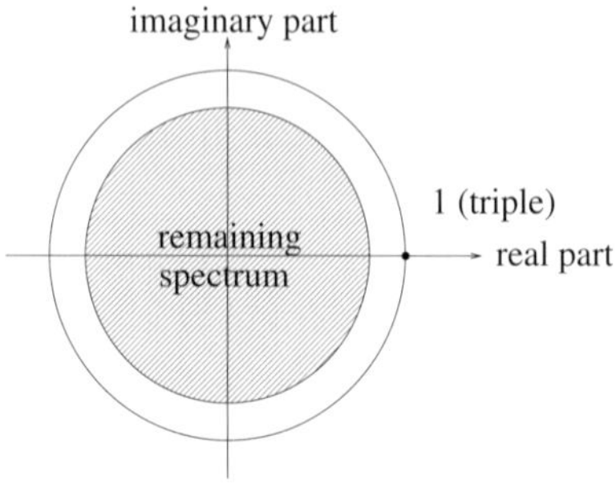

Fig. 3.2. Floquet multipliers μ of $D\phi_{2\pi/\omega_*}(u_*)$ in the complex plane.

Our spectral assumption can also be expressed in a rotating coordinate frame ([82], [72]). Let

$$L := D\Delta + \omega_* \frac{\partial}{\partial \varphi} + D_u F(u_*), \tag{3.7}$$

where $\frac{\partial}{\partial \varphi}$ denotes the partial derivative with respect to the angular component, i.e. $\frac{\partial}{\partial \varphi} = x_1 \partial_{x_2} - x_2 \partial_{x_1}$. In a co-rotating coordinate frame, the relative equilibrium u_* becomes an equilibrium. Eigenvalues of L are therefore Floquet exponents of the periodic solution $\phi_t(u_*)$. The equivalent spectral assumption is basically that zero is a triple eigenvalue of L and the remaining spectrum has negative real part, uniformly bounded away from the imaginary axis.

The symmetry of equation (3.1), which causes the non-hyperbolic eigenvalue zero to be triply degenerate, gives rise to a three-dimensional invariant manifold of $\phi_t(u_*)$. This is the center manifold which coincides with the group orbit of u_*, namely $SE(2)u_* := \{\rho_g u_*; g \in SE(2)\}$. Thus the center manifold of u_* is simply the set of all translations and rotations of the initial rotating wave u_*. See [72] for this result and some generalizations.

3.2.3. *Lattice symmetry*

The main mathematical feature of the above invariant center manifold $SE(2)u_*$ is its normal hyperbolicity. An invariant manifold is called *normally hyperbolic* if the linearized dynamics in the normal directions are of faster exponential rate than those in tangential direction. Normally hyperbolic invariant manifolds persist under small perturbations [30].

To account for static, lattice periodic light patterns in the photosensitive BZ reaction, we add a symmetry breaking perturbation $\varepsilon H(u)$ to equation (3.1) as follows:

$$u_t = D\Delta u + F(u) + \varepsilon H(u). \tag{3.8}$$

We assume that ε is small and $H \in C^{k+2}(X^\alpha, X)$ keeps only a translational lattice symmetry $\mathbb{Z}^2$, but breaks the rotational symmetry. Therefore the semiflow ϕ_t^ε of the perturbed equation (3.8) is only lattice equivariant:

$$\phi_t^\varepsilon(\rho_{(0,S)}u) = \rho_{(0,S)}\phi_t^\varepsilon(u) \quad \text{for all } S \in \mathbb{Z}^2. \tag{3.9}$$

For example H could be a superposition operator $(H(u))(x) = h(x, u(x))$, where $h : \mathbb{R}^2 \times \mathbb{R}^N \to \mathbb{R}^N$ is a C^{k+2}-function that satisfies mild growth conditions. Translational equivariance then requires $h(x + S, u) = h(x, u)$ for all $S \in \mathbb{Z}^2$.

In the photosensitive BZ reaction this corresponds to a lattice symmetric lighting pattern of the BZ medium. We recall our fundamental question: Which motions of the spiral tip can be realized by choosing an appropriate perturbation H? According to equivariance condition (3.9) the lighting pattern in the photosensitive BZ reaction can only be chosen freely on a unit square of the underlying media. The function H is then determined on the entire plane, by $\mathbb{Z}^2$-periodic extension. The main result below asserts, that for *every* prescribed ODE motion of the perturbed spiral wave tip there is a lighting pattern, represented by H, such that this prescribed motion can be realized, to leading order in the perturbation parameter ε. An example of an "illumination" functional H will be given such that, both, pinning and drifting of the perturbed spiral tip, coexist.

We now state the center manifold theorem of the perturbed system:

Theorem 1. [33] *Assume the existence of a rotating wave solution u_* of the unperturbed system. Suppose 1 is a triple Floquet multiplier of the linearized flow and the remaining spectrum lies strictly in the interior of the complex unit circle. Then, for ε sufficiently small, there exists a lattice- and flow-invariant three-dimensional manifold $\mathcal{M}^\varepsilon \subset X^\alpha$. The manifold $\mathcal{M}^\varepsilon$ is a locally exponentially attracting manifold. Differentiability of $\mathcal{M}^\varepsilon$ is $C^{k,1}$, including dependence on the parameter ε: the manifold can be represented by maps with Lipschitz continuous k-th derivatives.*

We briefly comment on the basic steps in the proof of theorem 1. For homogeneous lighting of intermediate strength a rigidly rotating spiral wave solution was assumed to be given by u_*. The manifold $\mathcal{M}^\varepsilon$ is close to the unperturbed normally hyperbolic center manifold $SE(2)u_*$ given by the translations and rotations of the spiral wave u_*. The existence of the manifold $\mathcal{M}^\varepsilon$ implies that the spiral shapes will stay close to that of u_*, even under slightly lattice periodic lighting patterns which destroy the full Euclidian equivariance. The symmetry breaking will cause the slightly deformed spirals to move. Projecting the dynamics of $\mathcal{M}^\varepsilon$ onto the 3-dimensional reference manifold $SE(2)u_*$ yields the dynamics of the perturbed spirals, via the rotational component R and the two translational components S of $SE(2)$. The rigid rotation of the unperturbed spiral u_* is replaced by a slow modulation of the rotation frequency. The translational component becomes time dependent and is responsible for non-trivial tip dynamics.

The center manifold in theorem 1 is constructed by applying the method of the graph transform. We give a brief summary here; for further background see Shub [67]. First we choose appropriate local coordinates in a neighborhood of the group orbit. The neighborhood of the group orbit $SE(2)u_*$ is a vector bundle over $SE(2)u_*$. Therefore every profile $u \in X^\alpha$ sufficiently close to the group orbit $SE(2)u_*$ can be written as the sum of an element of the group orbit $SE(2)u_*$ and a small element of an infinite-dimensional subspace V of X^α which corresponds to the stable eigenspace of the linearized flow. See also Figure 3.4 below. Thus u has a shape close to (a translated or rotated version of) the spiral wave solution u_*. A section of the vector bundle is a function mapping every element of $SE(2)u_*$ to the complementary infinite-dimensional stable eigenspace. The perturbed flow ϕ_T^ε applied to the graph of a section will again be a section of the bundle due to normal hyperbolicity of $SE(2)u_*$ if T is large enough. This graph transform, which maps sections to sections, becomes a contraction on the space of Lipschitz-continuous sections equipped with the metric induced by the sup-norm. Hence it has a unique fixed point and the graph of the fixed point is the perturbed center manifold $\mathcal{M}^\varepsilon$.

Using graph transforms, $\mathcal{M}^\varepsilon$ automatically becomes time- and flow-invariant. Smoothness of the manifold $\mathcal{M}^\varepsilon$ is a very delicate question and has been settled in [33].

As an example we state the two-component Oregonator model of the photosensitive BZ reaction with external forcing, see for example [84]:

$$\begin{aligned} \frac{\mathrm{d}u}{\mathrm{d}t} &= \Delta u + \frac{1}{\varepsilon}\left[u - u^2 - (fv + I(t))\frac{u-q}{u+q}\right], \\ \frac{\mathrm{d}v}{\mathrm{d}t} &= u - v. \end{aligned} \tag{3.10}$$

Here u corresponds to the concentration of the autocatalytic species, bro-

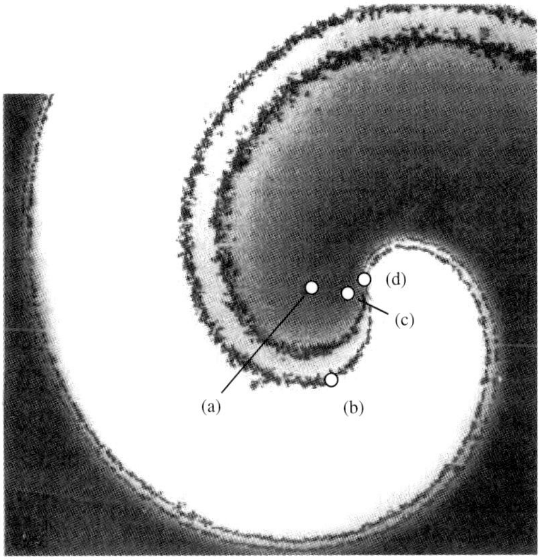

Fig. 3.3. "Tips" of spiral wave pattern in the BZ reaction: a) core center, b) maximum curvature, c) rotation center, d) inflection point; see [53].

mous acid $HBrO_2$, and v corresponds to the oxidized form of the catalyst. The parameters $\varepsilon = 0.05$, $q = 0.002$, and $f = 2.0$ are fixed and $I = I(t)$, induced by external light intensity, describes the bromide production.

3.2.4. *Spiral tip dynamics*

Many definitions of spiral tips are in use. One of them considers the spiral tip as "the" point of the spiral wave front with maximal curvature [8]. A practical way to determine the tip position in an experiment is to look at spiral wave fronts extracted from two consecutive frames of a digitized recording. The intersection of these lines is also considered a tip position [37]. For several other definitions see [35, 91] and Figure 3.3.

In this section we clarify the mathematical relation – and equivalence – of all this zoo of tip definitions; see (3.11)–(3.21) below. The main point is that any *tip function* $z^* : X^\alpha \to \mathbb{C}$ should associate a tip position $z^*(u)$ to a spatial profile $u \in X^\alpha$, such that shifted or rotated profiles u give rise to a correspondingly shifted and rotated tip position z^*. Any reasonable definition of a tip function z^* will certainly have to satisfy at least this minimal condition of equivariance with respect to the Euclidean group $SE(2)$. It turns out, below, that any tip definition then gives rise to the same basic form of the reduced ODE on a center manifold near the unperturbed reference spiral wave shape $SE(2)u_*$, as long as the tip faithfully represents the translation action in $SE(2)$.

We discuss the interpretation of the tip function z_* for the spatially homogeneous, fully $SE(2)$-equivariant case first. The modifications for the photosensitive case with lattice symmetry will be discussed in section 3.2.5 below.

From [72] we recall the existence of a center manifold $\mathcal{M}$ associated to a rigidly rotating reference spiral, alias a relative equilibrium $SE(2)u_*$; see also (3.1)–(3.7) above. In fact, the center manifold $\mathcal{M}$ accounts for all solutions which remain in a neighborhood of the relative equilibrium $SE(2)u_*$ for all positive and negative times. In particular, all bifurcations due to point spectrum on the imaginary axis of the linearization L from (3.7) occur in the finite-dimensional center manifold $\mathcal{M}$. For simplicity of presentation, however, we then suppress the dependence of $\mathcal{M}$, and of the flow on $\mathcal{M}$, on any extra bifurcation parameters.

To perform any specific calculations, and to clarify the zoo of possible tip functions z^*, we need good coordinates on $\mathcal{M}$: coordinate free abstract nonsense will not suffice here. Palais has constructed good coordinates near group orbits of proper, but not necessarily compact, group actions. We apply these coordinates to the group $SE(2)$ and the reference group orbit $SE(2)u_*$ of rigidly rotating waves. Let $g \in SE(2)$ and let $v \in V$ denote elements of a local section V, transverse to the group orbit $SE(2)u_*$ in the center manifold $\mathcal{M}$. Clearly gv, or more precisely $\rho_g v$, then cover a tubular neighborhood of $SE(2)u_*$, see Figure 3.4. Notationally, we may let $v = 0$ correspond to u_* itself here.

In [22] the resulting ODEs in the center manifold $\mathcal{M}$ of a relative equilibrium $SE(2)u_*$ were derived, in these Palais coordinates. Suppressing extra parameters, again, the result is

$$\begin{aligned} \dot{g} &= a(v)g \\ \dot{v} &= \varphi(v). \end{aligned} \tag{3.11}$$

Here $a : V \to \mathrm{alg}(SE(2))$ is a function from the section V to the Lie algebra of $SE(2)$, and $\varphi(v)$ denotes a vector field on V. Anyway, please read on. To be more specific, we rewrite (3.11) in terms of suitable coordinates $g = (R, S)$ on the Euclidean group $SE(2)$. Using complex notation, we write $g = (R, S) = (e^{i\alpha}, z)$. Here $R = e^{i\alpha}$, acting multiplicatively, denotes rotations of $x \in \mathbb{R}^2 \cong \mathbb{C}$. The translation component $S = z \in \mathbb{R}^2 \cong \mathbb{C}$, acting additively, denotes translations. In coordinates $(\alpha, z, v) \in (\mathbb{R}/2\pi\mathbb{Z}) \times$

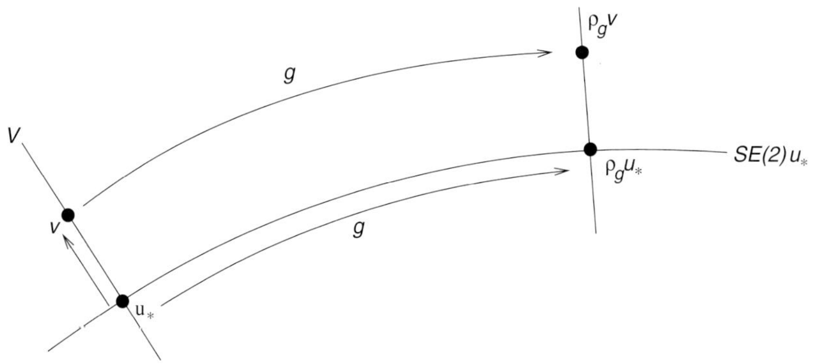

Fig. 3.4. The construction of local Palais coordinates (g, v) for a tubular neighborhood of a group orbit $SE(2)u_*$.

$\mathbb{C} \times V$, system (3.11) now demystifies and reads

$$\begin{aligned} \dot{\alpha} &= \omega(v) \\ \dot{z} &= e^{i\alpha}\sigma(v) \\ \dot{v} &= \varphi(v). \end{aligned} \tag{3.12}$$

with given functions $(\omega, \sigma) : V \to \mathbb{R} \times \mathbb{C}$.

How to interpret systems (3.11), (3.12)? First of all, let us interpret the Palais coordinates $(g, v) = (e^{i\alpha}, z, v) \in SE(2) \times V$ themselves. Obviously, the definition of Palais coordinates implies that elements u_1, u_2 in the same group orbit have the same component v. Therefore v parametrizes the *shape* of spatial profiles, in our setting. Indeed, spatial profiles $u_1, u_2 \in X^\alpha$ which have the same v-component differ only in g, viz. by a rotation and translation of their spatial profiles:

$$u_2(x) = (\rho_g u_1)(x) = u_1(e^{-i\alpha}(x - z)). \tag{3.13}$$

The group coordinates $g = (e^{i\alpha}, z)$ then correspond to a translation of the spatial profile by z, and a rotation by an angle $\alpha \in \mathbb{R}/2\pi\mathbb{Z}$.

In Palais coordinates $(g, v) = (e^{i\alpha}, z, v)$, systems (3.11), (3.12) have *skew product structure*: the pure shape dynamics $\dot{v} = \varphi(v)$ of the shape variable v is autonomous, and is not influenced by the group coordinates $g = (e^{i\alpha}, z)$ of rotation angle α and (complex) translation z. On the other hand, the shape variable v acts as a forcing on the group variables $(e^{i\alpha}, z)$.

We are now ready to clarify the role of tip functions $z^* : \mathcal{M} \to \mathbb{R}^2 \cong \mathbb{C}$. We only assume equivariance

$$z^*(\rho_g(u)) = gz^*(u) \tag{3.14}$$

for any $g \in SE(2)$, $u \in \mathcal{M}$. In fact we reduce attention to the center manifold $\mathcal{M}$ because all bifurcation phenomena must occur in $\mathcal{M}$. We claim that we may substitute $z^* = z^*(u)$ for the translation component z in the Palais coordinates $(e^{i\alpha}, z, v)$, without changing the skew product structure of the ODE (3.12) on the center manifold $\mathcal{M}$.

Indeed, equivariance condition (3.14) on the tip function implies that

$$z^* = z^*(u) = z^*(\rho_{(e^{i\alpha},z)}v) = (e^{i\alpha}, z)z^*(v) = e^{i\alpha}z^*(v) + z. \tag{3.15}$$

This defines a transformation

$$(e^{i\alpha}, z, v) \mapsto (e^{i\alpha}, z^*, v) \tag{3.16}$$

with inverse given explicitly by

$$z = z^* - e^{i\alpha}z^*(v). \tag{3.17}$$

Thus the tip position z^* may indeed be considered as the translation component z of the Palais coordinates, directly, up to a coordinate transformation (3.16). Let us calculate, briefly, the resulting skew product ODE component for z^*:

$$\begin{aligned} \frac{\mathrm{d}}{\mathrm{d}t}z^* &= \frac{\mathrm{d}}{\mathrm{d}t}(e^{i\alpha}z^*(v) + z) = i\dot{\alpha}e^{i\alpha}z^*(v) + e^{i\alpha}z^{*\prime}(v)\dot{v} + \dot{z} = \\ &= e^{i\alpha}(i\omega(v)z^*(v) + z^{*\prime}(v)\varphi(v) + \sigma(v)) =: e^{i\alpha}\sigma^*(v) \end{aligned} \tag{3.18}$$

with the obvious definition for the new function $\sigma^*(v)$. Therefore the tip function z^* satisfies a transformed ODE of the same structure as the translation component z itself.

For illustration purposes, let us now interpret meandering and drift in terms of the Palais coordinates $(e^{i\alpha}, z, v)$. First note that the relative equilibrium $v = 0$, which corresponds to the rigidly rotating spiral, becomes an equilibrium $\varphi(0) = 0$ of the shape ODE $\dot{v} = \varphi(v)$ in (3.12). Indeed the group orbit $SE(2)u_*$, which corresponds to one and the same shape variable $v = 0$, contains the time orbit of the rigidly rotating spiral u_*. Therefore

$$\begin{aligned} \dot{\alpha} &= \omega(0) \\ \dot{z} &= e^{i(\omega(0)t+\alpha(0))}\sigma(0) \end{aligned} \tag{3.19}$$

indicates a rigid rotation frequency $\omega_* = \omega(0)$ and a circular tip motion $z(t)$; see Figure 3.1.

Now assume the shape dynamics $\dot{v} = \varphi(v)$ undergoes a Hopf bifurcation from this trivial (relative) equilibrium $v = 0$, albeit with suppressed parameters. Then $v(t)$ will be periodic, say with "breathing" frequency ω_1 of the pure shape dynamics. Multiplying (3.12) by the nonzero Euler multiplier $\omega_*/\omega(v) \approx 1$ and redefining $\sigma(t) := \sigma(v(t))\ \omega_*/\omega(v(t))$ we obtain the tip equation

$$\dot{z} = e^{i\alpha(t)}\sigma(t) = e^{i\omega_* t}\sigma(t). \tag{3.20}$$

Obviously, $\sigma(t)$ has inherited minimal period $2\pi/\omega_1$ from the shape equation. Fourier expansion of $\sigma(t)$ and direct integration of (3.20) implies that $z(t)$ undergoes a 2-frequency bounded meandering epicycle motion, unless the frequencies

$$\omega_* = m\omega_1 \tag{3.21}$$

are in integer resonance, for some $m \in \mathbb{N}$. In the latter case, nonvanishing $(-m)$-th complex Fourier coefficients of $\sigma(t)$ will provide an unbounded drifting motion which is linear in t. Note that the rotation frequency ω_* must be an integer multiple of the breathing frequency ω_1 of the pure shape dynamics $\dot{v} = \varphi(v)$, for drifting tip motion to occur. Both, meandering and drifting spirals change their shape $v(t)$ periodically and are therefore also called *relative periodic orbits*.

By our above considerations, these meandering/drifting effects can be interpreted as tip motions, directly. The results are universal, at the same time, in the sense that the observed motions do not depend on the particular choice of a tip function $z^*(u)$ from the large zoo of $SE(2)$-equivariant possibilities

Nonautonomous time-periodic forcings with frequency ω_1 can be treated analogously. Indeed, $\omega = \omega(t)$ and $\sigma = \sigma(t)$ then depend on time t directly and our remarks on (3.20), (3.21) apply. We will return to this observation when we describe the alternative kinematic approach, in section 3.3.

Some arbitrariness is still involved in the choice of the transverse Palais section V in Figure 3.4. This geometric arbitrariness can in fact be used to further simplify the skew product form (3.11). Such an approach, which first simplifies $\dot{v} = \varphi(v)$ to Poincaré-Birkhoff normal form, and then redefines V to further simplify the cross term $a(v)$, has been pursued in [23].

3.2.5. *Reduced tip equations for the photosensitive system*

As above, we consider an unperturbed reference spiral wave $SE(2)u_*$ which is rigidly rotating with constant frequency ω_*. In this section we adapt the considerations of section 3.2.4 on the tip dynamics $z^*(t)$, and the skew product flow (3.12) in the center manifold $\mathcal{M}$, to lattice periodic perturbations $\varepsilon H(u)$; see (3.8). We recall that such perturbations account for slight, spatially periodic but temporally constant, variations of light intensity in the photosensitive BZ reaction.

Under the spectral assumptions of section 3.2.3, theorem 1, it is now possible to reduce the perturbed dynamics to a three-dimensional center manifold $\mathcal{M}^\varepsilon$ which is modeled over the group $SE(2)$ itself. In Palais coordinates

$$\begin{aligned}\dot{\alpha} &= \omega_* + \varepsilon\gamma(\alpha, z, \varepsilon)\\ \dot{z} &= \varepsilon h(\alpha, z, \varepsilon).\end{aligned} \tag{3.22}$$

As was justified in section 3.2.4, the angle α denotes the phase and z the position of the spiral tip. The Palais section coordinate $v \in V$ is absent here, because the critical spectrum is now three-dimensional, only, and is accounted for by the three-dimensional group $SE(2)$ itself. Therefore the center manifold $\mathcal{M}^\varepsilon$ is a graph over the group coordinates $(e^{i\alpha}, z) \in SE(2)$. A rigorous derivation of the reduced equation (3.22) has indeed been achieved in [25, 33], under the assumption that the unperturbed spiral wave $u_*(\cdot)$ is *spectrally* stable with the exception of a triple critical eigenvalue due to symmetry; see theorem 1. Note that the nonlinearities $\gamma(\alpha, z, \varepsilon)$ and $h(\alpha, z, \varepsilon)$ obey the lattice symmetry relic of full Euclidean symmetry, namely

$$\begin{aligned}\gamma(\alpha, z+k, \varepsilon) &= \gamma(\alpha, z, \varepsilon)\\ h(\alpha, z+k, \varepsilon) &= h(\alpha, z, \varepsilon)\end{aligned} \tag{3.23}$$

for every angle $\alpha \in S^1$, $z \in \mathbb{C} \cong \mathbb{R}^2$ and integer $k \in \mathbb{Z}^2 \subseteq \mathbb{R}^2$. See also (3.9).

In fact (3.22) therefore induces a (nonautonomous) flow on the 2-torus $z \in \mathbb{C}/\mathbb{Z}^2$, by reinterpreting the angle α as the new "time". This is possible because $\dot{\alpha}(t) \neq 0$ for ε sufficiently close to zero and all real t. Multiplying (3.22) by the nonzero Euler multiplier $1/(\omega_* + \varepsilon\gamma(\alpha, z, \varepsilon))$ therefore provides the nonautonomous tip equation

$$\dot{z} = \varepsilon\widetilde{h}(\alpha, z, \varepsilon) \tag{3.24}$$

Here we have replaced $h/(\omega_* + \varepsilon\gamma)$ by a new function $\widetilde{h}$, and we can consider α as new time because $\dot{\alpha} = 1$. The lattice symmetry (3.9), (3.23) is inherited by the new nonlinearity $\widetilde{h}$. Therefore (3.24) indeed defines a nonautonomous slow flow on the 2-torus $z \in T^2 = \mathbb{R}^2/\mathbb{Z}^2$. Passing to slow time $t = \varepsilon\alpha$, (3.24) transforms to the rapidly periodically forced system

$\dot{z} = \widetilde{h}(t/\varepsilon, z, \varepsilon)$, where h is 2π-periodic in its first argument $\alpha = t/\varepsilon$. Standard averaging procedures then reduce (3.24) to the autonomous system

$$\dot{z} = h(z, \varepsilon), \tag{3.25}$$

up to any finite order in ε, with a suitably defined new vector field h [74]. For example,

$$h(z, 0) = \frac{1}{2\pi} \int_0^{2\pi} \widetilde{h}(\alpha, z, 0) \mathrm{d}\alpha \tag{3.26}$$

turns out to simply be the average of the nonautonomous vector field $\widetilde{h}$ over its fast periodic forcing variable $\alpha = t/\varepsilon$ – hence the name *averaging* for this procedure. For more advanced results on averaging see for example [20] or [24, 28] and the references there.

As expected, one easily recovers the unperturbed spiral tip motion

$$\dot{z}(t) = 0 \tag{3.27}$$

in equation (3.24), by setting $\varepsilon = 0$. Of course, the precise form of $\widetilde{h}$ and h in (3.24) and (3.22) depends crucially on the exact choice of the original PDE perturbation H. We write $h = h(z, \varepsilon; H)$ in (3.22) to emphasize this important dependence. In particular, the choice of H determines the vector field (3.22) which itself determines the (averaged) spiral tip dynamics (3.25).

Expanding $h(z, \varepsilon)$ with respect to small ε yields

$$h(z, \varepsilon; H) = h_1(z; H) + \varepsilon h_2(z; H) + \ldots \tag{3.28}$$

We study the dependence of h_1 on H, because h_1 governs the dynamics (3.25) of the perturbed spiral wave tip, to leading order in ε. Therefore we consider the map

$$\mathcal{L} : H \mapsto h_1(H) \tag{3.29}$$

where $h_1(H) = h_1(\cdot, H)$. For the perturbation H we choose any bounded C^{k+2}-functional which satisfies the lattice symmetry. Then $\mathcal{L}$ maps H into the space of all vector fields h on $\mathbb{R}^2$ of class C^k which satisfy

$$h(z + k) = h(z), \tag{3.30}$$

for every $z \in \mathbb{C} \cong \mathbb{R}^2$ and $k \in \mathbb{Z}^2$.

The range of $\mathcal{L}$ has an interesting interpretation: It represents the vector fields of all possible motions of the perturbed spiral wave tip $z(t)$, to leading order in ε. Indeed, a vector field h in the range of $\mathcal{L}$ allows one to find a function H with $h = h_1(H)$ and thus determines the motion of the spiral tip to leading order; see (3.22).

The main result now states that the mapping $\mathcal{L}$ is linear, continuous and surjective [25, 27]. Note that this result does *not* depend on the particular coordinates which have to be introduced in order to obtain the reduced averaged equation (3.25). Continuity and linear dependence of $\mathcal{L}$ on the function H is very intuitive: By slightly changing the light intensity pattern one expects the motion of the spiral wave to change only slightly. Also a superposition of two different light intensity patterns H_1 and H_2 should result in a superposition of the corresponding motions of the spiral tip positions, at least to leading order. Although the surjectivity of $\mathcal{L}$ may sound just as reasonable, from a physical point of view, we have to keep in mind that $h = h_1(H)$ for some H is the leading order term which governs the time averaged dynamics of the perturbed spiral tip, see (3.28). From a mathematical point of view it could happen, that certain changes in the light intensity pattern, and thus in H, influence the spiral tip motion only in the second order terms of (3.28) with respect to ε. Moreover, local perturbations $H(u)(x) = H(x, u(x))$ which merely evaluate the pattern $u(x)$ at any given point x, may not suffice to ensure surjectivity of $\mathcal{L}$. Within the larger class of *functionals* H, rather than just functions, however, the strong property of surjectivity prevails.

We repeat that the construction of the functional H in [25] involves quantities such as the group orbit $SE(2)u_*$ of the unperturbed spiral wave. In particular, H will be a nonlocal functional, in general. Keeping in mind that the experimentalist has only a few control parameters at hand, it remains a challenging task to adjust the light intensity pattern in order to obtain the desired spiral tip motion. For specific results in this direction see [38, 84, 85, 87, 88], and our brief discussion in section 3.4.

Up to these difficulties, we have seen that there is a certain lattice periodic perturbation pattern, represented by a functional H, which realizes a prescribed averaged tip motion (3.25).

3.2.6. *Pinning versus drifting*

It was shown in the previous section that every vector field of the perturbed spiral tip motion can be realized, to leading order, by choosing an appropriate pattern which is represented by a lattice periodic, nonlinear perturbation functional H. We recall from (3.25) that the motion of the

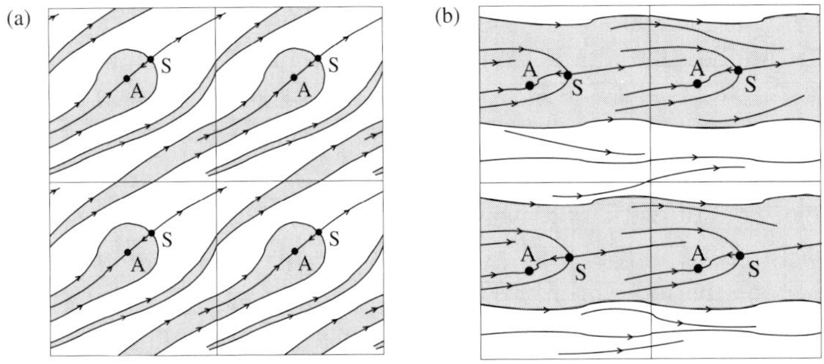

Fig. 3.5. Cherry flows: (a) irrational winding; (b) rational winding with stable and unstable periodic drifts. Basins of A = pinning (grey), and drifting region (white).

perturbed spiral wave tip is governed by the averaged differential equation

$$\dot{z} = h(z, \varepsilon) \tag{3.31}$$

on the 2-torus $z \in \mathbb{C}/\mathbb{Z}^2$

We now consider a particular such vector field, where pinning of the perturbed spiral tip occurs, and coexists with drifting motions. This particular flow on the torus is called the *Cherry flow*, see Figure 3.5(a) and, for further background, [57]. The Cherry flow features two equilibria, a saddle S and a sink A. Every solution $z(t)$ with initial tip position $z(0)$ in the open grey basin gets attracted to the sink A, eventually. The stable manifold $W^s(S)$ of the saddle S marks the boundary of the basin of attraction of A and denotes points which converge to S, eventually. For other initial values (white region) the solution neither gets attracted to the saddle S nor to the sink: the spiral keeps drifting.

We state an explicit example of such a vector field for $z = (z_1, z_2)$:

$$\begin{aligned} \dot{z}_1 &= c_1 - \sin(2\pi z_1) + c_2(1 - \cos(2\pi z_2)) \\ \dot{z}_2 &= -\sin(2\pi z_2) \end{aligned} \tag{3.32}$$

with constants $0 < c_1 < 1$ and $c_2 \neq 0$. Because the vector field (3.32) is 1-periodic in z_1,z_2, it indeed defines a vector field on the 2-torus.

Keeping in mind that the dynamics of (3.31) represents the motion of the perturbed spiral wave tip, this example shows that there are indeed open sets of initial positions $z(0)$ for which the spiral tip $z(t)$ finally gets pinned to the sink A. Coexisting are regions for which the spiral tip never gets attracted to any specific point and undergoes a drifting motion on the 2-torus, which is unbounded in $\mathbb{R}^2$. The regions are interwoven in a Cantor-like structure. It should be mentioned that the drifting region has zero

Lebesgue measure in the above example; see also [44, 49]. Zero Lebesgue measure implies that almost every spiral tip eventually becomes pinned, but possibly after a very long transient phase of drifting. In general, however, it seems that drifting can also prevail on regions of positive Lebesgue measure.

At least in presence of periodic orbits on the 2-torus, drifting and pinning can coexist with positive measure; see Figure 3.5(b). Indeed the pinning region of the attracting equilibrium (A) is now trapped between two unstable periodic orbits. The unstable periodic orbits also bound the basin of attraction of a stable periodic orbit. The stable periodic orbit signifies stable unbounded drift of the spiral tip across the lattice periodic perturbation pattern.

3.3. Kinematics

3.3.1. *Curves and tips*

The kinematic approach to the propagation of planar excitation waves idealizes the wave front location to be given by a one-dimensional curve

$$z = Z(t, s) \in \mathbb{R}^2 \cong \mathbb{C}. \tag{3.33}$$

The curve Z is parametrized by arc length s at any given time t. The basic modeling assumption is then the hypothesis that the curve $Z(t, s)$ moves in normal direction $\underline{n} := iZ_s$ with a normal velocity $U \in \mathbb{R}$ which depends on the curvature

$$\kappa\underline{n} := -Z_{ss}. \tag{3.34}$$

We abbreviate this modeling assumption as

$$U = U(t, \kappa), \tag{3.35}$$

where the explicit dependence on t allows for external forcing. We will obtain rigidly rotating curves as well as meanders and drifts in this setting.

In the mathematical literature, the special case of a power law dependence

$$U = U(\kappa) = -\kappa^\beta, \tag{3.36}$$

with $\beta, \kappa > 0$ has attracted much attention under the name of *"curve shortening"*; see for example [2, 3, 19, 26] and the references there. Closed convex curves Z, for example, shrink to a point in finite time, for $\beta = 1$, and become asymptotically circular before they disappear. A celebrated generalization of this fact to convex surfaces under mean curvature flow was obtained by [31]. Much excitement has been caused in the mathematical community, recently, by Perelman's contributions relating flows of compact three-dimensional manifolds under their Ricci curvature to the Poincaré conjecture [58–60].

Due to their parabolic PDE nature, curvature flows have a smoothing effect on the curves $Z(t, \cdot)$. Curvature flows are therefore also used in image processing to smoothen ragged boundaries [14, 47, 50]. Of particular interest in image processing is the case $\beta = 1/3$, which is in addition affine equivariant: initial curves which are affine images of each other retain this property under their evolution (3.36).

As we have noted in the introduction, experimental evidence for a kinematic description of excitable wave fronts is rich. Based on hyperbolic wave equations and the Huygens principle, Wiener and Rosenblueth [81] recommend the eikonal approach of geometric optics: waves propagate at a constant normal speed

$$U(t, \kappa) = c(t), \tag{3.37}$$

which does not depend on the curvature of the wave front. In section 3.3.3 below, we will describe meander and drift of spirals under this very restrictive assumption. More general affine linear dependencies

$$U(t, \kappa) = c(t) - D(t)\kappa, \tag{3.38}$$

albeit with constant coefficients c, D, have been proposed by several authors, both for BZ systems ([52, 76]) and for surface waves in catalysis ([32, 45]). Motivated by the photosensitive BZ reaction we modify these models to include time dependent coefficients $c(t)$, $D(t)$ which account for external forcing.

In section 3.2 we have seen how rigid rotations, meanders, and drifts are closely tied to equivariance with respect to the Euclidean group $SE(2)$ of planar rotations and translations. We therefore prefer to not describe our curvature flows in terms of the position vector $Z(t, s) \in \mathbb{C}$, directly. Instead, we work with the curvature scalar

$$\kappa = \kappa(t, s), \tag{3.39}$$

which eliminates the $SE(2)$-action. Indeed, curves $Z_1(t, s)$ and $Z_2(t, s)$ are Euclidean images of one another, under some fixed $(R, S) \in SE(2)$, if and only if their curvature functions κ coincide. Therefore the parametrization by κ can be viewed as a particularly suitable choice of a Palais shape coordinate v, in the terminology of section 3.2.4; see (3.11), (3.12) and Figure 3.4.

An important modification to the standard approaches concerns our treatment of the spiral tip. Traditionally, waves in excitable media are thought of as arising from a singular perturbation cycle in the reaction term. This view point readily identifies a "front" and "back" for the wave. Serious difficulties obstruct joining these curves in the sense of matched asymptotic expansions [36]. Instead, we merge "front" and "back" to be represented by a single curve. In other words, we view the excitation excursion as a single narrow pulse phenomenon, rather than a widely separated succession of first a front and then a back transition. The problematic core junction of front and back is then distilled into a single point, say $s = s_0(t)$, where the pulse curve $z = Z(t, s)$ terminates, or rather initiates. The above curvature flow then accounts for the dynamics of $Z(t, s)$, for $s \geq s_0(t)$. It is natural to call the end point $s = s_0(t)$ of the curve its *tip*. Certainly, such a definition is compatible with our discussion of tip choices in section 3.2.4. Moreover, the tip location can be seen as a degeneracy point of the singular perturbation excitable cycle, see [18] or, in the terminology of Winfree [80], as a phase singularity.

But what is the proper dynamics of the tip itself? Based on experimental evidence, we make the following three modeling assumptions:

(T1) the normal velocity of the tip is given by $U(t, \kappa)$, as everywhere else;

(T2) the tangential velocity of the tip is given by a function $G(t)$;

(T3) the curvature at the tip is given by a function $\kappa_0(t) > 0$.

Here (T1) is just a continuity assumption. Assumptions (T2), (T3) are based on observations of initial conditions which prepare an excitable cycle across an interval, in the x-plane, which terminates at two endpoints. The interval then propagates, while the two end points extend and curl inward to produce a pair of spiral-antispiral cores. See [7, 65, 86]. Careful examination of such experiments should reveal the values of G and κ_0, quantitatively. For a theoretical attempt at a derivation of tangential speeds $G(t)$ of spiral tips see also [62]. Although the more general case $G = G(t, \kappa)$ could be incorporated, we assume G, κ_0 to depend on t, only, for simplicity and for lack of experimental detail and mathematical derivation, alike.

Summarizing our discussion we arrive at the following PDE description of the dynamics of excitable media waves:

$$\kappa_t + U(t,\kappa)_{ss} + \left(\kappa \int_0^s \kappa U(t,\kappa)\mathrm{d}\sigma\right)_s + G(t)\kappa_s = 0 \tag{3.40}$$

for $s \geq 0$, with the boundary condition

$$\kappa = \kappa_0(t) \qquad \text{at } s = 0. \tag{3.41}$$

Here the Dirichlet boundary condition (3.41) follows from modeling assumption (T3). The drift term $G\kappa_s$ in (3.40) arises from (T2) when we normalize the tip to occur at $s = 0$ instead of $s_0(t)$. Likewise the integral term accounts for reparametrization by arc length, which is necessary as the propagating curve extends or contracts. See [13, 43, 45] for detailed derivations.

In section 3.3.2 we discuss rigidly rotating wave solutions for autonomous velocity functions $U = U(\kappa)$; see also [32, 43] for earlier partial analysis of the affine case $U(\kappa) = c - D\kappa$. In section 3.3.3 we discuss the κ-independent, but forced, eikonal case $U = U(t) = c(t)$ to obtain meanders, drifts and superspiral patterns in the kinematic setting.

3.3.2. *Rigidly rotating spirals*

In this section we discuss stationary solutions $\kappa = \kappa(s)$ of the curvature flow (3.35) for autonomous normal velocity functions $U = U(\kappa)$ and constant $\kappa_0 > 0$, $G \neq 0$. More precisely we solve (3.40), (3.41) for relative equilibria, alias rotating waves $\kappa = \kappa(s)$, in the integrated form

$$\begin{aligned} U(\kappa)_s + \kappa \int_0^s \kappa U(\kappa)\mathrm{d}\sigma + G\kappa &= \omega \\ \kappa(0) &= \kappa_0 \end{aligned} \tag{3.42}$$

with some suitable real integration constant ω. In (3.59), (3.60) below, we will see that ω is in fact the rotation frequency of the rotating Archimedean spiral given by the solution $\kappa(s)$ of (3.42). Differentiating (3.42) with respect to arc length s and using (3.42) itself to eliminate the integral term, we obtain the G-independent equation

$$U(\kappa)_{ss} + \frac{\kappa_s}{\kappa}\big(\omega - U(\kappa)_s\big) + \kappa^2 U = 0 \tag{3.43}$$

with parameter ω.

Our discussion of (3.43) follows the global and rather complete existence, multiplicity, and bifurcation analysis in [15, 16]. This analysis assumes $U'(\kappa) \neq 0$ and invertibility of the relation $U = U(\kappa)$ between normal velocity and curvature by an inverse function

$$\kappa = \Gamma(U). \tag{3.44}$$

Mimicking the affine case $U(\kappa) = c - D\kappa$, for which $\Gamma(U) = (c-U)/D$, we assume $\Gamma \in C^4$ satisfies $\Gamma'(U) < 0$ for all U, and $\Gamma(c) = 0$ for some $c > 0$. This allows us to rewrite (3.43) as a second order equation of Lienard type,

$$U_{ss} + \frac{\Gamma_U}{\Gamma} U_s(\omega - U_s) + \Gamma^2 U = 0, \tag{3.45}$$

with singular behavior at $U = c$ where $\Gamma(U) = 0$. The eikonal case $U(\kappa) \equiv c$, where such an inversion is not possible, will be addressed in section 3.3.3.

Our alert reader will have noticed how we "forgot" the Dirichlet boundary condition $\kappa(0) = \kappa_0$ in passing from (3.42) to (3.45). In fact we will supplement (3.45) by the Neumann boundary condition

$$U_s(0) = \omega - G\kappa_0, \tag{3.46}$$

which readily follows from (3.42). We now give a bifurcation result for the special case $G = 0$; see [16, Theorem 1.1] in terms of the bifurcation parameter $\omega > 0$ in (3.45), (3.46). We comment on the general case of nonzero G and the translation to the original problem (3.42) at the end of this section.

Theorem 2. [16, center manifold] *Under the above assumptions on the normal velocity $U = U(\kappa)$ as a function of curvature κ, all solutions $U = U(s)$ of (3.45), (3.46) with bounded normal velocity U and nowhere vanishing curvature, for all values of the rotation frequency $\omega > 0$, are contained in the bifurcation diagram of Figure 3.6.*

More precisely, there exists a strictly decreasing sequence $\omega_1 > \omega_2 > \cdots \searrow 0$ and associated functions $U_n^{\pm}(\omega)$, for $0 < \omega \le \omega_n$, with the following property. A C^3-solution $U = U(s)$, $s \ge 0$ of the curvature equation (3.45), (3.46) as above exists if, and only if, the tip velocity $U(0)$ and the derivative $U_s(0)$ at the tip $s = 0$ satisfy

$$U(0) \in \{U_n^-(\omega), U_n^+(\omega)\}, \quad U_s(0) = \omega \tag{3.47}$$

for some positive integer n.

The associated curves $z = Z(t,s) = r\exp(i\varphi)$ are right winding, left rotating spirals which are asymptotically Archimedean, for $s \to \infty$. The tip $Z(t,0)$ rotates on a circle of radius $\rho = |U(0)/\omega|$. The asymptotic wave length of the Archimedean spiral in the far field $r \to \infty$ is given by $2\pi c/\omega$.

The functions $\omega \mapsto U_n^{\pm}(\omega)$ possess the following properties, for every $n \in \mathbb{N}$:

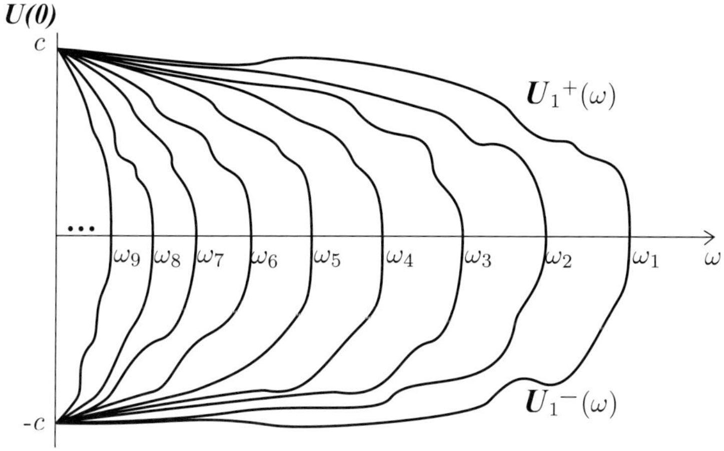

Fig. 3.6. A global bifurcation diagram of rotating Archimedean spirals with rotation frequency $\omega > 0$ and normal tip velocity $U(0)$.

(i) the union of the graphs of $U_n^\pm$ forms a C^3-curve, with nonvanishing curvature at $\omega = \omega_n$;

(ii) $U_n^\pm(\omega_n) = 0$, and $U_n^\pm(\omega) \to \pm c$ for $\omega \searrow 0$;

(iii) for all angular rotation speeds $0 < \omega < \omega_n$ we have the strict ordering

$$U_1^+(\omega) > U_2^+(\omega) > \ldots > U_n^+(\omega) > 0 > \\ > U_n^-(\omega) > \ldots > U_2^-(\omega) > U_1^-(\omega)$$

see Figure 3.6.

The results of theorem 2 are based on a center manifold analysis after regularization of the degenerate pendulum system (3.45). Rewriting (3.45) as a system for U and $V := U_s$, and multiplying the right hand side by the Euler multiplier $\kappa = \Gamma(U)$ to eliminate the vanishing denominator $\Gamma(c) = 0$, we obtain

$$\begin{aligned} \dot{U} &= \Gamma \cdot V \\ \dot{V} &= -\Gamma_U \cdot V(\omega - V) - \Gamma^3 U. \end{aligned} \tag{3.48}$$

Equilibria sit at $U = 0$ and at $U = c$, where $\Gamma = 0$, each together with $V = 0$. Under our assumptions $\Gamma_U < 0 < \omega$ the equilibrium at $(U, V) = (0, 0)$ is an unstable node or focus. The equilibrium at $(U, V) = (c, 0)$, in contrast, possesses one zero eigenvalue and one negative eigenvalue. Finite-dimensional center manifold analysis, much in the spirit of section 3.2.1

above, produces a center manifold $\mathcal{M}$ which is given as a graph

$$V = \Phi(\omega, U) = \frac{c}{\omega}\Gamma_U(c)^2(c-U)^3 + \dots \tag{3.49}$$

near $U = c$, $V = 0$.

It turns out that all solutions of theorem 2 must lie in this center manifold, under just the boundedness and sign conditions imposed there. As an easy consequence all such solutions are asymptotically Archimedean in the far field $s \to +\infty$. Indeed this follows directly from the first order ODE

$$U_s = V = \Phi(\omega, U) = \frac{c}{\omega}\Gamma_U(c)^2(c-U)^3 + \dots \tag{3.50}$$

Substituting $\kappa_s = \Gamma_U(c) \cdot U_s + \dots$ and $\big(\Gamma_U(c) \cdot (c-U)\big)^3 + \dots = -\Gamma^3 = -\kappa^3$ for U near c and for small κ, we obtain

$$\kappa_s = -\frac{c}{\omega}\kappa^3 + \dots \tag{3.51}$$

in the far field $\kappa \to 0$. For $s \to +\infty$ this of course implies

$$\kappa(s) = as^{-1/2} + \dots, \qquad \text{with} \qquad a = \left(\frac{\omega}{2c}\right)^{1/2}. \tag{3.52}$$

The asymptotics (3.52) for the curvature κ readily identifies Archimedean spirals. Indeed, arc length parametrization of the position vector $z = Z(t, s)$ in (3.33) implies that we can write the unit vector $Z_s(t, s)$ in the form

$$Z_s(t, s) = \exp(-iw(t, s))Z_s(t, 0) \tag{3.53}$$

for a suitable real angle variable $w(t, s)$. Recalling the definition $\kappa i Z_s = -Z_{ss}$ of curvature, from (3.34), we observe that

$$w(t, s) = \int_0^s iZ_{ss}/Z_s \mathrm{d}\sigma = \int_0^s \kappa \mathrm{d}\sigma. \tag{3.54}$$

In the present case, where κ does not depend on t, we obtain the explicit expansion

$$w(s) = 2as^{1/2} + \dots \tag{3.55}$$

for $s \to +\infty$, from (3.52). Elementary integration of (3.53), once again, provides the explicit expansion

$$Z(t, s) = Z_s(t, 0) \cdot \Big(\frac{i}{a}s^{1/2}e^{-2ias^{1/2}} + \dots\Big). \tag{3.56}$$

Writing $Z = r\exp(i\varphi)$ in polar coordinates (r, φ), expansion (3.56) readily identifies Z to be asymptotically Archimedean with

$$\lim_{s\to\infty} \frac{\mathrm{d}r}{\mathrm{d}\varphi} = -\frac{1}{2a^2} = -\frac{c}{\omega}. \tag{3.57}$$

In particular, this also determines the asymptotic wave length in the far field to be $2\pi c/\omega$, as claimed. The minus sign in (3.57) indicates that the Archimedean spiral Z is indeed right winding in outward direction.

We now show that the rotation speed $\tilde{\omega}$ of the rigidly rotating spiral $Z(t, s) = \exp(i\tilde{\omega}t)Z(s)$ indeed coincides with the integration constant ω introduced in (3.42) above. The normal velocity U is given by the scalar product with the normal vector $\underline{n}$, in complex notation and polar coordinates, as

$$U = (Z_t, \underline{n}) = (i\tilde{\omega}Z, iZ_s) = \tilde{\omega}\mathrm{Re}(Z \cdot \overline{Z}_s) = \tilde{\omega}rr_s. \tag{3.58}$$

We now argue in the far field $s \to \infty$, where $U \to c$ and $\kappa = \Gamma(U) \to 0$. In particular, (3.58) implies

$$\lim_{s\to\infty} rr_s = c/\tilde{\omega}, \tag{3.59}$$

and $r_s \to 0$ because $r \to \infty$. On the other hand, $(r\varphi_s)^2 + r_s^2 = 1$ holds for arc length parametrization, and implies $\lim_{s\to\infty} r\varphi_s = \pm 1$. Therefore the right winding $\varphi_s < 0$ and the asymptotics (3.57) imply that

$$\lim_{s\to\infty} rr_s = \lim_{s\to\infty} r\frac{\mathrm{d}r}{\mathrm{d}\varphi} \cdot \varphi_s = c/\omega. \tag{3.60}$$

Together, (3.59) and (3.60) prove $\omega = \tilde{\omega}$ is the rotation frequency, indeed.

We conclude this section with some remarks on the relation of our bifurcation analysis of (3.45) with the original problem (3.42). Expressing the left center manifold $\mathcal{M}$ of the equilibrium $(U, V) = (c, 0)$ of (3.48) by $V = \Phi(\omega, U)$, as in (3.49), the bifurcation diagram of Figure 3.6 is the solution set of the equation

$$\omega = \Phi(\omega, U(0)). \tag{3.61}$$

Indeed $V(0) = U_s(0) = \omega - G\kappa_0 = \omega$ for $G = 0$, by (3.46). The different branches of the bifurcation diagram correspond to the different intersection points of the line $V = \omega$ with the center manifold $\mathcal{M}$ in the phase portrait of Figure 3.7(a). The original boundary condition

$$\kappa(0) = \kappa_0 = \Gamma(U(0)) \tag{3.62}$$

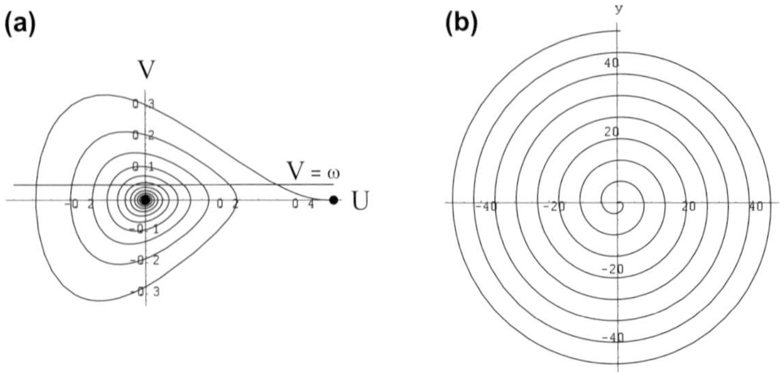

Fig. 3.7. Archimedean spiral (b) for $\Gamma(U) = 1 - 2U$, $\omega = 0.05$. For the underlying center manifold in the plane $(U, V) = (U, U')$ see (a).

can easily be satisfied: we only have to rescale the vertical U-axis of Figure 3.6 by the diffeomorphism Γ to represent κ_0. The horizontal ω-axis then provides all rotation frequencies, if any, of rigidly rotating wave solutions with prescribed Dirichlet boundary condition $\kappa = \kappa_0$ at the spiral tip. As a caveat, we add that some of the rigidly rotating spirals in the global bifurcation diagram of Figure 3.6, although asymptotically Archimedean in the far field $r, s \to \infty$, turn out to be self-intersecting. See [16] for further discussion.

For nonzero G, similarly, the relevant bifurcation diagram corresponds to the solution set of

$$\omega - G \cdot \Gamma(U(0)) = \Phi(\omega, U(0)), \tag{3.63}$$

again with $U(0) = \Gamma^{-1}(\kappa_0)$ and with Φ representing the various branches $V = \Phi(\omega, U)$ of the same center manifold $\mathcal{M}$ of (3.48). We omit a detailed analysis of this case. Instead, we note how G affects the angle γ at which the spiral emanates from the circle of its tip motion. An elementary trigonometric calculation at the tip shows

$$\tan\gamma = -U(0)/G = -\Gamma^{-1}(\kappa_0)/G. \tag{3.64}$$

The spiral therefore emanates perpendicularly if, and only if, $G = 0$.

3.3.3. *Eikonal meanders and drifts*

We return to Wiener and Rosenblueth [81] in this section to study the propagation of spirals under periodically forced normal velocity

$$U = U(t, \kappa) = c(t) > 0\,; \tag{3.65}$$

see (3.35), (3.37). We review [43] and present new results from [34]. Under the modeling assumptions (T1), (T2), (T3) of section 3.3.1 we obtain meandering and drifting motions of the spiral tip. For simplicity of presentation, we assume the normal velocity U to not depend on curvature. Differently from section 3.3.2, where the resulting equation (3.40) for our curvature flow was essentially parabolic, we now obtain an eikonal equation familiar from geometric optics. Throughout this section we consider the case of positive tangential tip velocity G, where the tangential tip motion extends the curve. Unlike [43] we do not impose restrictive assumptions like $G(t) = G_0(t) - \gamma\kappa_0(t)$, $\gamma = \text{const} > 0$. As we will see, spiral wave solutions then correspond to smooth regular solutions of the resulting nonlinear hyperbolic equation. We will have to consider growing solutions, however, and their response to periodic forcing.

Equation (3.40) now becomes

$$\kappa_t + c(t)\Big(\kappa \int_0^s \kappa \mathrm{d}\sigma\Big)_s + G(t)\kappa_s = 0, \tag{3.66}$$

again with Dirichlet boundary condition $\kappa(t,0) = \kappa_0(t) > 0$ at $s = 0$. As in our discussion (3.53)–(3.57) of resulting spiral shapes $Z(t,s)$, we introduce the negative tangent angle $w = \int_0^s \kappa \mathrm{d}\sigma$, and rewrite (3.66) as

$$\begin{aligned} w_t + (c(t)w + G(t))w_s &= G(t)\kappa_0(t) \\ w(t,0) &= 0. \end{aligned} \tag{3.67}$$

Note how the boundary condition $w_s(t,0) = \kappa(t,0) = \kappa_0(t)$ follows from (3.67), for $G(t) \neq 0$. We are interested in time-periodic forcing functions $c(t)$, $G(t)$, $\kappa_0(t)$. The characteristics of the nonlinear hyperbolic balance law (3.67) satisfy

$$\begin{aligned} \dot{s} &= c(t)w + G(t) \\ \dot{w} &= G(t)\kappa_0(t) \end{aligned} \tag{3.68}$$

with $\dot{} = \mathrm{d}/\mathrm{d}t$, $t \geq 0$, $s \geq 0$. In the following we discuss the case of positive c, G, κ_0. For other signs see section 3.4.

In case $G(t) > 0$, for all t, the characteristics at the boundary $s = 0$, where $w = 0$, are pointing inwards towards $s > 0$. We can therefore propagate the Dirichlet boundary data (3.67) to obtain the explicit solution

$$\begin{aligned} w(t, s(t,t_0)) &= \int_{t_0}^t G(\tau)\kappa_0(\tau)\mathrm{d}\tau, \qquad \text{at} \\ s(t,t_0) &= \int_{t_0}^t \Big(G(\tau) + c(\tau)\int_{t_0}^{\tau} G(\tau')\kappa_0(\tau')\mathrm{d}\tau'\Big)\mathrm{d}\tau, \end{aligned} \tag{3.69}$$

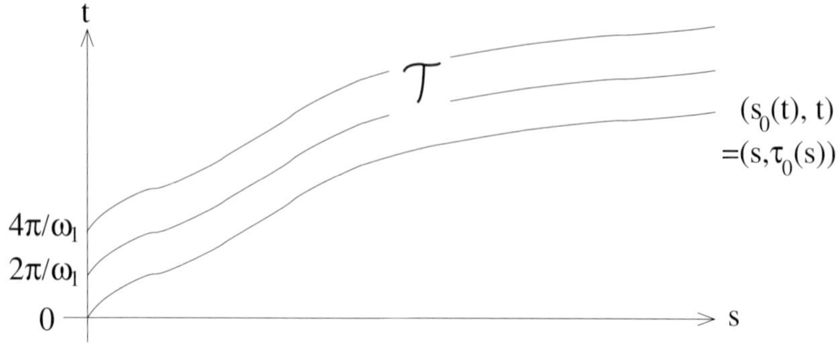

Fig. 3.8. Characteristics of (3.67) in $t \geq 0$, $s \geq 0$ for $G > 0$.

parametrized by $0 \leq t_0 \leq t$. Note that $(t, t_0) \mapsto (t, s(t, t_0))$ is a diffeomorphism onto its image, because $c, G, \kappa_0 > 0$. Hence (3.69) indeed defines w uniquely in the region $\mathcal{T}$ above the characteristic $s_0(t) := s(t, 0)$; see Figure 3.8. Moreover, T-periodicity of G and κ_0 implies $s(t + T, t_0 + T) = s(t, t_0)$, and therefore T-periodicity of the solution w holds in $\mathcal{T}$:

$$w(t + T, s) = w(t, s), \qquad \text{for } s \leq s_0(t). \tag{3.70}$$

Below the characteristic $s_0(t)$, only, the initial conditions $w(0, s) = \int_0^s \kappa(0, \sigma) d\sigma$ are relevant.

To avoid shock solutions, we assume

$$w_s(0, s) = \kappa(0, s) > 0. \tag{3.71}$$

It is easy to see that this condition, which prevents shocks to occur, locally, propagates globally. Indeed we only have to differentiate (3.67) with respect to s. We then observe that $y := w_s$ propagates along the same characteristics (3.68) as w itself, and satisfies $\dot{y} = -c(t)y^2$ with $c > 0$.

We conclude that *any solution* $w(t, s)$ *coincides with the unique periodic solution of (3.67), as soon as* $s_0(t) \geq s$, *independently of the initial condition.* This is a very strong stability statement of finite time convergence, uniformly on bounded subsets s. For further details see [34].

To render our convergence statement more quantitative and, at the same time, exhibit the Archimedean spiral character of our solutions, we now address the asymptotics of the characteristic $s_0(t)$. We claim

$$\tau_0(s) = \left(\frac{2}{\langle c \rangle \langle G\kappa_0 \rangle} \right)^{1/2} s^{1/2} + \ldots \tag{3.72}$$

for the inverse function $\tau_0 := s_0^{-1}$, where $\langle\cdot\rangle$ denotes time average over period T.

Indeed, $\int_0^t a(\tau)\mathrm{d}\tau = \langle a\rangle\, t + \ldots$ for any periodic function a. The remainder is bounded periodic. Applying this simple observation to $s_0(t) = s(t,0)$ in (3.69), twice, we obtain

$$s_0(t) = \frac{1}{2}\langle c\rangle \langle G\kappa_0\rangle t^2 + \ldots \tag{3.73}$$

with a remainder of order t. This proves (3.72).

To exhibit the Archimedean spiral character of the T-periodic solution w, we evaluate $w(t,s)$ along characteristics as in (3.69). Because $w(t + kT, s) = w(t,s)$, it suffices to restrict attention to the fundamental band $\tau_0(s) \leq t \leq \tau_0(s) + T$, where $0 \leq t_0 \leq T$. The relevant asymptotics of w is therefore given by

$$\begin{aligned} w(t,s) &= w(\tau_0(s), s) + \ldots = \langle G\kappa_0\rangle \tau_0(s) + \ldots \\ &= \left(2\langle G\kappa_0\rangle / \langle c\rangle\right)^{1/2} s^{1/2} + \ldots \end{aligned} \tag{3.74}$$

with bounded remainder. Extending the above estimate to include s-derivatives we obtain the curvature asymptotics

$$\kappa(t,s) = w_s(t,s) = \left(\frac{1}{2}\langle G\kappa_0\rangle / \langle c\rangle\right)^{1/2} s^{-1/2} + \ldots \tag{3.75}$$

By the curvature analysis of section 3.3.2, (3.52)–(3.60) this shows that our periodic solution $w(t,s)$ represents a periodically fluctuating Archimedean spiral of time independent asymptotic (average) rotation frequency

$$\omega = \langle G\kappa_0\rangle \tag{3.76}$$

and asymptotic wavelength $2\pi\langle c\rangle / \langle G\kappa_0\rangle$ in the far field.

It may be worth interpreting the characteristic front $s_0(t)$ in the Archimedean spiral geometry. Let $r(t) = r(t, s_0(t))$ denote the distance of the characteristic point on the spiral from its origin. For large s, t, we then have a radial propagation speed $\dot r$ of the stability zone which is given by

$$\frac{1}{2}\frac{\mathrm{d}}{\mathrm{d}t} r^2 = r\dot r = r r_s \dot s_0 + \ldots = \frac{\langle c\rangle^2}{\omega}\langle G\kappa_0\rangle t + \ldots \tag{3.77}$$

Here we have used (3.73) and, for the asymptotically rigidly rotating spiral, also (3.60). Moreover we have omitted the term rr_t which is zero under rigid rotation. With (3.76), integration of (3.77) yields

$$\dot r = \langle c\rangle + \ldots \tag{3.78}$$

This perhaps intuitive result asserts that *the stable core region of the spiral synchronizes the far field at the average radial speed of propagation of the spiral itself.* For further details see [34].

What is the resulting tip dynamics, then? In the notation of section 3.3.2, let $z(t) = Z(t,0)$ denote the tip position and $\exp(i\alpha(t)) := Z_s(t,0)$ the tip tangent. Then

$$\begin{aligned} \dot{\alpha} &= G(t)\kappa_0(t) \\ \dot{z} &= e^{i\alpha}(ic(t) - G(t)). \end{aligned} \tag{3.79}$$

Indeed, $\dot{z}$ results from the normal and tangential velocities c and G which are rotated by α into the appropriate coordinate frame. Since the normal velocity $c(t)$ is identically constant along the curve, the only contribution to the angle change $d\alpha$ comes from the extension of the curve by an arc of curvature radius $1/\kappa_0$ at the tip and of length Gdt. This provides the equation for $\dot{\alpha}$. It is perhaps helpful to compare this derivation with the rigidly rotating case of (3.46), where the rotation frequency ω – alias $\dot{\alpha}$ – was given by the sum of $G\kappa_0$ and the now vanishing term $U_s(0)$.

Comparing with section 3.2.4 we see how the tip dynamics (3.79) takes on the skew product form (3.12). Indeed we only have to assume the variable v to be T-periodic, as Hopf bifurcation had caused it to be there, and our current tip dynamics (3.79) fits right in. Defining $\omega_* := \langle G\kappa_0 \rangle$, the analysis (3.20), (3.21) of meanders and drifts applies with

$$\sigma(t) := (ic(t) - G(t)) \langle G\kappa_0 \rangle / (G(t)\kappa_0(t)) \tag{3.80}$$

of frequency $\omega_1 = 2\pi/T$.

In summary, we have shown that *the periodically forced eikonal flow (3.66) exhibits Archimedean spirals with very strongly stable meandering and drifting tip motions, if the normal velocity $c(t)$, the tangent tip speed $G(t)$, and the tip curvature $\kappa_0(t)$, are all positive* [34, 43, 61].

As mentioned in the introduction, meandering and resonant drift of spiral waves in the photosensitive BZ reaction can be achieved experimentally by a periodically changing light intensity [9, 90]. The Doppler effect imposes superspiral structures in the case of spiral wave dynamics other than rigid rotation. These superspiral structures have been observed in experiments; see for example [11, 41, 55, 56, 89]. See [70] for a mathematical analysis based on linearized analysis and eigenfunctions.

Numerical evidence for superspirals in the kinematic theory has first been presented in [61]. The normal velocity U and the tangential tip velocity G have been assumed to be curvature dependent and a meandering tip

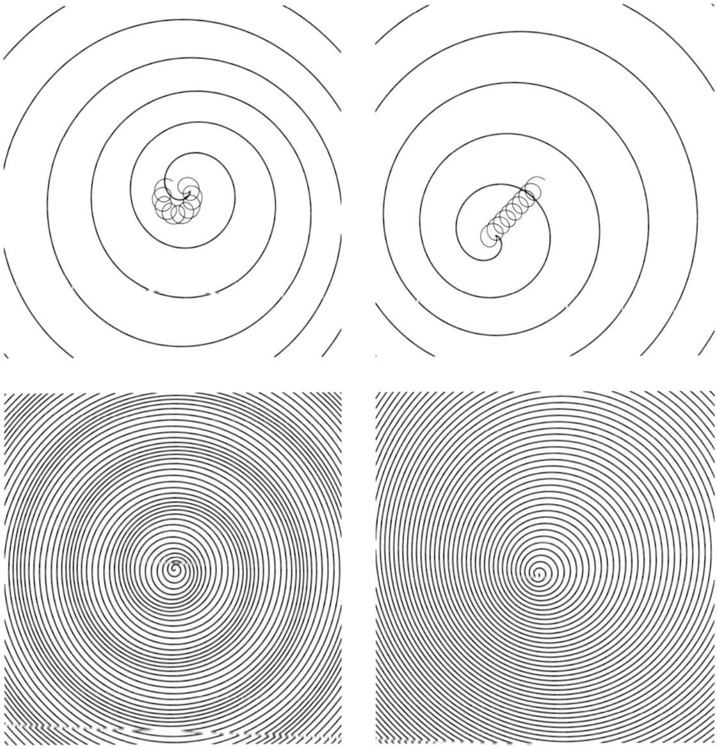

Fig. 3.9. Meandering superspiral for $\delta = 1.1$ (left) and drifting spiral for $\delta = 1$ (right) with tip dynamics in the close-up view (top) and Doppler effect (bottom). Forcing is at frequency 1, near-resonant to the rotation frequency δ.

trajectory was prescribed, a priori. Superspiral patterns have then been observed, numerically, via a stepwise reconstruction of the spiral wave shape.

In our setting of curvature independent but time dependent normal velocity $U = U(t) = c(t)$, as in (3.65), the analysis of the eikonal equation (3.66) reveals superspirals if the normal velocity $c(t)$, the tangent tip speed $G(t)$ or the curvature at the tip $\kappa_0(t)$ are positive and time periodic near resonance. To be explicit, we choose for example $c(t) = 1 + \varepsilon \sin(t)$, $G(t) = \delta\ c(t)$ and $\kappa_0(t) = 1/c(t)$ with $\varepsilon = 0.3$ and $\delta = 1.1$ or $\delta = 1$. Note that $\delta = G\kappa_0$ denotes the rotation frequency of the spiral, via $\dot{\alpha} = \delta$ in (3.79), and the forcing frequency is fixed to be 1, via the term $\sin(t)$. Then equations (3.66) and (3.79) can be solved analytically. For resulting tip motions and superspiral patterns see Figure 3.9.

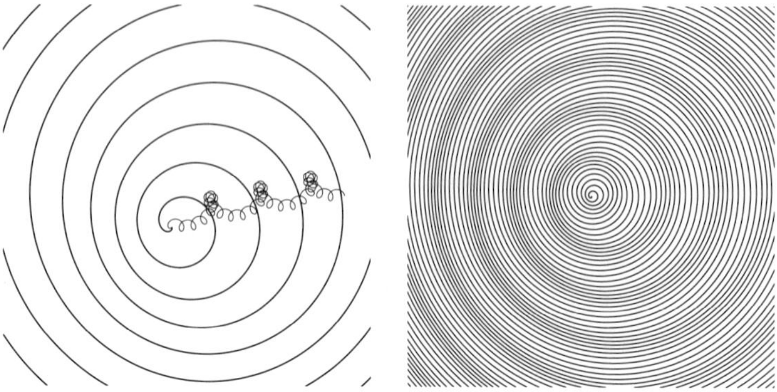

Fig. 3.10. Drifting superspiral: Meandering and drifting spiral with tip dynamics in the close-up view (left) and Doppler effect (right).

Even for periodic functions $c(t)$, $G(t)$ and $\kappa_0(t)$ more complicated tip motions and superspiral structures of higher order can be prescribed. Choose for example $c(t) = 1 + 2\ \varepsilon\ \sin(t) + 3\ \varepsilon\ \sin(1.1\ t)$, $G(t) = c(t)/5$, $\kappa_0(t) = 1/G(t)$ with $\varepsilon = 0.15$. The rotation frequency is thus fixed at 1, because $\dot{\alpha} = G\kappa_0 = 1$. The near resonant component of $c(t)$ with forcing frequency 1.1 is expected to produce a superspiral, as in the left part of Figure 3.9. The resonant component of $c(t)$ with frequency 1 superimposes a linear drift with an associated Doppler effect on, both, the spiral and the superspiral as in the right part of Figure 3.9. We choose initial conditions $z(0) = 0$ for the tip position and $\alpha(0) = 0$ for the tip tangent angle. Solving equation (3.79) yields exactly:

$$
\begin{aligned}
\alpha(t) &= t \\
z_1(t) &= -\frac{2081}{1400} - \frac{3}{20}\ t \\
&\quad - \frac{9}{4}\ \sin(0.1\ t) - \frac{1}{5}\ \sin(t) + \frac{3}{40}\ \sin(2\ t) + \frac{3}{28}\ \sin(2.1\ t) \\
&\quad + \frac{9}{20}\ \cos(0.1\ t) + \cos(t) + \frac{3}{200}\ \cos(2\ t) + \frac{3}{140}\ \cos(2.1\ t) \qquad (3.81) \\
z_2(t) &= \frac{125}{56} - \frac{3}{100}\ t \\
&\quad - \frac{9}{20}\ \sin(0.1\ t) + \sin(t) + \frac{3}{200}\ \sin(2\ t) + \frac{3}{140}\ \sin(2.1\ t) \\
&\quad - \frac{9}{4}\ \cos(0.1\ t) + \frac{1}{5}\ \cos(t) - \frac{3}{40}\ \cos(2\ t) - \frac{3}{28}\ \cos(2.1\ t).
\end{aligned}
$$

Fig. 3.11. Control of the spiral wave dynamics: Tip dynamics (left) induced by control function $h(t)$ (right).

An explicit expression for the evolution of the curve $Z(t, s(t, t_0))$ along the characteristics can be derived from (3.69); for a visualization see Figure 3.10. The two different Fourier frequencies 1 and 1.1 of the forcing in $c(t)$, indeed, cause superimposed meandering and drifting tip dynamics, see the tip trajectory in Figure 3.10. In the far field we recognize the superspiral structure induced by the meandering of the tip, which is drifting due to resonance. Similar observations apply to quasi-periodic superspirals supported by an incommensurate 2-frequency forcing. Superspirals of infinite order can then be achieved, due to the resulting arbitrarily small difference frequencies. For further analysis see [34].

Can the tip dynamics be designed, arbitrarily? Linearly dependent normal and tangential tip velocities, $c(t) = \text{const} \cdot G(t)$, and a constant tip curvature κ_0 admit rigid rotation, only. This follows by integration of (3.79), independently of the choices for $c(t)$, $G(t)$. For $\kappa_0(t) = 1/G(t)$, on the other hand, we can prescribe the spiral path, arbitrarily. For example, choose $c(t) = 1 + 0.1 \sin(\pi/2\ h(t) - t)$, $G(t) = 0.2 - 0.1 \cos(\pi/2\ h(t) - t)$. At any time t, the control function $h(t)$ assigns a tip drift which superimposes the rotational dynamics. For a control function $h(t)$ as sketched in Figure 3.11, we obtain a desired “SFB”.

3.4. Concluding remarks

We first comment on several aspects of eikonal meanders; see section 3.3.3. Specifically we mention the relation to tip control experiments [38, 84, 85, 87, 88], discuss the occurrence and interpretation of shock waves, the role of the signs of G and κ_0, possible generalizations to dynamics of scroll wave filaments in three dimensions, and the role of diffusion and viscous regularization in the sense of section 3.3.2. We then return to the reaction-diffusion view point of section 3.2 and discuss the chances of a reduced eikonal description of autonomous meanders by relative Hopf bifurcation in the Euclidean group $SE(2)$.

To describe *control of tip motion*, in the spirit of [38, 84, 85, 87, 88], we return to the tip dynamics

$$\begin{aligned} \dot{\alpha} &= G(t)\kappa_0(t) \\ \dot{z} &= e^{i\alpha}(ic(t) - G(t)). \end{aligned} \tag{3.82}$$

as derived in (3.79); see also (3.20), (3.21). Our derivation did not depend on time periodicity of c, D, κ_0, of course. Control of z can now be effected as follows. By external lighting, the triple $\lambda := (c, G, \kappa_0) \in \{\lambda_0, \lambda_1\}$ is set to two different values λ_0 and λ_1, depending on time. Mostly $\lambda = \lambda_0$ with corresponding rigid rotation of (α, z). When the wave front passes a sensor point z_0, lighting is changed to λ_1, briefly, and then resumes λ_0. As (α, z) pass through their λ_0 cycle, a small offset dz is thus effected during each period. The offsets add up to a superimposed drift

$$\frac{\mathrm{d}z}{\mathrm{d}\tau} = f(z - z_0) \tag{3.83}$$

on a slow time scale τ. Here $f(z) := \exp(ir + \beta\varphi + a_0)$ for polar coordinates $z = \exp(i\varphi)$. Indeed, f reflects the asymptotically Archimedean character of the spiral expressed by $G\kappa_0 r + c\varphi = \text{const.}$, with values λ_0; see also (3.57). The constant a_0 measures the resulting offset between lighting λ_0 and λ_1. Similarly, triggering lighting at several sensor positions z_k superimposes to a slow drift

$$\frac{\mathrm{d}z}{\mathrm{d}\tau} = \sum_k f(z - z_k) \tag{3.84}$$

with correspondingly richer possibilities for control. Alternatively, and more efficiently, active control can be based on the temporal pattern of sensor signals, within one period, to explicitly reconstruct – and control – position $z(t)$ and phase $\alpha(t)$ of the spiral tip.

Balance laws (3.67) are known to exhibit *shock waves*, i.e. solutions w with discontinuities of the (negative) tangent angle w at shock positions $s(t)$. See Figure 3.12 and, for some qualitative remarks, also [62]. The Rankine-Hugoniot condition specifies the shock speed $\dot{s}(t)$ to satisfy

$$\dot{s} = G + \frac{1}{2}c\frac{[w^2]}{[w]}, \tag{3.85}$$

where $[w] = w_+ - w_-$ measures the jump of w from the left value w_- to the right value w_+ at the discontinuity. The Lax entropy condition

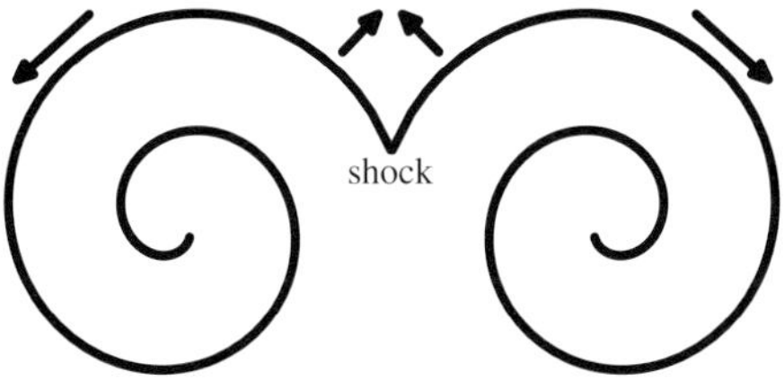

Fig. 3.12. Collision of spirals and shocks in solutions w of balance laws (3.67).

requires $[w] < 0$, for positive c. See for example [40, 68] and the references there. In (3.71) we have excluded shocks, imposing the initial condition $w_s > 0$. At collision of spirals, however, shocks may occur; see Figure 3.12. Indeed (3.53) then indicates a jump in the tangent direction $Z_s(t,s) = \exp(-iw(t,s))Z_s(t,0)$ of the joined spiral-antispiral curve $Z(t,s)$ at the interface of the two counter-rotating spirals. The interface evolves according to the Rankine-Hugoniot condition (3.85) until the shock strength $[w]$ has decayed to zero and the solution regains regularity.

In section 3.3 above we have imposed *positive sign conditions* $c, G, \kappa_0 > 0$. Other combinations of signs are of course conceivable. We only mention the case $c, G < 0 < \kappa_0$ briefly. Reversing time, $t \mapsto -t$, then converts the characteristics to the previous ones of section 3.3.3; see (3.68). Therefore asymptotically Archimedean spirals still exist, as a periodic response to the periodic forcing by c, G, k_0. However, they now rotate backwards compared to the previous case. Moreover, these spirals are now extremely unstable, just as they were extremely stable before. Indeed, any perturbation in the far field now propagates inwards at negative radial velocity $\dot{r} = \langle c \rangle + \ldots$, see (3.78), and destructively reaches the spiral tip after some finite time. Afterwards the Archimedean spiral disappears and the initial conditions reign. This may account for far-field break-up of spiral waves.

The eikonal approach of section 3.3.3 suggests generalizations to *scroll wave dynamics* in three space dimensions. For beautiful experimental results based on computer tomography see [12]. For numerical simulations in an excitable media reaction-diffusion setting see [18], for example, and the references there. For experimental and numerical investigations of chemical turbulence of scroll waves see [1, 4, 51]. Indeed we may consider an embedded oriented surface S in $\mathbb{R}^3$, called a scroll wave, with one-dimensional boundary curve γ, called a filament. Of course, the surface S generalizes our rotating planar curve $Z(t,s)$, and the curve γ corresponds to the tip z of Z. We may then propagate S and its boundary γ with speed c in the (oriented) normal direction. The propagation of the boundary curve γ will also occur with speed G along the outward tangent of S, and with prescribed sectional curvature of S. Several variants of mean or Gauss curvature flows may also be invoked to generalize the curvature dependence of normal velocity to the surface S, see also [4, 46]. As a result, meandering, drifting, and even colliding scroll wave filaments should be observed.

Eikonal curvature flow precludes *relative Hopf bifurcation.* Indeed the resulting tip dynamics

$$\begin{aligned} \dot{\alpha} &= G\kappa_0 \\ \dot{z} &= e^{i\alpha}(ic - G). \end{aligned} \tag{3.86}$$

of (3.79) generates purely exponential dynamics of $(\exp(i\alpha), z) \in SE(2)$ with circular motion, for $G\kappa_0 \neq 0$. This excludes meanders and drifts. At present, it is unknown whether the curvature dependent normal velocities $U = U(\kappa)$ of section 3.3.2 are able to remedy this modeling difficulty. It only seems clear that affine linear dependencies $U(\kappa) = c - D\kappa$ as in (3.38) will not remedy the problem, as long as $D > 0$ remains small. Indeed small D act as a viscous regularization in (3.40) and, analogously, for the balance law (3.67). For monotone solutions $w_s = \kappa > 0$ which are regular, however, small positive D act as a regular perturbation which should not be strong enough to produce meanders and drifts which are absent for $D = 0$.

An easy remedy for this problem would be the introduction of additional *"hidden" variables* which influence the dynamics of the curve, notably via c, D, G, or κ_0. A modification of G or κ_0 at the tip alone, for example, might be sufficient. Specifically we may introduce a hidden scalar variable v at the tip, alone, together with some hypothetical interaction dynamics

$$\begin{aligned} \dot{\kappa}_0 &= g(\kappa_0, v) \\ \dot{v} &= h(\kappa_0, v) \end{aligned} \tag{3.87}$$

which undergoes Hopf bifurcation as a (suppressed) parameter is varied. Here $\kappa_0 = \kappa_0(t)$ is supposed to provide the Dirichlet boundary condition (3.41) for the curvature PDE (3.40). The resulting periodic fluctuation $\kappa_0(t)$ would then drive and synchronize the Archimedean spiral, globally, into meandering and drifting motions.

For the moment, a coupling like (3.87) remains speculation. Only reaction-diffusion systems as studied in section 3.2 provide a modeling description which is based on reasonably "first" principles. The mathematically proper approach to autonomous meanders and relative Hopf bifurcation, which are present in the reaction-diffusion setting, would then be a derivation of reduced systems like (3.87), (3.40), (3.41) from reaction-diffusion systems which model excitable media. The relevant techniques, vaguely, are to include singular perturbation theory. But details still require much future effort.

References

[1] S. Alonso, R. Kähler, A. S. Mikhailov, and F. Sagués. Expanding scroll rings and negative tension turbulence in a model of excitable media. *Phy. Rev. E*, **70**(5):1–10, 2004.

[2] S. Angenent. Parabolic equations for curves on surfaces. I: Curves with p-integrable curvature. *Ann. Math.*, **132**:451–483, 1990.

[3] S. Angenent. Parabolic equations for curves on surfaces. II: Intersections, blow-up and generalized solutions. *Ann. Math.*, **133**:171–251, 1991.

[4] S. Alonso, F. Sagués, and A. S. Mikhailov. Periodic forcing of scroll rings and control of Winfree turbulence in excitable media. *Chaos*, **16**:1–11, 2006.

[5] D. Barkley. Euclidean symmetry and the dynamics of rotating spiral waves. *Phys. Rev. Lett.*, **72**:164–167, 1994.

[6] D. Barkley. Linear stability analysis of rotating spiral waves in excitable media. *Phys. Rev. Lett.*, **68**:2090–2093, 1994.

[7] H. Brandtstädter, M. Braune, I. Schebesch, and H. Engel. Experimental study of the dynamics of spiral pairs in light-sensitive Belousov-Zhabotinskii media using an open-gel reactor. *Chem. Phys. Lett.*, **322**:145–154, 2000.

[8] M. Braune and H. Engel. Compound rotation of spiral waves in a light-sensitive Belousov-Zhabotinsky medium. *Chem. Phys. Lett.*, **204**:257–264, 1993.

[9] M. Braune and H. Engel. Compound rotation of spiral waves in active media with periodically modulated excitability. *Chem. Phys. Lett.*, **211**:534–540, 1993.

[10] M. Bertram and A. Mikhailov. Pattern formation on the edge of chaos: mathematical modeling of CO oxidation on a Pt(110) surface under global delayed feedback. *Phys. Rev. E*, **67**:1–9, 2003.

[11] A. L. Belmonte, Q. Ouyang, and J.-M. Flesselles. Experimental Survey of Spiral Dynamics in the Belousov-Zhabotinsky Reaction. *Journal de Physique II*, **7**:1425–1468, 1997.

[12] T. Bánsági and O. Steinbock. Nucleation and Collapse of Scroll Rings in Excitable Media. Preprint, 2006.

[13] P. K. Brazhnik and J. J. Tyson. Nonspiral excitation waves beyond the eikonal approximation. *Phys. Rev. E*, **54**:4338–4346, 1996.

[14] F. Cao and L. Moisan. Geometric computation of curvature driven plane curve evolutions. *SIAM J. Numer. Anal.*, **39**(2):624–646, 2001.

[15] B. Fiedler, J.-S. Guo, and J.-C. Tsai. Multiplicity of rotating spirals under curvature flows with normal tip motion. *J. Differ. Equations*, **205**(1):211–228, 2004.

[16] B. Fiedler, J.-S. Guo, and J.-C. Tsai. Rotating spirals of curvature flows: a center manifold approach. To appear in Ann. Mat. Pura Appl., 2006.

[17] M. Falcke and H. Levine. Pattern Selection by Gene Expression in Dictyostelium Discoideum. *Phys. Rev. Lett.*, **80**:3875–3878, 1998.

[18] B. Fiedler and R. M. Mantel. Crossover collision of scroll wave filaments. *Doc. Math., J. DMV*, **5**:695–731, 2000.

[19] M. Fila and H. Matano. Blow-up in nonlinear heat equations from the dynamical systems point of view. In *B. Fiedler (ed.), Handbook of dynamical systems. Volume 2.* Amsterdam: Elsevier, 723–758, 2002.

[20] B. Fiedler and J. Scheurle. Discretization of homoclinic orbits, rapid forcing and "invisible" chaos. *Mem. Am. Math. Soc.*, **570**:79 p., 1996.

[21] B. Fiedler and A. Scheel. Spatio-temporal dynamics of reaction-diffusion patterns. In *M. Kirkilionis (ed.) et al., Trends in nonlinear analysis. On the occasion of the 60th birthday of Willi Jäger.* Berlin: Springer, 23–152, 411–417, 2003.

[22] B. Fiedler, B. Sandstede, A. Scheel, and C. Wulff. Bifurcation from relative equilibria of noncompact group actions: Skew products, meanders, and drifts. *Doc. Math., J. DMV*, **1**:479–505, 1996.

[23] B. Fiedler and D. Turaev. Normal forms, resonances, and meandering tip motions near relative equilibria of Euclidean group actions. *Arch. Ration. Mech. Anal.*, **145**(2):129–159, 1998.

[24] V. Gelfreich. Numerics and exponential smallness. In *B. Fiedler (ed.), Handbook of dynamical systems. Volume 2.* Amsterdam: Elsevier, 265–312, 2002.

[25] M. Georgi. Spiral waves in reaction-diffusion systems – the reduced vector field (Spiralwellen in Reaktions-Diffusions-Systemen – Das reduzierte Vektorfeld). *Diploma Thesis, Free University of Berlin*, 2003.

[26] M. Gage and R. Hamilton. The heat equation shrinking convex plane curves. *J. Differ. Geom.*, **23**:69–96, 1986.

[27] M. Georgi and N. Jangle. Spiral wave motion in reaction-diffusion systems. In F. Dumortier, H. Broer, J. Mawhin, A. Vanderbauwhede, and S. V. Lunel, editors, *Proceedings Equadiff 03, International Conference on Differential Equations, Hasselt 2003*, pages 651–656, Singapore, 2005. World Scientific.

[28] V. Gelfreich and V. Lazutkin. Splitting of separatrices: Perturbation theory and exponential smallness. *Russ. Math. Surv.*, **56**(3):499–558, 2001.

[29] D. Henry. *Geometric theory of semilinear parabolic equations.* Lecture Notes in Mathematics. 840. Berlin-Heidelberg-New York: Springer-Verlag, 348 p., 1981.

[30] M. Hirsch, C. Pugh, and M. Shub. *Invariant manifolds.* Lecture Notes in Mathematics. 583. Berlin-Heidelberg-New York: Springer-Verlag. 149 p., 1977.

[31] G. Huisken. Local and global behaviour of hypersurfaces moving by mean curvature. In *R. Greene (ed.) et al., Differential geometry. Part 1: Partial differential equations on manifolds. Proceedings of a summer research institute, held at the University of California, Los Angeles, CA, USA, July 8-28, 1990. Providence, RI: American Mathematical Society. Proc. Symp. Pure Math. 54, Part 1*, 175–191, 1993.

[32] R. Ikota, N. Ishimura, and T. Yamaguchi. On the structure of steady solutions for the kinematic model of spiral waves in excitable media. *Japan J. Ind. Appl. Math.*, **15**(2):317–330, 1998.

[33] N. Jangle. Center manifolds in reaction-diffusion systems on unbounded domains (Zentrumsmannigfaltigkeiten in Reaktions-Diffusions-Systemen auf unbeschränkten Gebieten). *Diploma Thesis, Free University of Berlin*, 2003.

[34] N. Jangle. Dynamics of spatio-temporal patterns in heterogeneous media (Dynamik räumlich-zeitlicher Muster in heterogenen Medien). Dissertation thesis in preparation, 2006.

[35] W. Jahnke, W. Skaggs, and A. Winfree. Chemical vortex dynamics in the Belousov-Zhabotinskii reaction and in the two-variable Oregonator model. *J. Chem. Phys.*, **93**:740–749, 1989.

[36] J. P. Keener. The core of the spiral. *SIAM J. Appl. Math.*, **52**(5):1370–1390, 1992.

[37] O. Kheowan, V. Gaspar, V. Zykov, and S. Müller. Measurements of kinematical parameters of spiral waves in media of low excitability. *Phys. Chem. Lett.*, **3**:4747–4764, 2001.

[38] O.-U. Kheowan and S. C. Müller. Control of spiral waves in excitable media. *Appl. Math. Comput.*, **164**(2):373–390, 2005.

[39] J. P. Keener and J. J. Tyson. The dynamics of scroll waves in excitable media. *SIAM Rev.*, **34**(1):1–39, 1992.

[40] P. D. Lax. *Hyperbolic systems of conservation laws and the mathematical theory of shock waves.* CBMS-NSF Regional Conference Series in Applied Mathematics. 11. Philadelphia, Pa.: SIAM, Society for Industrial and Applied Mathematics, 48 p. , 1973.

[41] G. Li, Q. Ouyang, V. Petrov, and H. L. Swinney. Transition from Simple Rotating Chemical Spirals to Meandering and Traveling Spirals. *Phys. Rev. Lett.*, **77**:2105–2108, 1996.

[42] R. Mantel and D. Barkley. Periodic forcing of spiral waves in excitable media. *Phys. Rev. E*, **54**:4791–4802, 1996.

[43] A. Mikhailov, V. Davydov, and V. Zykov. Complex dynamics of spiral waves and motion of curves. *Physica D*, **70**(1-2):1–39, 1994.

[44] P. Mendes. A metric property of Cherry vector fields on the torus. *J. Differ. Equations*, **89**(2):305–316, 1991.

[45] E. Meron. Pattern formation in excitable media. *Physics Reports*, **218**:1–66, 1992.

[46] A. S. Mikhailov. Three-dimensional Kinematics. *Chaos, Solitons and Fractals*, **5**:673–679, 1995.

[47] K. Mikula. Solution and application of anisotropic curvature driven evolution of curves (and surfaces). In *M. Falcone (ed.) et al., Numerical methods for viscosity solutions and applications.* Singapore: World Scientific. Ser. Adv. Math. Appl. Sci. 59, 173–196, 2001.

[48] A. S. Mikhailov. Modeling pattern formation in excitable media: The legacy of Norbert Wiener. In *J. Milton, P. Jung (eds.), Epilepsy as a dynamic disease. Biological and Medical Physics Series.* Springer, Berlin Heidelberg, 125–163, 2003.

[49] P. Moreira and A. Ruas. Metric properties of Cherry flows. *J. Differ. Equations*, **97**(1):16–26, 1992.
[50] R. Malladi and J. Sethian. Image processing via level set curvature flow. *Proc. Natl. Acad. Sci. USA*, **92**(15):7046–7050, 1995.
[51] A. S. Mikhailov and K. Showalter. Control of waves, patterns and turbulence in chemical systems. *Phys. Rep.*, **425**:79–194, 2006.
[52] A. S. Mikhailov and V. S. Zykov. Kinematical theory of spiral waves in excitable media: Comparison with numerical simulations. *Physica D*, **52**:379–397, 1991.
[53] S. C. Müller and V. S. Zykov. Simple and complex spiral wave dynamics. *Philos. Trans. R. Soc. Lond., Ser. A*, **347**:677–685, 1994.
[54] S. Nettesheim, A. von Oertzen, H. Rotermund, and G. Ertl. Reaction diffusion patterns in the catalytic CO-oxidation on Pt(110)-front propagation and spiral waves. *J. Chem. Phys.*, **98**:9977–9985, 1993.
[55] Q. Ouyang, H. L. Swinney, and G. Li. Transition from spirals to defect-mediated turbulence driven by a doppler instability. *Phys. Rev. Lett.*, **84**:1047–1050, 2000.
[56] V. Pérez-Muñuzuri, R. Aliev, B. Vasiev, V. Pérez-Vilar, and V. Krinsky. Super-spiral structures in an excitable medium. *Nature*, **353**:740–742, 1991.
[57] J. j. Palis and W. de Melo.S *Geometric theory of dynamical systems. An introduction. Transl. from the Portugese by A. K. Manning.* New York - Heidelberg - Berlin: Springer-Verlag, 198 p., 1982.
[58] G. Perelman. Finite extinction time for the solutions to the Ricci flow on certain three-manifolds. Preprint, 2006.
[59] G. Perelman. Ricci flow with surgery on three-manifolds. Preprint, 2006.
[60] G. Perelman. The entropy formula for the Ricci flow and its geometric applications. Preprint, 2006.
[61] V. Pérez-Muñuzuri, M. Gómez-Gesteira, and V. Pérez-Villar. A geometrical-kinematical approach to spiral wave formation: super-spiral waves. *Physica D Nonlinear Phenomena*, **64**:420–430, 1993.
[62] L. Pismen. *Patterns and Interfaces in Dissipative Dynamics.* Berlin: Springer-Verlag. 369 p., 2006.
[63] T. Plesser and K. Müller. Fourier analysis of the complex motion of spiral tips in excitable media. *Int. J. Bifurcation Chaos*, **5**:1071–1084, 1995.
[64] A. Panfilov, V. Zykov, S. Müller, and J. Keener. Defibrillation of cardiac tissue by multiple subthreshold shocks. *Phys. Rev. E*, **61**:4644–4667, 2000.
[65] M. Ruiz-Villarreal, M. Gómez-Gesteira, C. Souto, A. P. Muñuzuri, and V. Pérez-Villar. Long-term vortex interaction in active media. *Phys. Rev. E*, **54**:2999–3002, 1996.
[66] A. Scheel. Bifurcation to spiral waves in reaction-diffusion systems. *SIAM J. Math. Anal.*, **29**(6):1399–1418, 1998.
[67] M. Shub. *Global stability of dynamical systems. With the collab. of Albert Fathi and Remi Langevin. Transl. from the French by Joseph Christy.* New York etc.: Springer-Verlag, 150 p., 1987.
[68] J. Smoller. *Shock waves and reaction-diffusion equations. 2nd ed.* Grundlehren der Mathematischen Wissenschaften. 258. New York: Springer-

Verlag, 632 p., 1994.

[69] G. Skinner and H. Swinney. Periodic to quasiperiodic transition of chemical spiral rotation. *Physica D*, **48**:1–16, 1991.

[70] B. Sandstede and A. Scheel. Superspiral Structures of Meandering and Drifting Spiral Waves. *Phys. Rev. Lett.*, **86**:171–174, 2001.

[71] B. Sandstede and A. Scheel. Defects in oscillatory media: towards a classification. *SIAM J. Appl. Dyn. Syst.*, **3**(1):1–68, 2004.

[72] B. Sandstede, A. Scheel, and C. Wulff. Dynamics of spiral waves on unbounded domains using center-manifold reductions. *J. Differ. Equations*, **141**(1):122–149, 1997.

[73] B. Sandstede, A. Scheel, and C. Wulff. Bifurcations and dynamics of spiral waves. *J. Nonlinear Sci.*, **9**(4):439–478, 1999.

[74] J. Sanders and F. Verhulst. *Averaging methods in nonlinear dynamical systems*. Applied Mathematical Sciences, 59. New York etc.: Springer-Verlag, 247 p., 1985.

[75] O. Steinbock, V. Zykov, and S. C. Müller. Control of Spiral-Wave Dynamics in Active Media by Periodic Modulation of Excitability. *Nature*, **366**:322–324, 1993.

[76] J. J. Tyson and J. P. Keener. Singular perturbation theory of traveling waves in excitable media (A review). *Physica D*, **32**(3):327–361, 1988.

[77] A. Vanderbauwhede. Centre manifolds, normal forms and elementary bifurcations. *Dynamics Reported* **2**:89–169, 1989.

[78] A. Vanderbauwhede and G. Iooss. Center manifold theory in infinite dimensions. In *Dynamics Reported, New Series* **1**:125–163, 1992.

[79] A. Winfree. Spiral waves of chemical activity. *Science*, **175**:634–636, 1972.

[80] A. T. Winfree. *The geometry of biological time. 2nd ed.* Interdisciplinary Applied Mathematics. Berlin: Springer, 777 p., 2001.

[81] N. Wiener and A. Rosenblueth. The mathematical formulation of the problem of conduction of impulses in a network of connected excitable elements, specifically in cardiac muscle. *Arch. Inst. Cardiol. Mexico*, **16**:205–265, 1946.

[82] C. Wulff. Theory of Meandering and Drifting Spiral Waves in Reaction-Diffusion Systems. *PhD thesis, FU Berlin*, 1996.

[83] V. Zykov, G. Bordiougov, H. Brandtstädter, I. Gerdes, and H. Engel. Periodic forcing and feedback control of nonlinear lumped oscillators and meandering spiral waves. *Phys. Rev. E*, **68**:1–11, 2003.

[84] V. Zykov, G. Bordiougov, H. Brandtstädter, I. Gerdes, and H. Engel. Global control of spiral wave dynamics in an excitable domain of circular and elliptical shape. *Phys. Rev. Lett.*, **92**:1–4, 2004.

[85] V. Zykov, H. Brandtstädter, G. Bordiougov, and H. Engel. Interference patterns in spiral wave drift induced by a two-point feedback. *Phys. Rev. E*, **72**:1–4, 2005.

[86] V. Zykov and H. Engel. Experimental study of the dynamics of spiral pairs in light-sensitive BelousovZhabotinskii media using an open-gel reactor. *Chem. Phys. Lett.*, **323**:145–154, 2000.

[87] V. Zykov and H. Engel. Dynamics of spiral waves under global feedback in excitable domains of different shapes. *Phys. Rev. E*, **70**:1–9, 2004.

[88] V. Zykov and H. Engel. Unified approach to feedback-mediated control of spiral waves in excitable media. *This volume*, 2006.

[89] L. Q. Zhou and Q. Ouyang. Experimental studies on long-wavelength instability and spiral breakup in a reaction-diffusion system. *Phys. Rev. Lett.*, **85**:1650–1653, 2000.

[90] V. Zykov, O. Steinbock, and S. Müller. External forcing of spiral waves. *Chaos*, **4**:509–518, 1994.

[91] V. Zykov. *Simulation of wave processes in excitable media. Transl. from the Russian by Potrans, edited by A. T. Winfree.* Nonlinear Science: Theory and Applications. Manchester etc.: Manchester University Press, 235 p., 1987.

Chapter 4

Cellular calcium oscillations: From bifurcation analysis to experiment

Antonio Z. Politi[1], Lawrence D. Gaspers[2], Anneke Brümmer[1], Andrew P. Thomas[1,*] and Thomas Höfer[1]

[1] *Theoretical Biophysics, Humboldt University Berlin, Invaliden Str. 42, 10115 Berlin, Germany*

[2] *Department of Physiology and Pharmacology, University of Medicine and Dentistry of New Jersey, Newark*

[*] *thomas.hoefer@rz.hu-berlin.de*

Contents

4.1. Introduction

Cells translate extracellular signals in intracellular responses, such as changes in the metabolic activity or the expression of specific genes. The signaling substances bind to receptor proteins in the cell membrane, which transmit the incoming information into the inner of the cell using diffusible messengers. Ca^{2+} ions serve as a ubiquitous second messenger that control the activity of several key proteins that convey the desired cellular response. In unstimulated cells, the cytoplasmic Ca^{2+} concentration is kept low by the action of Ca^{2+} pumps ($\sim$ 100 nM, which is orders of magnitude below typical Ca^{2+} content of drinking water). Due to the energy-consuming transport of Ca^{2+} across the cell membrane and into an intracellular compartment, the endoplasmic reticulum (ER), large concentration gradients are generated. The binding of a hormone to its receptor at the plasma membrane initiates a signaling cascade that finally activates Ca^{2+} channels in the ER membrane, triggering Ca^{2+} release.

Interestingly, the presence of a constant external signal produces periodic increases in the intracellular Ca^{2+} concentration ($[Ca^{2+}]_c$). This phenomenon of Ca^{2+} oscillations has been intensively studied experimentally as well as theoretically in the last twenty years and the molecular key players have been identified. However, one of the central questions, namely the nature of the oscillatory mechanisms, is still controversial. On the one hand, it has been argued that the Ca^{2+} oscillations are generated by Ca^{2+} feedbacks on the Ca^{2+} channels. These channels are fast activated at low and slowly inhibited at high Ca^{2+} concentrations. On the other hand, recent studies have shown that inositol-1,4,5-trisphosphate (IP_3), the molecule activating the Ca^{2+} channels, also oscillates in time, indicating that the oscillatory mechanism may lie in the signaling cascade.

We will first introduce the biological system and briefly review previous work. Then an approach to identify the oscillatory mechanism(s) is developed. Chiefly, it consists in perturbing the IP_3 dynamics by introducing an IP_3 binding protein into the cell. The realization of this experiment shows that oscillations in the IP_3 signaling cascade are essential for $[Ca^{2+}]_c$ oscillations. Finally, we speculate on the physiological role of the proposed oscillatory mechanism. In particular, we show that it enables the frequency encoding of the hormone stimulus and facilitates cell-cell communication.

4.2. Background

Cells are composed of different intracellular compartments embedded in the cytoplasm, and are bounded by the plasma membrane. One of this compartments, the endoplasmic reticulum (ER), acts as intracellular store for Ca^{2+} ions. In a resting cell the Ca^{2+} concentration in the cytoplasm ($[Ca^{2+}]_c$) is kept low ($\sim$ 100 nM) and in the ER high ($\sim$ 20 μM) by active transport through the ER and plasma membrane. Binding of agonist to its receptor at the plasma membrane activates phospholipase C (PLC). This enzyme is responsible for the production of the second messenger IP_3, which can freely diffuse in the cytoplasm. IP_3 activates Ca^{2+} channels located in the ER membrane, the IP_3-receptors (IP_3R), and causes an outflow of Ca^{2+} into the cytoplasm increasing its concentration by about 10 fold ($\sim 1\ \mu M$). In a multitude of cell types one does not observe a simple increase of $[Ca^{2+}]_c$ to a new steady state but a sequence of $[Ca^{2+}]_c$ spikes, $[Ca^{2+}]_c$ oscillations. The strength of the extracellular stimulus is encoded primarily in the frequency of the $[Ca^{2+}]_c$ oscillations, which increases with the degree of stimulation. For example, in rat hepatocytes the periods of $[Ca^{2+}]_c$ oscillations range over one order of magnitude, from above 250 sec for low concentrations of hormones, such as vasopressin and noradrenalin, to about 30 sec for higher hormone doses [22]. Cells can be coupled with their next neighbors via channels, gap junctions, that allow for the diffusion of small molecules. In the liver, one observes that cells which are uncoupled oscillate with different periods, whereas they oscillate with equal period when coupled. This coordination becomes manifest as repetitive intercellular waves initiating from specific sites in the liver [21, 27].

A long-standing question has been whether the oscillations are generated by the cellular Ca^{2+} transporters and channels themselves or whether they originate upstream in the signal transduction machinery, between hormone binding to its receptor and the activation of Ca^{2+} fluxes. It has been proposed that the periodic release of Ca^{2+} ions from the ER can be brought about through the regulatory properties of the IP_3R [1, 6, 14, 33]. Mathematical models have demonstrated how fast activation and delayed inhibition of the IP_3R by cytoplasmic Ca^{2+} can drive repetitive Ca^{2+} spiking [4, 12, 23]. In these models, IP_3 is required to initially open the IP_3R and sensitize the channel toward feedback activation by cytoplasmic calcium. Therefore, Ca^{2+} oscillations can occur when IP_3 concentration is held at a constant value. However, models based on a simple description of the IP_3R dynamics generally produce $[Ca^{2+}]_c$ oscillations with short periods ($\sim$ 10-60 sec) and thus do not reproduce the long inter-spike intervals observed experimentally. Recently, it has become possible to monitor IP_3 changes in intact cells. These experiments have shown that the IP_3 concentration is highly dynamic and can oscillate together with cytoplasmic calcium [9, 17, 26, 32]. There exist both positive and negative feedbacks of Ca^{2+} on IP_3 metabolism which could mediate fluctuations in cellular IP_3 levels. The principal positive feedback is the activation of IP_3 production via a Ca^{2+} sensitive PLC [7, 20]. The principal negative feedback acts through Ca^{2+} dependent degradation of IP_3 via phosphorylation by the IP_3 3-kinase (IP_33K) [3, 15, 31]. All described interactions are shown in Fig. 4.1.

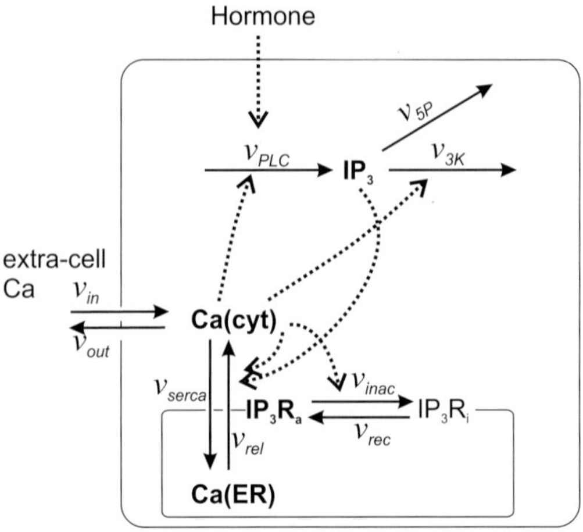

Fig. 4.1. Interactions between Ca^{2+} transport processes and IP_3 metabolism. The solid, and dashed arrows indicate transport/reaction steps and activations, respectively. The model variables are bold faced: IP_3, cytoplasmic IP_3; Ca(cyt), free cytoplasmic Ca^{2+}; Ca(ER), free Ca^{2+} in the ER; IP_3R_a, active conformation of the IP_3R. Reproduced from [18].

It is presently not clear what effects such feedbacks have on Ca^{2+} oscillations. Importantly, it is not known which of the feedbacks is predominant and whether the involvement of these Ca^{2+}-dependent feedbacks on IP_3 serves a physiological role. For the identification of Ca^{2+} feedbacks on IP_3 we suggest to perturb the IP_3 turnover time. This will first be illustrated using a simple model that allows to derive analytical results. Then a detailed model is presented and the results of slowing the IP_3 turnover with an IP_3 binding molecule are compared to experimental data. The physiological role of Ca^{2+} feedbacks on IP_3 will be studied with respect to how an IP_3-Ca^{2+} oscillator (i) generates long-period oscillations that underlie the efficient frequency-encoding of the hormone dose, and (ii) may favor the coordination of Ca^{2+} signals in a multicellular system.

4.3. Identification of feedbacks on IP_3

In order to understand the role of IP_3 oscillations we proposed to perturb the system by modifying the turnover time of IP_3. The turnover time can be accelerated by over-expressing IP_3 degrading enzymes or slowed by expressing IP_3 binding molecules (see section 4.3.2). We will first show, for a general model that captures the main properties of the Ca^{2+}-IP_3

oscillator and which allows the derivation of analytical results, that slowing the turnover time of IP_3 is the method of choice to unravel the Ca^{2+}-IP_3 feedback structure. In the second part, we present experiments with an IP_3 binding protein and use a detailed model to simulate its effects.

4.3.1. *General properties of Ca^{2+}-IP_3 oscillators*

A simple model for coupled Ca^{2+}-IP_3 oscillations contains three variables (see Fig. 4.1): cytoplasmic and total calcium concentration c and z, and the IP_3 concentration p. Here we assume that the (in)activation of the IP_3R by Ca^{2+} and IP_3 is fast compared to other processes. The Ca^{2+} concentration in the ER s is related to the total concentration by $s = (z - c)/\beta$, where β is the ratio of effective cytoplasmic volume to effective ER volume. The time changes of the three variables are described by

$$\frac{dc}{dt} = f(c, z, p), \quad \frac{dz}{dt} = g(c), \quad \text{and} \quad \frac{dp}{dt} = w(c, p). \tag{4.1}$$

The functions $f(c, z, p)$ and $g(c)$ include the transport fluxes of cytoplasmic Ca^{2+} across the ER and plasma membrane, $w(c, p)$ the production and degradation of IP_3. For realistic functions f, g and w the existence of a limit cycle must generally be shown numerically. However, the local stability properties of the steady state give an idea. One can show that, due to the Ca^{2+} transport across the plasma membrane, there is a unique steady state $(\bar{c}, \bar{z}, \bar{p})$ [23]. Therefore, changes in stability of the unique steady state are likely to be connected with an Hopf bifurcation and the birth/death of a limit cycle. Generally, if the steady state is unstable it is to be expected that the trajectories move toward a stable limit cycle. Note that a stable limit cycle and a stable steady state can coexist. Our analysis can make no predictions in this regard.

In the following we will focus on how the IP_3 dynamics can contribute to destabilizing or stabilizing the steady state and so possibly generate or abolish Ca^{2+} oscillations. The goal is to find the appropriate strategy to unmask the importance for Ca^{2+} oscillations of Ca^{2+} feedbacks on IP_3 metabolism in (4.1). The local stability of the steady state is determined by the Jacobian

$$J = \begin{pmatrix} f_c & f_z & f_p \\ g_c & 0 & 0 \\ w_c & 0 & w_p \end{pmatrix}. \tag{4.2}$$

The subscripts denote the partial derivatives at the steady state, e.g. $f_c = \frac{\partial f}{\partial c}\Big|_{(\bar{c},\bar{z},\bar{p})}$.

The properties of the IP_3R gating and of the transport processes allow to fix the signs of some of the terms

- $f_z > 0$, the Ca^{2+} concentration in the cytoplasm increases when the total concentration increases.
- $f_p > 0$, an increase in IP_3 causes the opening of the IP_3R and outflow of Ca^{2+} from the ER into the cytoplasm.
- $g_c < 0$, a high Ca^{2+} concentration in the cytoplasm favors the transport of Ca^{2+} outside the cell.
- $w_p < 0$, IP_3 is degraded proportional to its concentration.

The sign of w_c is positive when Ca^{2+} activates IP_3 production (positive feedback), negative in the presence of Ca^{2+} dependent degradation of IP_3 (negative feedback), and vanishes in the absence of feedbacks of Ca^{2+} on IP_3. When accelerating the IP_3 dynamics, e.g. by overexpressing IP_3 metabolizing enzymes, w_c and w_p increase in magnitude without changing sign. Conversely, slowing the IP_3 dynamics, e.g. by expressing an IP_3 binding protein, decreases w_c and w_p.

The stability of the steady state is determined using the Routh-Hurwitz criterion which states that the Eigenvalues of the Jacobian have all negative real parts when

$$
\begin{aligned}
a_0 &= f_z g_c w_p \\
a_1 &= -f_z g_c - f_p w_c + f_c w_p \\
a_2 &= -f_c - w_p \\
a_3 &= a_2 a_1 - a_0
\end{aligned}
$$

are all positive. Due to the previous assumptions $a_0 > 0$. The other terms simplify when the IP_3 dynamics is fast compared to the other processes $w_p, w_c \gg 1$, then $a_1 \approx -f_p w_c + f_c w_p$ and $a_2 \approx -w_p$. When the IP_3 dynamics is slow $w_p, w_c \ll 1$, then $a_1 \approx -f_z g_c$ and $a_2 \approx -f_c$. We compare three models: positive feedback, negative feedback and absence of feedbacks of Ca^{2+} on IP_3. If the three models have an identical steady state then the partial derivatives, except of course w_c and w_p, are identical for the three cases considered. Depending on the magnitude of the positive feedback of Ca^{2+} on the IP_3R and on the transport processes that remove Ca^{2+} f_c can be either positive or negative; both cases will be discussed.

The stability properties of the steady state when $f_c > 0$ are summarized

Table 4.1. Stability properties of the steady state when $f_c > 0$ in a model without feedback, positive or negative feedback of Ca^{2+} on IP_3. In the negative feedback, for fast IP_3 dynamics, the steady state is stable when $f_p w_c - f_c w_p < 0$.

If $f_c > 0$					
IP_3 dynamics	a_0	a_1	a_2	a_3	**stability**
no feedback, $w_c = 0$	+	±	±	−	unstable
positive feedback, $w_c > 0$					
slow dynamics	+	+	−	−	unstable
fast dynamics	+	−	+	−	unstable
negative feedback, $w_c < 0$					
slow dynamics	+	−	−	−	unstable
fast dynamics	+	±	+	±	stable/unstable

in Table 4.1. If the system can oscillate in the absence of Ca^{2+} feedbacks on the IP_3 metabolism, i.e. the steady state is unstable, slowing the IP_3 dynamics does not stabilize the system, i.e. abolish the oscillations, in either positive or negative feedback models. Overexpressing the enzyme $IP_3$3K, which corresponds to accelerating the IP_3 dynamics in the negative feedback model, can stabilize the steady state and so may abolish the oscillations.

The stability properties of the steady state when $f_c < 0$ are summarized in Table 4.2. When the steady state is stable in the absence of feedbacks, the presence of positive feedback on IP_3 can destabilize the system and so generate oscillations. For this to occur the IP_3 dynamics must be sufficiently fast and $f_p w_c - f_c w_p > 0$. In the last formula one recognizes that the mutual activation of IP_3 and Ca^{2+}, $f_p w_c$, needs to be strong compared to processes responsible for the clearance of Ca^{2+} and the degradation of IP_3, which enter $f_c w_p$. Slowing of the IP_3 dynamics stabilizes the system, and

Table 4.2. Stability properties of the steady state when $f_c < 0$. For fast IP_3 dynamics and in the presence of Ca^{2+} activation of IP_3 production the steady state is unstable when $f_p w_c - f_c w_p > 0$.

If $f_c < 0$					
IP_3 dynamics	a_0	a_1	a_2	a_3	**stability**
no feedback, $w_c = 0$	+	+	+	+	stable
positive feedback, $w_c > 0$					
slow dynamics	+	+	+	+	stable
fast dynamics	+	±	+	±	stable/unstable
negative feedback, $w_c < 0$					
any dynamics	+	+	+	+	stable

so abolishes the oscillations. The presence of negative feedback does not affect the stability of the steady state and so may not give rise to Ca^{2+} oscillations.

These conclusions are independent of specific expressions for the rates. In particular, they show that slowing the IP_3 dynamics may unmask a critical role of positive feedback of Ca^{2+} on IP_3 in the oscillatory mechanism. The slowing of the IP_3 dynamics is predicted to cause a transition from oscillatory to non-oscillatory behavior.

4.3.2. *A detailed model for Ca^{2+}-IP_3 interactions*

In a detailed model we consider the three variables mentioned before (Ca^{2+} and IP_3 in the cytoplasm, c and p, and total Ca^{2+} concentration z) and additionally the activity state of the IP_3R, denoted by r. The positive feedback model is described in detail in [18]. Here we give a summary of the rate expressions used. According to Fig. 4.1 the time changes in the four variables are given by

$$\frac{dc}{dt} = f(c, z, p, r) = v_{rel} - v_{serca} + v_{in} - v_{out} \tag{4.3}$$

$$\frac{dz}{dt} = g(c) = v_{in} - v_{out} \tag{4.4}$$

$$\frac{dp}{dt} = w(c, p) = v_{PLC} - v_{5P} \tag{4.5}$$

$$\frac{dr}{dt} = h(c) = v_{rec} - v_{inac}. \tag{4.6}$$

Where

$$v_{rel} - v_{serca} = \left(k_1 \left(r \frac{c}{K_a + c} \frac{p}{K_p + p}\right)^3 + k_2\right)(z - c(1+\beta))/\beta \\ -V_{serca} \frac{c^2}{K_{serca}^2 + c^2} \tag{4.7}$$

and

$$v_{in} - v_{out} = v_0 + \phi V_{PLC} - V_{pm} \frac{c^2}{K_{pm}^2 + c^2} \tag{4.8}$$

For the IP_3 dynamics we take

$$v_{PLC} - v_{5P} = \left(1 + \frac{BK_p}{(K_p + p)^2}\right)^{-1} \left(V_{PLC} \frac{c^2}{K_{PLC}^2 + c^2} - k_{5P} p\right). \tag{4.9}$$

Table 4.3. Parameters are taken from literature [2, 3, 5, 13, 24]. The maximal rate of PLC, V_{PLC}, is a stimulation-dependent control parameter. The half-activation constant of PLC, K_{PLC}, is used to tune the positive feedback of Ca^{2+} on IP_3. See also [18].

Reference parameters	
	Positive feedback
IP_3 dynamics parameters:	
IP_3 dephosphorylation rate constant, k_{5P}	1/s
Half-activation constant of PLC, K_{PLC}	0.2 μM
Ca^{2+} transport and structural parameters:	
Ratio of volumes ER/cytosol, β	0.185
Maximal SERCA pump rate, V_{serca}	0.9 μM/s
Half-activation constant, K_{serca}	0.1 μM
Maximal PMCA pump rate, V_{pm}	0.02 μM/s
Half-activation constant, K_{pm}	0.12 μM
Constant influx, v_0	0.004 μM/s
Stimulation-dependent influx, ϕ	0.002/s
IP_3R parameters:	
Maximal rate of Ca^{2+} release, k_1	1.11/s
Ca^{2+} leak, k_2	0.0203/s
Ca^{2+} binding to activating site, K_a	0.08 μM
Ca^{2+} binding to inhibiting site, K_i	0.4 μM
IP_3 binding, K_p	0.13 μM
Characteristic time of IP_3R inactivation, τ_r	6 s

The latter equation includes the effect of an IP_3 binding molecule, the IP_3 buffer. We assume that binding of the buffer to IP_3 is fast compared to other processes. Using a rapid-equilibrium approximation one obtains the first term of Eq. 4.9, where B is the IP_3 buffer concentration. The parameter K_{PLC} characterizes the sensitivity of PLC to Ca^{2+} and is used to tune the strength of the positive feedback. For the IP_3R, we assume that their activation by Ca^{2+} and IP_3 is fast, whereas the inactivation is slow [13]

$$v_{rec} - v_{inac} = \frac{1}{\tau_r}\left(1 - r\frac{K_i + c}{K_i}\right). \qquad (4.10)$$

This equation indicates that for high Ca^{2+} the fraction of activable receptors r decreases with a characteristic time τ_r. The parameter values are listed in Table 4.3.

4.3.3. *Expression of an IP_3 buffer suppresses Ca^{2+} oscillations*

To test experimentally the model predictions made in section 4.3.1, we used a molecular IP_3 buffer which consists of the N-terminal ligand binding domain of rat type 1 IP_3R linked to enhanced green fluorescent protein (EGFP-LBD). Chinese hamster ovary (CHO) cells were transiently transfected with EGFP (control) or EGFP-LBD then challenged with submaximal and maximal ATP concentrations. The subsequent $[Ca^{2+}]_c$ responses were monitored using a fluorescent probe, fura-2. EGFP fluorescence was utilized to distinguish transfected from non-transfected cells in a given field of view and to estimate the intracellular concentration of the transgene.

The addition of low ATP concentrations elicited periodic $[Ca^{2+}]_c$ spikes in $> 85\%$ of the CHO cells expressing EGFP (Fig. 4.2A, B) or non-expressing cells from cultures transfected with EGFP-LBD (not shown). The presence of EGFP-LBD had a dose-dependent effect on the agonist dependent Ca^{2+} oscillations in CHO cells (Fig. 4.2A). High levels of EGFP-LBD expression correlated with a loss of repetitive $[Ca^{2+}]_c$ spiking and the appearance of low amplitude $[Ca^{2+}]_c$ increases (Fig. 4.2A, B). Moreover, EGFP-LBD expression significantly slowed the rate of $[Ca^{2+}]_c$ rise (Fig. 4.2C; $p < 0.01$) and significantly broadened the width of the $[Ca^{2+}]_c$ spike (Fig. 4.2D; $p < 0.05$) compared to EGFP expressing cells.

According to the theoretical results, section 4.3.1, the disappearance of Ca^{2+} oscillations after slowing the IP_3 dynamics, may indicate that IP_3 oscillations, driven by positive feedback of Ca^{2+} on IP_3 production, are involved in this system. We performed simulations using the detailed model for the Ca^{2+}-IP_3 oscillator with positive feedback of Ca^{2+} on IP_3 (section 4.3.2). The concentration of the IP_3 buffer and the degree of stimulation are varied by changing B and the maximal rate of IP_3 production V_{PLC} (Eq. 4.9). Without IP_3 buffer ($B = 0$) the model exhibit Ca^{2+} oscillations where the frequency increases with increasing stimulation (Fig. 4.2E, Control). At high concentrations of IP_3 buffer (Fig. 4.2E, $B = 13$ μM), the model exhibits single transients (for lower agonist dose), and repetitive truncated spikes (for high agonist dose). Both responses closely resemble the experimentally observed patterns in cells expressing high amounts of EGFP-LBD. The Ca^{2+} oscillations at lower concentrations of IP_3 buffer (Fig. 4.2F) exhibit a broadening of the individual spikes, which is very similar to the experimental observation in cells expressing low amounts of EGFP-LBD. Also the observed decrease of the rate of Ca^{2+} rise is reproduced by the model (not shown). These results are corroborated by a more detailed analysis of the theoretical system. A detailed model with negative feedback through Ca^{2+} dependent degradation of IP_3 mediated by $IP_3$3K could not account for any of the experimental findings (see [18]).

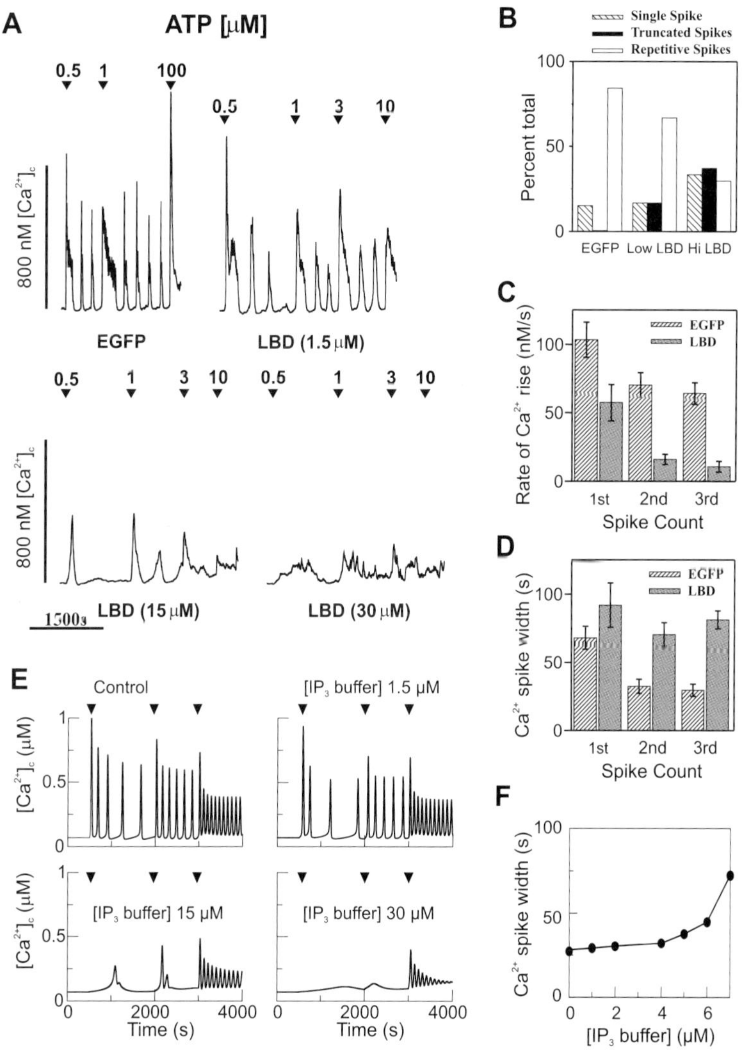

Fig. 4.2. Detailed figure description is in the text. Reproduced from [18].

In Fig. 4.2 the traces in **(A)** show typical ATP-evoked $[Ca^{2+}]_c$ spikes in CHO cells transiently expressing EGFP or different levels of EGFP-LBD (IP_3-buffer). **(B)** Data show the effects of increasing EGFP-LBD expression on ATP-evoked Ca^{2+}-signals. Truncated Spikes are defined as low amplitude $[Ca^{2+}]_c$ oscillations similar to those shown in the bottom traces of panel A. Data in **(C)** and **(D)** give the initial rate of $[Ca^{2+}]_c$ rise and the widths of the $[Ca^{2+}]_c$ spike respectively in control cells (expressing EGFP) and cells expressing low and high EGFP-LBD. **(E)** The model with Ca^{2+} activation of PLC (positive feedback) show good agreement with the EGFP-LBD experimental data shown in A. The IP_3 buffer concentration B is increased as indicated. An increase in ATP (arrowheads) is simulated by an increase in the maximal activity V_{PLC} of PLC, Eq. 4.9 . **(F)** We also observe a significant increase in spike width as shown in D. Parameters listed in Table 4.3. All rates have been slowed by a factor of 10. In (E) $V_{PLC} = 0.125$, 0.2, 0.4 μM/s. Initial condition at $V_{PLC} = 0.05$ μM/s. In (F) $V_{PLC} = 0.2$ μM/s.

4.4. Physiological role of Ca^{2+} feedbacks on IP_3 metabolism

4.4.1. *The wide range of oscillation periods is due to interactions of IP_3 and Ca^{2+} dynamics*

The experiments with the IP_3 buffer indicate that Ca^{2+} feedback on PLC plays a crucial role in the oscillatory process. To elucidate the possible physiological role of this feedback we varied its strength. This is done by changing the sensitivity of PLC for Ca^{2+} (i.e. changing K_{PLC}, see Eq. 4.9). For K_{PLC} being much lower than the basal $[Ca^{2+}]_c$, PLC is always saturated with Ca^{2+} and its activity is independent of variations in $[Ca^{2+}]_c$. In particular, by setting $K_{PLC} = 0$ positive feedback will effectively be eliminated. This model with constant IP_3 concentration shows fast calcium oscillations with a period of 10-15 sec (Fig. 4.3A, dashed line). Introducing positive feedback by setting $K_{PLC} > 0$ causes oscillations with long periods at low stimulation. The frequency encoding of the stimulus becomes very pronounced when the sensitivity of PLC to changes in $[Ca^{2+}]_c$ is just above basal $[Ca^{2+}]_c$ (Fig. 4.3A, solid lines, $K_{PLC} = 0.1$ and 0.2 μM). This indicates that positive feedback of Ca^{2+} on IP_3 may serve a physiological role by greatly enhancing the range of frequency-encoding of the agonist stimulus and promoting long period oscillations.

We also quantified the relative control of the IP_3 dynamics and the other

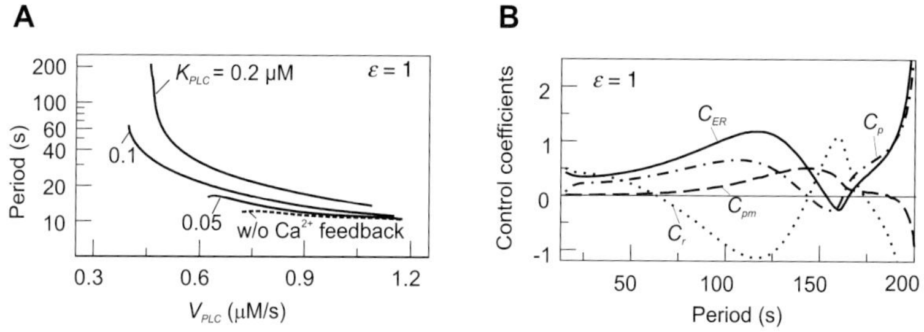

Fig. 4.3. Frequency encoding of agonist stimulus. **(A)** Oscillation periods observed at different stimulation strengths (varying V_{PLC}). Increasing the half-saturation constant of PLC for Ca^{2+}, K_{PLC}, from 0 (no functional positive feedback) to 0.2 μM (functional feedback) greatly enhances frequency encoding. **(B)** Control coefficients for the oscillation period of Ca^{2+} exchange across the ER membrane (C_{er}, solid line) and plasma membrane (C_{pm}, dashed line), IP_3 metabolism (C_p, dash-dotted line), and IP_3R dynamics (C_r, dotted line). A positive period control coefficient signifies that a slowing of the corresponding process increases the period. Parameter as in Table 4.3 except $\tau_r = 12.5$ s, $v_0 = 0.0004$ μM/s, $\phi = 0.004$, $k_{5P} = 0.66$/s. Reproduced from [18].

processes present in the model on the oscillation period. To this end, we used the following sensitivity measure

$$C_i = \frac{\tau_i}{T}\frac{\partial T}{\partial \tau_i} \quad \text{for } i = er,\, pm,\, p,\, r \tag{4.11}$$

which we refer to as period control coefficients (see also [30]). The C_i set the change of the oscillation period T in proportion to the change in a characteristic time τ_i of an individual process i. We analyzed the control of the following processes: IP_3 metabolism, the IP_3R dynamics, Ca^{2+} transport across the ER membrane, and Ca^{2+} transport across the plasma membrane. A positive period control coefficient implies that a slowing of the respective process (i.e. increase in τ_i) raises the period. At any point, the period control coefficients sum to unity, $C_p + C_r + C_{er} + C_{pm} = 1$, so that each coefficient quantifies the relative contribution of a single process to the oscillation period [8].

The control coefficients were calculated for various levels of stimulation. Because these levels correspond to different oscillation periods, we can plot the C_i against the period (Fig. 4.3B). There are several notable features. The control is distributed mainly between the IP_3 dynamics (dash-dotted line), IP_3R dynamics (dotted line), and the fluxes through the ER membrane (solid line). The fast oscillations are dominated by the exchange fluxes through the ER membrane and the IP_3R dynamics. Interestingly, there is an overall tendency that the IP_3 dynamics is more relevant for slow oscillations indicating its role for long-period oscillations. We also found a rather counter-intuitive behavior at intermediate periods where acceleration of the IP_3R dynamics would result in a slowing of the oscillations ($C_r < 0$).

This quantification of period control reveals that no process can be singled out as a unique period controlling factor. Depending on the oscillation mechanism and the reference period, the IP_3 turnover, the ER Ca^{2+} fluxes and the IP_3R dynamics can all exert strong control.

4.4.2. *Intercellular coupling*

In the intact liver, cells can communicate with their neighbors by diffusion of small molecules through intercellular channels (gap junctions). This is a prerequisite for a coordinated response to the external stimulation, where one observes repetitive Ca^{2+}-waves propagating through a large number of cells [21]. Isolated liver cells, conserving gap-junctional coupling with one or two of their neighbours, show upon stimulation synchronous (1:1 phase-locked) $[Ca^{2+}]_c$ oscillations. If the gap-junctional coupling is disrupted each cell oscillates independently and the oscillation frequencies can differ by a factor of up to two [27]. There are indications that this is due to variations in the amount of agonist receptors and so in the maximal IP_3 production rate. In models whithout feedbacks of Ca^{2+} on IP_3 (i.e. IP_3 is not oscillating), intercellular diffusion of Ca^{2+} alone is sufficient to account for the coordinated response observed in coupled cells [10, 11]. For this the gap-junctional Ca^{2+} permeability must be higher than a critical value. When IP_3 is not oscillating its intercellular diffusion smooth out the differences between cells. This decreases the value of the critical Ca^{2+} gap-junctional permeability and, if the Ca^{2+} permeability is to low, allows coordination for a limited number of spikes [4, 19].

In vivo, the theoretically estimated Ca^{2+} gap-junctional permeability necessary for intercellular coordination may not be attained as Ca^{2+} is strongly buffered in the cytoplasm so that its permeability is reduced by a factor of $\sim$ 10-20 [28]. Our results show that Ca^{2+} exerts feedback on IP_3 by activating its production, and we were interested how this influences the response to hormone of coupled cells.

We present the results for two coupled cells. The equation for IP_3 and Ca^{2+} dynamics in the first cell changes to

$$\frac{dc_1}{dt} = f(c_1, z_1, p_1, r_1) + \gamma_c(c_2 - c_1) \quad (4.12)$$

$$\frac{dz_1}{dt} = g(c_1) + \gamma_c(c_2 - c_1) \quad (4.13)$$

$$\frac{dp_1}{dt} = w(c_1, p_1) + \gamma_p(p_2 - p_1). \quad (4.14)$$

The gap-junctional diffusion rate constant of Ca^{2+} and IP_3 are denoted respectively by γ_c and γ_p. The subscripts 1 and 2 indicate the concentrations in the first and second cell respectively. The symmetric expression is used for the second cell.

The permeability for Ca^{2+}, γ_c, and IP_3, γ_p, are parameters which are difficult to measure. We use them as control parameters together with the maximal production rate V_{PLC} of IP_3. For two identical cells one can identify four different coupled oscillatory modes. The stimulatory region and stability of these modes is shown in Fig. 4.4A. Only two modes, regular synchronous (*diamonds*) and regular asynchronous (*triangles*) oscillations (Fig. 4.4B and C respetively), were found to be locally stable. We were not able to find a parameter combination where the two types of irregular asynchronous oscillations (*squares* and *circles* in Fig. 4.4A) are stable. These oscillatory modes are characterized by different amplitudes in the two cells (Fig. 4.4D-E) Interestingly, for two identical cells stable regular asynchronous oscillations have not been observed in a similar model without feedback on IP_3 [10]. In the region near the first hopf bifurcation HB_{s1}, opaque diamands, we found oscillations with very long interspike intervals characteristic for canards [29].

It turns out that the stable asynchronous oscillations crucially depend on the IP_3 and Ca^{2+} coupling. In Fig. 4.5 we systematically varied the two permeabilities γ_p and γ_c and looked whether, by changing V_{PLC}, one could find regular asynchronous oscillations. Only when the Ca^{2+} permeability is low is this oscillatory mode stable (region a). For larger Ca^{2+} permeabilities the regular asynchronous oscillations are unstable (region b) or even completely disappear (region c).

If the Ca^{2+}-IP_3 oscillators are not identical the possible collective dynamics becomes more complicated. We were interested in the parameter region where synchronous oscillations are observed, as this is the behavior observed for hepatocytes doublets [27]. The oscillations in the two cells are of course not identical but their phase difference is locked and the spike ratio is 1:1. The two cells differ in their intrinsic frequencies due to differences in

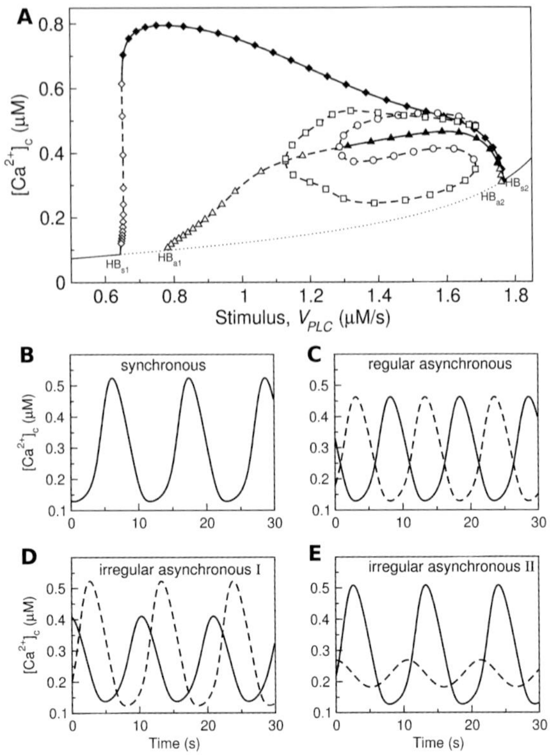

Fig. 4.4. Two identical cells are coupled via intercellular diffusion of IP_3 only ($\gamma_p = 0.2$/s, $\gamma_c = 0$). (**A**) Maxima of the $[Ca^{2+}]_c$ oscillations as function of the stimulus. One can distinguish between three types of coupled oscilatory behaviors: Regular synchronous oscillations (*diamond*, see B), regular asynchronous oscillations (*triangle*, see C), two branches of irregular asynchronous oscillations (*circles* and *square*, see D and E respectively). Both regular asynchronuous and synchronous oscillations have stable and unstable branches (*filled* and *opaque* symbols, respectively), irregular asynchronous oscillations have only unstable branches (*opaque* symbol). (**B**) Regular synchronous: no phase difference and equal amplitude. (**C**) Regular asynchronous: $\pi/2$ phase difference and equal amplitude. (**D**) and (**E**) Irregular asynchronous: phase difference $> \pi/2$, different amplitudes. In B–E $V_{PLC} = 1.55$ μM/s. Parameter as given in Table 4.3 except $V_{pm} = 0.18$ μM/s, $\phi = 0.084$, $v_0 = 0.0072$ μM/s, $\tau_r = 12.5$ s, $k_{5P} = 0.66$/s.

their IP_3 production rate V_{PLC}. As shown in Fig. 4.6 stable synchronous solutions are obtained either by increasing the IP_3 or Ca^{2+} intercellular permeability. We will denote by critical permeability the permeability over which the oscillations are synchronous. One notices that if the difference between cells is small ($V_{PLC1} = 0.8$ μM/s and $V_{PLC2} = 1$ μM/s) the critical

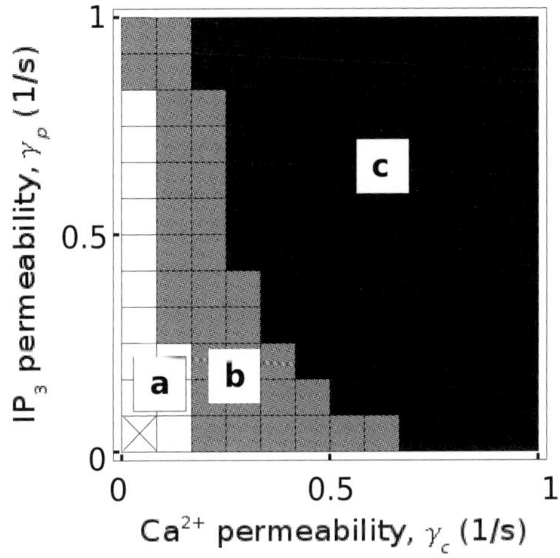

Fig. 4.5. Two identical cells coupled via intercellular diffusion of IP_3 and Ca^{2+}. Depending on the strength of Ca^{2+} and IP_3 coupling regular asynchronous oscillations as shown in Fig. 4.4C can be found. In region **a** this oscillatory mode is stable for certain values of V_{PLC} (see Fig. 4.4A), in region **b** it is unstable for any values of V_{PLC}, and in region **c** this oscillatory mode is not present. Parameters as in Fig. 4.4.

IP_3 permeability is always lower than the critical Ca^{2+} permeability. This relation inverses when the difference between cells is larger ($V_{PLC1} = 0.8$ μM/s and $V_{PLC2} = 1.5$ μM/s). This may be due to the fact that for the second case cells oscillate faster so that the amplitude in IP_3 oscillations is reduced, making it more difficult for the first cell coupled via IP_3 diffusion to sense the oscillation frequency in the second cell. Despite this variation in sensitivity toward Ca^{2+} or IP_3 coupling one should keep in mind that Ca^{2+} is strongly buffered so that its effective permeability is about 10 to 20 times lower. Therefore, for low gap-junctional coupling the synchronization of Ca^{2+} oscillations is more likely to be carried by IP_3 diffusion than by Ca^{2+} diffusion.

4.5. Conclusions

We have shown, by combining theoretical and experimental analysis, how it is possible to elucidate the feedback structure underlying hormone induced Ca^{2+} oscillations. IP_3 oscillations appear to play an important dynamic role for the Ca^{2+} oscillations. This result is also supported by a recent work of Sneyd et al. [25] where they perturb the system by releasing small

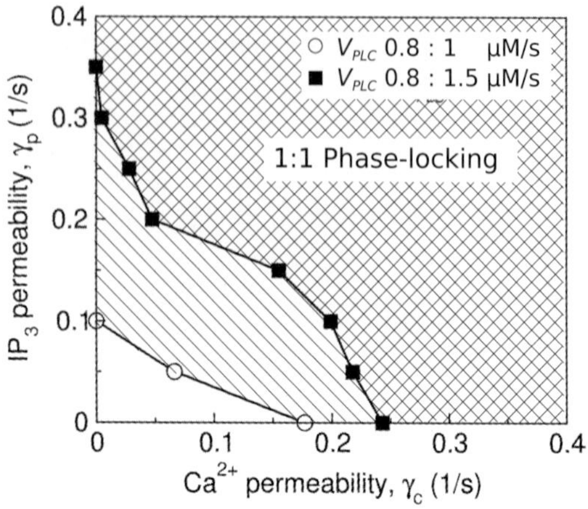

Fig. 4.6. Two cells which differ in their PLC activity are coupled. Uncoupled the cells oscillate with their own period (27, 18.8 and 11.6 sec for $V_{PLC} = 0.8$, 1 and 1.5 μM/s respectively). The region of synchronous, 1:1 phase-locked oscillations depends on the intercellular permeabilities for Ca^{2+} and IP_3 (γ_c and γ_p respectively). Parameters as in Fig. 4.4.

amounts of IP_3. Here, we concentrated on a positive feedback system, Ca^{2+} activation of IP_3 production via PLC. The physiological relevance of this interaction may lie in its ability to support efficient frequency encoding of the stimulus and in improving the coordination of cells coupled by intercellular diffusion.

Two other feedbacks of Ca^{2+} on IP_3 metabolism have been proposed, both of which are negative. The already mentioned Ca^{2+} activation of IP_3 degradation, via $IP_3$3K, and PKC induced inactivation of agonist receptors [16, 32]. Modelling of these two feedbacks show that both can improve frequency encoding of the stimulus. However, one cannot identify them by perturbing the IP_3 turnover as we have done for the positive feedback of Ca^{2+} on IP_3. Thus to asses their role in Ca^{2+} oscillations it is necessary to develop new theoretical and experimental approaches.

References

[1] I. Bezprozvanny and B.E. Ehrlich. The inositol 1,4,5-trisphosphate ($InsP_3$) receptor. *J. Membr. Biol.* , 145:205–216, 1995.

[2] J.L. Blank, A.H. Ross, and J.H. Exton. Purification and characterization of two G-proteins that activate the β1 isozyme of phosphoinositide-specfic phospholipase C. *J. Biol. Chem.*, 266:18206–18216, 1991.

[3] D. Communi, V. Vanweyenberg, and E. Christophe. D-myo-inositol 1,4,5-trisphosphate 3-kinase a is activated by receptor activation through a calcium:calmodulin-dependent protein kinase II phophorylation mechanism. *EMBO J.*, 16:1943–1952, 1997.
[4] G. Dupont, S. Swillens, C. Clair, T. Tordjamann, and L. Combettes. Hierarchical organization of calcium signals in hepatocytes: from experiments to models. *Biochem. Biophys. Acta*, 1498:134–152, 2000.
[5] C.C. Fink, B. Slepchenko, I.I. Moraru, J. Schaff, J. Watras, and LM. Loew Morphological control of inositol-1,4,5-trisphosphate-dependent signals. *J. Cell Biol.*, 147:929–935, 1999.
[6] C. Hajnoczky and A.P. Thomas. Minimal requirements for calcium oscillations driven by the IP_3 receptor. *EMBO J.*, 16:3533–3543, 1997.
[7] A. Harootunian, J. Kao, S. Paranjape, and R. Tsien. Generation of calcium oscillations in fibroblasts by positive feedback between calcium and IP_3. *Science*, 251:75–78, 1991.
[8] R. Heinrich and S. Schuster. *The regulation of cellular systems.* New York, NY, Chapman & Hall, 1996.
[9] K. Hirose, S. Kadowaki, M. Tanabe, H. Takeshima, and M. Lino. Spatiotemporal dynamics of inositol 1,4,5-trisphosphate that underlies complex Ca^{2+} mobilization patterns. *Science*, 284:1527–1530, 1999.
[10] T. Höfer. Model of intercellular calcium oscillations in hepatocytes: synchronization of heterogeneous cells. *Biophys. J.*, 77:1244–1256, 1999.
[11] T. Höfer, A. Politi, and R. Heinrich. Intercellular Ca^{2+} wave propagation through gap-junctional Ca^{2+} diffusion: a theoretical study. *Biophys. J.*, 80:75–87, 2001.
[12] J. Keizer, Y.X. Li, S. Stojilkovic, and J. Rinzel. $InsP_3$-induced ca^{2+} excitability of the endoplasmic reticulum. *Mol. Biol. Cell*, 6:945–951, 1995.
[13] Y. Li and J. Rinzel. Equations for $InsP_3$ receptor-mediated $[Ca^{2+}]_i$ oscillations derived from a detailed kinetic model: a hodgkin-huxley like formalism. *J. Theor. Biol.*, 166:461–473, 1994.
[14] J.S. Marchant and C.W. Taylor. Rapid activation and partial inactivation of inositol trisphosphate receptors by inositol trisphosphate. *Biochemistry*, 37:11524–11533, 1998.
[15] M.M. Nalaskowski and G.W. Mayr. The families of kinases removing the Ca^{2+} releasing second messenger Ins(1,4,5)P_3. *Curr. Mol. Med.*, 4:277–290, 2004.
[16] M.S. Nash, M.J. Schell, P.J. Atkinson, N.R. Johnston, S.R. Nahorski, and R.A. Challis. Determinants of metabotropic glutamate receptor-5-mediated Ca^{2+} and inositol 1,4,5-trisphophate oscillation frequency. *J. Biol. Chem.*, 277:35947–35960, 2002.
[17] M.S. Nash, K.W. Young, R.A. Challis, and S.R. Nahorski. Receptor-specific messenger oscillations. *Nature*, 413:381–382, 2001.
[18] A. Politi, L.D. Gaspers, A.P. Thomas, and T. Höfer. Model of IP_3 and calcium oscillations: frequency encoding and identification of underlying feedbacks. *Biophys. J.*, 90:3120–3133, 2006.
[19] A. Politi and T. Höfer. *Function and Regulation of Cellular Systems*, chapter

Modelling of periodic intercellular Ca^{2+} waves, pages 99–110. Birkhäuser-Verlag., 2004.

[20] M.J. Rebecchi and N.P. Pentyala. Structure, function, and control of phosphoinositide-specific phospholipase C. *Physiol. Rev.*, 80:1291–1335, 2000.

[21] L.D. Robb-Gaspers and A.P. Thomas. Coordination of Ca^{2+} signaling by intercellular propagation of Ca^{2+} waves in the intact liver. *J. Biol. Chem.* , 270:8102–8107, 1995.

[22] T.A. Rooney, E. Sass, and A.P. Thomas. Characterization of cytosolic calcium oscillations induced by phenylephrine and vasopressin in single fura-2-loaded hepatocytes. *J. Biol. Chem.*, 264:17131–17141, 1989.

[23] S. Schuster, M. Marhl, and T. Höfer. Modelling of simple and complex calcium oscillations. *Eur. J. Biochem.*, 269:1333–1355, 2002.

[24] C. Sims and N. Allbritton. Metabolism of inositol 1,4,5-trisphosphate and inositol 1,3,4,5-tetrakisphosphate by the oocytes of xenopus laevis. *J. Biol. Chem.*, 273:4052–4058, 1998.

[25] J. Sneyd, K. Tsaneva-Atanasova, V. Reznikov, Y. Bai, M.J. Sanderson, and D.I. Yule. A method for determining the dependence of calcium oscillations on inositol trisphosphate oscillations. *Proc. Natl. Acad. Sci. USA*, 103: 1675-1680, 2006.

[26] S. Thore, O. Dyachok, and A. Tengholm. Oscillations of phospholipase c activity triggered by depolarization and Ca^{2+} influx in insulin-secreting cells. *J. Biol. Chem.*, 279:19396–19400, 2004.

[27] T. Tordjmann, B. Berthon, M. Claret, and L. Combettes. Coordinated intercellular calcium waves induced by noradrenaline in rat hepatocytes: dual control by gap junction permeability and agonist. *EMBO J.* , 16:5398–5407, 1997.

[28] Z. Wang, M. Tymianski, O. Jones, and M. Nedergaard. Impact of cytoplasmic calcium buffering on the spatial and temporal characteristics of intercellular calcium signals in astrocytes. *J. Neurosci.*, 17:7359–7371, 1997.

[29] M. Wechselberger. Existence and bifurcation of canards in r3 in the case of a folded node. *SIADS*, 4:101–139, 2005.

[30] J. Wolf, S. Becker-Weimann, and R. Heinrich. Analysing the robustness of cellular rythms. *Systems Biol.*, 2:35–41, 2005.

[31] P.J. Woodring and J.C. Garrison. Expression, purification, and regulation of two isoforms of the inositol 1,4,5-trisphosphate 3-kinase. *J. Biol. Chem.*, 272:30447–30454, 1997.

[32] K.W. Young, M.S. Nash, A.J. Challiss, and S.R. Nahorski. Role of Ca^{2+} feedback on cell inositol 1,4,5-tisphosphate oscillations mediated by G-protein-coupled receptors. *J. Biol. Chem.*, 278:20753–20760, 2003.

[33] D.I. Yule, S.V. Straub, and J.I. Bruce. Modulation of Ca^{2+} oscillations by phosphorylation of Ins(1,4,5)P_3 receptors. *Biochem. Soc. Trans.*, 31:954–957, 2003.

Chapter 5

Pattern formation in semiconductors under the influence of time-delayed feedback control and noise

Eckehard Schöll*, Johanne Hizanidis, Philipp Hövel and Grischa Stegemann

Institut für Theoretische Physik, Technische Universität Berlin, Hardenbergstr. 36, D–10623 Berlin, Germany

**schoell@physik.tu-berlin.de*

Contents

5.1. Introduction

Over the past decade control of unstable states has evolved into a central issue in applied nonlinear science. This field has various aspects comprising stabilization of unstable periodic orbits embedded in a deterministic chaotic attractor, which is generally referred to as *chaos control* [1], stabilization of unstable fixed points, or control of the coherence and timescales of stochastic motion. Various methods of control, going well beyond the classical control theory [2, 3], have been developed since the ground-breaking work of Ott, Grebogi and Yorke [4]. One scheme where the control force is constructed from time-delayed signals [5] has turned out to be very robust and universal to apply. It has been used in a large variety of systems in physics, chemistry, biology, and medicine [6], in purely temporal dynamics as well as in spatially extended systems. Moreover, it has recently been shown to be applicable also to noise-induced oscillations and patterns [7–9].

This is an interesting observation in the context of ongoing research on the constructive influence of noise in nonlinear systems [10–12].

In this review we focus on modern semiconductor structures whose structural and electronic properties vary on a nanometer scale. They provide an abundance of examples for nonlinear dynamics and self-organized pattern formation [13–15]. In these nanostructures nonlinear charge transport mechanisms are given, for instance, by quantum mechanical tunnelling through potential barriers, or by thermionic emission of hot electrons which have enough kinetic energy to overcome the barrier. A further important feature connected with potential barriers and quantum wells in such semiconductor structures is the ubiquitous presence of space charges. This, according to Poisson's equation, induces a further feedback between the charge carrier distribution and the electric potential distribution governing the transport. This mutual nonlinear interdependence is particularly pronounced in the cases of semiconductor heterostructures (consisting of layers of different materials) and low-dimensional nanostructures where abrupt junctions between different materials on an atomic length scale cause conduction band discontinuities resulting in potential barriers and wells. The local charge accumulation in these potential wells, together with nonlinear transport processes across the barriers have been found to provide a number of nonlinearities in the current-voltage characteristics.

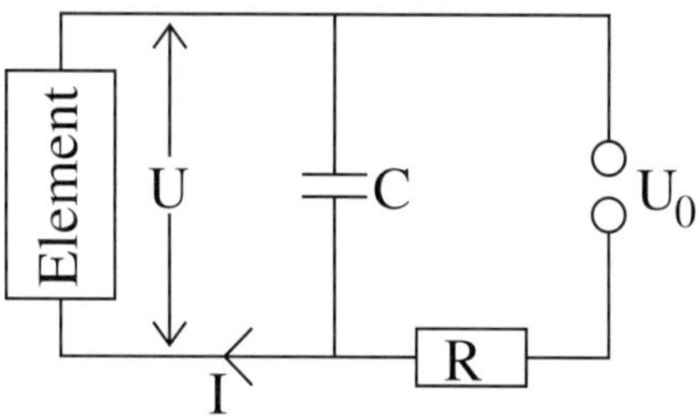

Fig. 5.1. A semiconductor element operated in a circuit with load resistor R and capacitor C, and applied bias voltage U_0.

In the case of negative differential conductance, the current I decreases with increasing voltage U, and vice versa, which normally corresponds to an unstable situation. The actual electric response depends upon the attached circuit which in general contains - even in the absence of external load resistors - unavoidable resistive and reactive components like lead resistances, lead inductances, package inductances, and package capacitances. These reactive components give rise to additional degrees of freedom which are described by Kirchhoff's equations of the circuit. If, for instance, a circuit is considered which contains a capacitance C parallel to the semiconductor device, and a load resistance R_L and a bias voltage U_0 in series with the device (Fig. 5.1), then Kirchhoff's laws lead to

$$U_0 = R_L I_0(t) + U(t) \tag{5.1}$$

$$I_0(t) = I(t) + C\frac{dU}{dt}. \tag{5.2}$$

Hence the temporal behavior of the voltage is determined by the circuit equation

$$\frac{dU(t)}{dt} = \frac{1}{C}\left(\frac{U_0 - U}{R_L} - I\right). \tag{5.3}$$

If a semiconductor element with negative differential conductance is operated in a reactive circuit, oscillatory instabilities may be induced by these reactive components, even if the relaxation time of the semiconductor is much smaller than that of the external circuit so that the semiconductor can be described by its stationary $I(U)$ characteristic and simply acts as a nonlinear resistor. Self-sustained semiconductor oscillations, where the semiconductor itself introduces an internal unstable temporal degree of freedom, must be distinguished from those circuit-induced oscillations. The self-sustained oscillations under time-independent external bias will be discussed in the following. Examples for internal degrees of freedom are the charge carrier density, or the electron temperature, or a junction capacitance within the device. Eq.(5.3) is then supplemented by a dynamic equation for this internal variable. It should be noted that the same class of models is also applicable to describe neural dynamics in the framework of the Hodgkin-Huxley equations [16].

Two important cases of negative differential conductivity (NDC) are described by an *N*-shaped or an *S*-shaped $j(F)$ characteristic, and denoted by *NNDC* and *SNDC*, respectively. However, more complicated forms like *Z*-shaped, loop-shaped, or disconnected characteristics are also possible [15]. NNDC and SNDC are associated with voltage- or current-controlled instabilities, respectively. In the NNDC case the current density is a single-valued function of the field, but the field is multivalued: the $F(j)$ relation has three branches in a certain range of j. The SNDC case is complementary in the sense that F and j are interchanged. In case of *NNDC*, the NDC branch is often but not always - depending upon external circuit and boundary conditions - unstable against the formation of nonuniform field profiles along the charge transport direction (*electric field domains*), while in the *SNDC* case *current filamentation* generally occurs, i.e., the current density becomes nonuniform over the cross-section of the current flow and forms a conducting channel. The elementary structures which make up these self-organized patterns are stationary or moving *fronts* representing the boundaries of the high-field domain or high-current filament. These primary self-organized spatial patterns may themselves become unstable in secondary bifurcation leading to periodically or chaotically breathing, rocking, moving, or spiking filaments or domains, or even solid-state turbulence and spatio-temporal chaos.

Chaotic oscillations should be avoided for a reliable operation of semiconductor devices. Therefore there is need for control of those. The important aspect of chaos control [1] is the emphasis of noninvasive control methods together with the observation that chaos supplies a huge number of unstable states that can be stabilized with tiny control power [4]. A particularly simple and efficient scheme uses time–delayed signals to generate control forces for stabilizing time periodic states [5] (*time–delay autosynchronization*, TDAS, or *Pyragas method*). Within this approach, an intrinsically unstable periodic orbit is stabilized using a feedback loop which couples back the difference of an output variable at the actual time t and the same variable at a delayed time $t - \tau$. This scheme is simple to implement, quite robust, and has been applied successfully in real experiments, e. g. Refs. 17–27. An extension to multiple time-delays (*extended time–delay autosynchronization*, ETDAS) has been proposed by Socolar et al [28], and analytical insight into those schemes has been gained by several theoretical studies [29–38] as well as by numerical bifurcation analysis [39]. Such self–stabilizing feedback control schemes (time–delay autosynchronization) with different couplings of the control force have also been applied to spatio-temporal patterns resulting from various models of semiconductor oscillators [40–48]. They should be easy to implement in practical semiconductor devices.

Time-delayed feedback control has also been applied to purely noise-induced oscillations in a regime where the deterministic system rests in a steady state. It has been shown that in this way both the coherence and the mean frequency of the oscillations can be controlled in simple models [7–9, 49, 50] as well as in spatially extended systems [51–53].

In the following, we use two paradigmatic models of semiconductor nanostructures which are of great current interest [15], see Fig. 5.2:

(i) Electron transport in semiconductor superlattices shows strongly

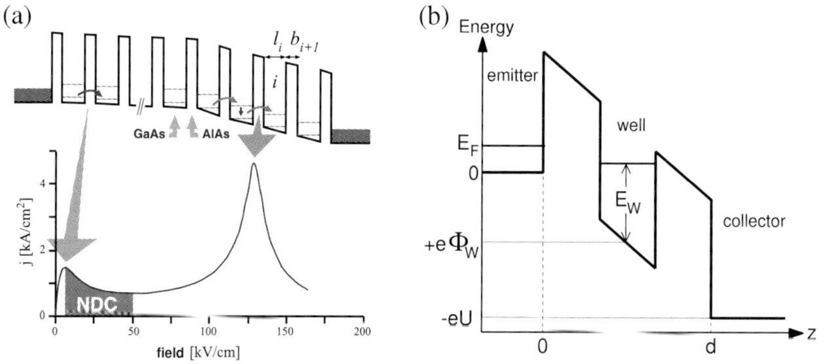

Fig. 5.2. **a)** Superlattice exhibiting domain formation. The associated current density (j) versus field (F) characteristic shows negative differential conductivity (NDC). The low-field domain corresponds to sequential tunnelling between equivalent levels of adjacent quantum wells (low-field peak of the $j(F)$ characteristic), while the high-field domain corresponds to resonant tunnelling between different levels of adjacent wells (high-field peak). **b)** Schematic potential profile of the double barrier resonant tunneling structure (DBRT). E_F and E_w denote the Fermi level in the emitter, and the energy level in the quantum well, respectively. U is the voltage applied across the structure.

nonlinear spatio-temporal dynamics. Complex scenarios including chaotic motion of multiple fronts have been found under time-independent bias conditions [54], showing signs of universal front dynamics [55, 56]. Unstable periodic orbits corresponding to travelling field domain modes can be stabilized by time delayed feedback control. A novel control scheme using low-pass filtering and allowing for control loop latency has been developed [48]. Noise-induced front patterns have also been found [57].

(ii) Charge accumulation in the quantum-well of a double-barrier resonant-tunneling diode (DBRT) may result in lateral spatio-temporal patterns of the current density and chaos [58]. We demonstrate that unstable current density patterns, e. g. periodic breathing or spiking modes, can be stabilized in a wide parameter range by a delayed feedback loop. We will compare the control domains of different control schemes, relating them to their Floquet spectrum [47]. Delayed feedback control of noise-induced patterns is also demonstrated [52, 59].

5.2. Chaos control of domains and fronts in superlattices

Semiconductor superlattices [60] have been demonstrated to give rise to self-sustained current oscillations ranging from several hundred MHz [61,

62] to 150 GHz at room temperature [63]. In any case, a superlattice constitutes a highly nonlinear system [15, 64–66], and instabilities are likely to occur. Indeed, chaotic scenarios have been found experimentally [67–69] and described theoretically in periodically driven [70] as well as in undriven systems [54]. For a reliable operation of a superlattice as an ultra-high frequency oscillator such unpredictable and irregular conditions should be avoided, which might not be easy in practice.

Here we focus on simulations of dynamic scenarios for superlattices under fixed time-independent external voltage in the regime where self-sustained dipole waves are spontaneously generated at the emitter. The dipole waves are associated with travelling field domains, and consist of electron accumulation and depletion fronts which in general travel at different velocities and may merge and annihilate. Depending on the applied voltage and the contact conductivity, this gives rise to various oscillation modes as well as different routes to chaotic behavior [54, 56].

We use a model of a superlattice based on sequential tunneling of electrons. In the framework of this model electrons are assumed to be localized at one particular well and only weakly coupled to the neighboring wells. The tunneling rate to the next well is lower than the typical relaxation rate between the different energy levels within one well. The electrons within one well are then in quasi–equilibrium and transport through the barrier is incoherent. The resulting tunneling current density $J_{m\to m+1}(F_m, n_m, n_{m+1})$ from well m to well $m+1$ depends only on the electric field F_m between both wells and the electron densities n_m and n_{m+1} in the wells (in units of cm^{-2}). A detailed microscopic derivation of the model has been given elsewhere [64]. A typical result for the current density vs electric field characteristic is depicted in Fig. 5.2 in the spatially homogeneous case, i.e. $n_m = n_{m+1} = N_D$, with donor density N_D.

In the following we will adopt the total number of electrons in each well as the dynamic variables of the system. The dynamic equations are then given by the continuity equation

$$e\frac{dn_m}{dt} = J_{m-1\to m} - J_{m\to m+1} \quad \text{for } m = 1, \dots N \tag{5.4}$$

where N is the number of wells in the superlattice, and $e < 0$ is the electron charge.

The electron densities and the electric fields are coupled by the following discrete version of Gauss's law

$$\epsilon_r \epsilon_0 (F_m - F_{m-1}) = e(n_m - N_D) \quad \text{for } m = 1, \dots N, \tag{5.5}$$

where ϵ_r and ϵ_0 are the relative and absolute permittivities, and F_0 and F_N are the fields at the emitter and collector barrier, respectively.

The applied voltage between emitter and collector gives rise to a global constraint

$$U = -\sum_{m=0}^{N} F_m d, \tag{5.6}$$

where d is the superlattice period.

The current densities at the contacts are chosen such that dipole waves are generated at the emitter. For this purpose it is sufficient to choose Ohmic boundary conditions:

$$J_{0\to 1} = \sigma F_0 \tag{5.7}$$

$$J_{N\to N+1} = \sigma F_N \frac{n_N}{N_D} \tag{5.8}$$

where σ is the Ohmic contact conductivity, and the factor n_N/N_D is introduced in order to avoid negative electron densities at the collector. Here we make the physical assumption that the current from the last well to the collector is proportional to the electron density in the last well. It is in principle possible to use a more realistic exponential emitter characteristic [71] or calculate the boundary conditions using microscopic considerations, but the qualitative behavior is not changed.

In our computer simulations we use a superlattice with $N = 100$ periods, $Al_{0.3}Ga_{0.7}As$ barriers of width $b = 5$nm and GaAs quantum wells of width $w = 8$nm, doping density $N_D = 1.0 \times 10^{11}\text{cm}^{-2}$ and scattering induced broadening $\Gamma = 8$meV at $T = 20$K. If the contact conductivity σ is chosen such that the intersection point with the homogeneous N–shaped current density vs. field characteristic is at a sufficiently low current value, accumulation and depletion fronts are generated at the emitter. Those fronts form a travelling high-field domain, with leading electron depletion front and trailing accumulation front. For fixed voltage U eq. (5.6) imposes constraints on the lengths of the high-field domains and thus on the front velocities. If N_a accumulations fronts and N_d depletion fronts are present, the respective front velocites v_a and v_d must obey $v_d/v_a = N_a/N_d$. Since the front velocities are monotonic functions of the current density [72], this also fixes the current. If the accumulation and depletion fronts have different velocities, they may collide in pairs and annihilate. With decreasing contact conductivity, or increasing voltage, chaotic scenarios arise, where the annihilation of fronts of opposite polarity occurs at irregular positions within the superlattice [54], leading to complex bifurcation diagrams.

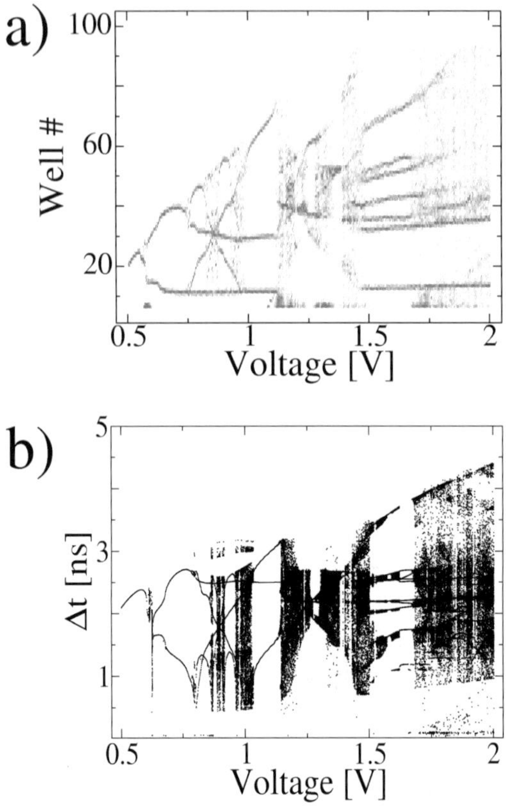

Fig. 5.3. (a) Positions where accumulation and depletion fronts annihilate vs. voltage at $\sigma = 0.5 \quad \Omega^{-1}\mathrm{m}^{-1}$. The greyscale indicates high (black) and low (white) numbers of annihilations at a given well. (b) Time differences between consecutive maxima of the electron density in well no. 20 vs. voltage. Time series of length 600 ns have been used for each value of the voltage. [54]

In Fig. 5.3(a) a density plot of the positions (well numbers) at which two fronts annihilate is shown as a function of the voltage. We see that for low voltage the annihilation takes place at a definite position in the superlattice with a variation of only a few wells. This distribution broadens for increasing voltage in characteristic bifurcation scenarios reminiscent of period doubling, leading to chaotic regimes. We note that in the chaotic regime periodic windows exist. A one-parameter bifurcation diagram is given in Fig. 5.3(b), obtained by plotting the time difference Δt between two consecutive maxima of the electron density in a specific well. Chaotic bands and periodic windows can be clearly seen.

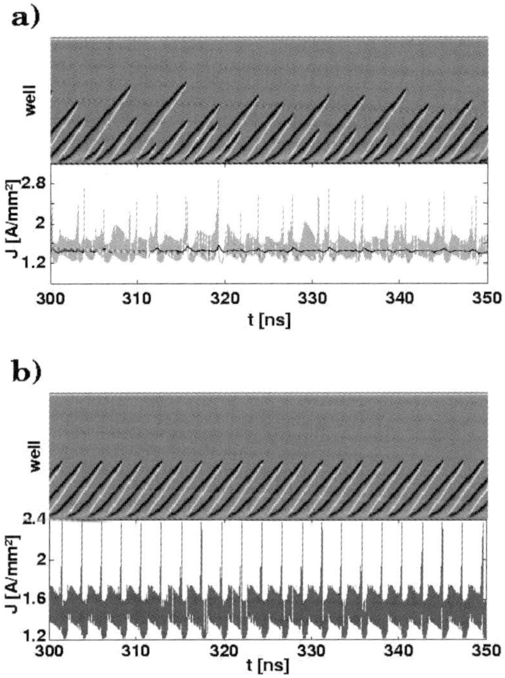

Fig. 5.4. Control of chaotic front dynamics by extended time-delay autosynchronization. a) Space-time plot of the uncontrolled charge density, and current density J vs. time. b) Same with global voltage control with exponentially weighted current density (denoted by the black curve). Parameters as in Fig. 1, $U = 1.15$ V, $\tau = 2.29$ ns, $K = 3 \times 10^{-6}\text{Vmm}^2/\text{A}$, $R = 0.2$, $\alpha = 10^9 s^{-1}$. [73]

The transition from periodic to chaotic oscillations is enlightened by considering the space-time plot for the evolution of the electron densities (Fig. 5.4(a)). At $U = 1.15$V chaotic front patterns with irregular sequences of annihilation of front pairs occur.

We shall now introduce a time-delayed feedback loop to control the chaotic front motion and stabilize a periodic oscillation mode which is inherent in the chaotic attractor [48, 73]. As a global output signal which is coupled back in the feedback loop, it is natural to use the total current density $J = \frac{1}{N+1}\sum_{m=0}^{N} J_{m\to m+1}$. For the uncontrolled chaotic oscillations, J versus time (grey trace in Fig. 5.4(a)) shows irregular spikes at those times when two fronts annihilate. Note that the grey current time trace is modulated by fast small-amplitude oscillations (due to well-to-well hopping of depletion and accumulation fronts in our discrete model) which are not resolved temporally in the plot. They can be averaged out by considering an exponentially weighted current density (black curve in Fig. 5.4(b)), which corresponds to a low–pass filter:

$$\overline{J}(t) = \alpha \int_0^t J(t')e^{-\alpha(t-t')}\mathrm{d}t', \tag{5.9}$$

with a cut-off frequency α.

The information contained in the low-frequency part of the current (Fig. 5.4(a), black curve) is then used as input in the extended multiple-time autosynchronization scheme. The voltage U across the superlattice is modulated by multiple differences of the filtered signal at time t and at delayed times $t - \tau$

$$U(t) = U_0 + U_c(t) \tag{5.10}$$

$$\begin{aligned} U_c(t) &= -K\left[\overline{J}(t) - \overline{J}(t-\tau)\right] + RU_c(t-\tau) \\ &= -K\sum_{\nu=0}^{\infty} R^{\nu}\left[\overline{J}(t-\nu\tau) - \overline{J}(t-(\nu+1)\tau)\right] \\ &= -K\left[\overline{J}(t) - (1-R)\sum_{k=1}^{\infty} R^{k-1}\overline{J}(t-k\tau)\right] \end{aligned} \tag{5.11}$$

where U_0 is a time–independent external bias, and U_c is the control voltage. K is the amplitude of the control force, τ is the delay time, and R is a memory parameter. A sketch of the whole control circuit is displayed in Fig. 5.5a. Such a global control scheme is easy to implement experimentally. It is non-invasive in the sense that the control force vanishes when the target state of period τ has been reached. This target state is an unstable periodic orbit of the uncontrolled system. The period τ can be determined *a priori* by observing the resonance-like behavior of the mean control force versus τ. The result of the control is shown in Fig. 5.4(b). The front dynamics exhibits annihilation of front pairs at fixed positions in the superlattice, and stable periodic oscillations of the current are obtained.

In Fig. 5.5(b) the control domain is depicted in the parameter plane of R and K. A typical horn–like control domain similar to the ones known from other coupling schemes [44] is found. Control is achieved in a range of values of the control amplitude K, which is widened and shifted to larger K with

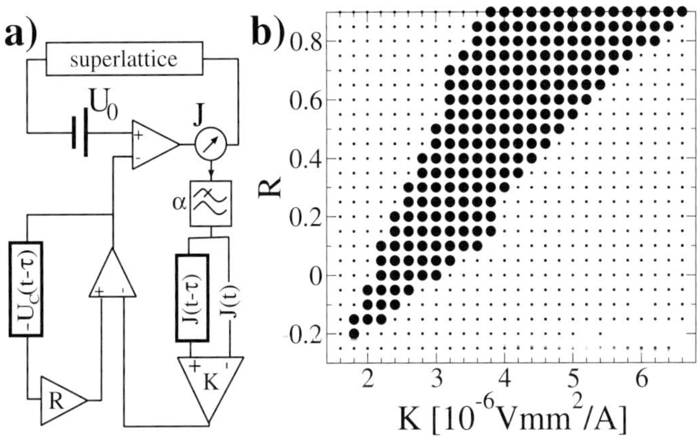

Fig. 5.5. a) Control circuit including the low-pass filter with cut-off frequency α and the time-delayed feedback loop (K) and its extension to multiple time delays (R). b) Control domain for global voltage control. Full circles denote successful control, small dots denote no control. Parameters as in Fig. 5.4. [73]

increasing memory parameter R. Typically, the left–hand control boundary corresponds to a period–doubling bifurcation leading to chaos for smaller K, while the right–hand boundary is associated with a Hopf bifurcation. The shape of our control domain and its size resemble the results obtained analytically for diagonal control schemes where observables are measured and controlled locally. In particular we do not observe the influence of other branches of the Floquet eigenvalue problem, which might reduce the size of the control domain severely [74]. Thus our control scheme is of similar control performance as local control.

In order to investigate the effect of the low-pass filtering on the frequency spectrum of the system, it is helpful to consider the transfer function formalism in the frequency domain for the ETDAS control scheme both with and without an additional low-pass filter.

In the frequency domain Eq. (5.11) reads

$$U_c(\omega) = -KT(\omega)J(\omega), \tag{5.12}$$

where $T(\omega)$ denotes the transfer function which can be calculated as

$$T(\omega) = T_{ETDAS}(\omega)T_{low-pass}(\omega). \tag{5.13}$$

$T_{ETDAS}(\omega)$ is the transfer function of the ETDAS control scheme [20] given by

$$T_{ETDAS}(\omega) = \frac{1 - e^{-i\omega\tau}}{1 - Re^{-i\omega\tau}} \tag{5.14}$$

and $T_{low-pass}(\omega)$ is the transfer function of the low-pass filter:

$$T_{low-pass}(\omega) = \frac{1}{1 + i\frac{\omega}{\alpha}}. \tag{5.15}$$

The shape of $|T_{ETDAS}(\omega)|$ is displayed in Fig. 5.6(a) for different values of R. As discussed by Sukow *et al* [20], the transfer function drops to zero at multiples of the frequency of the unstable periodic orbit (UPO), i.e., τ^{-1}. The notches at these frequencies become narrower for larger R. Due to the notches, the frequency of the UPO does not contribute to the control signal, so that the control force vanishes if stabilization is successful. The steeper notches for larger R indicate that the ETDAS feedback is more sensitive to frequencies different from the one to be controlled, so that more feedback is produced for these unwanted frequencies, which makes the control scheme more efficient.

The maximum value of $|T_{ETDAS}(\omega)|$ approaches unity for R close to 1 and the plateaus become flatter. Therefore, intermediate frequencies generate a smaller response for larger R and thus are less likely to destabilize the system.

The combined transfer function $|T(\omega)|$ for ETDAS and low-pass filtering is displayed in Fig. 5.6(b). As in Fig. 5.6(a), there are notches at multiples of the frequency of the UPO, which become narrower for increasing R. The amplitudes of frequencies larger than the cut-off frequency α are reduced and thus are only minor contributions to the feedback response. This is important to notice in order to understand how the low-pass filter improves the controllability of the system.

Consider a control signal that inhibits frequency components above the frequency of the unstable periodic orbit, $\omega_0 = 2\pi/\tau$. As discussed above, the ETDAS transfer function becomes zero at multiples of ω_0 so that these frequencies are stabilized since no feedback is generated. The harmonics of a small deviation from ω_0 are given by $m(\omega_0 + \epsilon) = m\omega_0 + m\epsilon$ with an integer number m. It is likely that special harmonics of the deviation coincide with one of the notches. In this case, ETDAS would generate a control force that stabilizes $\omega_0 + \epsilon$ and not only the desired frequency of the UPO ω_0. Inserting a low-pass filter overcomes this effect because higher frequency components are suppressed in the control signal.

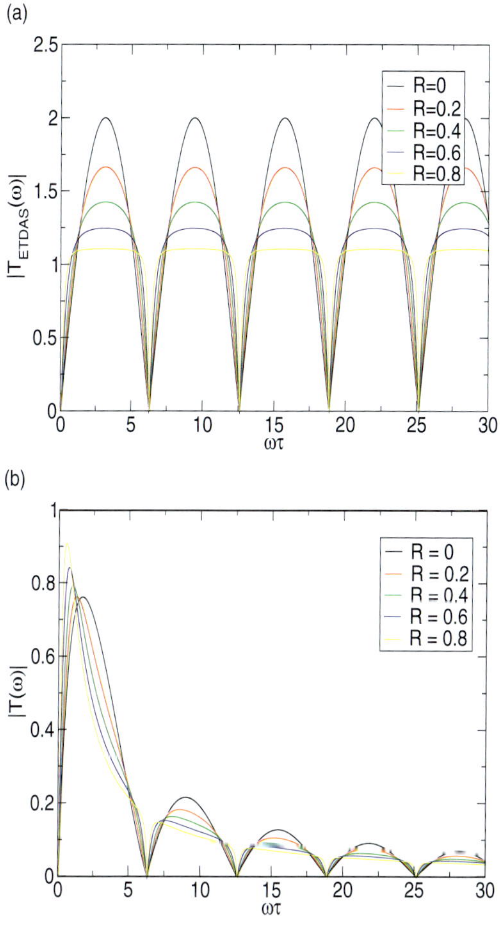

Fig. 5.6. (a) Absolute value of the ETDAS transfer function for different values of R =0 (TDAS), 0.2, 0.4 0.6, 0.8. (b) absolute value of the ETDAS transfer function including a low-pass filter with fixed $\alpha\tau = 1$ for the same values of R as in (a).

Another way to understand the influence of the low-pass filter is to take a look at frequencies which should be destabilized, i.e., suppressed by the control scheme. For this discussion see Fig. 5.7, which depicts the transfer function of the ETDAS method for $R = 0.2$ with and without a low-pass filter ($\alpha\tau = 1$) as the dashed and solid line, respectively. The circles and dots indicate an unwanted frequency ω_1 and its first three harmonics. Let us discuss first the case without a low-pass filter (solid line and black dots).

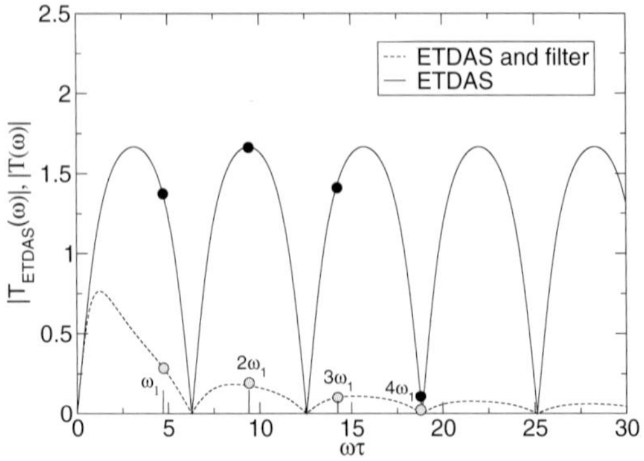

Fig. 5.7. Absolute value of the transfer function of the ETDAS method with and without low-pass filtering for $R = 0.2$: The solid line and black dots correspond to the case without a low-pass filter, the dashed line and the circles to the case with a low-pass filter ($\alpha\tau = 1$). The leftmost dot and circle depict an unwanted frequency ω_1, the three rightmost dots and circles at $2\omega_1$, $3\omega_1$, and $4\omega_1$ the first three harmonics.

Here we find that the third harmonic ($4\omega_1$) is almost located in the middle of a notch of the transfer function. Thus it will enter the generation of the control force with a high spectral weight so that the control scheme accidentally stabilizes its fundamental frequency ω_1. This effect can be overcome by an additional low-pass filter as shown by the dashed curve and the circles. Again the third harmonic is located near a minimum of the transfer function, but since the notch is not so steep due to the low-pass filter, the spectral weight of the harmonic is smaller. Therefore the component of the control force that supports the fundamental frequency ω_1 is reduced.

For a better understanding of the influence of the low-pass filter on the semiconductor superlattice, let us consider the Fourier power spectra of the global current density in the case with and without low-pass filter. Panels (a) and (b) of Fig. 5.8 depict the Fourier power spectra of the global current density J in the absence of a control scheme, i.e., $K = 0$, for both the unfiltered and filtered version of J where the red curve corresponds to J and the blue curve corresponds to its filtered counterpart. The parameters of the superlattice are as in Fig. 5.4 and the cut-off frequency of the filter is set to $\alpha = 1$GHz. Panel (b) is an enlargement of panel (a) to give details of the spectra in the range of 0 to 2GHz.

The spectra in panel (a) do not show sharp peaks, but a large band of frequencies indicating that the stabilization is not successful. In the

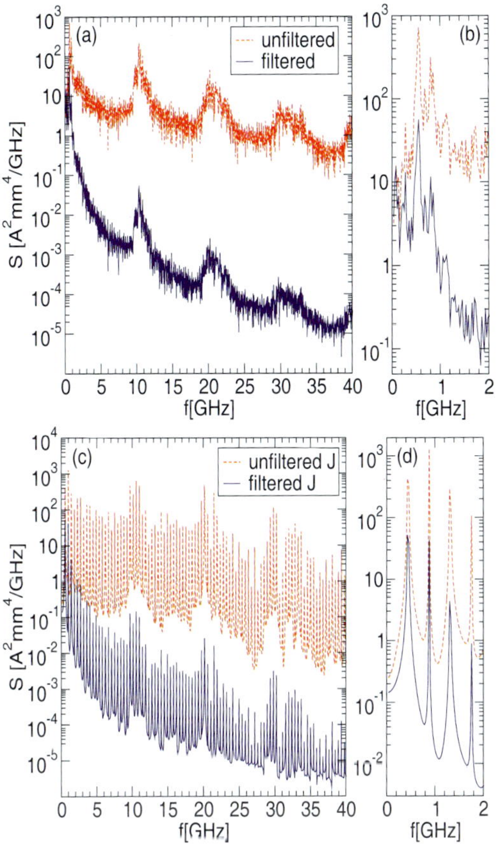

Fig. 5.8. Fourier power spectra of the global current density J: The red and the blue curve correspond to the system with and without low-pass filter (cutoff frequency $\alpha =$ 1GHz), respectively. (a) and (b) display the case of the absence of a control scheme, i.e., $K = 0$. (c) and (d) show the case of ETDAS control via the applied voltage for $K = 3 \times 10^{-6}$Vmm^2/A, $R = 0.2$, and $\tau = 2.29$ns. (b) and (d) are enlargements of (a) and (c) in the range of 0 to 2GHz, respectively.

spectra, there are some local maxima visible. Next to a maximum at about 0.5GHz [see panel (b)], there are additional local maxima at 10GHz and 20GHz, which can be identified with the high frequency of the well-to-well-hopping of the charge fronts and its first harmonic. As expected, the two spectra are similar in the range of low frequencies [see panel (b)], but the amplitudes of high frequency components are reduced in the filtered version of J. From this it can be predicted that an additional control voltage according to Eq. (5.11) will control the low frequency dynamics, which is subject of stabilization, and will not be sensitive to the disturbing high frequency parts.

In fact, panels (c) and (d) of Fig. 5.8 show exactly this result in the presence of ETDAS control with a low-pass filter ($\alpha = 1$GHz). The red and blue curves in the diagram refer again to the original unfiltered current density and its filtered counterpart, respectively. Panel (d) shows again an enlargement of panel (c) in the range of 0 to 2GHz.

Distinct peaks are visible in each spectrum. Panel (d) shows that these peaks belong to the fundamental frequency $f_0 = \tau^{-1} = 0.437$GHz and its harmonics. Thus, the control is successful. It can be seen from the spectrum of the unfiltered J in panel (c) that high frequencies are still present in the system as frequency components in the range of 10GHz, which corresponds to the well-to-well-hopping process of the fronts in the device (see also grey curve in the J vs. t plot of Fig. 5.4), but the oscillations are successfully stabilized because the control scheme is not sensitive to the disturbing high frequencies. This can be seen in the spectrum of the filtered current density. The frequencies of the well-to-well-hopping process of the electrons are strongly reduced. Since the control force is generated from the filtered J, there are no high frequency contributions in U_c which might destabilize the system.

In conclusion, time-delay autosynchronization represents a convenient and simple scheme for the self-stabilization of high-frequency current oscillations due to moving domains in superlattices under dc bias. This approach lacks the drawback of synchronization by an external ultrahigh-frequency forcing, since it requires nothing but delaying of the global electrical system output by the specified time lag.

5.3. Control of noise-induced oscillations in superlattices

Noise is an inevitable feature of physical models. Theoretical and experimental research has recently shown that noise can have surprisingly constructive effects in many nonlinear systems. In particular, an optimal noise level may give rise to ordered behavior and even produce new dynamical states [11]. Well-known examples are provided by stochastic resonance [75, 76] in periodically driven systems, and by coherence resonance [10, 77, 78] in autonomous systems. In spite of considerable progress on a fundamental level, useful applications of noise-induced phenomena in technologically relevant devices are still scarce. Here we will demonstrate that noise can give rise to oscillating current and charge density patterns in semiconductor nanostructures even if the deterministic system exhibits only a steady state, and that these space-time patterns can be controlled by the time-delayed feedback scheme applied to purely deterministic chaotic front patterns in a superlattice in the previous section.

We develop a stochastic model for the superlattice approximating the random fluctuations of the current densities by additive Gaussian white noise $\xi_m(t)$ with

$$\langle \xi_m(t) \rangle = 0, \quad \langle \xi_m(t)\xi_{m'}(t') \rangle = \delta(t-t')\delta_{mm'}, \tag{5.16}$$

in the continuity equation (5.4):

$$e\frac{dn_m}{dt} = J_{m-1\to m} + D\xi_m(t) - J_{m\to m+1} - D\xi_{m+1}(t), \tag{5.17}$$

where D is the noise intensity. Since the inter-well coupling in our superlattice model is very weak and the tunneling times are much smaller than the characteristic time scale of the global current, these noise sources can be treated as uncorrelated both in time and space. Charge conservation is automatically guaranteed by adding a noise term ξ_m to each current density $J_{m-1\to m}$. The physical origin of the noise may be, e. g. thermal noise, $1/f$ noise, or shot noise due to the randomly fluctuating tunneling times of discrete charges across the barriers [66, 79]. The latter is Poissonian and can be approximated by $D = (eJ_{m-1\to m}/A)^{1/2}$ which increases with decreasing current cross section A; thus this type of noise dominates for small devices. In the following we summarize the global effect of noise by a constant D. Note that the results coincide very well with those obtained for current-dependent shot noise.

We choose the control parameters U and σ such that the deterministic system exhibits no oscillations but is very close to a bifurcation thus yielding it very sensitive to noise. The transition from stationarity to oscillations in the system may occur either via a Hopf or via a saddle-node bifurcation on a limit cycle as depicted in the bifurcation diagram of Fig. 5.9. The different nature of these two bifurcations is reflected in the effect noise has in each case. The local character of the Hopf bifurcation is responsible for noise-induced high frequency oscillations of strongly varying amplitude around the stable fixed point. We try to characterize basic features of these oscillations such as coherence and time scales. The need to be able to adjust these features as one wishes will lead to the application of the time-delayed

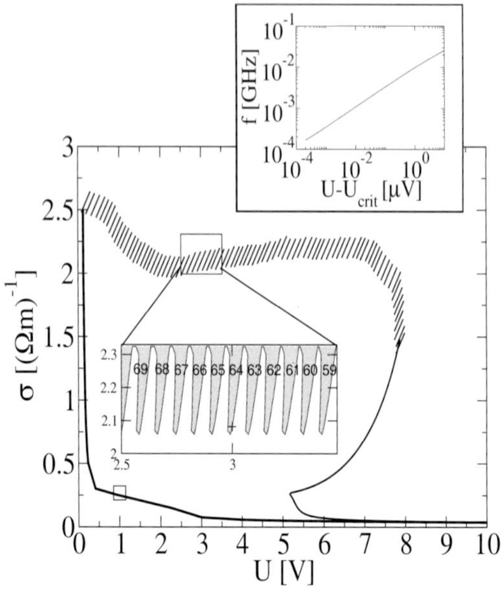

Fig. 5.9. Bifurcation diagram in the (σ, U) plane. Thick and hatched lines mark the transition from stationary to moving fronts via a Hopf or a saddle-node bifurcation on a limit cycle, respectively. The inset shows a blow-up of a small part of the hatched line revealing its saw-tooth-like structure. Dark and white correspond to stationary and moving fronts, respectively, where the numbers denote the positions of the stationary accumulation front in the superlattice. Upper inset shows the frequency f of the limit cycle which is born above the critical point (marked by a cross in the lower inset) as function of U. [57]

feedback scheme as introduced for the deterministic system in Sect. 2. But first we will be looking at the other dynamical regime [57], slightly below the global saddle-node bifurcation on a limit cycle (cross in the lower inset of Fig. 5.9).

There the scenario is quite different. As seen in Fig. 5.9, keeping σ fixed and increasing the voltage U, a limit cycle of approximately constant amplitude and increasing frequency is born. This happens through the collision of a stable fixed point and a saddle-point. Plotting the frequency of these oscillations *vs* the bifurcation parameter U, we obtain the characteristic *square-root scaling law* (upper inset of Fig. 5.9) that governs a saddle-node bifurcation on a limit cycle. At the critical point U_{crit} the frequency of the oscillations tends to zero. This corresponds to an infinite period oscillation and therefore this bifurcation is also known as saddle-node infinite period bifurcation or SNIPER [80, 81]. We now prepare the system at the stable fixed point, which corresponds to a stationary accumulation front (Fig. 5.10(a)), and introduce noise. As the noise intensity is increased, the behavior of the system changes dramatically (Fig. 5.10(b)): the accumulation front remains stationary only for a while, until a pair of a depletion and another accumulation front (i.e., a charge dipole with a high-field domain in between) is generated at the emitter. As is known from the deterministic system, this dipole injection depends critically upon the emitter current [56]. Here it is triggered by noise at the emitter (we have in fact verified that the same scenarios occur if noise is applied locally only to the wells near the emitter). Because of the global voltage constraint, Eq. (5.6), the growing dipole field domain between the injected depletion and accumulation fronts requires the high field domain between the stationary accumulation front and the collector to shrink, and hence that accumulation front starts moving towards the collector. For a short time there are two accumulation fronts and one depletion front in the sample, thereby forming a tripole [82], until the first accumulation front reaches the collector and disappears. When the depletion front reaches the collector, the remaining accumulation front must stop moving because of the global constraint, and this happens at the position where the first accumulation front was initially localized. After some time noise generates another dipole at the emitter and the same scenario is repeated.

There are two distinct time scales in the system. One is related to the time the depletion front takes to travel through the superlattice. The other timescale is associated with the time needed for a new depletion front to be generated at the emitter. These two time scales are also visible in the noise-induced current oscillations, see Fig. 5.11(a). The time series of the current density are in the form of a pulse train with two characteristic times: the activation time, which is the time needed to excite the system from this stable fixed point (time needed for a new depletion front to be generated at the emitter) and the excursion time which is the time needed to return from the excited state to the fixed point (time the depletion front needs to travel through the device). Low noise is associated with large activation times and small, almost constant, excursion times, while as the noise level increases activation times become smaller and at sufficiently large D vanish. At low D the spike train looks irregular, and the interval between excitations (mean interspike-interval $\langle T \rangle$) is relatively large and random in time. At moderate noise, the spiking is rather regular therefore suggesting that the mean interspike-interval does not vary substantially. Further increase of

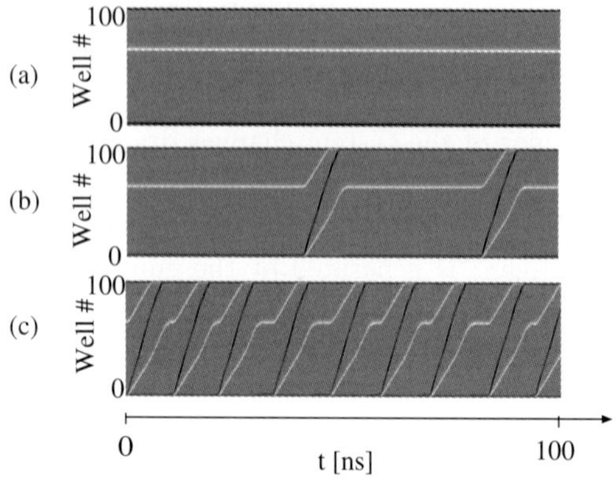

Fig. 5.10. Noise-induced front motion: Space-time plots of the electron density for (a) D=0 (no noise), (b) $D = 0.5As^{1/2}/m^2$, (c) $D = 2.0As^{1/2}/m^2$. Light and dark shading corresponds to electron accumulation and depletion fronts, respectively. The emitter is at the bottom. Parameters: $U = 2.99V$, $\sigma = 2.08210124888\Omega^{-1}m^{-1}$, other parameters as in Sect. 2. [57]

noise results in a highly irregular spike train with very frequent spikes.

To get deeper insight into the effect noise has on the time scales and coherence of the system we determine the interval between two consecutive excitations and calculate the mean interspike-interval $\langle T \rangle$. In Fig. 5.11(b) (top) the decrease of $\langle T \rangle$ as a function of D is shown thus demonstrating that the mean interspike-interval is strongly controlled by the noise intensity especially at lower values of the latter. This is very important in terms of experiments, where noise can induce oscillations by forcing stationary fronts to move. The corresponding spectral peak frequency f shows a linear scaling for small D. As a measure for coherence we use the normalized fluctuations of pulse duration [10]

$$R_T = \frac{(\langle T^2 \rangle - \langle T \rangle^2)^{1/2}}{\langle T \rangle} \tag{5.18}$$

This quantity, as seen in Fig. 5.11(b) (bottom), is a non-monotonic function of D, exhibiting a minimum at moderate noise intensity. This is the well phenomenon of *coherence resonance* and is strongly connected to excitability.

Next we prepare the system in the vicinity of the lower bifurcation line in Fig. 5.9, slightly below a Hopf bifurcation marked by the small rectangle in Fig.5.9 [51]. In the absence of noise the only stationary solution is a sta-

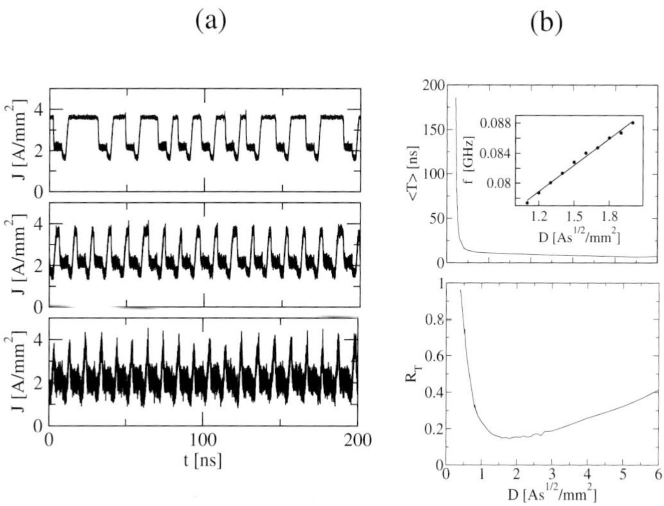

Fig. 5.11 (a) Three noise realizations of the current density $J(t)$. From top to bottom, $D = 0.8$, $D = 2.0$ and $D = 5.0As^{1/2}/m^2$. (b) Mean interspike-interval (top) and its normalized fluctuations R_T (bottom) *vs* noise intensity. The inset shows the peak frequency *vs.* D. [57]

ble fixed point which now corresponds to a stationary depletion front near the emitter, associated with a stationary current density flowing throughout the device (Fig. 5.12(a), top). In the phase space it corresponds to a stochastic oscillation around the stable fixed point. With increase of the noise intensity ($D > 0$), the current density starts oscillating in a rather regular way around the steady state (Fig. 5.12(a), middle). At larger noise intensities, the dynamics changes significantly (Fig. 5.12(a), bottom). Current oscillations become sharply peaked and spiky, and the average current is shifted towards larger values. In the spatiotemporal picture (omitted here), no significant front motion is observed. At low noise intensities the depletion front as a whole starts to "wiggle" around its deterministically fixed position. At higher noise intensities the depletion front exhibits a slightly more asymmetric motion and occasionally the onset of a tripole oscillation may be observed without, however, further development.

Although noise in this case fails to induce significant front motion, the associated current oscillations exhibit interesting features in dependence upon the noise intensity. To quantify the regularity of these oscillations a suitable and widely used measure is the correlation time t_{cor} given by the

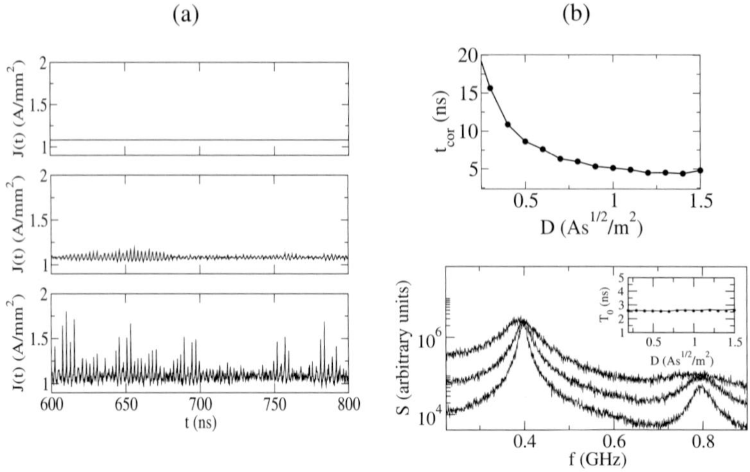

Fig. 5.12. (a) Current density times series. From top to bottom: $D = 0$, $D = 0.1$ and $D = 0.5 As^{1/2}/m^2$. (b) Correlation time *vs* D (upper plot). Lower plot: main and secondary spectral peak of the power spectral density $S(2\pi f)$ *vs* frequency f for increasing noise (from top to bottom: $D = 0.3, D = 0.5$ and $D = 1.0 As^{1/2}/m^2$). Inset shows basic period T_0 *vs* noise intensity D. Parameters: $U = 1V$, $\sigma = 0.266\Omega^{-1}m^{-1}$. [51]

formula [83]:

$$t_{cor} = \frac{1}{\psi(0)} \int_0^\infty |\psi(s)| ds, \tag{5.19}$$

where $\psi(s)$ is the autocorrelation function of the current density signal $J(t)$,

$$\psi(s) = \langle (J(t) - \langle J \rangle)(J(t-s) - \langle J \rangle) \rangle, \tag{5.20}$$

and $\psi(0)$ is its variance. Additionally we may consider the Fourier power spectral densities (shortly referred to as spectra) for different values of noise intensities D. They provide qualitative understanding of how the noise level affects basic characteristics of noise-induced oscillations (Fig. 5.12(b), lower panel). We see that an increase of the noise level broadens the spectral peak and suppresses secondary ones. At the same time the position of the main spectral peak, corresponding to the basic frequency of the oscillations, is almost unchanged. This is confirmed by the inset of the same figure, where the dependence of the basic period T_0 (the inverse of the frequency at which the spectral peak is centered) of the noisy oscillations versus the noise intensity is presented. This basic period is close to the period of self-oscillations above the Hopf bifurcation.

Next we are interested in the effect of time-delayed feedback on the coherence and time scales of the noise-induced oscillations. We will apply the scheme proposed by Eq. (5.10) but with $R = 0$ and cut-off frequency $\alpha = 1$ GHz. A natural choice for τ is the basic period of the Hopf oscillation (or integer multiples of it). As seen in Fig. 5.13(b) the application of control indeed improves the coherence of the current signal, since the main peak in the power spectrum becomes narrower. This improvement may also be expressed through the correlation time (Fig. 5.13a), where in the controlled system it acquires larger values especially at higher noise intensities. In order to study the influence of control on the time scales of the system, the parameter to be varied will be the time delay τ. Calculating spectra for increasing τ it was found that additional peaks appear, while the main (most pronounced) peak moves towards lower frequencies. Plotting the period of the resulting main peak as a function of τ, we see that $T_0(\tau)$ has an almost piecewise linear, oscillatory character (Fig 5.13c) in agreement with [7, 8]. Therefore, while the position of the main peak of the spectrum does not depend on the noise level in the case without control, it is indeed possible to shift its position by the proposed time-delayed feedback scheme.

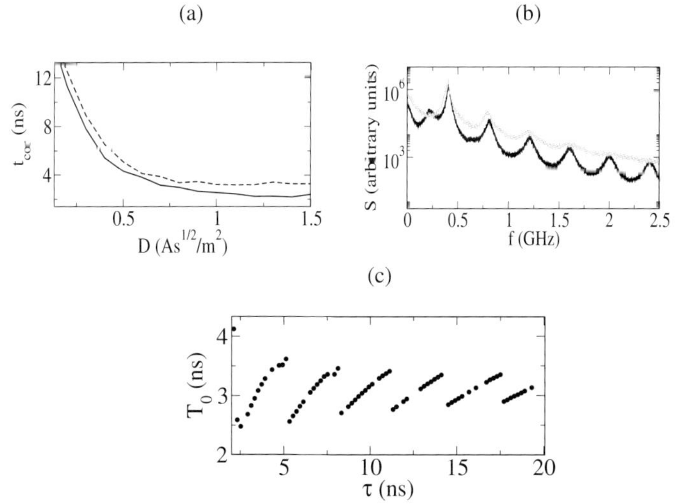

Fig. 5.13. (a) Correlation time of controlled system with $K = 3 \times 10^{-6} Vmm^2/A$, $\alpha = 10^9 s^{-1}$, and $\tau = 2.5ns$ (broken line) and (solid line) for uncontrolled system ($K = 0$). (b) Spectra with (black) and without control (gray) for $D = 0.5 As^{1/2}/m^2$. (c) Basic frequency T_0 *vs* time delay τ. [51]

To conclude, noise-induced front motion and oscillations have been observed in a spatially extended system. The former are induced in the vicinity of a global saddle-node bifurcation on a limit cycle where noise uncovers a mechanism of excitability responsible also for coherence resonance. In another dynamical regime, namely below a Hopf bifurcation, noise induces oscillations of decreasing regularity but with almost constant basic time scales. Applying time-delayed feedback enhances the regularity of those oscillations and allows to manipulate the time scales of the system by varying the time delay τ.

5.4. Chaos control of spatio-temporal oscillations in resonant tunneling diodes

Next we consider a double–barrier resonant tunneling diode (DBRT), which exhibits a Z-shaped (bistable) current–voltage characteristic [15]. We include the lateral re–distribution of electrons in the quantum well plane (x coordinate) giving rise to filamentary current flow [84, 85]. Complex chaotic scenarios including spatio-temporal breathing and spiking oscillations have been found in a simple deterministic reaction–diffusion model [58]. We extend this model to include control terms, and obtain the following equations [47] where we use dimensionless variables throughout:

$$\frac{\partial a}{\partial t} = f(a,u) + \frac{\partial}{\partial x}\left(D(a)\frac{\partial a}{\partial x}\right) - KF_a(x,t) \tag{5.21}$$

$$\frac{du}{dt} = \frac{1}{\varepsilon}(U_0 - u - rJ) - KF_u(t) \tag{5.22}$$

Here $u(t)$ is the inhibitor and $a(x,t)$ is the activator variable. In the semiconductor context $u(t)$ denotes the voltage drop across the device and $a(x,t)$ is the electron density in the quantum well. The nonlinear, nonmonotonic function $f(a,u)$ describes the balance of the incoming and outgoing current densities of the quantum well, and $D(a)$ is an effective, electron density dependent transverse diffusion coefficient. The local current density in the device is $j(a,u) = \frac{1}{2}(f(a,u) + 2a)$, and $J = \frac{1}{L}\int_0^L j dx$ is associated with the global current. Eq. (5.22) represents Kirchhoff's law of the circuit (5.3) in which the device is operated. The external bias voltage U_0, the dimensionless load resistance $r \sim R_L$, and the time-scale ratio $\varepsilon = R_L C/\tau_a$ (where C is the capacitance of the circuit and τ_a is the tunneling time) act as control parameters. The one–dimensional spatial coordinate x corresponds to the direction transverse to the current flow. We consider a system of width $L = 30$ with Neumann boundary conditions $\partial_x a = 0$ at $x = 0, L$ corresponding to no charge transfer through the lateral boundaries.

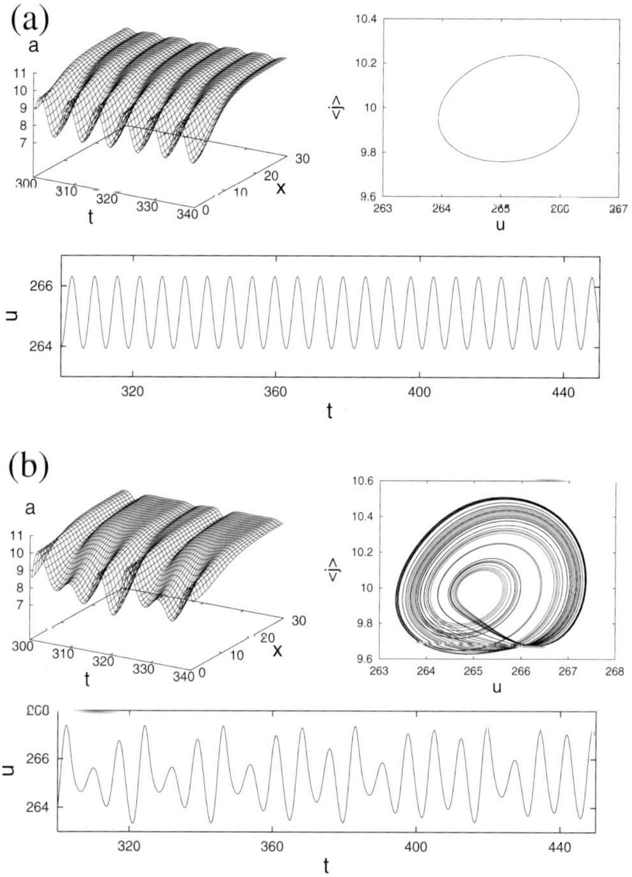

Fig. 5.14. Spatio-temporal breathing patterns of the DBRT: electron density evolution, phase portrait, and voltage evolution for (a) $\varepsilon = 7.0$: periodic breathing, (b) $\varepsilon = 9.1$: chaotic breathing ($r = -35$, $U_0 = -84.2895$, $K = 0$). Time t and space x are measured in units of the tunneling time τ_a and the diffusion length l_a, respectively. Typical values at 4 K are $\tau_a = 3.3$ ps and $l_a = 100$ nm [47].

Eqs. (5.21),(5.22) contain control forces F_a and F_u for stabilizing time periodic patterns. The strength of the control terms is proportional to the control amplitude K, which gives one important parameter of each control scheme. In the semiconductor context these forces can be implemented by appropriate electronic feedback circuits [43].

The dynamics of the free system, i.e. $K = 0$, develops temporally chaotic and spatially nonuniform states (spatio–temporal breathing or spiking) in appropriate parameter regimes [58], see Fig. 5.14. For any value of L the system, due to the global coupling, allows only single spikes at the boundary of the spatial domain [86]. In the semiconductor context the time and length scales of our dimensionless variables are typically given by 3.3 picoseconds (tunneling time) and 100 nanometers (diffusion length), respectively. Typical units of the electron density, the current density, and the voltage are $10^{10}\mathrm{cm}^{-2}$, $500\,\mathrm{A/cm}^2$, and $0.35\,\mathrm{mV}$, respectively. A characteristic bifurcation diagram exhibiting a period-doubling route to chaos is shown in Fig. 5.15.

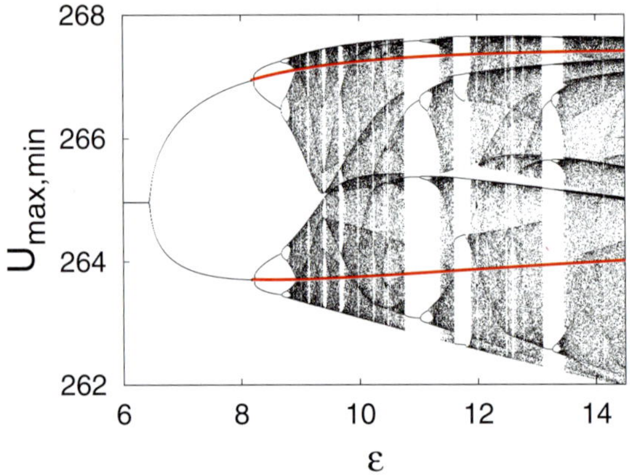

Fig. 5.15. Chaotic bifurcation diagram of the resonant tunneling diode. The maxima and minima of the voltage oscillations are plotted versus the time-scale parameter ε. [47]

We are concerned with controlling unstable time periodic patterns $u_p(t) = u_p(t+\tau)$, $a_p(x,t) = a_p(x, t+\tau)$ which are embedded in a chaotic attractor. For that purpose we apply control forces F_a and F_u which are derived from time–delayed differences of the voltage and the charge density. For example we may choose $F_u = F_{\mathrm{vf}}$ with the voltage feedback force

$$F_{\mathrm{vf}}(t) = u(t) - u(t-\tau) + RF_{\mathrm{vf}}(t-\tau) \tag{5.23}$$

(extended time–delay autosynchronization).

Here we concentrate on the question how the coupling of the control forces to the internal degrees of freedom influences the performance of the control. For our model we consider two different choices for the control force F_a. On the one hand we use a force which is based on the local charge density according to

$$F_{\text{loc}}(x,t) = a(x,t) - a(x,t-\tau) + RF_{\text{loc}}(x,t-\tau) \quad . \tag{5.24}$$

whereas on the other hand we propose a construction which is only based on its spatial average

$$F_{\text{glo}}(t) = \frac{1}{L}\int_0^L [a(x,t) - a(x,t-\tau)]\,dx + RF_{\text{glo}}(t-\tau) \quad . \tag{5.25}$$

We call the choice $F_a = F_{\text{loc}}$ a *local* control scheme in contrast to the *global* control scheme $F_a = F_{\text{glo}}$ which requires only the global average and does not depend explicitly on the spatial variable. The second option has considerable experimental advantages since the spatial average is related to the total charge in the quantum well and does not require a spatially resolved measurement.

In general the analysis of the control performance of time–delayed feedback methods results in differential–difference equations which are hard to tackle. Stability of the orbit is governed by eigenmodes and the corresponding complex valued growth rates (Floquet exponents). There exists a simple case (which we call *diagonal* control) where analytical results are available [30, 87], namely for $F_a = F_{\text{loc}}$ and $F_u = F_{\text{vf}}$. It is a straightforward extension to a spatially extended system of an identity matrix for the control of discrete systems of ordinary differential equations (cf. Ref. 29). Fig. 5.16 shows successful control of a chaotic breathing oscillation after the control force is switched on.

In Figure 5.17 the regime of successful control in the (K, R) parameter plane and the real part of the Floquet spectrum $\Lambda(K)$ for $R = 0$ is depicted. The control domain has its typical triangular shape bounded by a flip instability ($\Lambda = 0$, Imaginary part $\Omega = \pi/\tau$) to its left and by a Hopf (Neimark-Sacker) bifurcation to its right. Inclusion of the memory parameter R increases the range of K over which control is achieved. We observe that the numerical result fits very well with the analytical prediction.

To confirm the bifurcations at the boundaries we consider the real part of the Floquet spectrum of the orbit subjected to control. Complex conjugate Floquet exponents show up as doubly degenerate pairs. The largest nontrivial exponent decreases with increasing K and collides at negative values with a branch coming from negative infinity. As a result a complex conjugate pair develops and real parts increase again. The real part of the

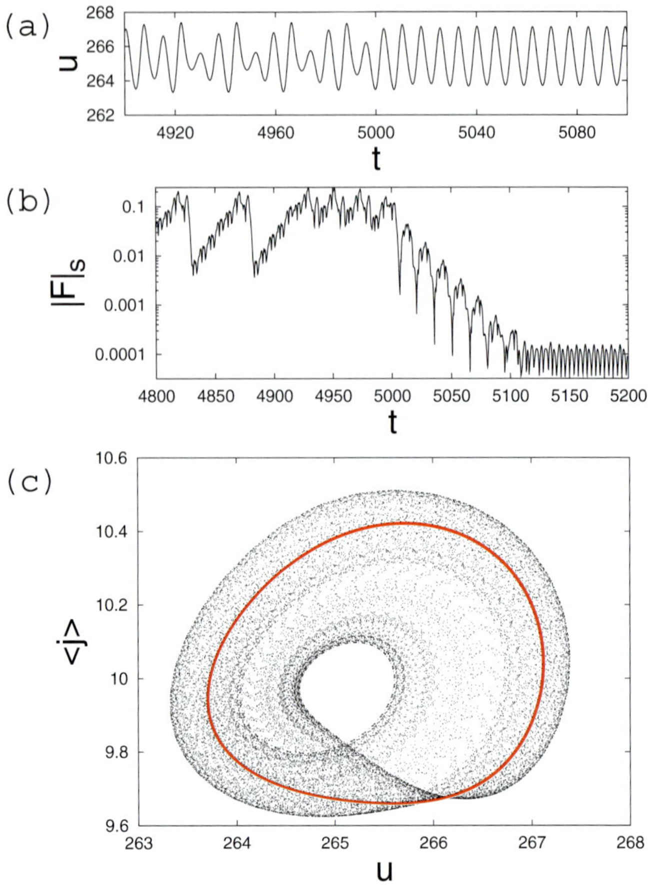

Fig. 5.16. Diagonal control in the DBRT, where the control force is switched on at $t = 5000$. (a) Voltage u vs. time, (b) Supremum of the control force vs. time, (b) Phase portrait (global current vs. voltage) showing the chaotic breathing attractor and the embedded stabilized periodic orbit (black cycle). Parameters: $r = -35$, $\varepsilon = 9.1$, $\tau = 7.389$, $K = 0.137$, $R = 0$. [47]

exponent finally crosses the zero axis giving rise to a Hopf bifurcation. Our numerical simulations are in agreement with the analytical result.

Let us now replace the local control force $F_a = F_{\text{loc}}$ by the global control $F_a = F_{\text{glo}}$. Fig. 5.18 shows the corresponding control regime and Floquet spectrum. The control domain looks similar in shape as for diagonal control, although the domain for the global scheme is drastically reduced. The shift in the control boundaries is due to different branches of the Floquet spectrum crossing the ($\Lambda = 0$)-axis.

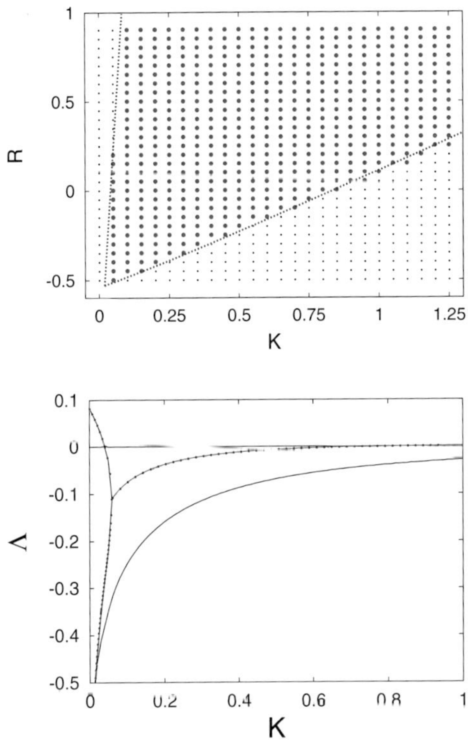

Fig. 5.17. Top: Control domains in the (K, R) parameter plane for diagonal control of the unstable periodic orbit with period $\tau = 7.389$. Large dots: successful control in the numerical simulation, small grey dots: no control, dotted lines: analytical result for the boundary of the control domain according to Ref. 87. Bottom: Leading real parts Λ of the Floquet spectrum for diagonal control in dependence on K $(R = 0)$. [47]

Finally, we note that the period-one orbit can be stabilized by our control scheme throughout the whole bifurcation diagram including chaotic bands and windows of higher periodicity, as marked by two solid lines in Fig. 5.15 for diagonal control. Thus our method represents a way of obtaining stable self-sustained voltage oscillations in a whole range of operating conditions, independently of parameter fluctuations.

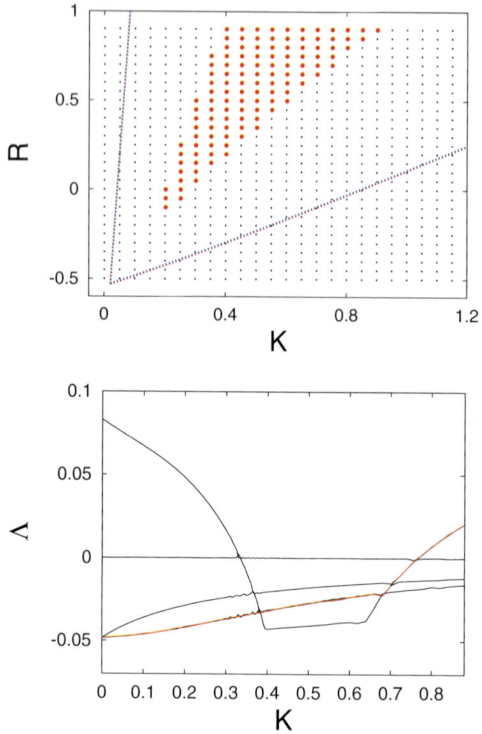

Fig. 5.18. Same as Fig. 5.17 for global control (Bottom: $R = 0.7$). [47]

5.5. Noise-induced spatio-temporal patterns in the DBRT

In the previous section we discussed the possibilities to control deterministic chaotic oscillations in the double barrier resonant tunneling diode (DBRT). Now we will study the effects of noise in this system and investigate whether we can control noise-induced spatio-temporal oscillations by the same method of time-delayed feedback [52, 59].

We use the same model equations (5.21) and (5.22). First we will consider the system without control ($K = 0$), but with two additional noise sources:

$$\begin{aligned} \frac{\partial a(x,t)}{\partial t} &= f(a,u) + \frac{\partial}{\partial x}\left(D(a)\frac{\partial a}{\partial x}\right) + D_a\xi(x,t) \\ \frac{du(t)}{dt} &= \frac{1}{\varepsilon}\left(U_0 - u - rJ\right) + D_u\eta(t) \end{aligned} \tag{5.26}$$

where $\xi(x,t)$ and $\eta(t)$ represent uncorrelated Gaussian white noise sources with noise intensities D_a and D_u, respectively:

$$\begin{aligned}\langle \xi(x,t)\rangle = \langle \eta(t)\rangle &= 0 \qquad (x \in [0,L]),\\ \langle \xi(x,t)\xi(x',t')\rangle &= \delta(x-x')\delta(t-t'),\\ \langle \eta(t)\eta(t')\rangle &= \delta(t-t').\end{aligned} \tag{5.27}$$

Here we concentrate on the effects of *external* noise modeled by the additional noise voltage $D_u\eta(t)$ in the current equation. This term is easily accessible in a real circuit and the noise intensity D_u can be adjusted in a large parameter range using a noise generator in parallel with the supply bias, as realized experimentally, e. g., in Ref. 88. In typical dimensional units of $\varepsilon k_B T/e$ [58] $D_u = 1$ corresponds to a parallel noise voltage of 2 mV at temperature $T = 4\,\mathrm{K}$. *Internal* fluctuations of the local current density on the other hand, e. g. shot noise [79], can not be tuned from the outside. Therefore in the following we keep this value fixed at a small noise amplitude of $D_a = 10^{-4}$, corresponding to a noise current density of the order of $50mA/cm^2$, which is within the range of Poissonian shot noise currents.

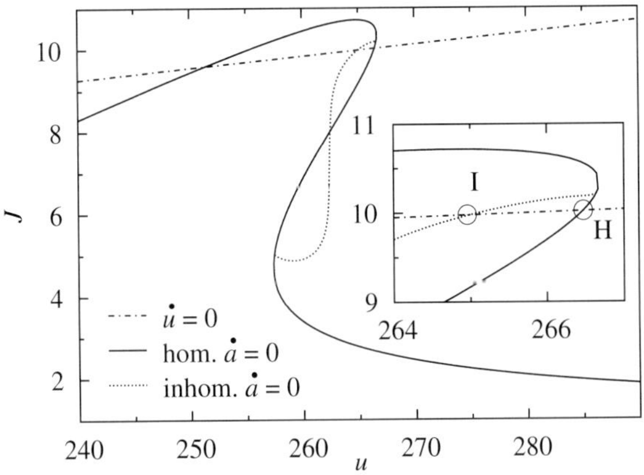

Fig. 5.19. Current-voltage characteristic of the DBRT model. The null isoclines for the dynamical variables u (which is the load line, dash-dotted) and a in the case of a homogeneous $a(x)$ (solid) and in the case of inhomogeneous $a(x)$ (dotted) are shown. The inset shows an enlargement, where I and H mark the inhomogeneous and the homogeneous fixed points of the system, respectively. $U_0 = -84.2895$, $r = -35$. Other parameters as in Refs. 47, 52, 59.

In the noise-free case, $D_u{=}D_a = 0$, one can calculate the null isoclines of the system. These are plotted in Fig. 5.19 using the current–voltage projection of the originally infinite-dimensional phase space. There are three curves, the null isocline $\dot{u} = 0$ (i. e., the *load line*) and two null isoclines $\dot{a} = 0$, one for a reduced system, including only spatially homogeneous states, and one for the full system. We call the system *spatially homogeneous* if the space dependent variable $a(x,t)$ is uniformly distributed over the whole width of the device, i. e. $a(x,t) = a(t)$ for all $x \in [0, L]$, otherwise it is called *spatially inhomogeneous.*

In Fig. 5.19 one can see the Z-shaped current-voltage characteristic of the DBRT (solid curve), and the inset represents our special regime of interest for the following investigations. We fix $\varepsilon = 6.2$ slightly below the Hopf bifurcation, which occurs at $\varepsilon_{\mathrm{Hopf}} \approx 6.469$ (cf. Fig. 5.15). In this regime we have a stable, spatially inhomogeneous fixed point marked 'I' in Fig. 5.19, which is determined by the intersection of the load line with the nullcline $\dot{a} = 0$ for inhomogeneous $a(x,t)$. The neighboring intersection of the load line with the homogeneous nullcline (marked 'H') defines another, spatially homogeneous fixed point which is a saddle-point. It is stable with respect to completely homogeneous perturbations but generally unstable against spatially inhomogenous fluctuations.

Finally, the system (5.26) has a stable homogeneous fixed point which is characterized by negative voltage u and almost zero current density J. This point corresponds to the non-conducting regime of the DBRT, which is beyond the scope of the present study.

In Fig. 5.20 one can see the rather rapid transition of the deterministic system from the slightly perturbed homogeneous fixed point (H) to the inhomogeneous filamentary one (I). This illustrates that for the given parameters the only stable solution, apart from a trivial, non-conducting fixed point, is an inhomogeneous steady state.

To quantify the degree of (in)homogeneity we use the measure of the *absolute spatial variation* $v(t)$ of $a(x,t)$ defined by

$$v(t) \equiv \int_0^L \left| \frac{\partial a(x,t)}{\partial x} \right| dx. \tag{5.28}$$

For completely homogeneous states $a(x,t) = a(t)$ the absolute spatial variation equals zero and the larger $v(t)$ grows, the more inhomogeneous the spatial charge carrier density distribution $a(x,t)$ appears. In Fig. 5.20(b) the spatial variation of $v(t)$ tends towards a fixed value of approximately 2.6, indicating the inhomogeneity of the corresponding fixed point.

In the following we will investigate the behavior of the system under variation of the noise intensity D_u. Note that this noise term does not have any space dependent influence upon a. Now we initialize the system at the

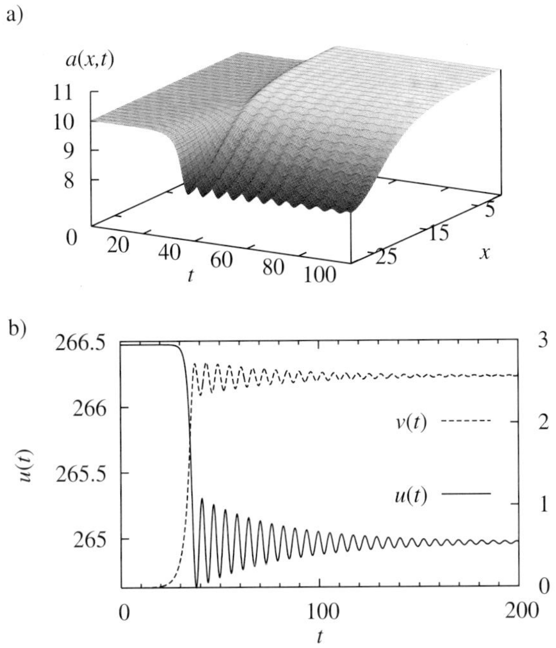

Fig. 5.20. Transition from the homogeneous initial state to the spatially inhomogeneous fixed point due to a small spatially inhomogeneous perturbation in the noise-free case. (a) Charge carrier density distribution $a(x,t)$, (b) Time series $v(t)$ (dashed) and $u(t)$ (solid). At $t = 0$ the system is prepared in the homogeneous fixed point $u(0) = 266.47$, $a(x, 0) = 10.02$ with a very small initial random perturbation. System parameters: $U_0 = -84.2895$, $r = -35$, $\varepsilon = 6.2$, $D_a = D_u = 0$. [59]

inhomogeneous fixed point and simulate it with different noise intensities D_u. The results can be seen in Fig. 5.21. While for small noise the system exhibits rather small oscillations around the inhomogeneous fixed point (topmost panel), with increasing noise intensity a transition to completely homogeneous oscillations occurs (bottom panel). For intermediate values of D_u one can see the competition between the inhomogeneous and the spatially homogeneous modes (middle panel).

Let us now quantify the spatial and the temporal ordering of the system. We call the system spatially coherent if the space dependent variable $a(x,t)$ is uniformly distributed over the whole length of the device. To reveal whether a particular state of the system is spatially homogeneous or not we use the simple measure of the *absolute spatial variation* defined in eq. (5.28)

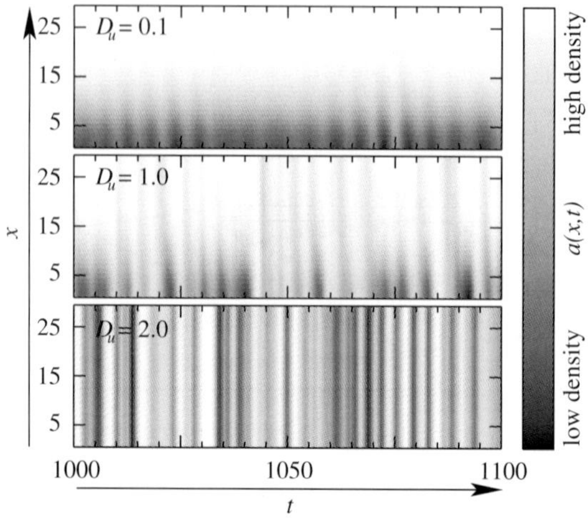

Fig. 5.21. Spatio-temporal patterns $a(x,t)$ induced by different noise intensities $D_u = 0.1, 1.0, 2.0$. At $t = 0$ the system is prepared in the spatially inhomogenous steady state 'I' and with the parameters of Fig. 5.19. The system is then simulated with $D_a = 10^{-4}$ and D_u as indicated. $U_0 = -84.2895$, $r = -35$, $\varepsilon = 6.2$. [52]

above.

The temporal ordering of the system, on the other hand, can be measured by the correlation time [83],

$$t_{\mathrm{cor}} \equiv \frac{1}{\sigma^2} \int_0^\infty |\Psi(s)|\, ds, \tag{5.29}$$

where $\Psi(s) \equiv \big\langle (u(t) - \langle u \rangle)\,(u(t-s) - \langle u \rangle)\big\rangle_t$ is the autocorrelation function of the variable $u(t)$ and $\sigma^2 = \Psi(0)$ is its variance.

By calculating the temporal mean values of $v(t)$ for different D_u we can characterize the shape of the dynamics in dependence on the noise intensity. In Fig. 5.22(a) these values are plotted versus the noise intensity and one can see that the mean value of v monotonically tends towards zero with increasing noise, indicating an increase in spatial coherence. The error bars in this plot show the standard deviation. In fact they reflect an essential feature of this transition, namely the competition between spatially inhomogeneous and homogeneous modes for intermediate values of D_u. The larger the standard deviation of v is, the more "mixed" the dynamics appears. Fig. 5.22(b) offers the same information showing the variance of v versus D_u. For noise close to zero only slight oscillations around the inhomogeneous fixed point with almost fixed spatial profile of $a(x,t)$ lead to

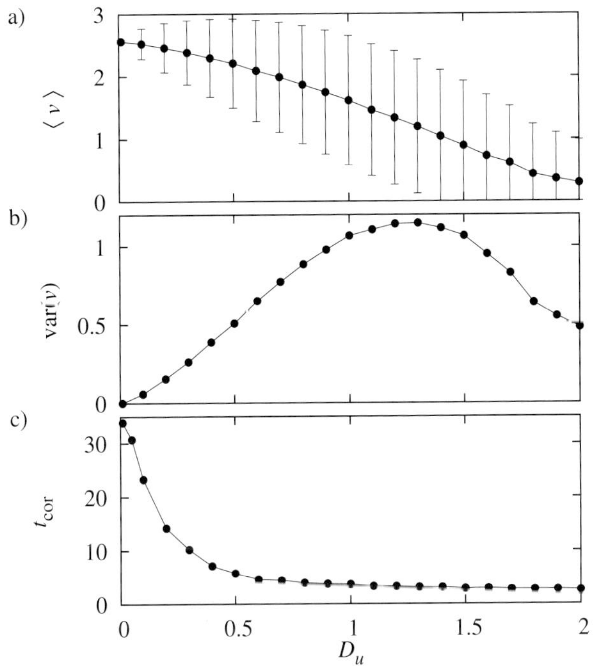

Fig. 5.22. Spatial and temporal ordering of the dynamics in dependence of noise intensity D_u for $D_a = 0.001$. (a) Time-average of the order parameter $v(t)$ defined in eq. (5.28), error bars correspond to the standard deviation. (b) Variance of the parameter v (corresponding to the square of the error bars from a)), (c) Correlation time (eq. (5.29)). [59]

a vanishingly small variance of v. With increasing noise more and more frequently the system tends to a homogeneous state. The variance exhibits a maximum around $D_u = 1.3$, indicating maximum fluctuations of the system between homogeneous and inhomogeneous modes. Thus, this value could be treated as a boundary between predominantly filamentary and predominantly homogeneous behavior. For even larger noise intensity the homogeneous mode is getting more and more dominant and therefore the variance of v again falls off towards zero.

On the other hand, the correlation time versus noise intensity in Fig. 5.22(c) shows that the temporal coherence of the system in contrast to the spatial ordering decreases rapidly with increasing noise.

In summary, noise induces oscillations in the system, which would otherwise rest in its inhomogeneous fixed point. With growing noise intensity the dynamics changes from small inhomogeneous oscillations which are quite coherent in time to spatially homogenous oscillations which on the other hand appear very irregular in time.

In order to control the noised-induced patterns, we will now use the method of time-delayed feedback which was previously applied successfully in deterministic chaos control of this particular system [47] as well as for control of noise-induced oscillations in simple models [7–9] without spatial degrees of freedom.

The voltage u is easily accessible in a real experiment. Therefore, as a simple and adaptive method of control we add the time-delayed feedback only to the voltage variable u in eq. (5.26):

$$\begin{aligned} \frac{\partial a(x,t)}{\partial t} &= f(a,u) + \frac{\partial}{\partial x}\left(D(a)\frac{\partial a}{\partial x}\right) + D_a\xi(x,t) \\ \frac{du(t)}{dt} &= \frac{1}{\varepsilon}(U_0 - u - rJ) + D_u\eta(t) \\ &\quad - K\left[u(t) - u(t-\tau)\right] \end{aligned} \tag{5.30}$$

By varying the control amplitude K we can adjust the strength of the control force; τ is the time delay of the feedback loop.

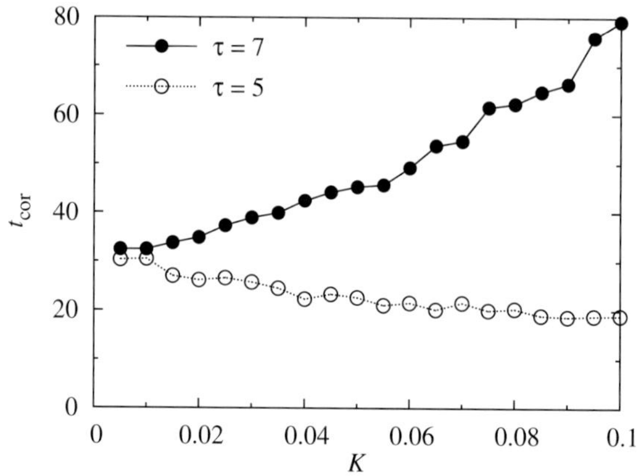

Fig. 5.23. Correlation time vs. feedback strength K for $\tau = 5$ and $\tau = 7$. $D_u = 0.1$, $D_a = 10^{-4}$. Averages from 100 time series of length $T = 10000$, parameters as in Fig. 5.21. [52]

To get a first impression whether or not this control force is able to change the temporal regularity of the noise-induced oscillations we fix $D_u = 0.1$, $D_a = 10^{-4}$, as in the upper panel of Fig. 5.21, and calculate the

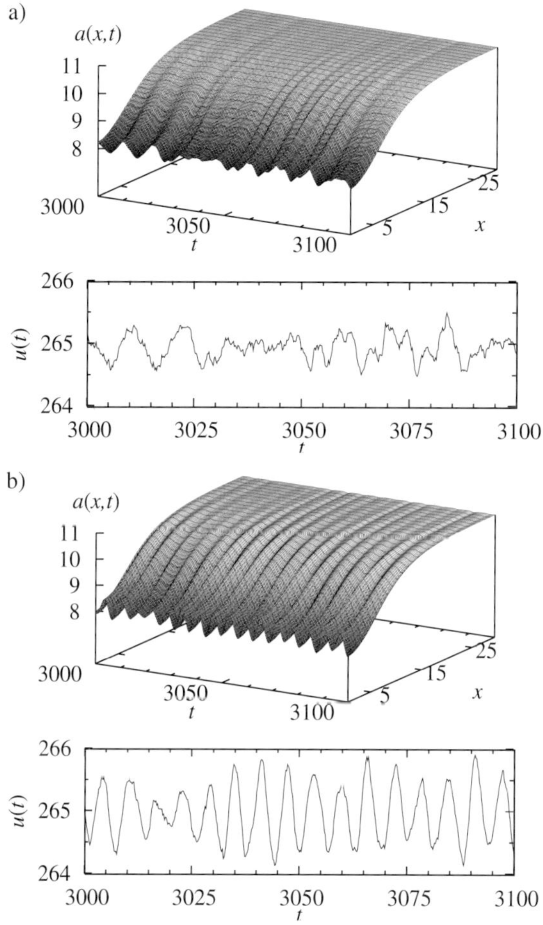

Fig. 5.24. Spatio-temporal patterns $a(x,t)$ and voltage time series $u(t)$ for different values of the control strength K and delay time τ: (a) $\tau = 4.0$, $K = 0.4$, (b) $\tau = 13.4$, $K = 0.1$. $D_u = 0.1$, $D_a = 10^{-4}$ and other parameters as in Fig. 5.21. [52]

correlation time in dependence of the feedback strength K for two different delay times τ. From Fig. 5.23 one can see that the qualitative result depends strongly upon the choice of the delay time. While for $\tau = 7$ the control loop strongly increases the correlation time with increasing K, it is on the other hand able to decrease it significantly for $\tau = 5$. The difference in regularity for different values of τ and K also shows up in the corresponding spatio-temporal patterns and voltage time series (Fig. 5.24), where (b) is clearly more regular than (a).

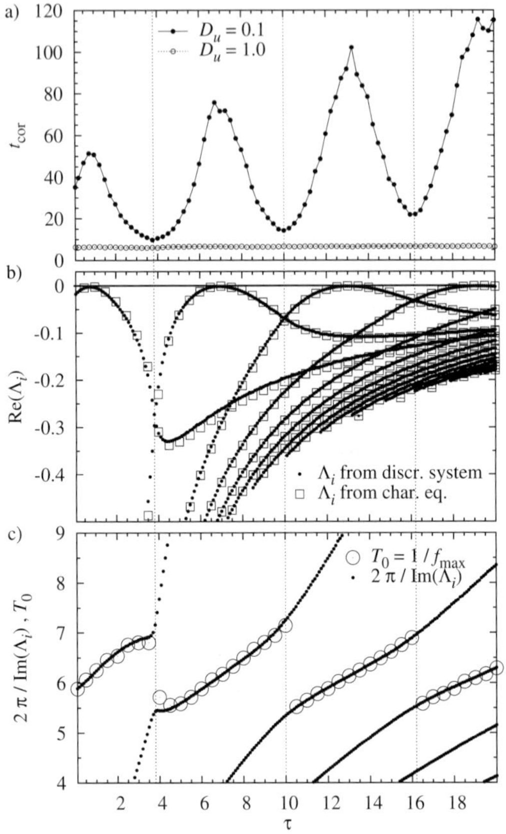

Fig. 5.25. (a) Correlation time (eq. (5.29)) for two different noise intensities in dependence of the feedback delay τ. (b) Real parts of the eigenvalues Λ_i of the linearized deterministic system ($D_a = D_u = 0$) calculated at the spatially inhomogeneous fixed point for $K = 0.1$. The black dots are calculated from the spatially discretized system (set of ODEs) whereas the squares are calculated from eq. (5.42) (see text). The vertical dotted lines mark values of τ at which the *leading* eigenvalue (i. e. the one with the largest real part) changes. (c) Eigenperiods $2\pi/\,\mathrm{Im}(\Lambda_i)$ of the deterministic system and basic periods $T_0 := 1/f_{\max}$ of the noise-induced oscillations, where $f_{\max}$ denotes the frequency of the highest peak in the Fourier power spectral density of the noisy system with $D_u = 0.1$, $K = 0.1$. [52]

The role of the appropriate choice of the control delay τ becomes even clearer if we keep K fixed and calculate the correlation time in dependence of τ. The result is plotted in Fig. 5.25(a) where one can clearly see the oscillatory character of the correlation time under variation of τ, which is characterized by the presence of "optimal" values of τ, corresponding to maximum regularity, and "worst" values of τ which are related to minimum regularity of the noise-induced dynamics. At the same time it is shown that the control with $K = 0.1$ produces no effect at all upon the correlation time if the noise is too large (lower curve for $D_u = 1.0$).

The fact that noise-induced oscillations take place in the vicinity of the spatially inhomogeneous fixed point gives us a hint that some properties of these oscillations could relate to the stability of the above mentioned fixed point. To gain some insight into how the control actually affects the systems dynamics around the spatially inhomogeneous fixed point we linearize the system equations (5.30) for $D_u = D_a = 0$ and calculate the complex eigenvalues Λ_i at the fixed point. First of all we calculate these eigenvalues from the spatially discretized system which we use for the numerical simulation. This discretized version is just a set of ordinary differential equations (ODEs) and the linearization and the eigenvalues can be computed easily.

In Fig. 5.25(b) one can see that the control with $K = 0.1$ does not change the stability of the inhomogenous fixed point since the real parts of all eigenvalues do not become positive within the given range of τ. Nevertheless with increasing τ the real parts of the eigenvalues intersect at particular values of τ (vertical dotted lines) and therefore the *leading* eigenvalue, i. e. the *least stable* one, or the one with the largest real part, changes at these values of τ. As one can see, these crossover points correspond to the minima of the correlation time in Fig. 5.25(a) whereas the local maxima of the real parts correspond to the maxima of the correlation time. This gives rise to a rather intuitive explanation for the behavior of the correlation time: The closer to zero the real part of an eigenvalue is, the weaker is the attracting stability of the fixed point and the easier it is for the noise to excite exactly the oscillating mode corresponding to this particular eigenvalue. On the other hand, at the intersection points of the real parts of the leading eigenvalue these values have the largest distance from zero, meaning that the attracting stability of the fixed point is stronger and in addition there are two different corresponding oscillating modes which are excited by the noise. Thus the control cannot reach its optimal effect.

As a direct consequence, the main frequency which is activated by the noise switches exactly at these values of τ to the eigenfrequency of the corresponding leading eigenvalue. In Fig. 5.25(c) the eigenperiods are plotted as black dots in dependence of τ. The circles mark the positions of the highest peak in the Fourier power spectrum for the corresponding noisy system with $D_u = 0.1$. One can see clearly that these main periods switch from one branch to another exactly at the positions where the real parts of two different eigenvalues cross over.

As we have already noted, the eigenvalues for the linearized deterministic system at the inhomogeneous fixed point, plotted as black dots in Figs. 5.25b,c are computed numerically for the system (5.30) in the deterministic case by using the spatially discretized set of ordinary differential equations.

To achieve a deeper understanding of the stability properties of the inhomogeneous fixed point under the influence of the control force and to obtain the general form of the characteristic equation which determines the eigenvalues of this linearized system, we perform the linearization of the original continuous system (5.30) at the spatially inhomogeneous fixed point $(a_0(x), u_0)$. Introducing

$$a_x \equiv \frac{\partial a}{\partial x}, \qquad a_{xx} \equiv \frac{\partial^2 a}{\partial x^2}, \qquad b\left(a, a_x, a_{xx}\right) \equiv \frac{\partial}{\partial x}\left(D(a)a_x\right), \tag{5.31}$$

and a linear operator

$$\mathcal{L} \equiv \left.\frac{\partial f}{\partial a}\right|_{a_0,u_0} + \left.\frac{\partial b}{\partial a}\right|_{a_0} + \left.\frac{\partial b}{\partial a_x}\right|_{a_0}\frac{\partial}{\partial x} + \left.\frac{\partial b}{\partial a_{xx}}\right|_{a_0}\frac{\partial^2}{\partial x^2}, \tag{5.32}$$

and using the ansatz $\delta a(x,t) = \mathrm{e}^{\Lambda t}\,\tilde{a}(x)$, $\delta u(t) = \mathrm{e}^{\Lambda t}\,\tilde{u}$ for the deviations from the fixed point we can write down the coupled eigenvalue problem:

$$\Lambda\tilde{a}(x) = \mathcal{L}\tilde{a}(x) + f_u(x)\tilde{u}, \tag{5.33}$$

$$\Lambda\tilde{u} = -\frac{r}{\varepsilon L}\int_0^L j_a(x)\tilde{a}(x)dx + \left[-\frac{1+rJ_u}{\varepsilon} + K\left(\mathrm{e}^{-\Lambda\tau} - 1\right)\right]\tilde{u}, \tag{5.34}$$

with $f_u \equiv \left.\frac{\partial f}{\partial u}\right|_{a_0,u_0}$, $\quad j_a \equiv \left.\frac{\partial j}{\partial a}\right|_{a_0,u_0}$, $\quad J_u = \frac{1}{L}\int_0^L \left.\frac{\partial j}{\partial u}\right|_{a_0,u_0} dx.$

For the case $K = 0$ this eigenvalue problem of the inhomogeneous filamentary fixed point has been analyzed generally [86]. In the voltage-clamped case ($\delta u = 0$), the Sturmian eigenvalue equation $\lambda\tilde{a} = \mathcal{L}\tilde{a}$ with Neumann boundary conditions (which can be shown to be self-adjoint) has solutions $\lambda_0 > \lambda_1 > \lambda_2 > \ldots$ where the corresponding eigenmode $\psi_n(x)$ has n nodes, and $\lambda_0 > 0$, while all other eigenvalues $\lambda_n < 0$ for $n \geq 1$ are stable. The eigenmodes of the full eqs. (5.33), (5.34) can be expanded in terms of the voltage-clamped eigenmodes,

$$\tilde{a}(x) = \sum_n (\tilde{a}, \psi_n)\, \psi_n(x), \tag{5.35}$$

where $(\tilde{a}, \psi_n) \equiv \frac{1}{L}\int_0^L \tilde{a}(x)\psi_n(x)dx$ denotes the usual scalar product in Hilbert space. Inserting this into eq. (5.33) yields

$$\Lambda \sum_n (\tilde{a}, \psi_n)\psi_n(x) = \sum_n \lambda_n (\tilde{a}, \psi_n)\psi_n(x) + f_u(x)\tilde{u}. \tag{5.36}$$

Forming the scalar product with ψ_m and using orthonormality gives the expansion coefficients

$$(\tilde{a}, \psi_m) = \frac{(f_u, \psi_m)}{\Lambda - \lambda_m}\tilde{u}. \tag{5.37}$$

The expansion (5.35) can be inserted into eq. (5.34):

$$\Lambda\tilde{u} = \left[-\frac{r}{\varepsilon}\sum_n \frac{(f_u, \psi_n)(j_a, \psi_n)}{\Lambda - \lambda_m} - \frac{1 + rJ_u}{\varepsilon} + K\left(\mathrm{e}^{-\Lambda\tau} - 1\right)\right]\tilde{u}. \tag{5.38}$$

We will now neglect the higher modes ψ_n because they oscillate fast whereas $a_0(x)$ varies slowly in space, and approximate the sum in eq. (5.35) by the dominant first term ψ_0 with $\lambda_0 > 0$. We obtain the characteristic equation for the eigenvalue Λ:

$$\Lambda^2 + \left(\frac{1 + rJ_u}{\varepsilon} - \lambda_0\right)\Lambda + (\lambda_0 - \Lambda)K\left(\mathrm{e}^{-\Lambda\tau} - 1\right) - \frac{\lambda_0}{\varepsilon}(1 + r\sigma_d) = 0, \tag{5.39}$$

where the static differential conductance at the inhomogeneous fixed point

$$\sigma_d \equiv \left.\frac{dJ}{du}\right|_{a_0, u_0} = J_u + \left(j_a, \frac{da}{du}\right) = J_u - (j_a, \psi_0)\frac{(f_u, \psi_0)}{\lambda_0} \tag{5.40}$$

has been introduced using eqs. (5.35), (5.37) in the static case $\Lambda = 0$ [15]. Without control, $K = 0$, eq. (5.39) reduces to a characteristic polynomial of second order, which gives the well-known conditions for stability of a filament [86]:

$$\begin{aligned} A &\equiv \frac{1 + rJ_u}{\varepsilon} - \lambda_0 > 0, \\ C &\equiv -\frac{\lambda_0}{\varepsilon}(1 + r\sigma_d) > 0. \end{aligned} \tag{5.41}$$

Without control a Hopf bifurcation on the two-dimensional center manifold occurs if $A = 0$. With control, eq. (5.39) can be expressed as

$$\Lambda^2 + A\Lambda + (B - \Lambda)K\left(\mathrm{e}^{-\Lambda\tau} - 1\right) + C = 0 \tag{5.42}$$

with $B \equiv \lambda_0 > 0$. The parameters A, B, C can be calculated directly from (5.41). For the inhomogeneous fixed point, $\lambda_0 = 1.0281$ has been calculated in [89]; $J_u = -0.1615$ can be obtained by using the condition for a Hopf bifurcation ($A = 0$) in eq. (5.41); $\sigma_d = 0.226$ can be estimated from the current-voltage characteristic shown in Fig. 5.19.

This yields $A = 0.0447$, $B = 1.0281$ and $C = 1.1458$. Note that in dimensional units the unstable eigenvalue of the voltage-clamped system $\lambda_0 = B$ is approximately equal to the inverse tunneling time $1/\tau_a$. With these values we can solve eq. (5.42). For $K = 0.1$ the real parts of this solution in dependence of τ are also shown in Fig. 5.25(b) as squares. They coincide with very good accuracy.

5.6. Conclusions

We have investigated the complex spatio-temporal behavior of two semiconductor nanostructures, viz. the superlattice and the double barrier resonant tunneling diode (DBRT). The first exhibits nonlinear dynamics of interacting fronts, while the second demonstrates breathing and spiking of filamentary current density patterns characteristic of globally coupled reaction–diffusion systems. Applying time-delayed feedback control of Pyragas type to both deterministic and stochastic oscillations, we have been able to suppress deterministic chaos and control the regularity and the mean period of noise-induced dynamics.

As an example for the constructive influence of noise in nonlinear systems, we have shown that random fluctuations are able to induce quite coherent oscillations of the current density in a regime where the deterministic system exhibits a stable fixed point, thereby demonstrating the phenomenon of coherence resonance for systems close to, but below, a Hopf bifurcation (superlattice and DBRT) as well as close to, but below, a global saddle-node bifurcation on a limit cycle (superlattice). This extends the phenomena of noise-induced oscillations from purely time-dependent generic models, e. g. Ref. 7, to space-time patterns. Moreover, we have shown for the DBRT that the noise which is applied globally to a space-independent variable determines the type of the spatio-temporal pattern of these oscillations. While for small noise intensity the system demonstrates oscillations which are quite correlated in time, but spatially inhomogeneous, with increasing noise intensity the shape of the spatiotemporal pattern changes qualitatively until the system reaches a highly homogeneous state. Thus the increase of spatial coherence is accompanied by the decrease of temporal correlation of the observed oscillations. In between these two situations for intermediate noise strength one can observe complex spatio-temporal behavior resulting from the competition between homogeneous and inhomogeneous oscillations.

We have seen that delayed feedback can be an efficient method for manipulation of essential characteristics of chaotic or noise-induced spatiotemporal dynamics in a spatially discrete front system and in a continuous reaction-diffusion system. By variation of the time delay one can stabilize particular unstable periodic orbits associated with space-time patterns, or deliberately change the timescale of oscillatory patterns, and thus adjust and stabilize the frequency of the electronic device. Moreover, with a proper choice of feedback parameters one can also effectively control the coherence of spatio-temporal dynamics, e. g. enhance or destroy it. Increase of coherence is possible up to a reasonably large intensity of noise. However, as the level of noise grows, the efficiency of the control upon the temporal coherence decreases.

The effects of the delayed feedback can be explained in terms of a Floquet mode analysis of the periodic orbits, or a linear stability analysis of the fixed point. For a better understanding of noise-induced patterns in the DBRT, we have derived the general form of the characteristic equation for the deterministic system (5.30) close to, but below, a Hopf bifurcation. Both dependences, coherence and timescale vs. τ, demonstrate an oscillatory character, which can be explained by oscillations of the real and imaginary parts of the eigenvalues of the linearized system at the fixed point, in the vicinity of which the noise-induced oscillations occur. The most coherent timescale corresponds to values of τ, for which the real parts of the eigenvalues attain a maximum. In some sense, the noise excites the least stable eigenmode: the less stable an eigenmode is, the greater is the coherence of the corresponding oscillations.

While these investigations have enlightened our basic understanding of nonlinear, spatially extended systems under the influence of time-delayed feedback and noise, they may also open up relevant applications as nano-electronic devices like oscillators and sensors.

Acknowledgment

This work was supported by DFG in the framework of Sfb 555. We are indebted to stimulating collaboration and discussion with A. Amann, N.

Baba, A. Balanov, O. Beck, H. Engel, B. Fiedler, B. Hauschildt, N. Janson, W. Just, K. Lüdge, J. Pomplun, S. Popovich, P. Rodin, L. Schimansky-Geier, J. Schlesner, J. E. S. Socolar, J. Unkelbach, A. Wacker, M. Wolfrum, H. J. Wünsche, S. Yanchuk.

References

[1] H. G. Schuster (Editor): *Handbook of Chaos Control* (Wiley-VCH, Weinheim, 1999).

[2] H. Nijmeijer and A. V. D. Schaft: *Nonlinear Dynamical Control Systems* (Springer, New York, 1996), 3rd ed.

[3] K. Ogata: *Modern Control Engineering* (Prentice-Hall, New York, 1997).

[4] E. Ott, C. Grebogi, and J. A. Yorke: *Controlling chaos*, Phys. Rev. Lett. **64**, 1196 (1990).

[5] K. Pyragas: *Continuous control of chaos by self-controlling feedback*, Phys. Lett. A **170**, 421 (1992).

[6] D. J. Gauthier: *Resource letter: Controlling chaos*, Am. J. Phys. **71**, 750 (2003).

[7] N. B. Janson, A. G. Balanov, and E. Schöll: *Delayed feedback as a means of control of noise-induced motion*, Phys. Rev. Lett. **93**, 010601 (2004).

[8] A. G. Balanov, N. B. Janson, and E. Schöll: *Control of noise-induced oscillations by delayed feedback*, Physica D **199**, 1 (2004).

[9] J. Pomplun, A. Amann, and E. Schöll: *Mean field approximation of time-delayed feedback control of noise-induced oscillations in the Van der Pol system*, Europhys. Lett. **71**, 366 (2005).

[10] A. Pikovsky and J. Kurths: *Coherence resonance in a noise-driven excitable system*, Phys. Rev. Lett. **78**, 775 (1997).

[11] J. García-Ojalvo and J. M. Sancho: *Noise in Spatially Extended Systems* (Springer, New York, 1999).

[12] B. Lindner, J. García-Ojalvo, A. Neiman, and L.Schimansky-Geier: *Effects of noise in excitable systems*, Phys. Rep. **392**, 321 (2004).

[13] E. Schöll: *Nonequilibrium Phase Transitions in Semiconductors*spat (Springer, Berlin, 1987).

[14] M. P. Shaw, V. V. Mitin, E. Schöll, and H. L. Grubin: *The Physics of Instabilities in Solid State Electron Devices* (Plenum Press, New York, 1992).

[15] E. Schöll: *Nonlinear spatio-temporal dynamics and chaos in semiconductors* (Cambridge University Press, Cambridge, 2001).

[16] A. L. Hodgkin and A. F. Huxley: *A quantitative description of membrane current and its application to conduction and excitation in nerve*, J. Physiol. **117**, 500 (1952).

[17] S. Bielawski, D. Derozier, and P. Glorieux: *Controlling unstable periodic orbits by a delayed continuous feedback*, Phys. Rev. E **49**, R971 (1994).

[18] T. Pierre, G. Bonhomme, and A. Atipo: *Controlling the chaotic regime of nonlinear ionization waves using time-delay autsynchronisation method*, Phys. Rev. Lett. **76**, 2290 (1996).

[19] K. Hall, D. J. Christini, M. Tremblay, J. J. Collins, L. Glass, and J. Billette: *Dynamic control of cardiac alterans*, Phys. Rev. Lett. **78**, 4518 (1997).
[20] D. W. Sukow, M. E. Bleich, D. J. Gauthier, and J. E. S. Socolar: *Controlling chaos in a fast diode resonator using time-delay autosynchronisation: Experimental observations and theoretical analysis*, Chaos **7**, 560 (1997).
[21] O. Lüthje, S. Wolff, and G. Pfister: *Control of chaotic taylor-couette flow with time-delayed feedback*, Phys. Rev. Lett. **86**, 1745 (2001).
[22] P. Parmananda, R. Madrigal, M. Rivera, L. Nyikos, I. Z. Kiss, and V. Gáspár: *Stabilization of unstable steady states and periodic orbits in an electrochemical system using delayed-feedback control*, Phys. Rev. E **59**, 5266 (1999).
[23] H. Benner and W. Just: *Control of chaos by time delayed feedback in high power ferromagnetic resonance experiments*, J. Kor. Phys. Soc. **40**, 1046 (2002).
[24] C. Beta, M. Bertram, A. S. Mikhailov, H. H. Rotermund, and G. Ertl: *Controlling turbulence in a surface chemical reaction by time-delay autosynchronization*, Phys. Rev. E **67**, 046224 (2003).
[25] C. von Loewenich, H. Benner, and W. Just: *Experimental relevance of global properties of time-delayed feedback control*, Phys. Rev. Lett. **93**, 174101 (2004).
[26] J. Schlesner, V. Zykov, H. Engel, and E. Schöll: *Stabilization of unstable rigid rotation of spiral waves in excitable media*, Phys. Rev. E **74**, 046215 (2006).
[27] S. Schikora, P. Hövel, H.-J. Wünsche, E. Schöll, and F. Henneberger: *All-optical noninvasive control of unstable steady states in a semiconductor laser*, Phys. Rev. Lett. (2006), in print.
[28] J. E. S. Socolar, D. W. Sukow, and D. J. Gauthier: *Stabilizing unstable periodic orbits in fast dynamical systems*, Phys. Rev. E **50**, 3245 (1994).
[29] M. E. Bleich and J. E. S. Socolar: *Stability of periodic orbits controlled by time-delay feedback*, Phys. Lett. A **210**, 87 (1996).
[30] W. Just, T. Bernard, M. Ostheimer, E. Reibold, and H. Benner: *Mechanism of time-delayed feedback control*, Phys. Rev. Lett. **78**, 203 (1997).
[31] H. Nakajima: *On analytical properties of delayed feedback control of chaos*, Phys. Lett. A **232**, 207 (1997).
[32] K. Pyragas: *Analytical properties and optimization of time-delayed feedback control*, Phys. Rev. E **66**, 26207 (2002).
[33] W. Just, H. Benner, and E. Schöll: *Control of chaos by time–delayed feedback: a survey of theoretical and experimental aspects*, in *Advances in Solid State Phyics*, edited by B. Kramer (Springer, Berlin, 2003), vol. 43, pp. 589–603.
[34] P. Hövel and J. E. S. Socolar: *Stability domains for time-delay feedback control with latency*, Phys. Rev. E **68**, 036206 (2003).
[35] P. Hövel and E. Schöll: *Control of unstable steady states by time-delayed feedback methods*, Phys. Rev. E **72**, 046203 (2005).
[36] S. Yanchuk, M. Wolfrum, P. Hövel, and E. Schöll: *Control of unstable steady states by long delay feedback*, Phys. Rev. E **74**, 026201 (2006).

[37] A. Amann, E. Schöll, and W. Just: *Some basic remarks on eigenmode expansions of time-delay dynamics*, Physica A **373**, 191 (2007).

[38] B. Fiedler, V. Flunkert, M. Georgi, P. Hövel, and E. Schöll: *Refuting the odd number limitation of time-delayed feedback control*, http://arxiv.org/abs/nlin.CD/0609056 (2006).

[39] A. G. Balanov, N. B. Janson, and E. Schöll: *Delayed feedback control of chaos: Bifurcation analysis*, Phys. Rev. E **71**, 016222 (2005).

[40] E. Schöll and K. Pyragas: *Tunable semiconductor oscillator based on self-control of chaos in the dynamic Hall effect*, Europhys. Lett. **24**, 159 (1993).

[41] D. Reznik and E. Schöll: *Oscillation modes, transient chaos and its control in a modulation-doped semiconductor double-deterostructure*, Z. Phys. B **91**, 309 (1993).

[42] D. P. Cooper and E. Schöll: *Tunable real space transfer oscillator by delayed feedback control of chaos*, Z. f. Naturforsch. **50a**, 117 (1995).

[43] G. Franceschini, S. Bose, and E. Schöll: *Control of chaotic spatiotemporal spiking by time-delay autosynchronisation*, Phys. Rev. E **60**, 5426 (1999).

[44] O. Beck, A. Amann, E. Schöll, J. E. S. Socolar, and W. Just: *Comparison of time-delayed feedback schemes for spatio-temporal control of chaos in a reaction-diffusion system with global coupling*, Phys. Rev. E **66**, 016213 (2002).

[45] N. Baba, A. Amann, E. Schöll, and W. Just: *Giant improvement of time-delayed feedback control by spatio-temporal filtering*, Phys. Rev. Lett. **89**, 074101 (2002).

[46] W. Just, S. Popovich, A. Amann, N. Baba, and E. Schöll: *Improvement of time–delayed feedback control by periodic modulation: analytical theory of Floquet mode control scheme*, Phys. Rev. E **67**, 026222 (2003).

[47] J. Unkelbach, A. Amann, W. Just, and E. Schöll: *Time–delay autosynchronization of the spatio-temporal dynamics in resonant tunneling diodes*, Phys. Rev. E **68**, 026204 (2003).

[48] J. Schlesner, A. Amann, N. B. Janson, W. Just, and E. Schöll: *Self-stabilization of high frequency oscillations in semiconductor superlattices by time–delay autosynchronization*, Phys. Rev. E **68**, 066208 (2003).

[49] E. Schöll, A. Balanov, N. B. Janson, and A. Neiman: *Controlling stochastic oscillations close to a Hopf bifurcation by time-delayed feedback*, Stoch. Dyn. **5**, 281 (2005).

[50] B. Hauschildt, A. G. Balanov, N. B. Janson, and E. Schöll: *Control of noise-induced cooperative dynamics in coupled neural models*, Phys. Rev. E **74**, 051906 (2006).

[51] J. Hizanidis, A. G. Balanov, A. Amann, and E. Schöll: *Noise-induced oscillations and their control in semiconductor superlattices*, Int. J. Bifur. Chaos **16**, 1701 (2006).

[52] G. Stegemann, A. G. Balanov, and E. Schöll: *Delayed feedback control of stochastic spatiotemporal dynamics in a resonant tunneling diode*, Phys. Rev. E **73**, 16203 (2006).

[53] A. G. Balanov, V. Beato, N. B. Janson, H. Engel, and E. Schöll: *Delayed feedback control of noise-induced patterns in excitable media*, Phys. Rev. E

74, 016214 (2006).
[54] A. Amann, J. Schlesner, A. Wacker, and E. Schöll: *Chaotic front dynamics in semiconductor superlattices*, Phys. Rev. B **65**, 193313 (2002).
[55] A. Amann, K. Peters, U. Parlitz, A. Wacker, and E. Schöll: *A hybrid model for chaotic front dynamics: From semiconductors to water tanks*, Phys. Rev. Lett. **91**, 066601 (2003).
[56] A. Amann and E. Schöll: *Bifurcations in a system of interacting fronts*, J. Stat. Phys. **119**, 1069 (2005).
[57] J. Hizanidis, A. G. Balanov, A. Amann, and E. Schöll: *Noise-induced front motion: signature of a global bifurcation*, Phys. Rev. Lett. **96**, 244104 (2006).
[58] E. Schöll, A. Amann, M. Rudolf, and J. Unkelbach: *Transverse spatio-temporal instabilities in the double barrier resonant tunneling diode*, Physica B **314**, 113 (2002).
[59] G. Stegemann, A. G. Balanov, and E. Schöll: *Noise-induced pattern formation in a semiconductor nanostructure*, Phys. Rev. E **71**, 16221 (2005).
[60] L. Esaki and R. Tsu: *Superlattice and negative differential conductivity in semiconductors*, IBM J. Res. Develop. **14**, 61 (1970).
[61] J. Kastrup, R. Klann, H. T. Grahn, K. Ploog, L. L. Bonilla, J. Galán, M. Kindelan, M. Moscoso, and R. Merlin: *Self-oscillations of domains in doped GaAs-AlAs superlattices*, Phys. Rev. B **52**, 13761 (1995).
[62] K. Hofbeck, J. Grenzer, E. Schomburg, A. A. Ignatov, K. F. Renk, D. G. Pavel'ev, Y. Koschurinov, B. Melzer, S. Ivanov, S. Schaposchnikov, and P. S. Kop'ev: *High-frequency self-sustained current oscillation in an Esaki-Tsu superlattice monitored via microwave emission*, Phys. Lett. A **218**, 349 (1996).
[63] E. Schomburg, R. Scheuerer, S. Brandl, K. F. Renk, D. G. Pavel'ev, Y. Koschurinov, V. Ustinov, A. Zhukov, A. Kovsh, and P. S. Kop'ev: *InGaAs/InAlAs superlattice oscillator at 147 GHz*, Electronics Letters **35**, 1491 (1999).
[64] A. Wacker: *Semiconductor superlattices: A model system for nonlinear transport*, Phys. Rep. **357**, 1 (2002).
[65] L. L. Bonilla: *Theory of nonlinear charge transport, wave propagation, and self-oscillations in semiconductor superlattices*, J. Phys.: Condens. Matter **14**, R341 (2002).
[66] L. L. Bonilla and H. T. Grahn: *Non-linear dynamics of semiconductor superlattices*, Rep. Prog. Phys. **68**, 577 (2005).
[67] Y. Zhang, J. Kastrup, R. Klann, K. H. Ploog, and H. T. Grahn: *Synchronization and chaos induced by resonant tunneling in GaAs/AlAs superlattices*, Phys. Rev. Lett. **77**, 3001 (1996).
[68] K. J. Luo, H. T. Grahn, K. H. Ploog, and L. L. Bonilla: *Explosive bifurcation to chaos in weakly coupled semiconductor superlattices*, Phys. Rev. Lett. **81**, 1290 (1998).
[69] O. M. Bulashenko, K. J. Luo, H. T. Grahn, K. H. Ploog, and L. L. Bonilla: *Multifractal dimension of chaotic attractors in a driven semiconductor superlattice*, Phys. Rev. B **60**, 5694 (1999).
[70] O. M. Bulashenko and L. L. Bonilla: *Chaos in resonant-tunneling superlat-*

tices, Phys. Rev. B **52**, 7849 (1995).

[71] A. Amann, J. Schlesner, A. Wacker, and E. Schöll: *Self-generated chaotic dynamics of field domains in superlattices*, in *Proc. 26th International Conference on the Physics of Semiconductors (ICPS-26), Edinburgh 2002*, edited by J. H. Davies and A. R. Long (2003).

[72] A. Amann, A. Wacker, L. L. Bonilla, and E. Schöll: *Dynamic scenarios of multi-stable switching in semiconductor superlattices*, Phys. Rev. E **63**, 066207 (2001).

[73] J. Schlesner, A. Amann, N. B. Janson, W. Just, and E. Schöll: *Self-stabilization of chaotic domain oscillations in superlattices by time–delayed feedback control*, Semicond. Sci. Technol. **19**, S34 (2004).

[74] W. Just, E. Reibold, K. Kacperski, P. Fronczak, J. A. Holyst, and H. Benner: *Influence of stable floquet exponents on time-delayed feedback control*, Phys. Rev. E **61**, 5045 (2000).

[75] L. Gammaitoni, P. Hänggi, P. Jung, and F. Marchesoni: *Stochastic resonance*, Rev. Mod. Phys. **70**, 223 (1998).

[76] V. S. Anishchenko, A. B. Neiman, F. Moss, and L. Schimansky-Geier: *Stochastic resonance: noise-enhanced order*, Phys. Usp. **42**, 7 (1999).

[77] Hu Gang, T. Ditzinger, C. Ning, and H. Haken: *Stochastic resonance without external periodic force*, Phys. Rev. Lett. **71**, 807 (1993).

[78] O. V. Ushakov, H.-J. Wünsche, F. Henneberger, I. A. Khovanov, L. Schimansky-Geier, and M. A. Zaks: *Coherence resonance near a Hopf bifurcation*, Phys. Rev. Lett. **95**, 123903 (2005).

[79] Y. M. Blanter and M. Büttiker: *Shot noise in mesoscopic conductors*, Phys. Rep. **336**, 1 (2000).

[80] A. Andronov, E. A. Leontovich, I. Gordon, and A. G. Maier: *Vol.2: Theory of bifurcations of systems on a plane* (Israel program for scientific translation, Jerusalem, 1971).

[81] J. Guckenheimer and P. Holmes: *Nonlinear Oscillations, Dynamical Systems, and Bifurcations of Vector Fields* (Springer, Berlin, 1986).

[82] A. Amann, A. Wacker, and E. Schöll: *Tripole current oscillations in superlattices*, Physica B **314**, 404 (2002).

[83] R. L. Stratonovich: *Topics in the Theory of Random Noise*, vol. 1 (Gordon and Breach, New York, 1963).

[84] M. Meixner, P. Rodin, E. Schöll, and A. Wacker: *Lateral current density fronts in globally coupled bistable semiconductors with S- or Z-shaped current voltage characteristic*, Eur. Phys. J. B **13**, 157 (2000).

[85] V. Cheianov, P. Rodin, and E. Schöll: *Transverse coupling in bistable resonant-tunneling structures*, Phys. Rev. B **62**, 9966 (2000).

[86] A. Alekseev, S. Bose, P. Rodin, and E. Schöll: *Stability of current filaments in a bistable semiconductor system with global coupling*, Phys. Rev. E **57**, 2640 (1998).

[87] W. Just, E. Reibold, H. Benner, K. Kacperski, P. Fronczak, and J. Holyst: *Limits of time-delayed feedback control*, Phys. Lett. A **254**, 158 (1999).

[88] V. V. Sherstnev, A. Krier, A. G. Balanov, N. B. Janson, A. N. Silchenko, and P. V. E. McClintock: *The stochastic laser: mid-infrared lasing induced*

by noise, Fluct. Noise Lett. **1**, 91 (2003).

[89] J. Unkelbach: *Rückkopplungsgesteuerte transversale Dynamik der resonanten Tunneldiode*, Master's thesis, TU Berlin (2002).

Chapter 6

Dynamics of coupled semiconductor lasers

Lutz Recke[1,*], Matthias Wolfrum[2,+] and Serhiy Yanchuk[1,!]

[1] *Humboldt-University of Berlin, Institute of Mathematics, Rudower Chaussee 25, 12489 Berlin, Germany*

[2] *Weierstraß Institute for Applied Analysis and Stochastics, Mohrenstraße 39, 10117 Berlin, Germany*

[*] *recke@math.hu-berlin.de*, [+] *wolfrum@wias-berlin.de*, [!] *yanchuk@wias-berlin.de*

Contents

6.1. Coupled rate equations model

Semiconductor lasers are key components in modern telecommunication systems. Coupling in lasers may occur due to different reasons, for example via signal transmission [2] or via the interaction of different sections in a multisection laser [10].

On the other hand, semiconductor lasers exhibit rich nonlinear behavior and a variety of collective synchronization phenomena. Among these phenomena, one can mention complete [11, 18, 27, 28] and phase synchronization [13, 26], retarded and anticipated synchronization [9, 19, 23], inverse synchronization [20], etc. The corresponding evidences exist for both: experimental systems and mathematical models.

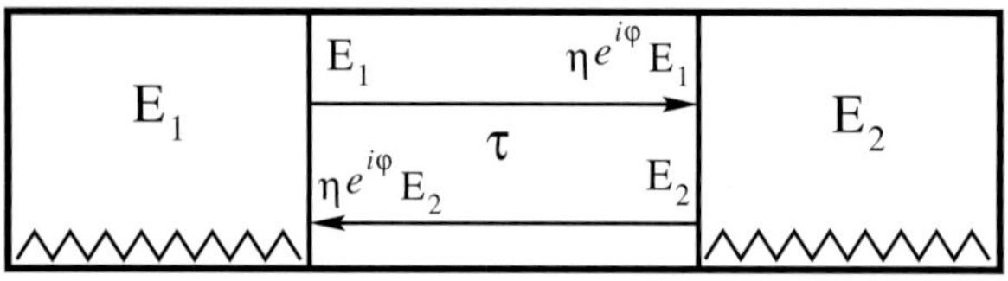

Fig. 6.1. Schematic view of two optically coupled lasers

The following system is widely used (cf. Refs. [6, 7, 12]) for describing the dynamics of two face to face coupled semiconductor lasers:

$$\begin{aligned} E_j' &= i\delta_j E_j + (1+i\alpha)N_j E_j + \eta e^{-i\varphi} E_{3-j}(t-\tau), \\ N_j' &= \varepsilon[J - N_j - (2N_j+1)|E_j|^2], \end{aligned} \tag{6.1}$$

where $j = 1, 2$, $E_{1,2}$ and $N_{1,2}$ denote the complex optical fields and the carrier densities of the lasers, respectively. The term $i\delta_j E_j$ accounts for the frequency detuning. Without loss of generality, we assume $\delta_1 = 0$. By J we denote the pumping current, and $\varepsilon = \tau_p/\tau_c$ is a small parameter given by the ratio of the photon to carrier lifetimes. The parameters η, φ and τ characterize injection rate, injection phase, and injection delay-time, respectively. System (6.1) is already presented in a dimensionless form. In particular, the delay τ is rescaled by the photon lifetimes.

Recently, there has been appeared new interest in lasers with a short cavity [5], motivated by several reasons: First, the study of the dynamics in this regime has become experimentally accessible. Also, this regime is very interesting from the dynamical point of view, since it has an intermediate complexity, allowing to analyze directly the mechanisms of either synchronization or the appearance of pulsations and chaotic dynamics. The same is true when the delay in the coupling is small, i.e. there is a *short external cavity*. For instance, this is the case in a two-section monolitically integrated device [10, 22]. The instantaneous coupling limit may serve as an appropriate starting point for the study of such systems. When neglecting the delay in the amplitude terms, it turns out to be important to take into account the propagation phase φ caused by the delay [29]. We now give an overview of the dynamical regimes arising in a model of instantaneously coupled rate equations

$$\begin{aligned} E_j' &= i\delta_j E_j + (1+i\alpha)N_j E_j + \eta e^{-i\varphi} E_{3-j}, \\ N_j' &= \varepsilon[J - N_j - (2N_j+1)|E_j|^2], \end{aligned} \tag{6.2}$$

where $\delta_1 = 0$, $\delta_2 := \delta$. In Sec. 6.5 we compare some of the obtained results with the model (6.1) including small delay. Sec. 6.6 considers the synchronization transition in an array of three coupled lasers. Finally, we discuss n coupled systems in Sec. 6.7.

6.2. Structural properties of the model

1. The optical phase-shift invariance $(E_1, N_1, E_2, N_2) \to (E_1 e^{i\psi}, N_1, E_2 e^{i\psi}, N_2)$ is frequently met in many optical devices, and, in particular, in system (6.1). This symmetry implies, that for suitable laser parameters there exist continuous wave (CW) solutions, i.e. solutions of the type $E_j(t) = E_{0j} e^{i\omega t}, N_j(t) = N_{0j}$ $(j = 1, 2; \omega \in \mathbb{R})$. These solutions correspond to laser regimes with stationary intensity $|E_{01}|^2 + |E_{02}|^2$ and fixed optical frequency ω. Moreover, this invariance implies that for suitable laser parameters there exist modulated wave (MW) solutions, i.e. solutions of the type $E_j(t) = E_{0j}(t) e^{i\omega t}, N_j(t) = N_{0j}(t)$ with T-periodic functions E_{0j} and N_{0j}. These solutions correspond to laser regimes with periodic intensity. Remark, that generically CW and MW solutions survive under small perturbations of the laser parameters. In particular, the optical frequency ω and the power period T depend smoothly on the laser parameters and are not locked (see, e.g. Refs. [14, 15]).
2. If $\delta_1 = \delta_2$, then (6.1) is invariant with respect to the reflections $(E_1, N_1, E_2, N_2) \to (E_2, N_2, E_1, N_1)$ and $(E_1, N_1, E_2, N_2) \to (-E_2, N_2, -E_1, N_1)$. This implies that the subspaces of synchronous states $E_1 = E_2$, $N_1 = N_2$ and of antisynchronous states $E_1 = -E_2$, $N_1 = N_2$ are flow invariant. In particular, for suitable laser parameters there exist synchronous and

antisynchronous CW solutions (see Sec. 6.3), which can bifurcate into synchronous and antisynchronous MW solutions (Hopf bifurcations) or into asynchronous CW solutions (spontaneous symmetry breaking in a pitchfork bifurcation). Moreover, for $\delta_1 \approx \delta_2$ they survive as "nearly synchronous" and "nearly antisynchronous" CW and MW solutions (see Sec. 6.4).

3. The following symmetry allows us to establish a one-to-one correspondence between synchronous and antisynchronous solutions in the case $\delta_1 = \delta_2$. If $(E_1(t), N_1(t), E_2(t), N_2(t))$ is a solution to (6.1) then $(E_1(t), N_1(t), -E_2(t), N_2(t))$ is also a solution to (6.1) with φ replaced by $\varphi + \pi$.

4. Suppose $\tau = 0$ and introduce polar coordinates $E_j = a_j e^{i\psi_j}$, $a_j, \psi_j \in \mathbb{R}$. Then, because of the phase shift invariance, system (6.1) in the new coordinates can be decomposed into the five dimensional system

$$\begin{aligned}
a_1' &= N_1 a_1 + \eta a_2 \cos(\Delta\psi + \varphi) \\
N_1' &= \varepsilon(J - N_1 - (2N_1 + 1)a_1^2), \\
a_2' &= N_2 a_2 + \eta a_1 \cos(\varphi - \Delta\psi), \\
N_2' &= \varepsilon(J - N_2 - (2N_2 + 1)a_2^2), \\
\Delta\psi' &= \delta + (N_1 - N_2)\alpha - \eta\frac{a_2}{a_1}\sin(\Delta\psi + \varphi) + \eta\frac{a_1}{a_2}\sin(\varphi - \Delta\psi)
\end{aligned} \tag{6.3}$$

and a one dimensional equation. Here $\Delta\psi = \psi_1 - \psi_2$. CW solutions to (6.1) correspond to stationary solutions of (6.3), and MW solutions to (6.1) correspond to periodic solutions of (6.3) (if $\Delta\psi$ is considered modulo 2π).

5. If $\tau = 0$ and $\varphi = 0$ or $\varphi = \pi$, then the subspace of amplitude synchronized states $a_1 = a_2$, $N_1 = N_2$ is invariant with respect to the flow of (6.3). In particular, for suitable laser parameters there exist stationary and periodic solutions to (6.3) in this subspace, corresponding to so-called amplitude synchronized CW and MW solutions to (6.1) with $|E_1(t)| = |E_2(t)|$. They survive for $\tau \approx 0$ and $\varphi \approx 0$ or $\varphi \approx \pi$ as nearly amplitude synchronized CW and MW solutions (see Sec. 6.4.5).

6. Suppose $\tau = \varepsilon = 0$ and $\varphi = \pi/2$ or $\varphi = 3\pi/2$. Then for any r the submanifolds $a_1^2 + a_2^2 = r$, $N_1 = N_2 = 0$ are invariant with respect to the flow of (6.3). Moreover, if $(a_1(t), a_2(t), \Delta\psi(t))$ is a solution on such an invariant manifold, then $(a_2(-t), a_1(-t), \Delta\psi(-t))$ is also a solution there. It seems that this time reversibility property is the reason for the existence of T-periodic solutions with the "inverse amplitude synchronization property" $a_1(t+T/2) = a_2(t)$ (see Sec. 6.4.6 and Refs. [16, 20]). Such solutions survive for $\tau \approx 0$, $\varepsilon \approx 0$ and $\varphi \approx \pi/2$ or $\varphi \approx 3\pi/2$ and correspond to nearly "inverse amplitude synchronized" MW solutions to (6.1) with $|E_1(t+T/2)| \approx |E_2(t)|$ and $|E_1(t)|^2 + |E_2(t)|^2 \approx \text{const}$.

6.3. CW solutions in the case of no detuning

Let us first examine the model (6.2) without detuning

$$\begin{aligned} E_j' &= (1+i\alpha)N_jE_j + \eta e^{-i\varphi}E_{3-j}, \\ N_j' &= \varepsilon[J - N_j - (2N_j+1)|E_j|^2]. \end{aligned} \tag{6.4}$$

6.3.1. *Dynamics in the synchronization subspace*

After substituting $N_1 = N_2 =: N$ and $E_1 = E_2 =: E$ into (6.4), we obtain the following equations for the dynamics in the synchronization subspace

$$\begin{aligned} E' &= (1+i\alpha)NE + \eta e^{-i\varphi}E, \\ N' &= \varepsilon[J - N - (2N+1)|E|^2]. \end{aligned} \tag{6.5}$$

For $J > 0$, $\varepsilon > 0$, and $0 < \eta < 1/2$ there exists the CW solution $E(t) = E_0 e^{i\omega_0 t}$, $N(t) = N_0$ with $\omega_0 = -\eta(\alpha\cos\varphi + \sin\varphi)$, $N_0 = -\eta\cos\varphi$, $E_0^2 = (J + \eta\cos\varphi)/(1 - 2\eta\cos\varphi)$, which is globally stable in the synchronization subspace. Similar results hold for the subspace of antisynchronous solutions.

6.3.2. *Transverse stability of the synchronous CW solutions*

Since the synchronous CW solution is stable within the synchronization subspace, its stability in the whole phase space is determined by its *transverse* stability, i.e. the stability with respect to perturbations of the initial data transverse to the synchronization subspace. The analysis of the transverse stability of synchronous CW solutions can be carried out by inspecting the transverse characteristic equation [29]

$$\begin{aligned} \chi_T^0(\Lambda) &= [\Lambda^2 + 4\Lambda\eta\cos\varphi + 4\eta^2](\Lambda + \varepsilon(1+S_0)) \\ &+ 2\varepsilon S_0(0.5 - \eta\cos\varphi)[\Lambda + 2\eta(\cos\varphi - \alpha\sin\varphi)] = 0, \end{aligned} \tag{6.6}$$

where $S_0 := (J - \eta\cos\varphi)/(0.5 + \eta\cos\varphi)$. A transverse pitchfork bifurcation takes place if there is a zero eigenvalue, i.e. $\chi_T^0(0) = 0$, and a transverse Hopf bifurcation corresponds to the existence of pure imaginary eigenvalues, i.e. $\chi_T^0(i\Omega) = 0$, where $\Omega \neq 0$ is some real parameter. These bifurcations can be identified and path-followed with respect to the system parameters. Here we choose the coupling strength η and injection phase φ to be the bifurcation parameters. The typical bifurcation diagram is shown in Fig. 6.2(a). Note that in order to obtain the result for antisynchronous solutions, we used the symmetry arguments of Sec. 6.2, i.e. the region 'A' is an image of the region 'S', which is shifted by π along the parameter axis φ. The transverse bifurcations that mediate the loss of synchronization are marked as 'P' for pitchfork and 'H' for Hopf, respectively. Note that we do not show in Fig. 6.2 all the bifurcation lines, but only those which mediate the stability loss of CW solutions.

There are also small regions where stable synchronous and stable anti-synchronous CW solutions coexist. They are located at $\varphi = \arctan(1/\alpha)$ and $\varphi = \arctan(1/\alpha) + \pi$, cf. Fig 6.2(b).

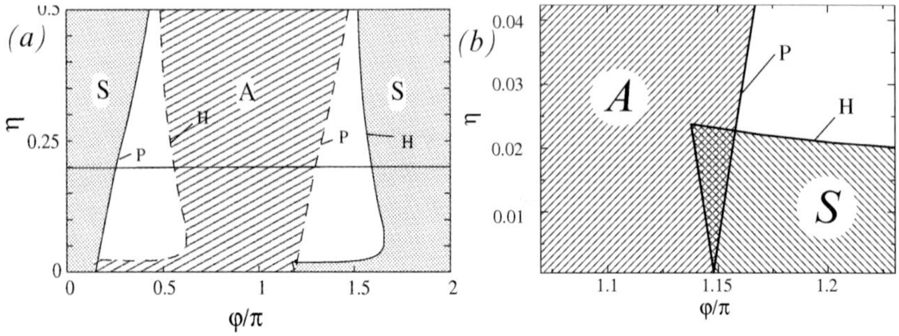

Fig. 6.2. Region of transverse stability for synchronous 'S' and antisynchronous 'A' CW solutions, respectively. 'P' denotes the curves of transverse pitchfork bifurcations and 'H' Hopf bifurcations for the parameters $\alpha = 2$, $J = 1$, $\varepsilon = 0.03$. (b) Zoom of the small part of (a) shows overlapping of the regions.

The bifurcation diagrams from Figs. 6.2 also reveals that the first destabilization threshold, i.e. destabilization of the CW solutions with increasing of the coupling η for fixed φ, may occur already for coupling strength of order ε via Hopf bifurcation.

6.3.3. *Asynchronous CW solutions*

CW solutions to (6.4), which are neither synchronous nor antisynchronous will be called asynchronous. They are of the type $E_1(t) = a_1 e^{i(\omega t + \psi)}$, $N_1(t) = N_1$, $E_2(t) = a_2 e^{i\omega t}$, $N_2(t) = N_2$, where $a_1, a_2, N_1, N_2, \omega, \psi$ are real constants to be determined. After substituting it into (6.4), we obtain a set of nonlinear equations, which afterwards can be effectively studied numerically. As a result we present the one-dimensional bifurcation diagram in Fig. 6.3, which corresponds to the parameters as in Fig. 6.2 with fixed $\eta = 0.2$. In addition to the synchronous and antisynchronous solutions, we observe branches of unstable asynchronous orbits connecting synchronous and antisynchronous CW solutions. These branches emerge from the subcritical pitchfork bifurcations P_s and P_a, respectively.

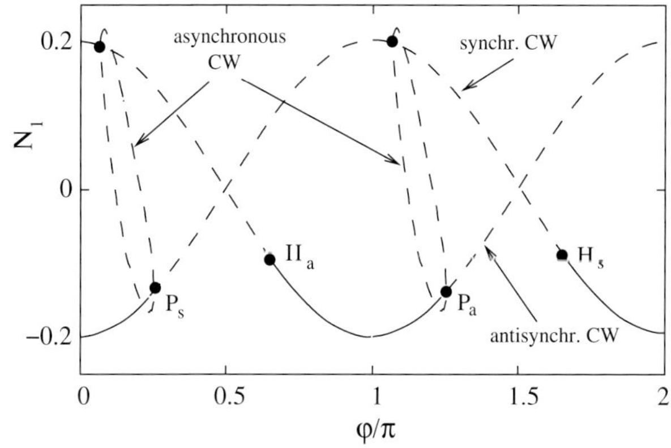

Fig. 6.3. Two branches of synchronous and antisynchronous CW solutions and the connecting branches of asynchronous solutions. P_a and P_s are pitchfork and H_a, H_s are Hopf bifurcations. Index 's' stands for the synchronous and 'a' for antisynchronous solutions, respectively. $\eta = 0.2$

6.4. Influence of the detuning

In this section, we study system (6.3) with detuning, i.e. $\delta \neq 0$. The following quantity (if it exists) is often used to determine the locking between two weakly coupled oscillators:

$$\overline{\Delta\Omega} := \lim_{t\to\infty} \frac{\Delta\psi(t)}{t}.$$

$\overline{\Delta\Omega}$ can be treated as averaged frequency difference between two weakly coupled lasers. Figure 6.4 shows results of computation of $\overline{\Delta\Omega}$ depending on the detuning δ. Three different curves are obtained for different values of φ with fixed $\eta = 0.3$. At each point, we averaged over the time interval $T_{\text{av}} = 1000$. One clearly observes the locking intervals $\overline{\Delta\Omega} = 0$ as plateaus near $\delta = 0$. Moreover, the width of these intervals strongly depends on the phase parameter φ. We will inspect the dependence of the locking on φ in more details in the next sections by studying bifurcations that are involved in the loss of locking and appearance of pulsations.

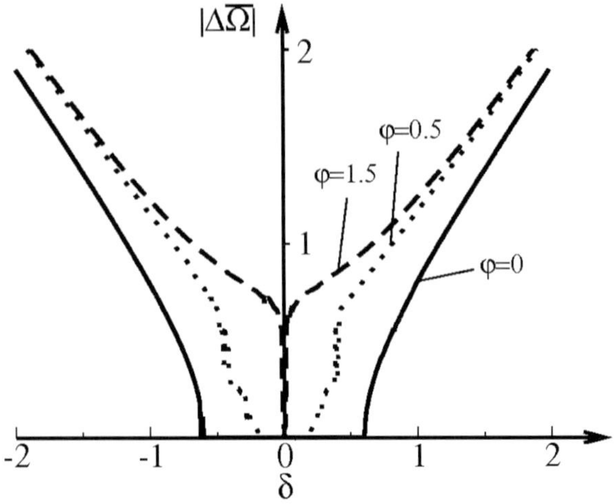

Fig. 6.4. Averaged frequency difference versus detuning for different values of φ ($\eta = 0.3$).

6.4.1. *Stationary states in the presence of detuning*

The CW solutions of system (6.2) are equilibria of system (6.3). Hence, one can use the standard path-following technique to follow their dependence on the parameters. The resulting bifurcation diagram is shown in Fig. 6.5, which is computed for the detuning $\delta = 0.1$. Remark that detuning breaks the first two symmetries in our system (cf. Sec. 6.2). Therefore, exactly synchronous and exactly antisynchronous solutions do not exist anymore. Moreover, the pitchfork bifurcations are no longer admissible for $\delta \neq 0$. Two questions arise naturally: What happens with the synchronous and antisynchronous solutions after the symmetry breaking by detuning? How is the pitchfork bifurcation perturbed in this case? The observed scenario, cf. Fig. 6.5, clarifies the situation. In particular, instead of the pitchfork bifurcation we have a saddle-node bifurcation (denoted as 'LP'). In the nonsymmetric case this saddle-node bifurcation connects the previously synchronous solutions via the unstable branch of asynchronous solutions to the antisynchronous, cf. Fig. 6.5. As a result, instead of the separate branches of synchronous and antisynchronous solutions, for $\delta \neq 0$ there are closed branches of solutions, which do not posses these symmetry properties. Nevertheless, as we shall see in Sec. 6.4.3, some parts of these branches still keep being close to the synchronous state and some to the

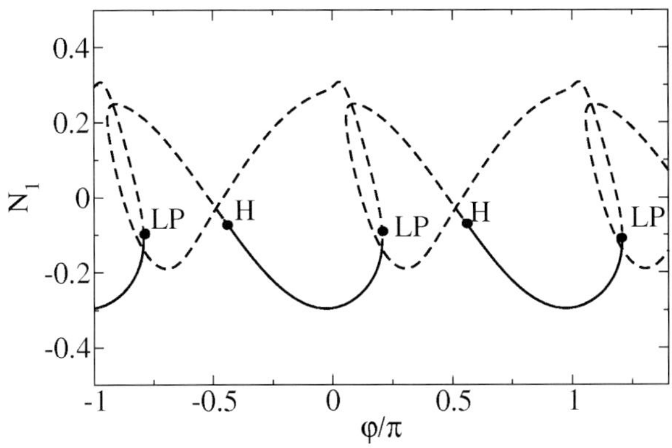

Fig. 6.5. Stationary states for system (6.2) with detuning. $\delta = 0.01$, $\eta = 0.2$. Stable branches are depicted by solid lines, unstable by dashed.

antisynchronous.

Comparing the bifurcation diagram in Fig. 6.5 and its symmetric counterpart in Fig. 6.3 we note that similar stability regions for stationary states, which are limited by the Hopf (H) and saddle-node (LP) bifurcations, exist in both cases. In fact, they can be obtained from each other by continuation along the parameter δ. Moreover, as we will see in Sec. 6.4.3, the corresponding branches are close to the synchronous (those that contain $\varphi = 0$) and to the antisynchronous one (containing $\varphi = \pm\pi$).

6.4.2. *Regions of locking*

In the following, we want to regard the existence of a stable CW solution as a locked state of the coupled laser system Θ. In order to study the influence of the detuning parameter δ, we investigate the boundaries of the stability region, i.e. the bifurcation points 'LP' and 'H' in Fig. 6.5, depending on δ. The resulting bifurcation diagram is shown in Fig. 6.6. There we denote by D_s and D_a two regions, corresponding to the existence of stable stationary states. We distinguish between these two regions because the first one is connected to the synchronous stationary states and the second one to the antisynchronous at $\delta = 0$. For more details about these states we refer to the next section.

Note that the diagram in Fig. 6.6 is obtained for fixed coupling strength η. Fig. 6.7 shows the bifurcation diagrams for fixed values of φ with varying δ and η. The different values of φ correspond to qualitatively different

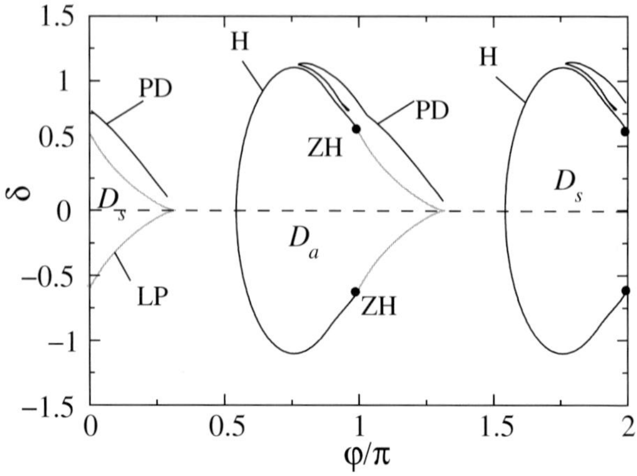

Fig. 6.6. Stability regions for the stationary states of coupled systems with detuning. 'LP' and 'H' denotes saddle-node and Hopf bifurcations, respectively. 'ZH' is a codimension 2 Zero-Hopf (or Guckenheimer-Gavrilov) bifurcation point. 'PD' - period-doubling bifurcation curve.

bifurcation diagrams. As it is expected, for sufficiently small coupling η in Fig. 6.7, the stability region is bounded by saddle-node bifurcation lines. In addition to the saddle-node bifurcation mechanism, Hopf bifurcation lines appear to confine partially the locking regions for larger values of η. The codimension two bifurcation points (ZH) appear where Hopf and saddle-node bifurcation lines meet.

6.4.3. *Stationary states after the symmetry breaking*

By definition, CW solutions have the form $E_1(t) = a_1 e^{i(\omega t + \Delta\psi)}$, $E_2(t) = a_2 e^{i\omega t}$, $N_1(t) = N_1$, $N_2(t) = N_2$. Now we show that their particular shape, i.e. the values of $a_1, a_2, \Delta\psi$, is influenced by the symmetry that is broken by the detuning δ. Particularly, in the region D_s (see Fig. 6.6) we have stationary solutions that are close to synchronous and in the region D_a close to antisynchronous states. In other words, as $\delta \to 0$ the stable locked solutions from the region D_s approach the synchronous CW states, and from D_a the antisynchronous, respectively. Figure 6.8 shows the ratio of amplitudes a_1/a_2 and a phase-shift $\Delta\psi$ as a function of φ for $\delta = 0.2$, i.e. parameters belong to D_s.

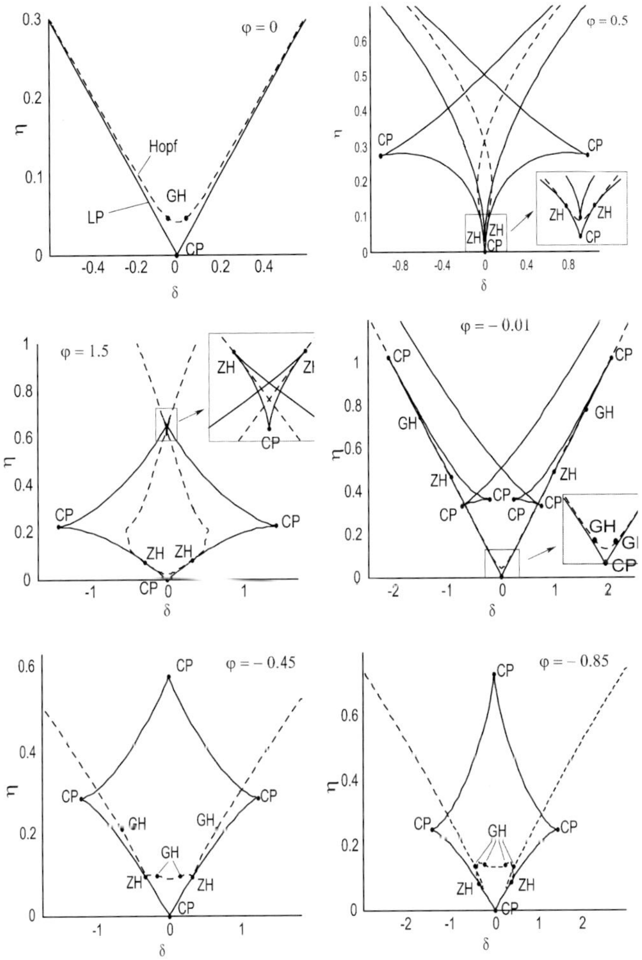

Fig. 6.7. Bifurcation diagrams for stationary solutions. Detuning δ and coupling strength η are plotted. The figures differ by the propagation phase parameter φ. Hopf bifurcations are denoted by dashed lines and saddle-nodes by solid lines. ZH — Zero-Hopf bifurcation; CP — cusp; GH — generalized Hopf bifurcation (where the Hopf bifurcation changes its criticality)

6.4.4. *Self-pulsations*

Self-pulsations, i.e. periodic oscillations of the field intensity, appear as a result of a Hopf bifurcation of the stationary locked states. In terms of

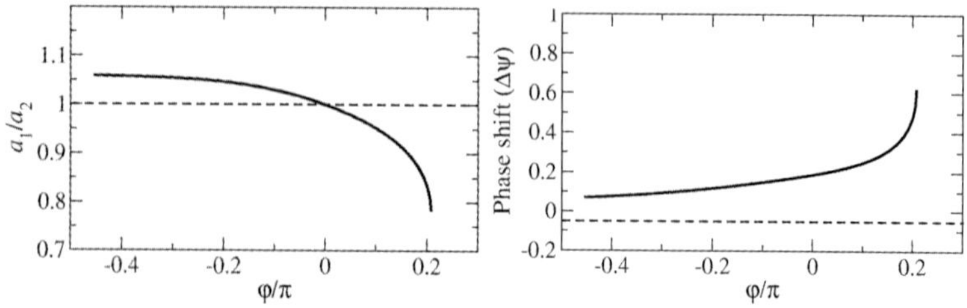

Fig. 6.8. Characteristics of some locked solutions, which are close to synchronous for the system with detuning. Solid lines: $\delta = 0.2$, dashed line: $\delta = 0$.

the original system (6.2) they are modulated waves lying on invariant tori, while for system (6.3) these solutions become periodic solutions.

We have noticed, that near the period-doubling bifurcation self pulsations appear, which are close to the diagonal in the space (a_1, a_2), cf. Fig. 6.9, orbit A. Such pulsations appear when there is no phase shift between the amplitudes a_1 and a_2 of the lasers, cf. Fig. 6.9(A). On the contrary, near the Hopf bifurcation, we observe that self pulsations are close to the "antidiagonal". Such a phenomenon was reported in Ref. [20] and called "inverse synchronization". In this case, a_1 is shifted with respect to a_2 by a half of the period, cf. Fig. 6.9(C). The orbit B in Fig. 6.9 corresponds to the intermediate regime. In the following two sections we consider these phenomena in more detail and show that the phase propagation parameter φ determines the possibility to observe identical or inverse amplitude synchronization.

6.4.5. *Identical amplitude synchronization*

It is a remarkable fact that coupled systems with detuning (6.3) still admit solutions for which the amplitudes behaves identically, i.e. $a_1 = a_2$. An example of such a solution is shown in Fig. 6.9(A). The necessary condition for the existence of such solutions is $\varphi = 0$ or $\varphi = \pi$. In this case, substituting $N := N_1 = N_2$ and $a := a_1 = a_2$ into (6.3) we obtain

$$a' = Na + \eta a \cos \Delta\psi \tag{6.7}$$
$$N' = \varepsilon(J - N - (2N + 1)a^2), \tag{6.8}$$
$$\Delta\psi' = \delta - 2\eta \sin \Delta\psi. \tag{6.9}$$

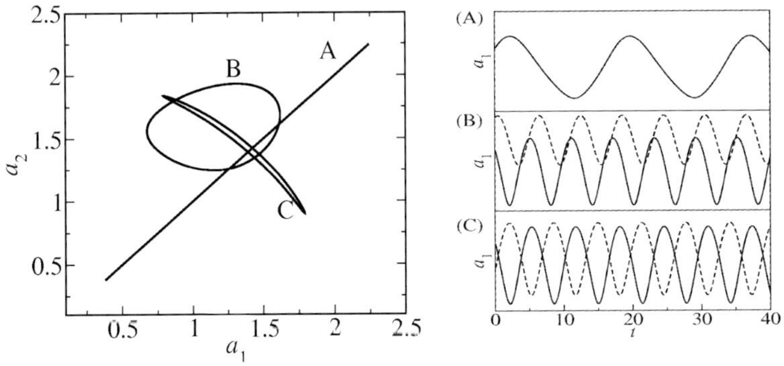

Fig. 6.9. Different types of self-pulsations. (A), (B), and (C) show time dependence of the amplitudes a_1 and a_2 of both lasers. The corresponding parameter values: $\delta = 0.7$, $\varphi = 0$ (A), $\varphi = \pi/4$ (B), and $\varphi = \pi/2$ (C).

The phase difference $\Delta\psi(t)$ can be explicitly found from (6.9). Namely, we have

$$\Delta\psi = 2\arctan\left[\frac{2\eta}{\delta} - \tanh\left(\frac{t}{2}\sqrt{4\eta^2-\delta^2}\right)\sqrt{\left(\frac{2\eta}{\delta}\right)^2 - 1}\right] \tag{6.10}$$

for $\delta < 2\eta$ and

$$\Delta\psi = 2\arctan\left[\frac{2\eta}{\delta} + \tan\left(\frac{t}{2}\sqrt{\delta^2-4\eta^2}\right)\sqrt{1 - \left(\frac{2\eta}{\delta}\right)^2}\right] + 2\pi n \tag{6.11}$$

for $\delta > 2\eta$. Here $n = [t\sqrt{\delta^2-4\eta^2}/2\pi]_i$, and $[\cdot]_i$ denotes an integer part. Solution (6.10) corresponds to a relatively small detuning. In this case, $\Delta\psi$ converges to a constant for $t \to \infty$, and the system settles on a stable CW solution. When the detuning δ becomes larger 2η, then $\Delta\psi$ is the periodic function (considered modulo 2π) (6.11) with the period $T_{syn} = 2\pi/\sqrt{\delta^2-4\eta^2}$. The growth rate of $\Delta\psi$ is determined by the term $2\pi[t\sqrt{\delta^2-4\eta^2}/2\pi]_i$, i.e. the averaged frequency of self-pulsations is $\overline{\Delta\Omega} = \sqrt{\delta^2-4\eta^2}$. Note that their period tends to infinity as $\delta \to 2\eta$.

To summarize, there exists for $\varphi = 0$ a set of amplitude-synchronous periodic solutions for $\delta > 2\eta$. We believe that approximate chaotic synchronization of intensities occurs as a result of further bifurcations of these solutions when φ is close to 0. As we will see in Sec. 6.4.7 there is a complicated dynamical mechanisms leading to the destabilization of self-pulsations. An example of such chaotic motion is shown in Fig. 6.4.6.

6.4.6. *Inverse amplitude synchronization*

When the propagation phase φ is close to $\pi/2$, we numerically observe MW solutions for which a_1 and a_1 behaves in antiphase, cf. Fig. 6.9(C). Such a phenomenon was reported also in Ref. [20] and called inverse synchronization. To start with, let us substitute [a] $\varphi = \pi/2$ into (6.3). We obtain the following system

$$\begin{aligned} a_1' &= N_1 a_1 - \eta a_2 \sin\Delta\psi, \\ N_1' &= \varepsilon(J - N_1 - (2N_1 + 1)a_1^2), \\ a_2' &= N_2 a_2 + \eta a_1 \sin\Delta\psi, \\ N_2' &= \varepsilon(J - N_2 - (2N_2 + 1)a_2^2), \\ \Delta\psi' &= \delta + (N_1 - N_2)\alpha + \eta(\frac{a_1}{a_2} - \frac{a_2}{a_1})\cos\Delta\psi, \end{aligned} \tag{6.12}$$

which can be considered as a perturbed uncoupled system with coupling parameter η acting as a perturbation parameter. The unperturbed system (i.e. $\eta = 0$) has an asymptotically stable fixed point $\hat{N}_1 = \hat{N}_2 = 0$, $\hat{a}_1 = \hat{a}_2 = \sqrt{J}$. The perturbation term $\eta(a_2 \sin\Delta\psi, 0, a_1 \sin\Delta\psi, 0)^T$ is tangent to the arc $N_1 = C$, $N_2 = C$, $a_1^2 + a_2^2 = R^2$ with some fixed R and C. We numerically observe that the resulting oscillations are close to the arc with $R = \sqrt{J}$, and $C = 0$ i.e. they appear around the point $\hat{a}_i, \hat{N}_i$, cf. Fig. 6.4.6. Note that the set $N_1 = N_2 = 0$, $a_1^2 + a_2^2 = J$ is not invariant for (6.12) with nonzero ε.

6.4.7. *Chaotic oscillations near Zero-Hopf bifurcation point*

In the vicinity of the Zero-Hopf bifurcation point, cf. Fig. 6.12, there is a branch of Neimark-Sacker bifurcations emerging from ZH point. When crossing this curve from above, the stable limit cycle undergoes a Neimark-Sacker bifurcation. It is a general observation (cf. Ref. [8], page 302) that the torus created by the Neimark-Sacker bifurcation exists only for parameter values near the corresponding bifurcation curve. If one moves away from the curve, the torus loses its smoothness and will be destroyed. The complete sequence of events is likely to involve an infinite number of bifurcations, since any weak resonance point on the Neimark-Sacker curve is the root of an Arnold phase-locking tongue. In view of this fact, we did not try to resolve the bifurcations numerically below the curve 'T' in Fig. 6.12(a). Instead, for randomly chosen initial conditions, we calculated

[a]similarly also in the case $\varphi = 3\pi/2$

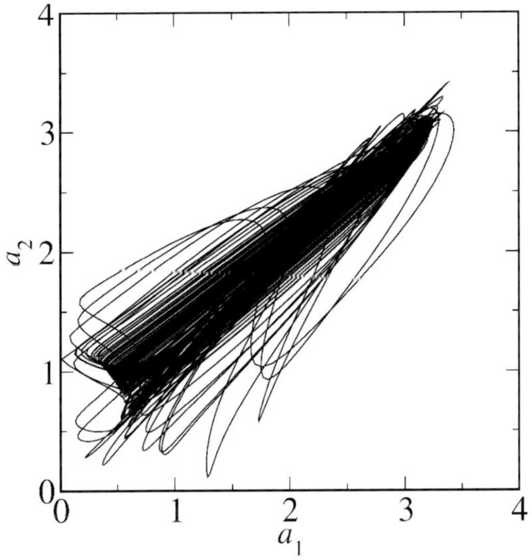

Fig. 6.10. Nearly amplitude-synchronous state for $\varphi = 0.01\pi$, $\delta = 0.6$.

Lyapunov exponents for different parameter values. Figure 6.12(b) shows the parameter values for which the largest Lyapunov exponent is positive, i.e. complex dynamics is present. In particular, we can see that this region comes arbitrary close (with the given precision) to the ZII point.

6.5. The case of a small delay

In this section we discuss some properties of the symmetric system with small delay

$$\begin{aligned} E_j'(t) &= (1 + i\alpha)N_j(t)E_j(t) + \eta e^{-i\varphi}E_{3-j}(t-\tau), \\ N_j'(t) &= \varepsilon[J - N_j(t) - (2N_j(t)+1)|E_j(t)|^2], \quad j = 1, 2 \end{aligned} \tag{6.13}$$

and compare them with the corresponding properties of the instantaneously coupled system (6.4). The zero-delay approximation (6.4) gives a correct representation for the dynamics of the delayed system (6.13), provided that the delay "is small enough". Mathematically, it can be regarded as existence of a stable finite-dimensional invariant manifold in the phase space

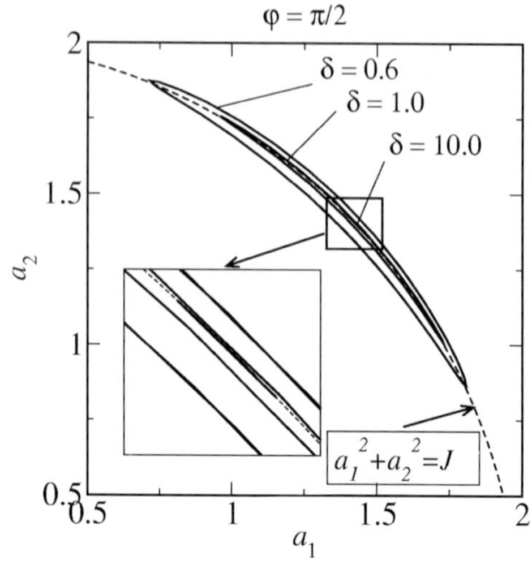

Fig. 6.11. "Inverse synchronization" for $\varphi = \pi/2$, $\eta = 0.3$, $J = 1$, $\varepsilon = 0.03$, $\alpha = 2$.

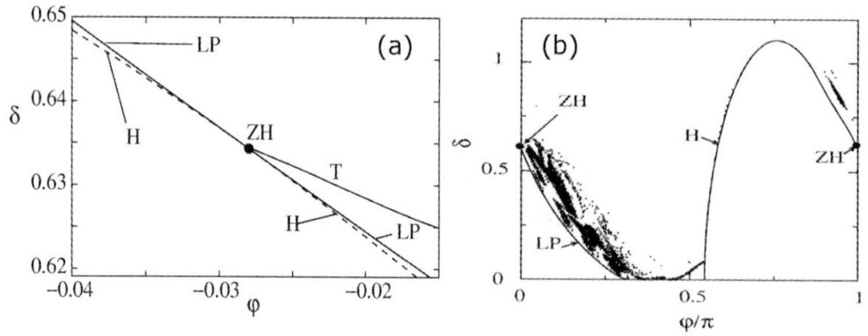

Fig. 6.12. (a) Neighborhood of the Zero-Hopf bifurcation. 'T' denotes the Neimark-Sacker bifurcation curve emerging from the Zero-Hopf point. (b) Parameter values, for which an attractor with positive largest Lyapunov exponent exists.

of (6.13) for small enough τ such that the dynamics on this manifold is described by the system of ODEs (6.4) [1]. This implies that self-pulsations as well as more complicated dynamical regimes (bounded chaotic attractors) that have been discovered in (6.4) have their counterparts in (6.13). The main problem-specific question is then for which values of τ system (6.4) approximates reasonably (6.13). In this section, we give some partial answer to this problem concerning CW solutions.

The dynamics of (6.13) within the synchronization subspace $E_1 = E_2 :=$ E, $N_1 = N_2 := N$ is governed by the Lang-Kobayashi system

$$\begin{aligned} E' &= (1+i\alpha)NE + \eta e^{-i\varphi}E(t-\tau), \\ N' &= \varepsilon[J - N - (2N+1)|E|^2]. \end{aligned} \tag{6.14}$$

The parameters of the synchronous CW solutions $E(t) = ae^{i\omega t}$, $N(t) = N$ of (6.14) satisfy the following set of equations

$$\begin{aligned} N &= -\eta\cos(\varphi + \omega\tau), \\ \omega - \alpha N &= -\eta\sin(\varphi + \omega\tau), \\ a^2 &= (J - N)/(2N+1). \end{aligned} \tag{6.15}$$

One can obtain sufficient conditions for system (6.14) to have only one external cavity mode, i.e. a unique solution of (6.15). For this, we write the equation for ω as

$$\omega = -\eta(\alpha\cos(\varphi + \omega\tau) + \sin(\varphi + \omega\tau)). \tag{6.16}$$

The saddle-node bifurcation, which gives rise to additional external cavity modes, can be identified [25] as a double root of (6.16). Hence, differentiating it with respect to ω, we obtain $1 = \tau\eta(\alpha\sin(\varphi + \omega\tau) - \cos(\varphi + \omega\tau))$. It is clear that the condition

$$\tau\eta < 1/\sqrt{1+\alpha^2} \tag{6.17}$$

guarantees that a double root does not exist. Hence, the inequality (6.17) roughly provides the limit within which one might expect that the delay τ does not qualitatively change the dynamics within the synchronization (antisynchronization) subspace.

The transverse stability of the unique synchronous CW solution is determined by the solutions of the characteristic equation

$$\begin{aligned} \chi_T^{\tau}(\Lambda) = {} & [\Lambda^2 + 2\eta\cos\theta(e^{-\Lambda\tau}+1)\Lambda + \eta^2(e^{-\Lambda\tau}+1)^2] \\ & \times(\Lambda + \varepsilon(1+S)) + 2\varepsilon S(1 - 2\eta\cos\theta) \\ & \times[\Lambda + \eta(\cos\theta - \alpha\sin\theta)(e^{-\Lambda\tau}+1)] = 0, \end{aligned} \tag{6.18}$$

where $S := (J - \eta\cos\theta)/(1 + 2\eta\cos\theta)$ and $\theta := \omega\tau + \varphi$. It turns out that for the values up to $\tau = 2$ the regions in the (φ, δ) parameter plane for the transverse stability of synchronous CW solution of (6.13) are qualitatively the same as in the case of zero delay (6.4). In Fig. 6.13 we plot the curves which delineate this stability region. All the remaining parameters are taken to be the same as in Fig. 6.2. The effect of delay for this range of τ can be seen to be only a continuous changing of the slope of the curves.

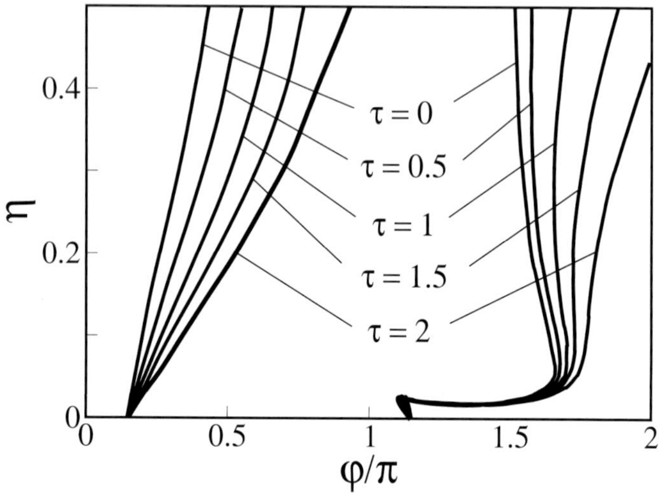

Fig. 6.13. Boundaries of the region for transverse stability of synchronous CW solutions of system (6.13) for different values of delay (similarly as in Fig. 6.2 without delay). The corresponding boundaries for antisynchronous CW solutions can be obtained by shifting along the φ axis by π.

6.6. Three coupled identical lasers

Arrays of three coupled lasers where studied by Winful and Rahman [24] for the case of semiconductor lasers coupled by the evanescent field, and by Terry et al. [21] for Nd:YAG lasers with either modulated pump or loss. In particular, Ref. [24] demonstrates the possibility of chaotic synchronization between the outer lasers in the array, and Ref. [21] presents a detailed analysis of different symmetries, which occur in the array of three coupled identical lasers. This section gives an overview of the results presented in Ref. [30].

The system of n coupled lasers (see Fig. 6.14) can be modeled by the following coupled rate equations

$$\begin{aligned}
E_1' &= (1+i\alpha)N_1E_1 + \eta e^{-i\varphi}E_2,\\
N_1' &= \varepsilon[J - N_1 - (2N_1+1)|E_1|^2],\\
E_j' &= (1+i\alpha)N_jE_j + \eta e^{-i\varphi}(E_{j-1}+E_{j+1}),\\
N_j' &= \varepsilon[J - N_j - (2N_j+1)|E_j|^2], \quad j = 2,\ldots,n-1,\\
E_n' &= (1+i\alpha)N_nE_n + \eta e^{-i\varphi}E_{n-1},\\
N_n' &= \varepsilon[J - N_n - (2N_n+1)|E_n|^2]
\end{aligned} \tag{6.19}$$

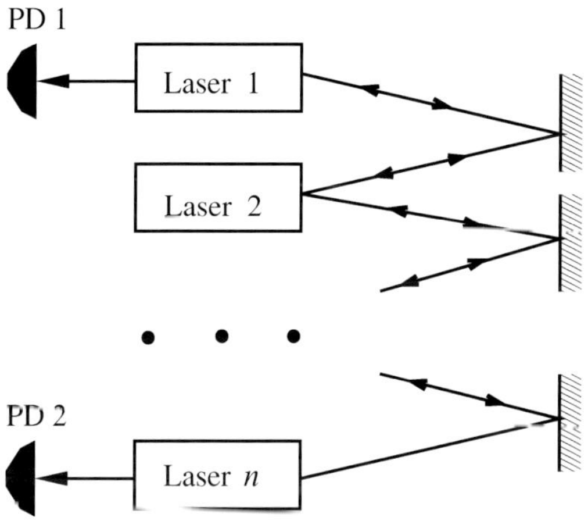

Fig. 6.14. Schematic representation of the array of linearly coupled lasers. PD1 and PD2 are the photodiodes, which detect the output from the outer lasers.

6.6.1. *Conditions for synchronization of the outer lasers*

Let us consider (6.19) with $n = 3$. Remark that the subspace $E_1 = E_2, N_1 = N_2$ is not invariant with respect to (6.19). Instead, the 6-dimensional subspace $M_s = \{E_1 = E_3, N_1 = N_3\}$, which corresponds to the synchronization of the outer lasers, is invariant.

We perform a stability analysis of the subspace M_s. First, we present the two-dimensional bifurcation diagram for the stability of CW solutions in Fig. 6.15(a). One can identify different bifurcations: Hopf, Fold, and pitchfork, which lead to the destabilization of synchronous stationary states.

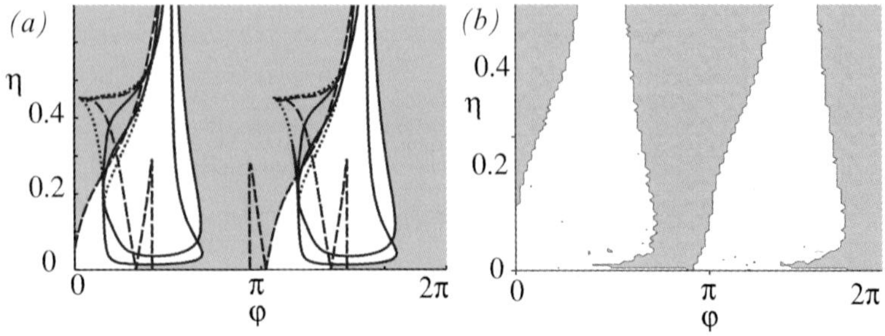

Fig. 6.15. (a) Two-dimensional bifurcation diagram for the stability of stationary states within the subspace M_s. The grey shaded regions show the parameter values, for which there exists a stable stationary state. Solid lines denote Hopf bifurcations, dotted lines Fold, and broken lines the pitchfork bifurcations. Fixed parameters are $\alpha = 2, J = 1, \varepsilon = 0.03$. (b) Numerically obtained parameter region for the stability of the synchronization subspace M_s.

Note, that Fig. 6.15(a) reveals the local stability properties of the synchronous stationary states. In order to have an insight into the transverse stability of the whole subspace M_s, we perform also a numerical analysis and show the results in Fig. 6.15(b). A grid 100x100 was introduced to discretize a square region of the parameters η and φ. From each obtained cell we have chosen a pair of parameters φ, η, for which an orbit was computed starting from some initial condition near the subspace, but not exactly symmetric. The gray area shows the regions, for which the orbit was attracted to the subspace, i.e. for which it was synchronized.

One can conclude, comparing Figs. 6.15(a) and (b), that the stationary synchronization dominates among all possible synchronization regimes. The difference in sizes of the shaded regions in Figs.6.15(a) and (b) indicate that there is a coexistence of the synchronous and non-synchronous regimes. In addition, one can find parameter values, where periodic or chaotic synchronization takes place. The detailed one-dimensional bifurcation diagram for $\varphi = \pi/2$ in Fig. 6.16 reveals these regimes. Comparing the bifurcation diagrams (a) and (b) in Fig. 6.16 we can see that for $0 < \eta < 0.0156$ a stable CW solution is an attractor of the system and stationary synchronization occurs. As a result of the Hopf bifurcation at $\eta = 0.0156$, a stable MW solution within the subspace M_s appears, but the synchronized state is still retained. Periodic synchronization loses its stability at $\eta = 0.022$ via transverse Hopf bifurcation, even though the limit cycle in the synchronization subspace is still stable. Further increase of the parameter η leads to the appearance of a chaotic invariant set within the synchronization subspace via a cascade of period doubling bifurcations at $\eta = 0.0275$. This set becomes again transversely stable at $\eta = 0.0284$, implying chaotic synchronization. For some narrow range around $\eta \approx 0.03$ (see Fig. 6.16(c-d)) a coexistence of the synchronous and asynchronous attractors in the phase space occurs. When the parameter η increases, the asynchronous attractor becomes chaotic via period-doubling (Fig. 6.16d) bifurcations. This multistability can be observed for $0.0284 < \eta < 0.0338$. The asynchronous solution becomes unstable and stable chaotic synchronization is dominant for $0.0338 < \eta < 0.0515$, since at $\eta = 0.0515$ the chaotic attractor loses its stability. The new asynchronous attractor can be shown to be hyperchaotic, because its spectrum of Lyapunov exponents contains two positive exponents ($\lambda_1 = 0.0442$ and $\lambda_2 = 0.0035$ for $\eta = 0.052$).

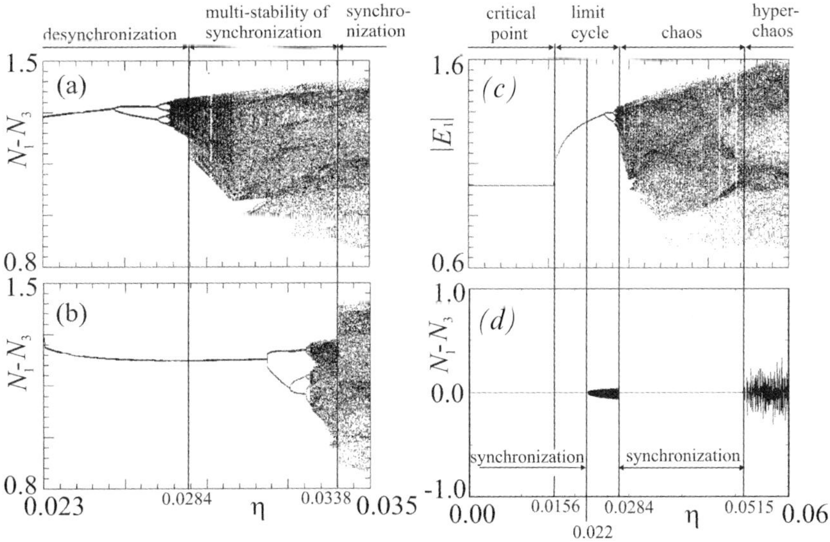

Fig. 6.16. One-dimensional bifurcation diagram of the synchronous system attractor (a) and corresponding ranges of synchronization (b). $\varphi = \pi/2$ is fixed. Bifurcation diagrams for synchronous (c) and asynchronous (d) attractors. Zoom of the interval $0.022 < \eta < 0.035$. The region of multistability is indicated.

A confirmation of the observations described above is given in Figure 6.17. This Figure shows the behavior of Lyapunov exponents corresponding to the subspace M_s. One observes the following points, where the behavior changes qualitatively: A is a transverse Hopf bifurcation. As a result of this bifurcation, a stable periodic solution appears, which is not located in the subspace M_s. At the point B the chaotic attractor in the subspace M_s becomes stable. It loses its stability at the point C. Therefore, we have stationary synchronization for $0 < \eta < 0.0156$, periodic synchronization for $0.0156 < \eta < 0.022$, and chaotic one for $0.0284 < \eta < 0.0515$.

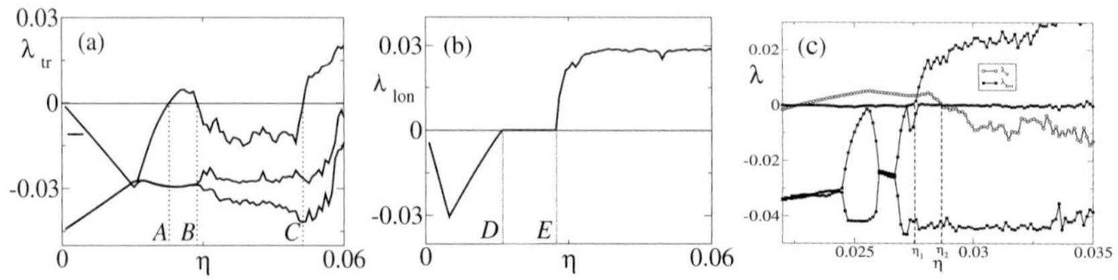

Fig. 6.17. Behavior of the Lyapunov exponents. (a) Transverse Lyapunov exponents and (b) the largest longitudinal Lyapunov exponent. The parameters are the same as in Fig. 6.16. Bifurcation points A-E are explained in the text. (c) Zoom of the interval $0.022 < \eta < 0.035$.

6.6.2. *Transition to chaotic synchronization via blowup of a transversely unstable synchronous invariant set*

Let us consider in detail the synchronization mechanism for the parameter value $\eta \approx 0.028$. One can observe that shortly after the synchronous attractor becomes chaotic at $\eta_1 = 0.0275$, it becomes transversely stable at $\eta_2 = 0.0284$, cf. Figs. 6.17,6.16.

As follows from the bifurcation diagram in Fig. 6.16, at the moment when the synchronous attractor becomes transversely stable at $\eta_2 = 0.0284$, it undergoes a blowup, i.e. the size of the attractor rapidly increases. The coincidence of these two effects seems to be not occasional. First, we suggest an explanation, which is based on the concept of local Lyapunov exponents [17]. Figure 6.18 shows regions within the chaotic attractors, which are transversely stable. Namely, we computed the local contraction/expansion ratio by means of the local transverse Lyapunov exponents $\lambda_{loc}^{(i)} = \ln|\mu_i(\Phi(T,x))|, i = 1,2,3$, where $\Phi(T,x)$ is the fundamental matrix corresponding to the linearized dynamics of the transverse perturbations. Here, $\mu_i(\cdot)$ denotes the i-th eigenvalue of a matrix. $\Phi(T,x)$ depends on the integration interval T, which is chosen to be small in our

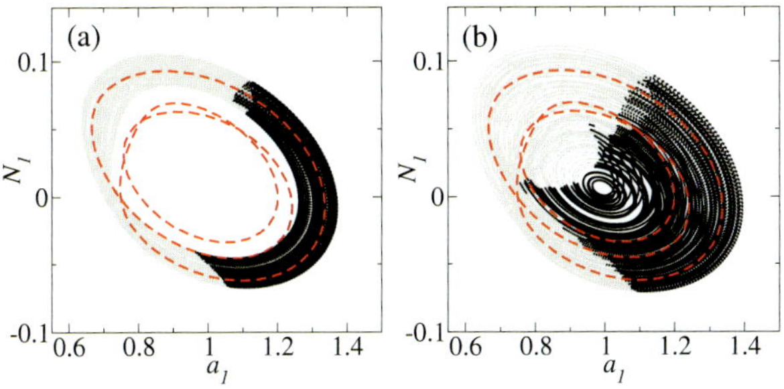

Fig. 6.18. The phase portraits of the synchronous attractor before the blowup (a) at $\eta = 0.0282$ and after (b) at $\eta = 0.0287$. Those parts of the attractors, which possesses negative local Lyapunov exponents are marked by the black points. Period-3 unstable periodic orbit is shown by broken red line.

case ($T = 1.0$) in order to reveal the local expansion properties, and x, which is a point on the synchronous attractor. We estimated numerically $\lambda_{loc}^{(i)}$ and marked in Fig. 6.18 those points on the attractor, for which all $\lambda_{loc}^{(i)}$ are negative. This method allows us to find transversely attracting regions on the attractor. The final transverse stability is then deduced by averaging over the whole attractor. It can be seen that the major part of the attractor after the blowup in Fig. 6.18 is covered by transversely stable points, i.e. the "new" part of the attractor brings additional transverse stability. This fact allows us to assume that the blowup of the attractor comes together with its transverse stabilization.

Let us clarify the underlying bifurcation mechanisms, which cause the observed phenomenon. The inspection of the system dynamics and the continuation technique reveal that the blowup of the synchronous attractor at $\eta = 0.0284$ occurs via a collision with a chaotic saddle. The chaotic saddle is located in the phase space as shown in Fig. 6.19(b). In order to detect the geometrical place of this unstable chaotic saddle, we used the continuation technique. We followed the unstable low-periodic orbits, which were born in the period-doubling cascade.

In the following, we will use the abbreviations $\mathcal{A}$ for the synchronous chaotic attractor (e.g. in Fig. 6.18), and CS for the chaotic saddle (e.g. in Fig. 6.19(b)).

Figure 6.19(a) shows the main bifurcation branch, which leads to the appearance of the CS. Particularly, at $\eta = 0.02762$ (LP point in the figure), a pair of period-3 orbits appears via a fold bifurcation. The branch with the smaller period leads then to the period-doubling bifurcation at $\eta = 0.02783$

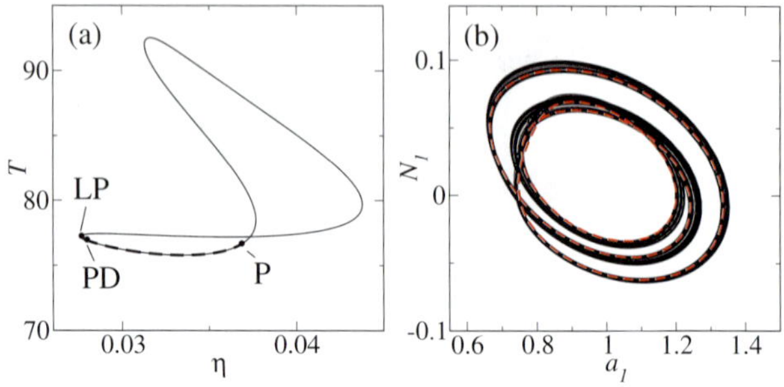

Fig. 6.19. (a) The bifurcation diagram for the period-3 unstable periodic orbit. LP – Fold bifurcation ($\eta = 0.02762$), PD – period-doubling bifurcation ($\eta = 0.02783$). T denotes the actual period of the orbit. (b) The three-band chaotic saddle (CS), which is created through the period-doubling cascade of period-3 orbit (broken red line) at $\eta = 0.0280$.

(PD), where period-6 orbit is created. We were able to follow a few consecutive period-doubling bifurcations, which apparently lead to the appearance of three-band CS in the vicinity of the period-3 orbit. We also found that this set is stable with respect to transverse perturbations. Since CS is longitudinally unstable, we observe the following scenario: when the system starts close to the CS, it undergoes a relatively long synchronous transient along the CS and then approaches the longitudinally stable chaotic set $\mathcal{A}$. Afterwards, it is repelled in the transverse direction exhibiting desynchronization.

The collision of the sets CS and $\mathcal{A}$ occurs at $\eta = 0.0284$, producing a larger attractor. This collision can already be observed from Fig. 6.18, however, only in a projection. To be moer correct, we plot in Fig. 6.20 the distance $d(\mathcal{A}, p3)$ between the attractor $\mathcal{A}$ and the period-3 orbit embedded in the CS. We define

$$d(\mathcal{A}, p3) = \inf_{t_1 \in [0,T_1]; t_2 \in [0,T_2]} |x(t_1) - y(t_2)|,$$

where $x(t_1)$ is the period-3 orbit with the actual period T_1 and $y(t_2)$ is the orbit on the attractor $\mathcal{A}$ of large enough length T_2. In our simulations we used $T_2 = 5 \cdot 10^4$. We clearly observe the transition at $\eta = 0.0284$ when the period-3 orbit becomes a part of a synchronous attractor. Small deviations from zero for $\eta > 0.0284$ are due to the finiteness of T_2.

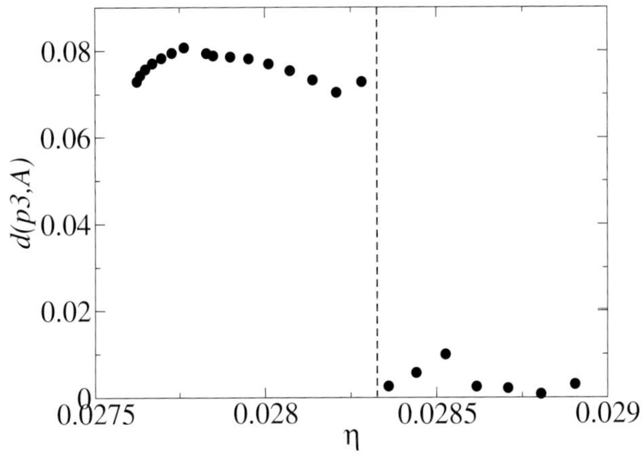

Fig. 6.20. Distance between the synchronous attractor $\mathcal{A}$ and the period-3 orbit, which is embedded into the chaotic saddle. At $\eta = 0.0284$ the distance rapidly decreases, indicating that the period-3 orbit becomes a part of the attractor for larger η.

6.7. Arrays of coupled oscillators

Having computed the bifurcation diagram in Fig 6.15 for the system of three coupled lasers, one can compare it with the similar stability diagram for two face-to-face coupled lasers. The qualitative similarity of the results is evident. In the both cases there are two large regions of synchronization of the outer lasers, which are located near $\varphi \approx 0$ and $\varphi \approx \pi$.

In order to check the universality of this observation, we computed numerically similar synchronization diagrams for larger arrays. The results are shown in Fig. 6.21 for 5 and 10 oscillators. One can recognize the same qualitative trend. Note, that the synchronization considered here corresponds only to the asymptotic coincidence of the variables of the outer lasers, i.e. $E_n = E_1, N_n = N_1$. This is not sufficient for the complete synchronization of the whole array. In general, the synchronization of the outer lasers may include some cluster states as well [3, 4]. Figure 6.21 has been obtained by computing asymptotic behavior of orbits starting from randomly chosen initial conditions.

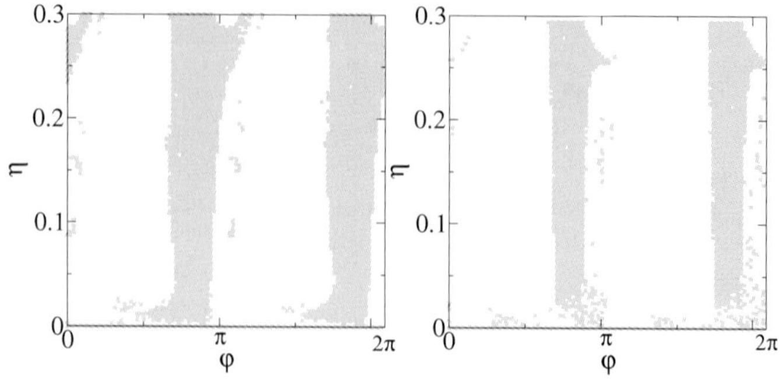

Fig. 6.21. The regions in the $\varphi - \eta$ parameter space, for which an orbit started from randomly chosen initial condition is attracted to the synchronization subspace M_s. This corresponds to the asymptotic synchronization of the outer lasers. (a) 5 coupled lasers, (b) 10 coupled lasers.

References

[1] C. Chicone. Inertial and slow manifolds for delay equations with small delay. *J. Differ. Equations*, 190:364–406, 2003.

[2] I. Fischer, Y. Liu, and P. Davis. Synchronization of chaotic semiconductor laser dynamics on subnanosecond time scales and its potential for chaos communication. *Phys. Rev. A*, 62:011801, 2000.

[3] M. Golubitsky and I. Stewart. Synchrony versus symmetry in coupled cells. In F. Dumortier, H.W. Broer, J. Mawhin, A. Vanderbauwhede, and S.M. Verduyn Lunel, editors, *Equadiff 2003: Proceedings of the International Conference on Differential Equations*, pages 13–24. World Scientific, 2005.

[4] M. Golubitsky, I. Stewart, and A. Torok. Patterns of synchrony in coupled cell networks with multiple arrows. *SIAM J. Appl. Dynam. Sys*, 4:78–100, 2005.

[5] T. Heil, I. Fischer, W. Elsäßer, B. Krauskopf, K. Green, and A. Gavrielides. Delay dynamics of semiconductor lasers with short external cavities: Bifurcation scenarios and mechanisms. *Phys. Rev. E*, 67:066214, 2003.

[6] J. Javaloyes, P. Mandel, and D. Pieroux. Dynamical properties of lasers coupled face to face. *Phys. Rev. E*, 67:036201, 2003.

[7] I. V. Koryukin and P. Mandel. Two regimes of synchronization in unidirectionally coupled semiconductor lasers. *Phys. Rev. E*, 65:026201, 2002.

[8] Y. Kuznetsov. *Elements of Applied Bifurcation Theory*, volume 112 of *Applied Mathematical Sciences*. Springer-Verlag, 1995.

[9] C. Masoller. Anticipation in the synchronization of chaotic semiconductor lasers with optical feedback. *Phys. Rev. Lett.*, 86:2782, 2001.

[10] M. Möhrle, B. Sartorius, S. Bauer, O. Brox, A. Sigmund, R. Steingrüber, M. Radziunas, and H.J. Wünsche. Detuned grating multisection-rw-dfb

lasers for high-speed optical signal processing. *IEEE J. Selected Topics Quantum Electron.*, 7:217, 2001.

[11] L. M. Pecora, T. L. Carroll, G. A. Johnson, D. J. Mar, and J. F. Heagy. Fundamentals of synchronization in chaotic systems, concepts, and applications. *Chaos*, 7:520–543, 1997.

[12] M. Peil, T. Heil, I. Fischer, and W. Elsäßer. Synchronization of chaotic semiconductor laser systems: a vectorial coupling-dependent scenario. *Phys. Rev. Lett.*, 88:174101, 2002.

[13] A. Pikovsky and M. Rosenblum. Phase synchronization of regular and chaotic self-sustained oscillators. In A. Pikovsky and Yu. Maistrenko, editors, *Synchronization: Theory and Application*, pages 187–219. Kluwer, Dordrecht, 2003.

[14] D. Rand. Dynamics and symmetry. Predictions for modulated waves in rotating fluids. *Arch. Rat. Mech. Anal.*, 79:1–37, 1982.

[15] L. Recke and D. Peterhof. Abstract forced symmetry breaking and forced frequency locking of modulated waves. *Journal of Differ. Equations*, 144:233–262, 1998.

[16] L. Recke and S. Yanchuk. On inverse amplitude synchronized modulated wave solutions in time reversable systems. *in preparation.*

[17] N. F. Rulkov and M. M. Sushchik. Robustness of synchronized chaotic oscillations. *Int. J. Bifurcation Chaos Appl. Sci. Eng.*, 7:625–643, 1997.

[18] K. R. Schneider and S. Yanchuk. Complete synchronization of symmetrically coupled autonomous systems. *Applicable Analysis*, 82:1127–1143, 2003.

[19] E. M. Shahverdiev, S. Sivaprakasam, and K. A. Shore. Parameter mismatches and perfect anticipating synchronization in bidirectionally coupled external cavity laser diodes. *Phys. Rev. E*, 66:017206, 2002.

[20] S. Sivaprakasam, I. Pierce, P. Rees, P. S. Spencer, K. Alan Shore, and A. Valle. Inverse synchronization in semiconductor laser diodes. *Phys. Rev. A*, 64:013805, 2001.

[21] J. R. Terry, K. S. Thornburg, D. J. DeShazer, G. D. VanWiggeren, S. Zhu, P. Ashwin, and R. Roy. Synchronization of chaos in an array of three lasers. *Phys. Rev. E*, 59:4036, 1999.

[22] H. Wenzel, U. Bandelow, H.J. Wünsche, and J. Rehberg. Mechanisms of fast self pulsations in two-section DFB lasers. *IEEE Journal of Quantum Electronics*, QE-32:69–79, 1996.

[23] J. K. White, M. Matus, and J. V. Moloney. Achronal generalized synchronization in mutually coupled semiconductor lasers. *Phys. Rev. E*, 65:036229, 2002.

[24] H. G. Winful and L. Rahman. Synchronized chaos and spatiotemporal chaos in arrays of coupled lasers. *Phys. Rev. Lett.*, 65:1575, 1990.

[25] M. Wolfrum and D. Turaev. Instabilities of lasers with moderately delayed optical feedback. *Opt. Commun.*, 212:127–138, 2002.

[26] H.-J. Wünsche, S. Bauer, J. Kreissl, O. Ushakov, N. Korneyev, F. Henneberger, E. Wille, H. Erzgräber, M. Peil, W. Elsäßer, and I. Fischer. Synchronization of delay-coupled oscillators: A study of semiconductor lasers. *Phys. Rev. Lett.*, 94:163901, 2005.

[27] S. Yanchuk, Yu. Maistrenko, and E. Mosekilde. Loss of synchronization in coupled Rössler systems. *Physica D*, 154:26–42, 2001.
[28] S. Yanchuk, Yu. Maistrenko, and E. Mosekilde. Synchronization of time-continuous chaotic oscillators. *Chaos*, 13:388–400, 2003.
[29] S. Yanchuk, K. R. Schneider, and L. Recke. Dynamics of two mutually coupled semiconductor lasers: Instantaneous coupling limit. *Phys. Rev. E*, 69:056221, 2004.
[30] S. Yanchuk, A. Stefanski, T. Kapitaniak, and J. Wojewoda. Dynamics of an array of coupled semiconductor lasers. *Phys. Rev. E*, 73, 2006.

Chapter 7

Trapping of phase fronts and twisted spirals in periodically forced oscillatory media

Oliver Rudzick and Alexander S. Mikhailov*

Abteilung Physikalische Chemie, Fritz-Haber-Institut der Max-Planck-Gesellschaft,
Faradayweg 4-6, 14195 Berlin, Germany
* http://www.fhi-berlin.mpg.de/complsys/

Contents

7.1. Periodically forced oscillatory media

The control of pattern formation is an important problem in engineering of spatially extended self-organized systems. A wide class of systems showing spontaneously formed waves and patterns can be summarized under the generic term "oscillatory media". They may be considered as a composition of a large number of coupled subsystems. The dynamics of each subsystem is oscillatory. Complex phenomena like the formation of patterns and waves or spatio-temporal chaos comes through the interaction between the subsystems. Despite such complex collective behavior the dynamics is very sensitive to feedback or to periodic forcing with frequencies close to an integer ratio of the oscillation frequency of the subsystems.

A well-known example is the Belouzov-Zhabotinsky-Reaction. involving the the oxidation of an organic compound by bromate in acidic solution. It is a robust oscillatory reaction with striking color changes. Spatiotemporal

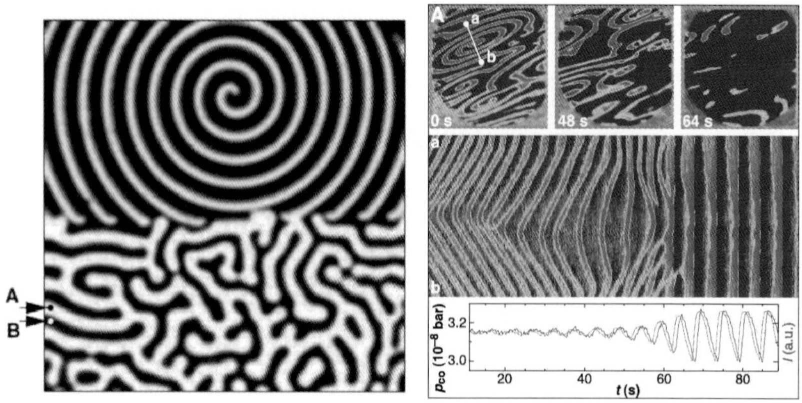

Fig. 7.1. Left: Effect of 2:1 resonant periodic forcing on the photosensitive BZ-reaction. A spiral wave is formed without forcing (upper part). Under the influence of illumination with light pulsed at twice the natural frequency of the reaction a labyrithine pattern appears. [3]. Right: Suppression of chemical turbulence in the catalytic CO oxidation on Pt(110) under global feedback. Upper row: subsequent PEEM images of the Pt surface illustrating the transition from spiral wave turbulence to homogeneous oscillations. Bright regions are mainly CO-covered. Middle row: Space-time diagram showing the evolution along the line AB. Bottom: Time series of the CO partial pressure [8].

wave behavior is exhibited in unstirred reaction mixtures. [1, 2]. For this reaction, application of global periodic forcing was shown to produce various cluster patterns [3–5] and induce turbulent regimes [6].

Another example which has been extensively studied is the catalytic CO oxidation on Pt(110). The interplay between desorption and surface diffusion of CO, reaction between the two adsorbed species, and an adsobate-driven structural change of the platinum surface can lead to oscillations of the CO and oxygen coverage [7]. In experiments where bulk oscillations were unstable and spatiotemporal chaos spontaneously developed, application of periodic forcing allowed to suppress chemical turbulence, produce intermittent regimes with cascades of amplitude defects, and generate oscillating cellular and labyrinthine patterns [8, 9]. Recently, front explosions have been predicted under periodic forcing [10].

7.2. The forced complex Ginzburg-Landau equation

The universal description of reaction-diffusion systems near a supercritical Hopf bifurcation is provided by the complex Ginzburg-Landau equation [11]. Action of global periodic forcing on the systems described by this equation has been first considered by Coullet and Emilsson [12, 13]. Under sufficiently strong resonant n:1 forcing, oscillations are entrained and stationary or traveling $2\pi/n$-fronts become possible. The 2π phase fronts for the 1:1 forcing are known as kinks (or phase slips). They represent traveling localized structures, because the states differing by the phase of 2π are physically identical. Therefore, as noticed in Ref. [13], they bear similarity with pulses in excitable media. Traveling π-fronts under 2:1 forcing represent nonequilibrium Bloch walls [14]. Kinks and traveling Bloch walls are elementary wave patterns under forcing conditions. Instabilities of kinks lead to backfiring and development of intermittent regimes with reproduction of amplitude defects [15–17]. Transverse instabilities of nonequilibrium planar Bloch walls give origin to the Bloch turbulence [6]. In heterogeneous media near a Bloch-Ising transition, complex behavior due to reflections of Bloch waves on Ising domains has been found [18].

Under global resonant n:1 forcing, the complex Ginzburg-Landau equation (CGLE) for the slow complex oscillation amplitude η is [12]

$$\dot{\eta} = (1 + i\nu)\eta - (1 + i\alpha)|\eta|^2\eta + (1 + i\beta)\nabla^2\eta + B\,(\eta^*)^{n-1}, \tag{7.1}$$

where detuning $\nu = \omega_0 - \omega_e/n$ is determined by the natural (ω_0) and forcing (ω_e) frequencies and B is the forcing amplitude. Oscillations are entrained by forcing in the parameter region known as the Arnold tongue (Fig. 7.2 (a)). Inside this region, kinks ($n = 1$) and Bloch walls ($n = 2$) traveling at a constant velocity are possible (see [12, 15, 16]). Moreover, wave trains formed by periodic sequences of such phase fronts can also be observed there.

7.3. Phase front propagation reversal

Our attention is focused on the properties of periodic trains formed by kinks or traveling Bloch walls. Our analysis reveals that, depending on the parameters of the oscillatory medium and the spatial period of a train, it can undergo a reversal of its propagation direction [19]. We show how this phenomenon can be used to design traps for traveling kinks and Bloch walls. Furthermore, we find that a new kind of patterns - twisted rotated spiral waves - exist in oscillatory media under the conditions of front propagation reversal.

Any traveling phase front is characterized by its chirality: "right" if the phase increases after front propagation and "left" if it decreases after that. A similar definition can be accepted for traveling wave trains. It is convenient furthermore to define the front velocity V in such a way that it is always positive ($V > 0$) if a front propagates to increase the oscillation phase and negative ($V < 0$) otherwise. With this convention, all "right" fronts move at a positive velocity, while the velocity of any "left" phase front is negative.

The velocity of an individual phase front is uniquely determined by the properties of the medium and the forcing parameters. For wave trains, it additionally depends on the spatial period λ of a train. Figure 7.2b shows dependences $V(\lambda)$ for two different values of the coefficient β, obtained by numerical continuation of wave train solutions of equation (7.1) with $n = 1$. When $\beta = 5.0$, velocity V remains positive for all spatial periods. This means that both a solitary kink and any kink train in such a medium possess the "right" chirality. In contrast to this, kinks move at a positive velocity (and have the "right" chirality) only for sufficiently short spatial periods at $\beta = 1.8$. At a critical spatial period λ_c, the propagation velocity of the train vanishes and $V(\lambda) < 0$ when $\lambda > \lambda_c$. Thus, solitary kinks and kink trains with large periods have the opposite "left" chirality in the latter case.

To illustrate the difference in the properties of wave patterns in such two media, we consider the following example. Suppose that the local oscillation frequency is increased in the center of a one-dimensional medium. If periodic 1:1 forcing is applied, the local frequency increase can still be so large that oscillators in the central region are not entrained and perform autonomous oscillations. This region acts then as a pacemaker which periodically generates phase slips propagating away as a kink train with the "right" chirality . Suppose now that this heterogeneity is removed and the activity of the pacemaker is terminated. When $\beta = 5.0$, generated kinks continue to move away from the center (Fig. 7.2c). The situation is however different, if $\beta = 1.8$ (see Fig. 7.2d). As spatial intervals between the kinks get larger, they subsequently reverse their propagation direction and move into the central region where repeated annihilations take place. This is because single kinks and the kink trains with sufficiently large periods are characterized by the "left" chirality in this medium and propagate in such a way that the oscillation phase becomes decreased.

7.4. Phase approximation

For sufficiently small forcing amplitudes B, the train velocity $V(\lambda)$ can be analytically estimated. In this parameter region, the dynamics is ap-

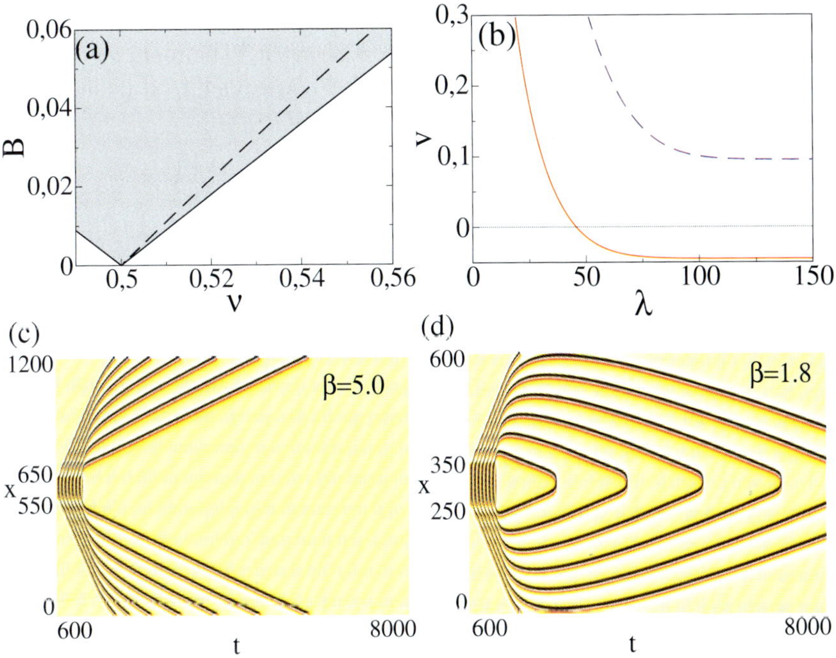

Fig. 7.2. Front propagation reversal ($n = 1, \alpha = 0.5$). (a) Standing kinks are found along the dashed line ($\beta = 1.8$) in the Arnold tongue. (b) Dependences of velocity V on spatial train period λ for $\beta = 1.8$ (solid line) and $\beta = 5.0$ (dashed line); here $\nu = 0.5525$, $B = 0.053$. Space-time diagrams showing behavior of wave patterns after termination of a pacemaker in media with (c) $\beta = 5.0$ and (d) $\beta = 1.8$; the same other parameters as in part b, local values of $\mathrm{Re}\eta(x,t)$ are shown in gray scale.

proximately described [12] by the reduced equation for the local oscillation phase φ,

$$\dot{\varphi} = \nu - \alpha - B\sqrt{1+\alpha^2}\sin n\varphi + a\left(\nabla\varphi\right)^2 + b\nabla^2\varphi, \tag{7.2}$$

where $\varphi = \phi + \arctan\alpha$ and $\eta = \rho\exp(i\phi)$. For brevity, we have introduced here notations $b = 1 + \alpha\beta$ and $a = \alpha - \beta$. Note that such reduced phase description is justified, when $b > 0$ so that uniform oscillations are modulationally (Benjamin-Feir) stable. The kinks exist for $B > B_\mathrm{A}$, where $B_\mathrm{A}(\nu) = |\nu - \alpha|/\sqrt{1+\alpha^2}$.

Applying the Cole-Hopf transformation $\varphi = (b/a)\ln u$, this phase dynamics equation is transformed to a simple form analogous to the equation for front propagation in one-component bistable media [20],

$$\partial_t u = Q(u) + b\partial_{xx}u \tag{7.3}$$

with the nonlinear function
$Q(u) = (a/b)u \left[\nu - \alpha - B\sqrt{1+\alpha^2}\sin\left(n\,(b/a)\ln u\right)\right]$. The roots u_j of equation $Q(u) = 0$ under the condition $Q'(u_j) < 0$ correspond to stable uniform locked states of the system. Explicitely, we have

$$u_j = \exp\left\{\frac{a}{nb}\left[2\pi j + \arcsin\left(\frac{\alpha-\nu}{B\sqrt{1+\alpha^2}}\right)\right]\right\}. \tag{7.4}$$

Although the system has an infinite sequence $j = 1, 2, 3, ...$ of such roots, only n of them represent physically different phase-locked states.

A front train with spatial period λ is a solution of equation (7.3) satisfying periodicity conditions $\varphi(x+\lambda/n) = \varphi(x) + 2\pi/n$ (for $n > 1$ one spatial period of the pattern consists of n subsequent $2\pi/n$-fronts). In terms of the variable u, these conditions take the form

$$u(x+\lambda/n) = \exp\left(2\pi a/nb\right)u(x). \tag{7.5}$$

Thus, the train solutions for u are not periodic, but grow exponentially with x.

When propagation reversal occurs, a stationary train is possible. In the stationary case, equation (7.3) has the first integral $(1/2)b(\partial_x u)^2 + W(u) = E$. Using the periodicity condition (7.5) and the property $W(u_{j+1}) = \exp\left(4\pi a/nb\right)W(u_j)$, we find that $E = 0$ for any stationary train. Thus, the wavelength λ_{st} of the stationary train is given by

$$\lambda_{st} = \int_{u_j}^{u_{j+1}} \sqrt{\frac{-bn^2}{2W(u)}}\,du. \tag{7.6}$$

This result does not depend on the choice of the root j.

Solitary $2\pi/n$ phase fronts are front solutions of equation (7.3), such that $u(x,t) \to u_j$ for $x \to -\infty$ and $u(x,t) \to u_{j\pm1}$ for $x \to \infty$. They can be also viewed as a limit of a periodic train with $\lambda \to \infty$. According to equation (7.6), the wavelength of a stationary train diverges, if $W(u_j) = 0$. Solving this equation, we find that stationary solitary $2\pi/n$ fronts exist along the line $B = B_{st}(\nu)$ given by

$$B_{st} = \frac{1}{2a}\left(\alpha - \nu\right)\sqrt{\frac{a^2+n^2b^2}{1+\alpha^2}}. \tag{7.7}$$

Note that, in the phase approximation, the boundaries of the Arnold tongue are $B = B_A(\nu)$. The line $B = B_{st}(\nu)$ is shown as the dashed line in Fig. 7.2a. Along this line, the reversal of the propagation direction of solitary phase fronts occurs. Phase fronts with the "right" chirality ($V > 0$) are found on the left side of this line, if $\beta > \alpha$.

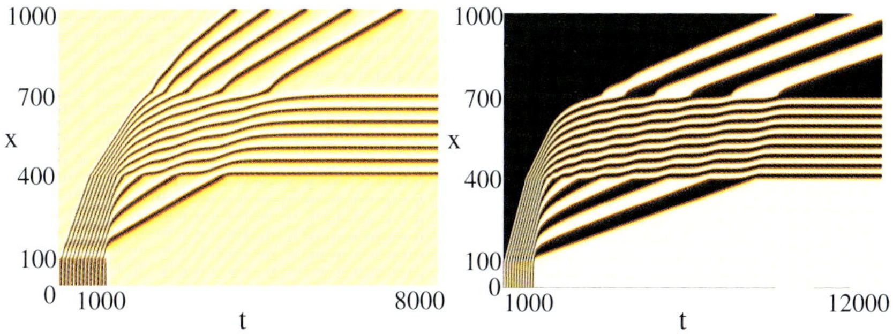

Fig. 7.3. Trapping of kinks and Bloch fronts in the 1-d medium. The coefficient β is set to $\beta = 5$ and decreased to $\beta = 1.8$ inside the central region of width 300. *Left:* Kink trap under the 1:1 resonance, the same parameters as in Fig. 1b. *Right:* Bloch front trap under the 2:1 resonance ($B = 0.061$, the same other parameters as in Fig. 1b).

As follows from (7.6), the condition for existence of stationary periodic trains is $W(u_j) < 0$. This means that they are found inside the region of the Arnold tongue, lying between the line $B = B_{st}(\nu)$ and the (nearest) boundary of the tongue (see Fig. 1a). Such a region always exists if $b > 0$. For any given set of parameters, the wavelength of the stationary train can be computed by numerical evaluation of the integral in equation (7.6).

The above analysis shows that front propagation reversal occurs near any n:1 resonance. For $n = 1$, stationary kinks (i.e., 2π-fronts) and periodic sequences of standing kinks are possible (the existence of stationary solitary kinks under global feedback conditions has previously been shown [15]). For $n = 1$, stationary π-fronts represent standing Bloch walls or their periodic sequences. Such standing structures are different from Ising walls, because the oscillation amplitude does not vanish here.

7.5. Trapping of phase fronts

The wave propagation reversal can be induced by varying parameters of the medium. Most conveniently, this can be done by changing the coefficient β in the CGLE, since this coefficient does not affect uniform oscillations and is only important for propagating waves.

The dependence of the wave propagation direction on the coefficient β can be used to trap kinks and Bloch fronts. Such traps can be designed by creating spatial regions, where the coefficient β is locally changed to reverse the propagation velocity. The left panel in Fig. 7.3 shows an example of a kink trap in the one-dimensional medium at the 1:1 resonance. The value

Fig. 7.4. Trapping of kinks and Bloch fronts in the two-dimensional medium. The coefficient β is set to $\beta = 5$ and decreased to $\beta = 1.8$ in the rectangular central region. The medium parameters are the same as in Fig. 7.2. The system size is 1000×1000. *Upper panel:* Kink trap in 1:1 resonance. The snapshots of the spatial distribution of $\mathrm{Re}\eta$ are taken at $t = 6000$, $t = 19000$ and $t = 32000$. *Lower panel:* Bloch wave trap in 2:1 resonance. Snapshots the spatial distribution of $\mathrm{Re}\eta$ at $t = 2000$, $t = 16000$ and $t = 46800$. ($B = 0.061$)

of β is decreased in the central region. No-flux boundary conditions are used in all our simulations. Initially, a rapid pacemaker operates at the left end of the medium. This pacemaker produces a kink train with a short spatial period. The train enters the modified central region and passes it with some deceleration. When the pacemaker is terminated, further kinks are not produced. However, the kinks inside the central modified region become trapped inside it and form a stationary pattern with a period corresponding to the velocity reversal. If the central heterogeneity is removed, the stored pulses would propagate out of it. The right panel in Fig. 7.3 demonstrates the trapping of Bloch fronts at the 2:1 resonance. The pacemaker at the left end of the medium produces a train of Bloch fronts.

Similar traps for kinks and Bloch fronts can also be constructed in two-dimensional media. The upper panel of Fig. 7.4 (see also video 1 [21]) shows a series of snapshots where trapping of kinks by a central modified region is seen. A pacemaker in the lower left corner emits a kink train with a short spatial period. The first snapshot shows the kink train passing the rectangular-shaped inhomogeneity. Inside the inhomogeneity, the velocity of the kinks is decreased, resulting in a delay of the kinks at the upper right corner of the rectangle. In the second snapshot, the rear end of the kink train has reached the lower left boundary of the inhomogeneity. The kinks inside the inhomogeneity close to this boundary can no longer propagate and the kinks outside propagate around the central region. This leads to the formation of ring-shaped kinks in the upper right part of the inhomogeneity. As long as new kinks arrive, these rings collapse. When the kink train has passed the inhomogeneity, the stored kinks form stationary ring-shaped structures. In the lower panel of Fig. 7.4 (see also video 2 [21]), trapping of traveling Bloch fronts at the 2:1 resonance is demonstrated. It proceeds similar to the respective process for the kinks near the 1:1 resonance. In our simulations, wave traps with various sizes and with complicated geometries could be created.

In the above discussion, we assumed that the condition $\beta > \alpha$ was satisfied. If the opposite condition $\beta < \alpha$ holds, waves in the unforced CGLE have negative dispersion (see, e.g., [22]). In this case, inwardly rotating spirals ("antispirals") and inwardly propagating target patterns are possible. In such media, reversal of front propagation also takes place and the wavelength of a stationary train is again given by equation (7.6). However, the line $B = B_{st}(\nu)$ lies now on the left side of the Arnold tongue and stationary front trains are found in the region between this line and the left tongue boundary.

7.6. Twisted spirals

A special effect, related to kink propagation reversal, is the formation of twisted spirals near the 1:1 resonance in two-dimensional media (Fig. 7.5a). The central and outer parts in such a spiral are wound in opposite directions. These structures are stable, they are observed in numerical simulations starting with various initial conditions. A twisted spiral rigidly rotates as a whole, retaining its shape. In Fig. 7.5b, three subsequent snapshots of the spiral, separated by a third of the rotation period each, are superimposed (see also video 3 [21]). We see that the instantaneous rotation center does not coincide with the location of the spiral tip. Instead, the oppositely wound central part of the spiral is steadily rotating. Thus, this regime can also be characterized as a kind of meandering. Qualitatively, the development of twisted spirals can be understood by noticing that the waves are

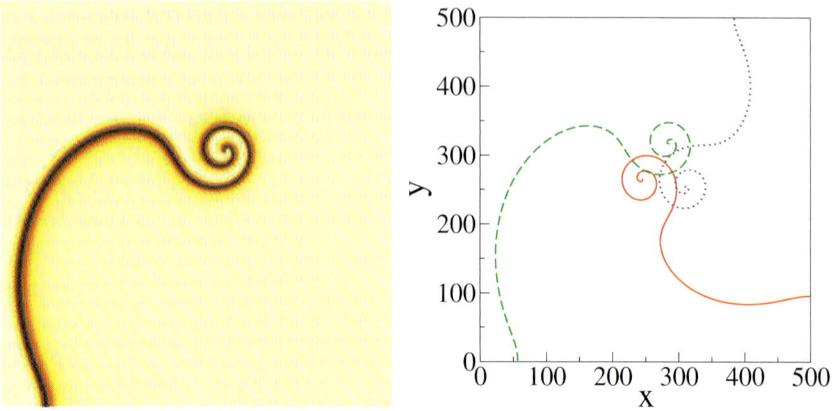

Fig. 7.5. Rotation of a twisted spiral. Left: Spatial distribution of Reη. Right: Position of the spiral at times $t = 0$ (solid), $t = 340$ (dashed), $t = 680$ (dotted). Parameters are $\alpha = 4.19$, $\beta = 0.992$, $\nu = 3.9895$, $B = 0.0455$, the system size is 500×500. Numerical integration using the explicit Euler scheme with $\Delta x = 0.2$ and $\Delta t = 0.0025$.

tightly wound near the center and, therefore, their propagation direction should be reversed there. In the displayed simulation, the medium was characterized by negative dispersion ($\beta < \alpha$). Similar behavior has, however, been found by us in the simulations for the media with positive wave dispersion ($\beta > \alpha$) [19]. By changing the forcing intensity and frequency, winding and unwinding of the central part of the spiral can be controlled.

Our theoretical study has shown that, applying periodic forcing, one can induce propagation reversal of kinks, Bloch walls and $2\pi/n$ phase fronts for higher resonances with $n > 2$. Using this effect, traps for propagating kinks and other phase fronts can be designed by creating appropriate heterogeneities in the medium. In our simulations, such heterogeneities were introduced by spatial variation of the coefficient β in the CGLE, but similar effects can be achieved by varying other parameters of the medium or by applying inhomogeneous forcing. We have also shown that, in uniform media with 1:1 forcing, steadily rotating twisted spirals can develop. Though our results have been obtained only for the CGLE, we expect that they should be characterictic for a class of media where oscillations are not strongly relaxational. Our simulations using a realistic model of the catalytic surface reaction of CO oxidation on platinum have shown that the wave propagation reversal under periodic forcing takes place near a supercritical Hopf bifurcation in this reaction and that the wave traps can be constructed there [19]. Another experimental system where the predictions of our theory can be tested is the oscillatory photosensitive Belousov-Zhabotinsky reaction.

References

[1] A. N. Zaikin and A. M. Zhabotinsky. Concentration wave propagation in two-dimensional liquid-phase self-oscillating system. *Nature*, 225:535, 1970.

[2] Arthur T. Winfree. Scroll-shaped waves of chemical activity in three dimensions. *Science*, 7:937–939, 1973.

[3] V. Petrov, Q. Ouyang, and H. L. Swinney. Resonant pattern formation in a chemical system. *Nature*, 388:655 – 657, 1997.

[4] A. L. Lin, M. Bertram, K. Martinez, H. L. Swinney, A. Ardelea, and G. F. Carey. Resonant phase patterns in a reaction-diffusion system. *Phys. Rev. Lett.*, 84:4240, 2000.

[5] V. K. Vanag, A. M. Zhabotinsky, and I. R. Epstein. Oscillatory Clusters in the Periodically Illuminated, Spatially Extended Belousov-Zhabotinsky Reaction. *Phys. Rev. Lett.*, 86:552, 2001.

[6] B. Marts, A. Hagberg, E. Meron, and A. L. Lin. Bloch-Front turbulence in a periodically forced Belousov-Zhabotinsky reaction. *Phys. Rev. Lett.*, 93:108305, 2004.

[7] K. Krischer, M. Eiswirth, and G. Ertl. Oscillatory CO oxidation on Pt(110): modelling of temporal self-organization. *J. Chem. Phys.*, 96(12):9161–9172, 1992.

[8] M. Kim, M. Bertram, M. Pollmann, A. v. Oertzen, A. S. Mikhailov, H. H. Rotermund, and G. Ertl. Controlling chemical turbulence by global delayed feedback: pattern formation in catalytic co oxidation reaction on pt(110). *Science*, 292:1357–1360, 2001.

[9] M. Bertram, C. Beta, H. H. Rotermund, and G. Ertl. Complex patterns in a periodically forced surface reaction. *J. Phys. Chem. B*, 107:9610, 2003.

[10] J. Davidsen, A. S. Mikhailov, and R. Kapral. Front explosion in a periodically forced surface reaction. *Phys. Rev. E*, 72:046214, 2005.

[11] I. S. Aranson and L. Kramer. The world of the complex Ginzburg-Landau equation. *Rev. Mod. Phys.*, 74(1):99–143, 2002.

[12] P. Coullet and K. Emilsson. Strong resonances of spatially distributed oscillators: a laboratory to study patterns and defects. *Physica D*, 61:119–131, 1992.

[13] P. Coullet and K. Emilsson. Pattern formation in the strong resonant forcing of spatially distributed oscillators. *Physica A*, 188:190–200, 1992.

[14] P. Coullet, J. Lega, B. Houchmanzadeh, and J. Lajzerowicz. Breaking chirality in nonequilibrium systems. *Phys. Rev. Lett.*, 65:1352, 1990.

[15] F. Mertens, R. Imbihl, and A. S. Mikhailov. Breakdown of global coupling in oscillatory chemical reactions. *J. Chem. Phys.*, 99(11):8668–8671, 1993.

[16] H. Chaté, A. Pikovsky, and O. Rudzick. Forcing oscillatory media: Phase kinks vs. synchronization. *Physica D*, 131(1-4):17–30, 1999.

[17] M. Argentina, O. Rudzick, and M. G. Velarde. On the Back-firing instability. *Chaos*, 14(3):777 – 783, 2004.
[18] C. Hemming and R. Kapral. Phase front dynamics in inhomogeneously forced oscillatory systems. *Physica D*, 306:199–210, 2002.
[19] O. Rudzick and A. S. Mikhailov. unpublished.
[20] A. S. Mikhailov. *Foundations of Synergetics I. Distributed Active Systems.* Springer, Berlin, 2nd ed., 1994.
[21] Supporting online material can be found at `http://www.fhi-berlin.mpg.de/complsys/cs_videos/twisted_spirals/`.
[22] M. Stich and A. S. Mikhailov. Complex pacemakers and wave sinks in heterogeneous oscillatory chemical systems. *Z. Phys. Chem.*, 216:521–533, 2002.

Chapter 8

Visualizing pitting corrosion on stainless steel

Monika Dornhege, Christian Punckt* and Harm H. Rotermund

[1] *Fritz-Haber-Institute of the Max-Planck-Society, Department of Physical Chemistry, Faradayweg 4-6, D-14195 Berlin, Germany*

**punckt@fhi-berlin.mpg.de*

Contents

8.1. Introduction

Corrosion is a phenomenon affecting all of us in everyday life; just imagine rusting cars or water pipes. In order to develop a finer grasp for the processes leading to the corrosive destruction of materials, a great effort is usually made to study the chemistry of corrosion in detail. However, in our group we follow a different approach: The transition from negligibly small and localized damage to a highly affected corroding surface is described using the conceptual framework of nonlinear dynamics and pattern formation. In order to motivate this approach, first the necessary background is provided (Chapter 8.2). On this basis a stochastic reaction-diffusion model is introduced (Chapter 8.3). To gain corresponding experimental support for this model, the methods described in Chapter 8.4 can be utilized. Finally, the results of our experimental investigations are described and discussed in Chapter 8.5.

8.2. Background

Stainless steel is corrosion resistant because a protective oxide layer naturally forms on top of the surface in the presence of oxygen and humidity. This protective oxide layer typically has a thickness in the order of nanometers, depending on the present environmental conditions. XPS studies of oxide films formed in air on AISI 316 revealed that not only oxidation of the material takes place, but also chromium and metallic nickel accumulate at the interface between oxide layer and bulk material [1]. The protective film is, of course, not perfect but contains defects like inclusions and grain boundaries. At these defects the film may locally break down and dissolution of the bulk material may start [2]. This kind of corrosion is called pitting corrosion and is estimated to cause a third of all chemical plant failures in the United States [3].

The onset of pitting corrosion occurs suddenly: If one performs electrochemical experiments with stainless steel, e. g. by applying a constant electrical potential to a sample immersed in dilute NaCl solution, the electrical current – which is an indicator for chemical activity (corrosion) on the metal surface – is low over a wide parameter range. But if critical parameters like temperature, potential, or electrolyte concentration exceed a certain critical value, the current rises abruptly and the metal surface is severely affected by pitting corrosion. The transition to high corrosion rates is preceded by the appearance of metastable corrosion pits.

A detailed mechanism for the formation of metastable pits was proposed by Pistorius and Burstein [4]: Electrochemically active inclusions are spread over the surface of stainless steel. These sites can be sulfide inclusions [2, 5], but also other kinds of electrochemically active defects associated with metal dissolution may be responsible for pit nucleation. They are attacked by the electrolyte (e. g. NaCl) while the protective oxide film remains stable. In other words, a surface defect may act as a local anode. This means that an electrochemical reaction is initiated and the bulk metal beneath the surface defect starts to be eroded. Chloride ions are attracted to maintain charge neutrality, and dissolution of sulphide inclusions may lead to the formation of thiosulphate ions. These ions, together with protons which are formed during hydrolysis of the metal, accelerate further dissolution of the metal beneath the protective oxide layer, which itself is hardly affected. A small pit forms beneath the oxide layer, filled with an aggressive solution of low pH-value (see Fig. 8.1A and B).

Since the dissolution process is diffusion controlled and the pit cover is

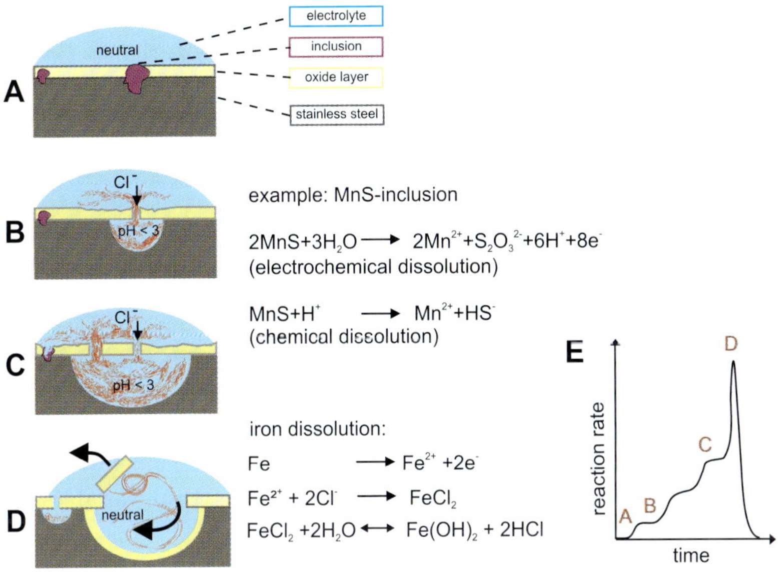

Fig. 8.1. Illustration of activation and repassivation of a pitting site, a detailed explanation is given in the text.

the major diffusion barrier, ruptures and additional holes in the cover are assumed to occur through mechanical stress caused by osmotic pressure due to the concentration gradient between the electrolyte within the pit and the bulk electrolyte [4] (see Fig. 8.1C). Each rupture is associated with a stepwise increase in the reaction rate, hence, with a stepwise increase of the electrical current flowing through the surface. But the breakdown of the protective oxide layer also leads to a dilution of the aggressive electrolyte inside the pit and the pH-value increases. The dilution can reach such an extend, that active dissolution of the metal bulk is no longer sustained, the protective oxide layer rebuilds inside the pit (self healing), and the pit passivates (see Fig. 8.1D). This is accompanied by a final sharp spike of the electrical current (cf. Fig. 8.1E).

Typically, the lifetime of metastable pits is in the order of a few seconds. Pits which exhibit electrochemical metal dissolution are called active. The remaining small holes in the metal surface after passivation of a pit are called inactive pits. Pits can also undergo reactivation.

The appearance of corrosion pits has up to now been described by stochastic processes, and the transition to higher corrosion rates was explained by the nucleation and stabilization of only a few corrosion pits with high activity [6, 7]. In our investigations we take a closer look at the transition from low to high corrosion rates. We analyze the spatiotemporal dynamics of this process both in numerical simulations and in the experiment and demonstrate that the onset of pitting corrosion is a cooperative critical process that precedes like a chain reaction.

8.3. Mathematical model

Recently, there has been growing evidence that metastable pits have effects both in time [6, 8] and space [9] on each other. As indicated in Fig. 8.1C and D, it is believed that active pits release aggressive ions which weaken the protective oxide layer. Thus, the nucleation rate of new metastable pits is enhanced in the vicinity of active pits. Based on these assumptions, a two-dimensional stochastic reaction-diffusion model was developed by A. Mikhailov, J. Hudson, and coworkers [10]. Four variables play an important role in this model:

(1) $s(x, y, t)$ denotes the oxide film damage (the intact oxide layer corresponds to $s = 0$).
(2) $c(x, y, t)$ is the accumulated concentration of aggressive species.
(3) $I_k(t)$ describes the total current flowing through the k-th pit.
(4) $\Phi(x, y, t)$ is the potential drop due to the current associated with active pits.

The nucleation rate of new pits increases with the oxide film damage s and the concentration of aggressive species c. In contrast to that, the ohmic potential drop Φ in the vicinity of an active pit inhibits pit nucleation. The influence of these three variables is combined in an auxiliary variable M:

$$M = \alpha_s \cdot s + \alpha_c \cdot c - \alpha_\Phi \cdot \Phi \,, \tag{8.1}$$

from which the local pit generation rate w is calculated as follows:

$$w(M) = \frac{w_0}{1 + \exp\left((M_0 - M)/H\right)} \,. \tag{8.2}$$

The influence of varying temperature or electrolyte concentration can be simulated by tuning w_0 which is the maximum pit generation rate. The graph of eq. (8.2) is displayed in Fig. 8.2. Below $M_0 = 50$ the rate is small, increases then rapidly and finally saturates. The width of the transition regime is specified by the parameter H.

Numerical simulations are performed on a quadratical grid with discrete time steps, thus the nucleation probability at a given point is calculated by

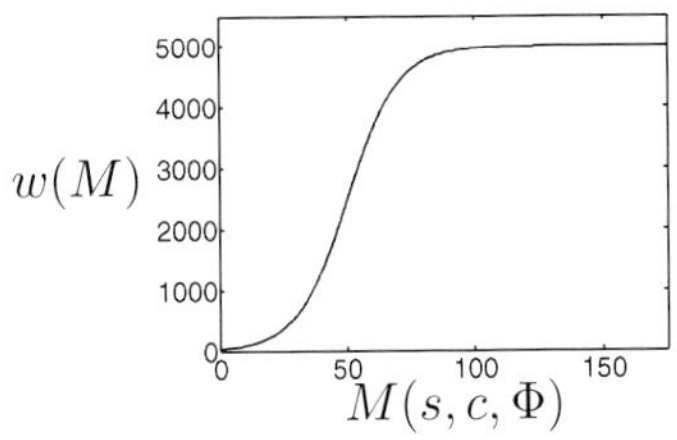

Fig. 8.2. Local generation rate as a function of M. $M_0 = 50$, $w_0 = 5000$, $H = 10$. Reproduced from [11].

multiplying the local value of w with the time step and the area associated with every point of the grid. A pit is initiated if a random number, which is calculated for every grid point and time step, is less than the local nucleation probability.

Once a pit is initiated, a current is flowing locally. Numerical simulations have shown, that the exact shape of the current spike produced by an active metastable pit does not qualitatively change the results. Here, a sharp rise followed by an exponential decay is used for each pit:

$$I_k = \begin{cases} 0 & t < t_k \\ I_0 \exp\left[-\dfrac{t - t_k}{\tau}\right] & t \geq t_k \end{cases} \tag{8.3}$$

with

I_0 : peak current,
τ : time constant for current decay,
t_k : time of initiation of a pit.

Aggressive ions are released by active pits proportional to the current. These ions can diffuse laterally within the thin diffusion boundary layer on top of the metal surface and into the bulk electrolyte. Two coupled partial differential equations describe the spatiotemporal evolution of s and c:

$$\partial_t s = \mu \cdot c - \nu \cdot s \,, \tag{8.4}$$

$$\partial_t c = -\gamma \cdot c + D \cdot \nabla^2 c + \beta \cdot \sum_k j_k \tag{8.5}$$

with

$$\gamma = \frac{2D}{d^2}, \quad \beta = \frac{2}{dnF} \quad \text{and} \quad j_k = \frac{I_k}{\pi a^2} \,.$$

The first term of eq. (8.4) describes the increase of oxide film damage in the presence of aggressive species, whereas the second term corresponds to the self-healing of the oxide film. The diffusion of aggressive species out of the boundary layer into the electrolyte is considered in the first term of eq. (8.5). The second term describes lateral diffusion of aggressive species (diffusion constant D). The concentration of aggressive ions which is released by active pits is taken into account in the third term and is calculated from the thickness of the boundary layer d, the oxidation state n of the released metal cations, the Faraday constant F, the pit radius a, and the local current contributions I_k of all active pits. In the model the release of aggressive species corresponds to the amount of released metal cations.

Local mean field approximation: By temporal and spatial averaging of the equations a local mean-field approximation was derived, from which qualitative aspects of the model can be gathered [12]. The expression "local" means, that averaging takes place not over the whole surface, but only over a small area. However, the area should be big enough to allow for a reasonable definition of pit density. Since the influence of the ohmic potential drop has only a small range it is neglected in the following. First, an equation for the current density i is derived from eq. (8.3) by summing up the current of all active pits k with nucleation times t_k and locations r_k:

$$\partial_t i = -\frac{i}{\tau} + I_0 \sum_k \delta(\mathbf{r} - \mathbf{r}_k) \cdot \delta(t - t_k) \,. \tag{8.6}$$

Temporal averaging of the two Dirac-functions δ yields the nucleation rate $\langle w \rangle$. Thus, using temporally and spatially averaged local variables the model equations can be written as follows:

$$\partial_t \langle i \rangle = -\frac{\langle i \rangle}{\tau} + I_0 \langle w \rangle \,, \tag{8.7}$$

$$\partial_t \langle s \rangle = \mu \cdot \langle c \rangle - \nu \cdot \langle s \rangle \,, \tag{8.8}$$

$$\partial_t \langle c \rangle = -\gamma \cdot \langle c \rangle + D \cdot \nabla^2 \langle c \rangle + \beta \cdot \langle i \rangle \,, \tag{8.9}$$

$$\langle M \rangle = \alpha_s \cdot \langle s \rangle + \alpha_c \cdot \langle c \rangle \,, \tag{8.10}$$

$$\langle w \rangle = \frac{w_0}{1 + \exp\left((M_0 - \langle M \rangle)/H\right)} \,. \tag{8.11}$$

Adiabatic elimination of eq. (8.7) is possible, since the timescale of the damage of the oxide layer is much slower than the timescale of the electric current density:

$$\langle i \rangle \approx I_0 \tau \langle w \rangle \, . \tag{8.12}$$

Hence, the local mean-field approximation can be summarized in the following two equations:

$$\partial_t \langle s \rangle \; = \; \mu \cdot \langle c \rangle \; - \; \nu \cdot \langle s \rangle \, , \tag{8.13}$$

$$\begin{aligned} \partial_t \langle c \rangle \; = \; & \frac{\beta I_0 \tau w_0}{1 + \exp\left(\dfrac{M_0 - \alpha_s \cdot \langle s \rangle - \alpha_c \cdot \langle c \rangle}{H} \right)} \\ & - \gamma \cdot \langle c \rangle \; + \; D \cdot \nabla^2 \langle c \rangle \, . \end{aligned} \tag{8.14}$$

Equations (8.13) and (8.15) reveal the autocatalytic nature of the model, which was before hidden in the stochastic part of the full model (eq. (8.1) and (8.2)): Aggressive ions and a high oxide film damage have an activating effect on the pit nucleation rate. In particular, the presence of active pits increases the nucleation rate in a diffusion-limited area around the active site. Thus, the model contains an autocatalytic component. Further details about the model can be found in [13].

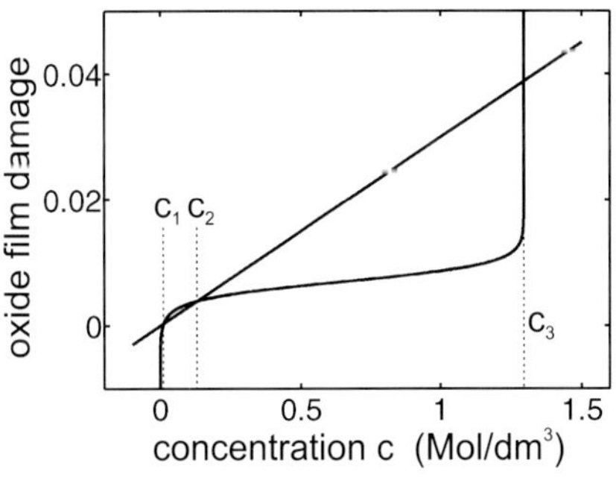

Fig. 8.3. Nullclines of eqs. (8.13) and (8.15). Reproduced from [11].

Choosing suitable parameters (corresponding to the ones used in [14], Text S1), the nullclines of these equations, i. e. $\partial_t \langle s \rangle = 0$ and $\partial_t \langle c \rangle = 0$, show three intersection points (cf. Fig. 8.3). There are two stable fixed points, indicating a bistable system, one corresponding to an intact oxide layer with low concentration of aggressive species (c_1), the other one corresponding to a damaged oxide layer with high concentration of aggressive species (c_3). A third, unstable fixed point is found at intermediate values

(c_2). Further analysis of the local mean-field approximation reveals the existence of propagating fronts between the two stable states. These results are confirmed by numerical simulations using the full model. Choosing the same parameters as for the nullclines, an autocatalytic reproduction of pitting sites is found. As shown in Fig. 8.4, the local film damage expands front-like. This process is accompanied by an exponential growth of the accumulated number of pits and the current.

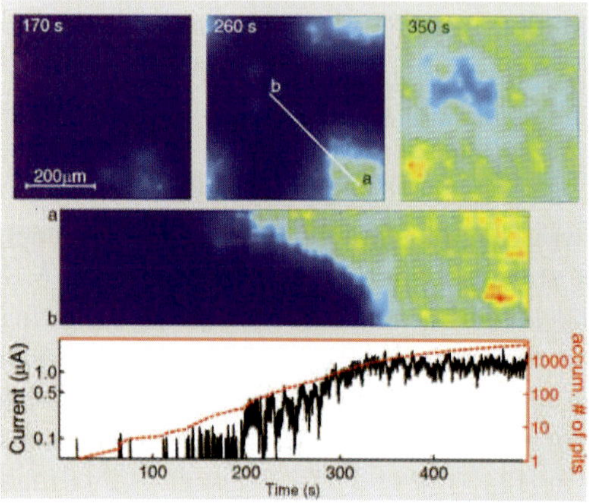

Fig. 8.4. Simulation of corrosion onset with periodic boundary conditions. (*Top*) Snapshots showing the local film damage at the indicated time moments. Blue corresponds to low, orange to high film damage. (*Middle*) Space/time diagram along the line marked with *ab*. (*Bottom*) *Red line*: Accumulated total number of pitting sites. *Blue line*: Total current. Reproduced from [11]

8.4. Experimental methods

The observation of propagating fronts in numerical simulations were the starting point of our experimental studies. Of course, microscopic investigations of corrosion at surfaces have been conducted, largely after the termination of the corrosion experiment [4, 15]. Optical microscopy has been applied in situ to observe relatively large, stable pits [16]. Individual metastable pits have been visualized using pH sensitive agar gels [17, 18]. By applying scanning methods changes in the thickness of the oxide layer in the vicinity of an active pit and the topography of surfaces before and after pitting events were analyzed [19, 20]. All these methods suffer from low temporal or spatial resolution, so that it was impossible to examine the interdependency between oxide layer weakening and occurrence of pits described above. In our group we gather information about the corrosion process by employing two different complementary microscopic techniques: **E**llipso-**M**icroscopy for **S**urface **I**maging (EMSI) [21, 22] and specially adapted contrast-enhanced optical microscopy. Both techniques are accompanied by parallel monitoring of the current. Using contrast-enhanced microscopy the temporal and spatial development of metastable pits is followed during the onset of pitting corrosion and active and inactive pits are differentiated. EMSI yields complementary information about changes of the protective oxide layer [14].

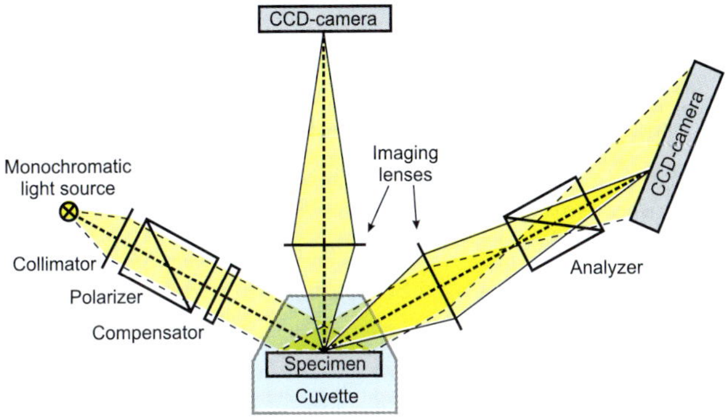

Fig. 8.5. Sketch of the experimental setup. Reproduced from [23]

Our experimental studies were conducted using an 8 mm diameter AISI 316 stainless steel disk electrode. The nominal composition was 0.13% C, 0.31% Co, 18.18% Cr, 0.38% Cu, 1.75% Mn, 2.40% Mo, 12.25% Ni, 0.016% S, 0.35% Si, with the balance Fe. The sample was polished up to 600 grit polishing paper followed by a 1 µm diamond paste polish. The surface was then cleaned with ethanol in an ultrasonic bath and mounted with silver lacquor onto a holder. The sample was placed vertically in a cuvette to serve as the working electrode of a three electrode configuration. In front of the sample a platinum wire was employed as a counter electrode, and a silver/silver chloride electrode acted as the reference electrode and was placed between the sample and the platinum wire. A potentiostat controlled the potential and also recorded the current flowing through the counter electrode.

The cell is a cuvette with specially arranged windows, which allow the light of a HeNe-Laser (20 mW) to pass perpendicularly through the glass. Light propagates through the electrolyte, is reflected at about 70° off the sample, and exits the system perpendicularly through the opposite window, as indicated in Fig. 8.5. The laser beam is elliptically polarized by combining a Glan-Thompson prism and a quarter wave plate in such a way that after reflection from the sample only linearly polarized light is leaving the cuvette. A second Glan-Thompson prism is oriented such that nearly all intensity is removed from the imaging path. The sample is magnified by a lens onto a CCD chip. The CCD chip is tilted about 75° away from the optical axis, thereby compensating for most of the distortion. A plane homogeneous sample will display a uniform blank image, the "null" image. However, local changes on the surface, e. g. areas with a different oxide layer thickness, are detected as bright spots by the CCD camera. In order to improve quality and contrast of the images we employ an Argus 20 image processing unit from Hamamatsu. It is capable of real time background subtraction, so that most of the inhomogeneities of the surface and interference patterns originating from the laser light are not visible.

The spatial resolution of EMSI is limited by geometrical restrictions to about 12 µm, and it cannot resolve individual pits. Therefore, we developed a contrast-enhanced optical microscope that images the sample with 8-times magnification onto a CCD chip. By using a Schwarzschild objective we were able to provide both the necessary large working distance ($\approx$ 20 mm) and a diffraction limited spatial resolution of 2 µm at a field of view of approx. 200 µm. In order to improve the contrast we again used a downstream Argus 20 image processing unit.

The subtracted and enhanced ellipsomicroscopic and microscopic images were stored on a DVD-recorder. Great care was taken to synchronize the video and the current measurements. For this purpose a flashlight was activated which could be easily identified on the videos. Simultaneously, a signal was send to the computer storing the temporal evolution of the current.

To ensure that current peaks can be associated with events detectable by EMSI and microscopy, the entire sample was coated – excluding the imaged area varying between 0.13 and 1 mm diameter – with insulating lacquer and Apiezon wax.

8.5. Experimental results and discussion

Fig. 8.6A displays four snapshots from an EMSI video sequence recorded at low pitting activity. They show bright areas with diameters up to 100 µm, with the brightness gradually fading towards the periphery. The space-time diagram in the middle panel displays the evolution along the line *ab* in the first snapshot. The bottom diagram displays the parallel recording of the current. All observed bright areas are unambiguously associated with current spikes. The lifetimes of the bright areas are 10–15 s, while the spikes in the current have a shorter duration of 4–10 s. EMSI is capable of directly

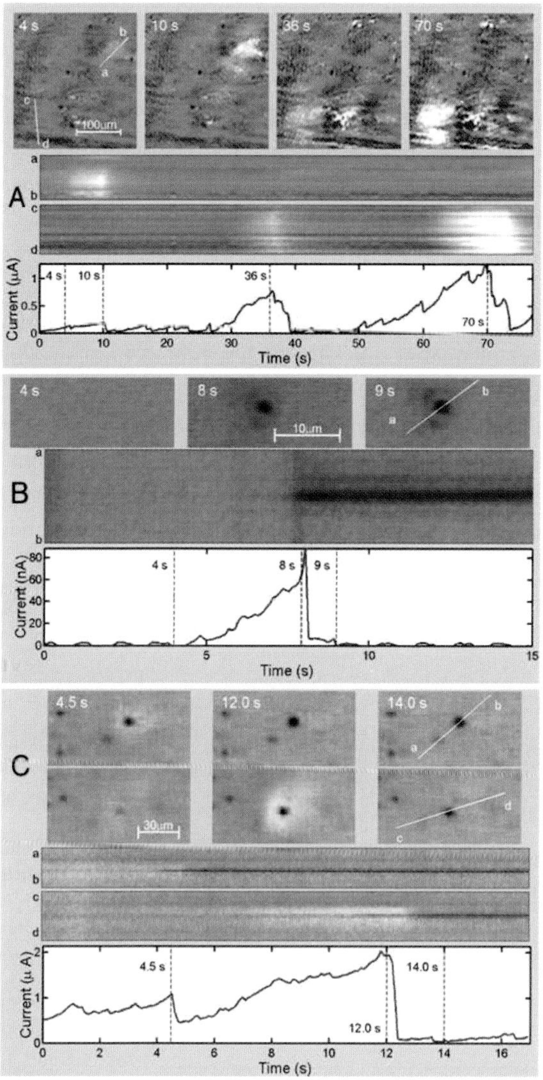

Fig. 8.6. Microscopic observations of individual pitting events. (**A**) EMSI visualizing oxide film damage. (**B**) Nucleation of a single pit seen by contrast-enhanced optical microscopy. (**C**) Reactivation of a single pit seen by contrast enhanced optical microscopy. Below the snapshots in each part, space-time diagrams showing evolution along the lines marked *ab* or *cd* in the respective images and parallel current recordings are displayed (dashed lines indicate snapshot moments). The reaction occurred in 0.05 M NaCl at 22 °C. The potential was scanned from 771 mV versus normal hydrogen electrode (mV_{NHE}) at 1 mV/s (A), from 542 mV_{NHE} (B) and held at 607 mV_{NHE} (C). Reproduced from [14].

imaging the damage of the oxide layer. In earlier scanning ellipsometric observations of localized corrosion, a decrease of a few angstrom in the oxide layer thickness was reported [19]. Hence, we believe that the observed bright areas reveal the local damage of the protecting oxide film around an active pit. However, individual pits are not visible in the EMSI images.

Fig. 8.6B shows a small metastable pit observed with the optical microscope. The formation of the pit is accompanied by a spike in the current (with an amplitude of only 80 nA), indicating an electrochemical reaction. About three seconds later, the pit abruptly passivates and the electrical current drops to its noise level. The pit remains seen as a dark spot.

The activity of metastable pits with a higher current can be identified with optical microscopy. In Fig. 8.6C, two pits become active at $t = 4.5$ s and $t = 12$ s showing peak currents of 0.6 μA and 2 μA, respectively. During their active states, each of the pits is surrounded by a bright "halo" in the image. This may be caused by a hemispherical concentration gradient of ions surrounding the active site, which form an effective micro lens illuminating the surface and lead to the observed bright halo. After the current has dropped to its noise level, the halo remains for ≈ 0.6 s, which is about the time needed for diffusion of the ions away from the pit.

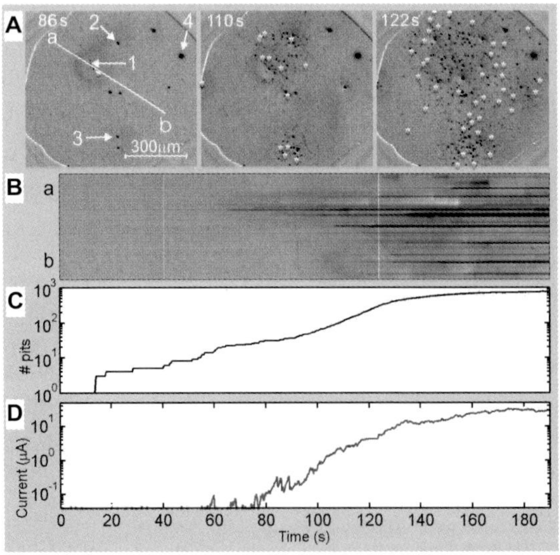

Fig. 8.7. Onset of pitting corrosion observed with optical microscopy. **(A)** Snapshots of a computer processed video sequence obtained with optical microscopy; green stars mark the nucleation of new pits. **(B)** Space-time plot along the line *ab* in A. **(C)** Total number of pits on a logarithmic scale. **(D)** Total current as a function of time. The reaction conditions were $T = 20.3$ °C with the potential held at 615 mV_{NHE}. Reproduced from [23].

To summarize, using EMSI and contrast enhanced microscopy we succeeded to visualize both the oxide film damage and the nucleation and activity of individual microscopic corrosion pits. The results described so far were obtained after scanning the potential from -93 mV_{NHE} with a velocity of 1 mV/s in the positive direction immediately after immersing the sample into the electrolyte. Using this procedure, we reproduced the exponential increase of the number of pitting sites, as had been found in the simulations, but no fronts were observed [14].

Obviously, the oxide layer thickness should be of great importance for the possibility to observe any fronts experimentally. It is known from XPS studies of stainless steel that the properties of the protective oxide layer change with time while being immersed in an electrolyte [1]. In the following we kept the sample for 90 min at open circuit potential after it was immersed. Then the potential was rapidly scanned to a pre-defined value at which the experiment should been conducted and held constant.

Results of one of our measurements using this method are displayed in Fig. 8.7. The first panel of Fig. 8.7A displays a snapshot of the electrode surface at $t = 86$ s. A few black spots have already appeared, each of them corresponding to the nucleation of a corrosion pit. Around the pits labeled 1–3, dark clouds have developed, which first appeared at around $t = 76$ s. The clouds reach an extent of almost 200 µm and evolve presumably due to the release of aggressive ions from the active pits. This process is only visible due to strong contrast enhancement. In the second panel of Fig. 8.7A, more pits are present on the surface. New pits have nucleated mainly in the vicinity of those pits with clouds. These pits obviously have changed their respective surroundings, such that nucleation of new pits is enhanced. This eventually leads to a spreading of the nucleation zones (see Fig. 8.7A, second and third panel and Fig. 8.7B) until the whole surface becomes covered with pits. Inspection of the microscopic images in Fig. 8.7 reveals, that not all initial pits on the surface initiated a pit nucleation zone. For example, around Pit 4 no dark cloud developed and no further pits appeared, even though it is much bigger than Pits 1, 2, and 3. This might indicate that not only the size, but also the chemistry of the dissolution site determines, whether the change in the environment is strong enough that pits nucleate preferentially in its vicinity. This, of course, has to be studied in future work.

The finding of a spreading nucleation zone encouraged us to use EMSI and optical microscopy simultaneously. In Fig. 8.8 one result of these measurements is displayed. Two bright regions are visible in the EMSI image at $t = 90$ s (first panel of Fig. 8.8A), one at the bottom (marked with letter a) and one at the top right corner of the image (marked with letter b).

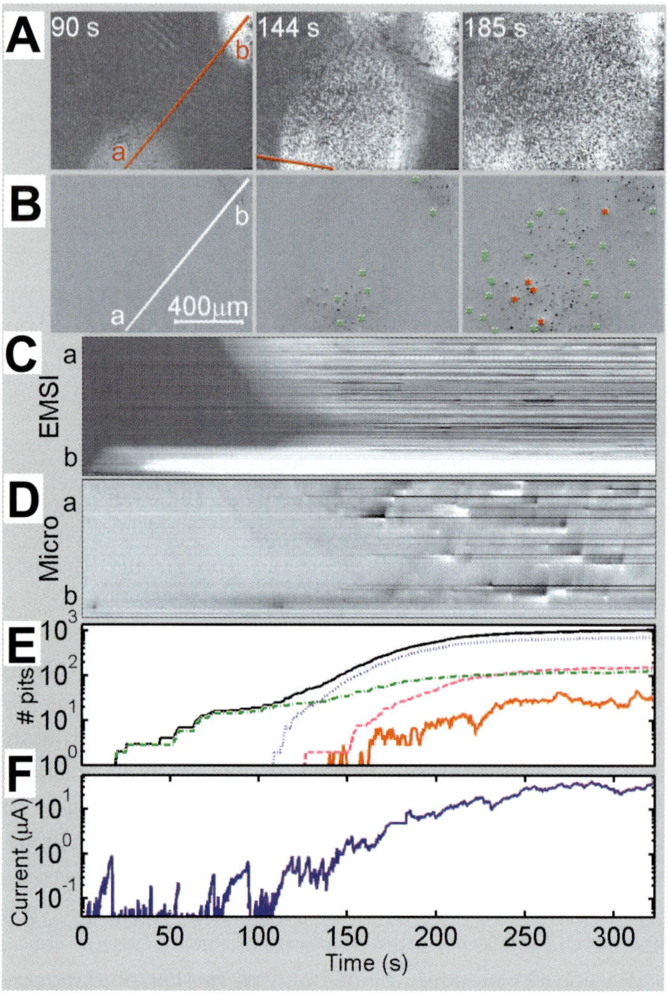

Fig. 8.8. Onset of pitting corrosion observed with EMSI and optical microscopy simultaneously. **(A)** Snapshots of a computer processed video sequence obtained with EMSI. **(B)** Snapshots of a computer processed video sequence obtained with optical microscopy corresponding to the snapshots in (A), green stars mark the nucleation of new pits whereas red stars mark highly active pits which develop a halo. **(C)** Space time diagram calculated along the line *ab* in (A). **(D)** Space time diagram along the line *ab* in (B). **(E)** Number of pits on a logarithmic scale. *Solid black line*: total number of pitting sites; *chain dotted green line*: number of pitting sites which occur in an area comparable to the bright region at the top-right of the middle panel in (a); *dotted blue line*: number of pitting sites which occur in an area comparable to the bright region at the bottom middle of the middle panel in (a); *dashed magenta line*: number of pitting sites which occur in an area comparable to the bright region of the top middle of the middle panel in (a); *solid red line*: number of the highly active pits with halo. **(F)** Total current as a function of time. The reaction conditions were $T = 21.9\,°C$ with the potential held at $637\,mV_{NHE}$. Reproduced from [23].

The bright region around b originates from the location of a pit that has nucleated at $t = 2\,\mathrm{s}$. It grows for approximately 15 s and reaches an extend of $\approx 200\,\mu\mathrm{m}$ diameter as can be estimated from the space time plot of Fig. 8.8C. New pits nucleate in the vicinity of the first pit, all of them in an area corresponding to the bright region observable with EMSI. The bright region around a originates from an additional nucleating pit at $t = 75\,\mathrm{s}$ at a different location near the lower rim. Here, a front-like spreading of the pit nucleation zone starts, clearly preceded by the propagation of a second bright region in the EMSI video sequence (see Fig. 8.8A-D).

A third pit nucleation zone forms at $t = 117\,\mathrm{s}$ at the top of the unsealed area and is visible in the second panel of Fig. 8.8A. Again it is accompanied by a bright region observed with EMSI. Eventually all three bright regions expand until the whole surface appears bright (as seen in the third panel of Fig. 8.8A). In the end the surface is covered with more than 1000 randomly distributed pits, as can be determined by the microscopic picture (see Fig. 8.9, where a microscopic image of the surface after termination of the experiment is shown).

An analysis of the growing bright region at the bottom of the EMSI image in Fig. 8.8 reveals that the initial propagation velocity is independent of the direction during the first 20 s and has a value of about $15\,\mu\mathrm{m/s}$. This value is of the same order of magnitude as that obtained with the stochastic reaction diffusion model mentioned above [9]. In a later stage of the experiment an elliptical shape of the bright region occurred.

Convection cannot be fully suppressed in our experiment. In the direction of convection the diffusive transport of ions is sustained by a flow of the electrolyte and the front propagates continuously. Perpendicular to this direction (indicated by the red line in Fig. 8.8A, second panel) the ions are spread only by diffusion. Thus after initial homogeneous front spreading, an asymmetry in growth occurs.

The measurements displayed in Fig. 8.7 and Fig. 8.8 exhibit both an exponential increase in the total number of pitting sites followed by saturation, consistent with numerical simulations and former experimental results [14, 15]. This was also found for each of the three growing bright regions individually (see Fig. 8.8E).

The gain in pit number was accompanied by a growth of the total current. Via an exponential increase a transition from sporadic individual pitting events, i.e., small spikes in the current, to a high permanent current of about 20–30 μA took place (see Fig. 8.7D and Fig. 8.8F). Initially, only newly nucleated pits contribute to the current. They generally passivate after a short time (approximately 2 s) and each makes only a small contribution to the overall current [14]. With time an increasing number of pits develops a bright halo, indicating higher activity (Fig. 8.8E). Typically, these pits are found to be sites which reactivate after they have been passivated.

By detecting the halo around highly active pits their lifetime can be determined. Most pits have a lifetime of less than 20 s, only a few of them show a lifetime up to 55 s. In our experiments we observed no stable pits, although we cannot rule out their formation at a later stage of the experiment. After approximately 250 s the number of highly active pits exceeds the number of newly nucleating pits. At that time the overall current is mainly due to the highly active pits. Although only metastable pits were observed, the overall current had reached an almost constant high value. Hence, the strong increase of the current results only from the rapid increase in the number of active pitting sites. This observation is contrary to the results reported by many other authors, who attributed the strong increase of current to the formation of stable pits [6, 7].

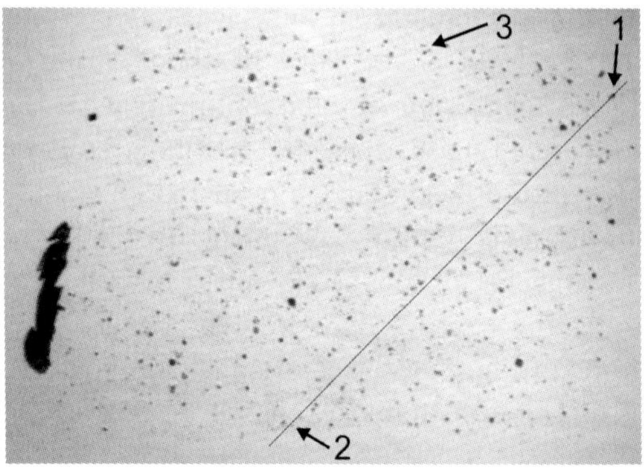

Fig. 8.9. Microscopic image of the sample in Fig. 8.8 after termination of the experiment and after removal of the lacquer. Reproduced from [23].

In the experiment shown in Fig. 8.8 the propagating fronts all have their origin at the rim of the unsealed surface. Thus, also crevice corrosion beneath the insulating lacquer and the associated acidification of the chemical environment could have affected our results [24]. To exclude this possibility the stainless steel sample was taken out of the electrolyte and imaged ex situ after termination of the experiment and careful removal of the insulating lacquer (see Fig. 8.9). The pits identified by arrows and numbers 1, 2, and 3 are the initial pits which released enough aggressive ions to weaken the oxide layer. It is possible that the first pits nucleated near the rim of the unsealed area because there the lacquer partly shielded potential pitting sites and diffusion takes place only in one direction, thus the probability for pitting corrosion is enhanced [4]. Clearly no sign of crevice corrosion is visible at these sites and can therefore be ruled out as source for the observed spreading of the pits. An example of crevice corrosion, which developed at a later stage of the experiment, is visible at the left side of Fig. 8.9. As evident from the original video, it started at $t = 348\,\mathrm{s}$ and grew for the remaining 140 s until the potential was set back to $-93\,\mathrm{mV}$.

8.6. Conclusion

Our findings can be interpreted as follows: Due to the release of aggressive species by metastable pits the oxide layer in their vicinity is weakened. Here, the probability for nucleation of pits is dramatically enhanced. Each new pit releases additional aggressive ions, leading to further weakening of the oxide layer and hence an expansion of the weakened area. This causes chain-reaction-like development of pits and a spreading of the active surface area.

Our results and their good agreement with the theoretical model strongly corroborate our believe that oxide layer weakening and nucleation rate of pits depend on each other and that the onset of pitting corrosion in our experiments can be regarded as an autocatalytic process. We hope that our approach gives new impulses for preventing corrosion damage.

In general, the experimental methods described here are applicable for the investigations of many materials protected by thin films.

References

[1] C. M. Abreu, M. J. Cristóbal, R. Losada, X. R. Nóvoa, G. Pena, and M. C. Pérez. *J. Electroanal. Soc.* **572** (2004) 335.
[2] G. S. Frankel. *J. Electrochem. Soc.* **145** (1998) 2186.
[3] J. A. Collins and M. L. Monack. *Mater. Prot. Perform.* **12** (1973) 11.
[4] P. C. Pistorius and G. T. Burstein. *Philos. T. Roy. Soc. A* **341** (1992) 531.
[5] M. P. Ryan, D. E. Williams, R. J. Chater, B. M. Hutton, and D. S. McPhail. *Nature* **415** (2002) 770.
[6] U. Bertocci, M. Koike, S. Leigh, F. Qiu, and G. Yang. *J. Electrochem. Soc.* **133** (1986) 1782.
[7] D. E. Williams, C. Westcott, and M. Fleischmann. Corrosion Chemistry within Pits, Crevices and Cracks. A. Turnbull (Editor) (1987).
[8] T. T. Lunt, S. T. Pride, J. R. Scully, J. L. Hudson, and A. S. Mikhailov. *J. Electrochem. Soc.* **144** (1997) 1620.

[9] T. T. Lunt, V. Brusamarello, J. R. Scully, and J. L. Hudson. *Electrochem. Solid St.* **3** (2000) 271.

[10] T. T. Lunt, J. R. Scully, V. Brusamarello, A. S. Mikhailov, and J. L. Hudson. *J. Electrochem. Soc.* **149** (2002) B163.

[11] C. Punckt. *On the optical investigation of pattern forming catalytic surface reactions.* Ph.D. thesis, Technische Universität Berlin (2005).

[12] L. Organ. *Spatiotemporal Patterns and Cooperative Behavior in Localized Corrosion.* Ph.D. thesis, University of Virginia (2004).

[13] L. Organ, J. R. Scully, A. S. Mikhailov, and J. L. Hudson. *Electrochim. Acta* **51** (2005) 225.

[14] C. Punckt, M. Bölscher, H. H. Rotermund, A. S. Mikhailov, L. Organ, N. Budiansky, J. R. Scully, and J. L. Hudson. *Science* **305** (2004) 1133.

[15] M. Reuter and K. E. Heusler. *Electrochim. Acta* **35** (1990) 1809.

[16] Y. S. Lim, J. S. Kim, S. J. Ahn, H. S. Kwon, and Y. Katada. *Corros. Sci.* **43** (2001) 53.

[17] H. S. Isaacs, G. Adzic, and C. S. Jeffcoate. *Corrosion* **56** (2000) 971.

[18] K. Sasaki and H. S. Isaacs. *J. Electrochem. Soc.* **151** (2004) B124.

[19] K. Sugimoto, S. Matsuda, Y. Ogiwara, and K. Kitamura. *J. Electrochem. Soc.* **132** (1985) 1791.

[20] L. F. Garfias and D. J. Siconolfi. *J. Electrochem. Soc.* **147** (2000) 2525.

[21] H. H. Rotermund, G. Haas, R. U. Franz, R. M. Tromp, and G. Ertl. *Science* **270** (1995) 108.

[22] J. Dicke, H. H. Rotermund, and J. Lauterbach. *J. Opt. Soc. Am. A* **17** (2000) 135.

[23] M. Dornhege, C. Punckt, J. L. Hudson, and H. H. Rotermund. *J. Electrochem. Soc.* **154** in press.

[24] N. J. Laycock, J. Stewart, and R. C. Newman. *Corros. Sci.* **39** (1997) 1791.

Chapter 9

Unified approach to feedback-mediated control of spiral waves in excitable media

Vladimir S. Zykov and Harald Engel*

Institut für Theoretische Physik, Technische Universität Berlin, Hardenbergstrasse 36, D-10623 Berlin, Germany

**h.engel@physik.tu-berlin.de*

Contents

9.1. Introduction

Rotating spiral waves represent very typical examples of self-organized spatio-temporal patterns in quite different reaction-diffusion systems. They have been observed in social amoeba colonies [1], in the Belousov-Zhabotinsky (BZ) reaction [2, 3], in heart muscle [4, 5], in the retina of the eye [6], during the oxidation of CO to CO_2 on platinum single crystal surfaces [7], as intracellular calcium waves in frog eggs [8], and in yeast extracts during glycolysis [9]. From a theoretical point of view, all these experimentally available systems belong to a single broad class known as excitable media [10–13].

Excitable media are essentially nonlinear dissipative dynamical systems with spatially distributed internal energy sources. Wave phenomena in active media differ strongly from optical or acoustic waves. Due to a balance between energy influx and dissipation the undamped propagation of a solitary wave is possible. Such a wave includes the transition from a stable rest state to a metastable excited state followed by a recovery transition back into the rest state. During the recovery transition the medium is refractory. Therefore, two colliding excitation waves usually annihilate each other in contrast to the emergence of interference patterns formed by waves in linear systems. Due to refractoriness there is a minimal wavelength of periodic wave trains in excitable media.

A spiral wave is created if some special initial conditions are used or if a single propagating wave front is broken down. Although the spatial nonuniformity of the medium can simplify the emergence of spiral waves, they exist in completely homogeneous media as well. The influence of the boundary conditions decreases very fast with the distance from the boundary and practically vanishes if this distance exceeds the spiral wavelength. Thus, once created spiral waves represent very robust sources of wave activity and oscillations in excitable media.

Sometimes such self-sustained activity is undesirable as it destroys the normal functioning of biological systems. One such complicated biological wave system is the heart. A disruption in waves of electrical activity propagating in heart tissue causes the formation of spiral waves. Spiral waves can lead to tachycardia, or rapid heart beat, which often is a precursor to a heart attack. A high percentage of cardiac disease deaths are sudden and result from a spatio-temporally irregular wave activity that prevents the main chambers of the heart from pumping blood normally. The role of spiral waves in cardiac arrhythmias is one of the most important motivations for the study of spiral wave dynamics, especially for the elaboration of gentle and efficient control methods.

One obvious way to suppress a spiral wave is to apply a huge perturbation in excitability to the whole system forcing it to return into the rest state. Obviously, such a "defibrillation shock" can be rather dangerous since it can damage other important living functions of the biological system. The main goal of our study is to propose basic ideas for control methods applying relatively weak perturbations.

The control methods we have in mind base on the resonant drift of spiral waves in response to a periodic change in the excitability of the whole medium exactly at the spiral rotation period [14]. Theoretical [14–17] and experimental [18, 19] studies show that a resonant parametric modulation induces a drift of the spiral core along a straight line. Thus, the spiral wave can be shifted to the boundary of the medium, where it hopefully should disappear. However, a more detailed analysis [20] reveals a repulsion of resonantly drifting spiral waves from no-flux boundaries. The influence of the boundary changes the rotation period and results in a reflection of the spiral wave. In addition, any variations of the rotation period due to temporal parameter variations or spatial inhomogeneity, for example, restrict the applicability of resonant drift at an a priori given frequency as a control tool to suppress undesirable self-sustained oscillations associated with the rotating spiral wave.

In this respect, feedback-mediated parametric modulation seems to be a more promising control strategy, since in this case the modulation period always coincides exactly with the actual rotation period of the spiral wave [20, 21]. Another important motivation to study feedback-mediated dynamics of spiral waves is related to the fact that a feedback is either naturally present or can be easily implemented in many excitable media [22–25]. For example, recent experimental investigations performed with the BZ medium [26, 27] and during the catalytic CO oxidation on platinum single crystal surfaces [28] reveal that global feedback can provide an efficient tool for the control of pattern formation.

In this paper we demonstrate that all known feedback-mediated control methods based on the phenomenon of resonant drift can be considered in the framework of a unified theoretical approach. This approach allows to analyze existing methods of discrete and continuous control, and helps to elaborate novel control algorithms. The theoretical predictions are confirmed by numerical computations and experimental studies.

9.2. Spiral waves under periodic parameter modulation

9.2.1. *Experimental system and underlying mathematical model*

The light-sensitive BZ reaction demonstrates all basic features of excitable media of quite different nature and represents a very suitable experimental system to study controlled motion of spiral waves. In the experiments reported below an open gel reactor is used [29–31]. The catalyst is immobilized in a silicahydrogel layer of 0.5 mm thickness prepared on a plate of depolished glass (diameter 63 mm). The active layer is in diffusive contact with a feeding solution prepared from stock solutions containing $[NaBrO_3]_0$=2.06×10^{-1} M (Aldrich, 99 %), $[H_2SO_4]_0 = 3.1 \times10^{-1}$ M (Aldrich, 95-98%), malonic acid $[CH_2(COOH)_2]_0 = 1.86 \times10^{-1}$ M (Aldrich, 99%), and $[NaBr]_0 = 4.12\times10^{-2}$ M (Fluka, 99%). This solution is pumped continuously through the reactor at a rate of 120 ml/h. To protect the active layer from stirring effects, it is covered by an inactive gel layer of 0.5 mm thickness not loaded with the catalyst.

The active layer is illuminated by a video projector (Panasonic PT-L555E) with intensity controlled by a computer via a frame grabber (Data Translation, DT 2851). The light is filtered with a bandpass filter (BG6, 310-530 nm). Every one second the wave pattern appearing in the gel layer is detected in transmitted light by a CCD camera (Sony AVC D7CE) and digitized with a frame grabber (Data Translation, DT 3155) for immediate processing by the computer. During the same time step the signal $I(t)$ controlling the light intensity generated by the projector can be changed in accordance with the feedback scheme under consideration.

A single spiral wave, which constitutes the initial condition for all experiments, is created in the center of the gel disk by breaking a wave front with an intense light spot. The location of the spiral wave tip is defined online as the intersection point of contour lines (0.6 $\times$ amplitude) extracted from two digitized images taken with time interval 2.0 s. The tip trajectory, the control signal and the wave activity at an arbitrary detection point can be visualized online by the computer.

For the chosen recipe concentrations the wave length of the unperturbed spiral is $\lambda \approx 2.0\ mm$. The spiral tip describes a meandering trajectory with about four outward pointing lobes. The rotation period measured far from the core center is $T_\infty \approx 40\ s$.

Basic features of the light-sensitive BZ reaction are reproduced by the Oregonator model given as a system of two coupled reaction-diffusion equations:

$$\frac{\partial u}{\partial t} = \nabla^2 u + \frac{1}{\epsilon}\left[u - u^2 - (fv + I)\frac{u-q}{u+q}\right], \qquad (9.1)$$

$$\frac{\partial v}{\partial t} = u - v.$$

Here, the variables u and v correspond to the concentrations of the autocatalytic species $HBrO_2$ and the oxidized form of the catalyst, respectively. In the following we keep the parameters $\epsilon = 0.05$, $q = 0.002$, and $f = 2.0$ fixed.

The term $I = I(t)$ describes the additional bromide production that is induced by external illumination [32]. Below all numerical simulations with the Oregonator model are performed using an explicit Euler method on a 380×380 array with a grid spacing $\Delta x = 0.14$ and time steps $\Delta t = 0.002$.

The autonomous system with $I(t) = 0.06$ has a steady state which is stable with respect to small perturbations. However, a supra-threshold perturbation, once locally applied, gives rise to a concentric wave propagating through the medium. Near to the center of the simulated domain a spiral wave rotating counterclockwise was created by a special choice of initial conditions. The spiral tip performed a compound rotation (meandering motion) including at least two different frequencies. The oscillation period measured far enough from the core center of unperturbed trajectory was $T_\infty = 3.6$.

9.2.2. *Archimedean spiral approximation*

Up to now there is no general theory which allows to predict the parameters and the shape of a spiral wave in a two-dimensional excitable medium specified by a reaction-diffusion model like (9.1), though this problem has been the subject of numerous studies [14–16, 33].

The very first attempt to construct a simplified kinematical description of a rotating spiral wave has been done in the classical paper of N. Wiener and A. Rosenblueth [33]. This description is based on the assumption that wave fronts propagate in a uniform and isotropic medium with equal velocity from any stimulated points into a region where the medium is in the rest state. Due to Huygens' principle, successive wave fronts are perpendicular to a system of rays which represent the position which may be assumed by stretched cords starting from the stimulated point. The back of the wave is another curve of the same form, which follows the wave front at a fixed distance λ_e measured along these rays.

Under these assumptions one can imagine a spiral wave rotating around a circular hole with no-flux boundary, as shown in Fig. 9.1(a). The perimeter λ of this hole should be, of course, larger than the distance λ_e. The shape of the wave front in this case coincides with the involute of the hole, that is, for any points B of the front the segment AB of a tangent to the hole has the same length as the arc length AC. The wavelength of the rotating spiral is equal to the hole perimeter λ and the rotation period T is determined as $T = \lambda/c_0$, where c_0 is the front velocity. In parametric

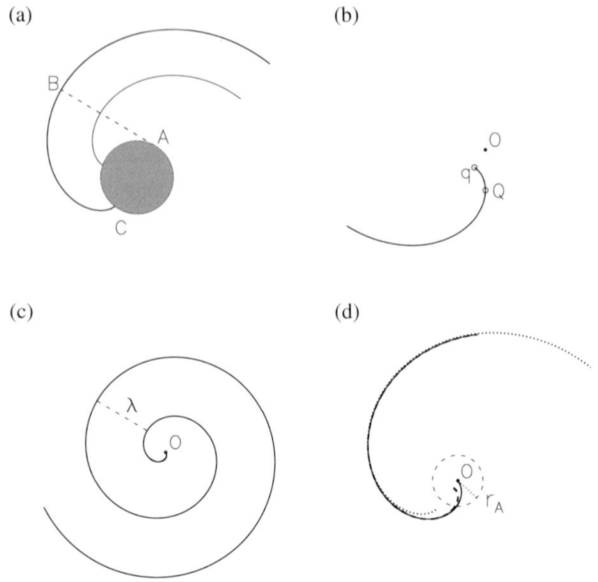

Fig. 9.1. Different representations of a rigidly rotating spiral wave. (a) Involute of a hole, (b) Solution of the kinematical equation with linear velocity-curvature relationship, (c) Archimedean spiral, (d) Superposition of the three wave fronts where the dotted, dashed and solid lines correspond to (a), (b), and (c), respectively. Far from the rotation center the fronts practically coincide.

representation the shape of the involute reads

$$x(\theta) = R_h \cos(\theta + \theta_0) + R_h\theta \sin(\theta + \theta_0),$$
$$y(\theta) = R_h \sin(\theta + \theta_0) - R_h\theta \cos(\theta + \theta_0),$$

where $R_h = \lambda/(2\pi)$ is the hole radius, and θ_0 specifies the orientation of the wave front.

The main disadvantage of this kinematical construction is that it corresponds to spiral waves pinned to a hole but fails to describe freely rotating spiral waves in uniform excitable media which are frequently observed both experimentally as well as in numerical simulations of the underlying reaction-diffusion equations.

A more elaborated kinematical description of a freely rotating spiral wave in a uniform medium is based on the assumption that the normal velocity c of a curved wave front is not a constant, but depends on its curvature [34]. The simplest approximation of this relationship is a linear dependence

$$c(K) = c_0 - DK, \tag{9.2}$$

where the local curvature K is considered positive for a convex wave front. This relationship can be derived directly from the reaction-diffusion system (9.1) assuming widely separated time scales of the variables u and v with $\epsilon \ll 1$ [35, 36]. To this aim the isoconcentration lines $u(x, y, t) = u_e$ are considered as a boundary of an excited region.

For a freely rotating spiral, wave front and wave back should coincide at one site, called phase change point [38], see point q in Fig. 9.1(b). In q the normal velocity of the wave front vanishes. This point lies at the shortest distance to the rotation center O. At another interesting point, Q, the radial direction is tangent to the isoconcentration line.

To determine the shape of a rigidly rotating spiral wave it is convenient to specify the curvature K as a function of the arc length s. A purely geometrical consideration shows that the function $K = K(s)$, the so-called natural representation of a curve, satisfies the following integro-differential equation [39]:

$$K \int_0^s Kc(K)ds' - D\frac{dK}{ds} = \omega. \tag{9.3}$$

Here, the arc length s is measured from the point Q, and $\omega = 2\pi/T$ denotes the rotation frequency of the spiral wave. It was shown that two boundary conditions $K(0) = K_Q$ and $K(\infty) = 0$ specify a unique value for the rotation frequency ω, which can be expressed as

$$\omega = c_0 K_Q \Omega(\eta), \tag{9.4}$$

where $0 \leq \eta \leq 1$ is a dimensionless parameter $\eta = DK_Q/c_0$ and the function $\Omega(\eta)$ can be well approximated as

$$\Omega(\eta) = 0.685\eta^{1/2} - 0.06\eta - 0.293\eta^2.$$

The wave front computed from Eq. (9.1) with $c_0 = 0.363$, $D = 1.0$, and $K_Q = 0.2$ is shown in Fig. 9.1(b). The rotation frequency of this spiral wave is $\omega = 0.028$, and its wavelength is $\lambda = 81.7$.

Since with increasing arc length the front curvature goes rapidly to zero, it is not necessary to use this rather complicated approach to determine the front shape far away from the rotation center. A. Winfree was the first who suggested to approximate the spiral front by an Archimedean spiral [2]

$$\Theta(r, t) = \Theta_0 - \frac{2\pi}{\lambda} r + \omega t, \tag{9.5}$$

where Θ and r are polar coordinates with origin at the rotation center. This simple analytical form determines the asymptotic of spiral waves in the complex Ginzburg-Landau equation as well [41, 42]. In Cartesian coordinates the front shape with $\lambda = 81.7$ is shown in Fig. 9.1(c).

It is important to stress that the above three descriptions of the spiral wave practically coincide far away from the rotation center. Moreover, already at a relatively small distance r_A from the rotation center, the Archimedean spiral becomes very close to the curvature affected spiral obtained from Eq. (9.3), as can bee seen in Fig. 9.1(d). In this example r_A can be estimated as $r_A \approx 9.0 \ll \lambda$. Recent computations performed with the Oregonator model [40] and experiments with the BZ reaction [43] also confirm that an Archimedean spiral provides a suitable approximation of the wave front except in a relatively small region of radius $r_A \ll \lambda$ near the rotation center. Even the shape of a slightly meandering spiral waves exhibits only small oscillations around an Archimedean shape, and the amplitude of these oscillations vanishes very quickly with r [44]. Therefore, the Archimedean spiral approximation will be used below to specify the shape of the wave front.

9.2.3. *Resonant drift of a spiral wave under periodic parameter modulation*

Resonant drift of a spiral wave is a displacement of the spiral wave center induced by a periodic modulation of the medium excitability. This phenomenon has been predicted for a kinematical model of weakly excitable media [14], confirmed experimentally as well as in numerous numerical simulations [16, 19, 20, 45], and explained as a generic property of excitable media [15, 17].

Let us assume that an unperturbed spiral wave is rotating rigidly at a frequency ω_0 around a center located at a site $z_0 = x_0 + iy_0$. One has to approximate the shape of this spiral by expression (9.5) to determine the spiral phase Θ_0. After a short perturbation, modelled as $I(t) = A_I\delta(t)$ in Eq. (9.1), the rotation center, generally speaking, is shifted to a new site

$$z_1 = z_0 + h\exp(i\Theta_0 + i\varphi), \tag{9.6}$$

where h is the magnitude of this displacement whose direction is determined by the spiral phase Θ_0 and a constant φ characterizing the excitable medium. Additionally the applied perturbation changes the rotation phase of the spiral by $\delta\Theta$. Changes in the spiral shape induced by the perturbation are assumed to relax on a time scale T_{rel} much smaller than the rotation period $T_0 = 2\pi/\omega_0$.

A second perturbation applied after the time interval T_m should produce a similar displacement. For $T_{rel} \ll T_m$ the location of the spiral center after the second perturbation can be written as

$$z_2 = z_1 + h\exp(i\Theta_1 + i\varphi), \tag{9.7}$$

where

$$\Theta_1 = \Theta_0 + \delta\Theta + \omega_0 T_m. \tag{9.8}$$

Since the direction of the displacement remains the same after subtracting 2π on the right hand side of Eq. (9.8), the last expression can be written as

$$\Theta_1 = \Theta_0 + (\omega_0 + \frac{\delta\Theta}{T_m} - \omega_m)T_m. \tag{9.9}$$

Consequently, the spiral displacement after k similar perturbations according to $I(t) = A_I \sum_{l=0}^{k-1} \delta(t - lT_m)$ reads

$$z_k = z_{k-1} + h\exp(i\Theta_{k-1} + i\varphi), \tag{9.10}$$

$$\Theta_{k-1} = \Theta_{k-2} + (\bar{\omega} - \omega_m)T_m, \tag{9.11}$$

where $k \geq 2$ and

$$\bar{\omega} = \omega_0 + \delta\Theta/T_m \tag{9.12}$$

is the averaged rotation frequency of the spiral wave perturbed by the applied modulation.

The phenomenon of resonant drift is induced under modulation with frequency $\omega_m = \bar{\omega}$. In this case, from Eq. (9.11) follows $\Theta_k = \Theta_{k-1} = \ldots = \Theta_0$. All particular displacements occur in the same direction and the total shift of the spiral location after k perturbations is given by

$$z_k = z_0 + kh\exp(i\Theta_0 + i\varphi). \tag{9.13}$$

It is important to stress, that if $\omega_m = n\bar{\omega}$, where n is an integer, $n > 1$, the result of n subsequent displacements, i.e, after one rotation period of the spiral wave, is equal to zero, because $z_{k+n} = z_k$. A long term drift is absent in this case.

Another interesting consequence from Eqs. (9.10)-(9.12) can be obtained in the case when ω_m is close to the averaged frequency of the perturbed spiral wave, $|\omega_m/\bar{\omega} - 1| \ll 1$. Under the assumption that the displacement h is small, the discrete map (9.10),(9.11) can be transformed into an ordinary differential equation for the spiral location $z(t)$:

$$\dot{z} = \frac{h}{T_m} \exp[i((\bar{\omega} - \omega_m)t + \Theta_0 + \varphi)]. \tag{9.14}$$

Thus, the velocity of the resonant drift induced by the periodic modulation is determined by the ratio h/T_m. Under resonant forcing, $\omega_m = \bar{\omega}$, the drift is along a straight line whose direction depends on the initial orientation of the spiral wave Θ_0 and on the constant φ. More generally, if the parameter modulation is given by

$$I(t) = A_I \sum_{l=0}^{k} \delta(t - lT_m - t_0) \tag{9.15}$$

with arbitrary $t_0 \geq 0$, the drift direction γ should depend also on the modulation phase $\phi_{mod} = \bar{\omega} t_0 = \omega_m t_0$, i.e.,

$$\gamma = \varphi + \Theta_0 + \phi_{mod}. \tag{9.16}$$

Note that Eq. (9.14) obtained for the discrete modulation (9.15) practically coincides with induced drift equations known for a continuous periodic modulation [14–16]. The only difference is that the constant φ in (9.14) specifies a displacement direction induced by a sequence of δ-perturbations, but not by a harmonic perturbation $I(t) = A\cos(\bar{\omega}t)$. In both cases this angle, which determines the drift direction for $\Theta_0 = \phi_{mod} = 0$ is a characteristic parameter that depends on the properties of the excitable medium and on the applied modulation method.

9.3. Discrete feedback control

9.3.1. *Resonance attractor for spiral waves subjected to one-channel feedback*

One-channel feedback has been first applied to control meandering spiral waves in experiments with the light-sensitive BZ medium [21]. Later a theory of this control method has been elaborated for rigidly rotating [40, 43] and for meandering [30] spiral waves. In accordance with this control algorithm, the wave activity (e.g. the value of the variable v in Eq. (9.1)) is measured at a particular detection point as a function of time. This value oscillates with time and exceeds the value v_e every time instant t_l, when a wave front touches the detector point. A short perturbation is applied to the system immediately at t_l or after some time delay τ, i.e. at instances $t_l' = t_l + \tau$. The frequency of the periodic sequence of generated perturbations is practically equal to the rotation frequency of the spiral wave and induces a resonant drift.

Let us assume that the drift is slow and determine its direction. Obviously, the phase of the pulse sequence depends on the spiral location. Indeed, if the spiral wave center is placed at the point $z = x + iy$, the wave front specified by Eq. (9.5) crosses the detector point located at the origin of the coordinate system each time t_l satisfying the following equation:

$$\bar{\omega}t_l' + \Theta_0 - \frac{2\pi}{\lambda}|z| = \pi + arg(z) + 2\pi l + \bar{\omega}\tau. \tag{9.17}$$

Hence the stimulating sequence will be generated as

$$I(t) = A_I \sum_{l=0}^{\infty} \delta(t - lT_m - \phi_{mod}/\omega_m). \tag{9.18}$$

This expression coincides with Eq. (9.15), where $T_m = 2\pi/\bar{\omega}$ and the phase shift ϕ_{mod} is determined as

$$\phi_{mod} = \pi + arg(z) - \Theta_0 + \frac{2\pi}{\lambda}|z| + \bar{\omega}\tau. \tag{9.19}$$

To determine the drift direction expression (9.19) should be substituted into Eq. (9.16). This yields

$$\gamma = \varphi + \pi + arg(z) + \frac{2\pi}{\lambda}|z| + \bar{\omega}\tau. \tag{9.20}$$

It is important to stress that under the described one-channel feedback the drift direction does not depend on the initial orientation Θ_0 of the spiral. The location z of the spiral center and the time delay τ completely determine the drift angle γ. Thus, the drift velocity field induced by one-channel feedback control follows from Eq. (9.14):

$$\dot{z} = \frac{h}{T_m}\exp(i\gamma) \tag{9.21}$$

with γ defined by Eq. (9.20).

This drift velocity field is shown in Fig. 9.2 corresponding to $\tau = 0$. The constant φ is taken as $\varphi = -1.8$ in accordance with the chosen parameters for the Oregonator model (9.1). The field has rotational symmetry, but the drift angle γ monotonously increases with the distance $|z|$ from the detector point. Hence, there is a discrete set of sites along any radial direction, where the drift direction is orthogonal to the radial one i.e.

$$\gamma = arg(z) + \pi/2 + \pi n, \tag{9.22}$$

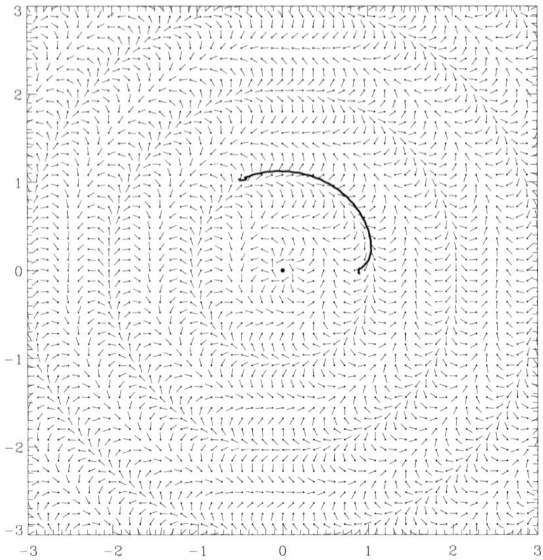

Fig. 9.2. Drift velocity field under one-channel feedback for $\tau = 0$. The thick solid line shows the trajectory of the spiral center computed from the Oregonator model (9.1) .

where n is an arbitrary integer. The sites corresponding to fixed n form circular orbits around the detector. It follows from Eq. (9.20) that the motion along this orbit will be stable only if n is an even number $n = 2m$. Indeed, according to Eq. (9.20) γ increases with $|z|$. Therefore, small deviations from the circular orbit are damped out (amplified) for $n = 2m$ ($n = 2m+1$). The radius R of a stable orbit can be found from Eqs. (9.20) and (9.22):

$$R/\lambda = m - 0.25 - \frac{\varphi}{2\pi} - \tau\bar{\omega}/(2\pi). \tag{9.23}$$

These stable orbits represent the attractors of the spiral drift under one-channel feedback. The basins of attraction are bounded by unstable orbits, which correspond to $n = 2m + 1$ and have radii

$$R/\lambda = m + 0.25 - \frac{\varphi}{2\pi} - \tau\bar{\omega}/(2\pi). \tag{9.24}$$

An example of feedback-induced drift computed for the Oregonator model (9.1) is shown by the solid line in Fig. 9.2. It can be seen that the drift field determined by Eqs. (9.20),(9.21) describes not only the attractor location, but the transient motion starting somewhere within the basin of attraction, too.

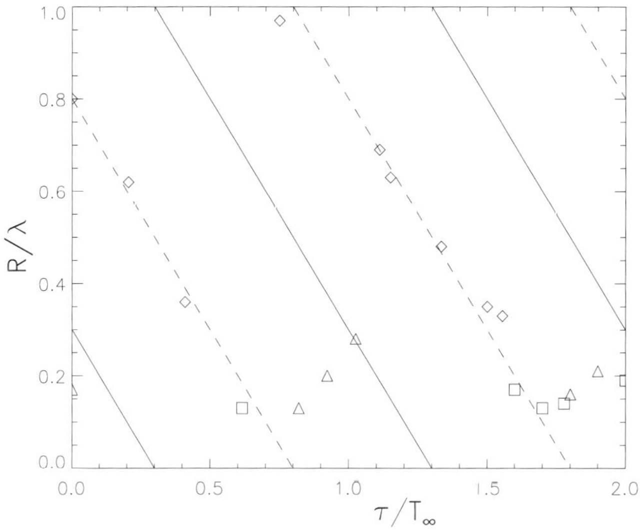

Fig. 9.3. Radii of resonance attractors (diamonds) determined experimentally for meandering spiral waves in the light-sensitive BZ medium vs the time delay in the feedback loop. The dashed lines show the theoretical predictions obtained from Eq. (9.23), the solid lines are the boundaries of the basins of attraction according to Eq. (9.24). Radii of observed entrainment and asynchronous attractors are shown by triangles and squares, respectively [47].

Application of one-channel feedback control to spiral waves in the light-sensitive BZ system allows to observe the discrete set of stable resonant attractors experimentally [21, 30, 43, 46]. Note, that Eq. (9.23) for the radius of the resonance attractor contains only one medium dependent parameter φ, which specifies the direction of the resonance drift. To avoid a rather complicated experimental procedure to determine this value, the obtained experimental data were fitted to the theoretically predicted linear dependence (9.23) using $\varphi = -0.31$. The results are shown in Fig. 9.3 by dashed lines. Then, the boundaries of the basin of attraction were specified in accordance with Eq. (9.24) (solid lines in Fig. 9.3).

In the region of small radii, where $R < r_A \approx 0.2\lambda$, the dependence (9.23) becomes violated. This is quite natural, since close to the rotation center the spiral wave deviates considerably from the Archimedean shape, and Eqs. (9.17)-(9.20) are not applicable. When the distance between the core center and the measuring point is small the entrainment attractor or the asynchronous attractor can be observed (for details see [30]). The radii of these attractors grow with the time delay until the basin of attraction of the resonance attractor is reached. With further increase of the time delay the resonance attractor is observed.

9.3.2. *Spiral wave drift near a straight line detector*

In another feedback algorithm short perturbations are generated each time when the rotating spiral wave front is tangent to a straight line. Fig. 9.4(a) illustrates the location of the spiral wave observed during an experiment with the light-sensitive BZ medium exactly at the instance when it touches a line detector. Of course, wave activity cannot be traced along the detector continuously but rather at a discrete set of sites approximating the virtual line detector.

Fig. 9.4(b) shows the spiral tip trajectory obtained experimentally under this feedback control. After a short transient the spiral core center drifts in parallel to the line detector. The asymptotic drift trajectory reminds the resonance attractor observed under one-channel control, because a small variation of the initial location of the spiral wave does not change the final distance between the detector and the drift line. To construct the drift velocity field for this control algorithm an Archimedean spiral approximation is used again. Assume the detector line is given as $x = 0$ and an Archimedean spiral described by Eq. (9.5) is located at a site (x, y) with $x > 0$. A pure geometrical consideration shows that the spiral front touches the detector each time t_l satisfying the following equation:

$$\bar{\omega} t_l + \Theta_0 - \frac{2\pi}{\lambda} r = \pi - \arctan(\frac{\sigma}{x}) + 2\pi l, \tag{9.25}$$

where

$$r(x) = x\sqrt{0.5 + \sqrt{0.25 + (\frac{\lambda}{2\pi x})^2}}, \tag{9.26}$$

and

$$\sigma = \sqrt{r^2 - x^2} = \frac{\lambda x}{2\pi r}. \tag{9.27}$$

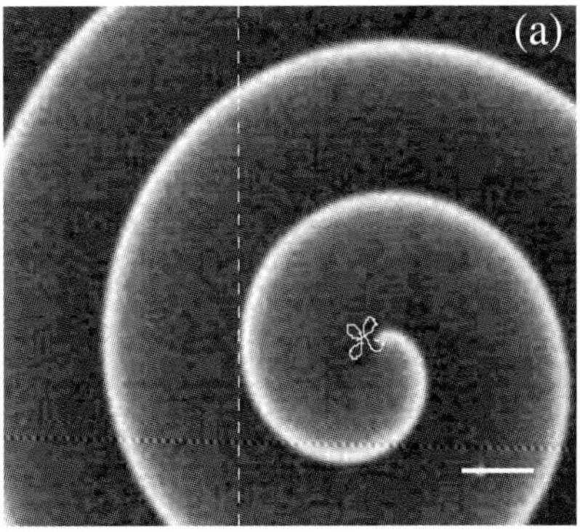

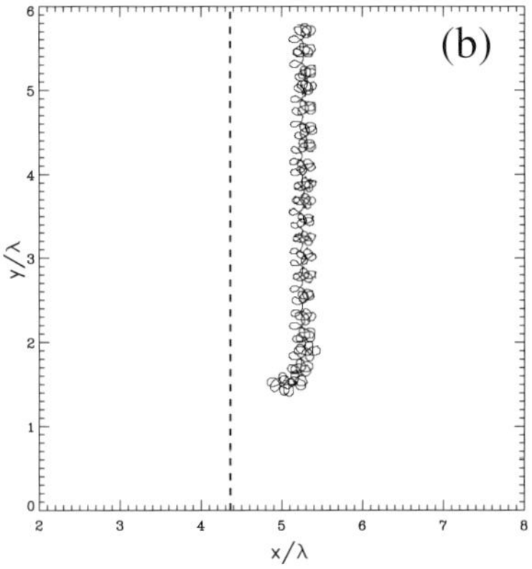

Fig. 9.4. Spiral wave drift near a virtual line detector. (a) Snapshot of a spiral wave in a thin layer of the light-sensitive BZ reaction. The overlaid dashed line represents the detector. The white curve shows the spiral tip trajectory in the absence of feedback. (b) Trajectory of the spiral wave tip induced by the feedback.

Eq. (9.25) determines a modulating sequence similar to Eq. (9.15) with the phase shift

$$\phi_{mod} = \pi - \Theta_0 + \frac{2\pi}{\lambda} r - \arctan(\frac{\sigma}{x}) + \bar{\omega}\tau \tag{9.28}$$

which takes into account a possible time delay τ. Substituting Eq. (9.28) into Eq. (9.16) we obtain the direction γ of the induced drift as a function of the coordinate $x > 0$:

$$\gamma(x) = \varphi + \pi + \frac{2\pi}{\lambda} r(x) - \arctan(\frac{\sigma}{x}) + \bar{\omega}\tau. \tag{9.29}$$

On the left hand side from the detector line the drift direction is specified by a similar expression, but an additional rotation by the angle π should be included. This gives for $x < 0$

$$\gamma(x) = \varphi + \frac{2\pi}{\lambda} r(x) + \arctan(\frac{\sigma}{x}) + \bar{\omega}\tau. \tag{9.30}$$

Finally, the drift of the spiral center will be specified by Eq. (9.21) with γ determined by Eqs. (9.29) and (9.30).

The obtained drift velocity field is shown in Fig. 9.5(a). In analogy to one-channel feedback, there is a set of stationary trajectories given here by the condition

$$\gamma(x) = \pi/2 + \pi n, \tag{9.31}$$

where n is an arbitrary integer. A stable stationary trajectory corresponds to an even number $n = 2m$, while an odd number $n = 2m + 1$ characterizes an unstable one. Solid lines in Fig. 9.5(a) show three trajectories of spiral wave centers placed initially at different distances from the line detector. Asymptotically they are attracted by stable stationary trajectories with basins of attraction separated by unstable stationary trajectories. The similarity between this attractor structure and that described for one-

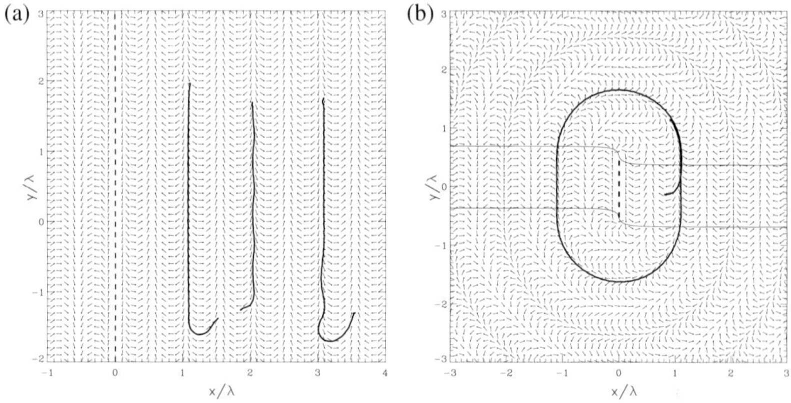

Fig. 9.5. Velocity field for spiral wave drift induced near a straight line detector (a) and a line segment (b). Solid lines show trajectories of the spiral center computed for the Oregonator model (9.1). $\tau = 0$.

channel feedback becomes even more obvious, when a line segment is used as a detector. Indeed, if short perturbations are generated each time the wave front is tangent to a segment of a straight line or touches its open end, then the drift velocity field consists of three regions, as shown in Fig. 9.5(b). In the central region the drift direction is determined by Eqs. (9.29) and (9.30) for the line detector. However, for a spiral center located in the top or bottom region the front can never become tangent to the segment. Instead, it will touch the nearest open end. Thus, in these regions the drift direction follows from Eq. (9.21) valid for one-channel feedback control. These three regions are separated by two boundaries $y_b = y_b(x_b)$ given in the case of a vertical line segment as

$$y_b = y_{tip} - \frac{\lambda x_b}{2\pi r(x_b)}, \tag{9.32}$$

where y_{tip} specifies the y-coordinates of the open ends, and $r(x)$ is given by Eq. (9.26). The drift velocity field is smooth at these boundaries, and the attractor structure existing in one region continuously transforms into another one. Direct integrations of the Oregonator model (9.1) demonstrate a continuous drift of a spiral center along a closed trajectory surrounding the line segment.

9.3.3. *Spiral wave drift near a curved one-dimensional detector*

Drift induced near a line detector represents an efficient method to move a spiral wave along a straight line and, hence, to shift it from a given initial location A to a desirable site B along the shortest pathway. Now let us

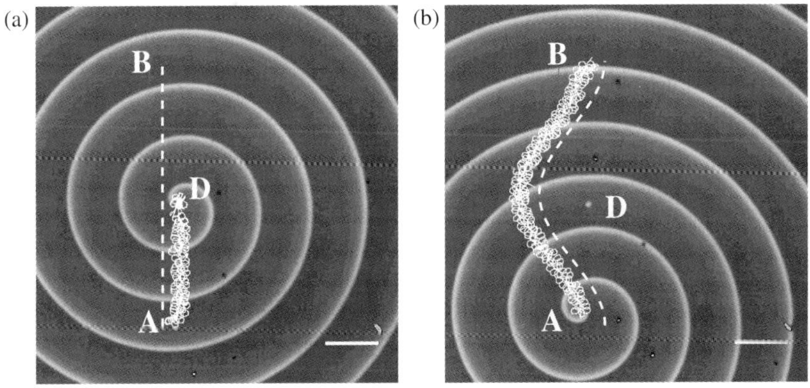

Fig. 9.6. Feedback-mediated drift of a spiral wave observed in a thin layer of the BZ reaction. (a) Spiral wave captured by a heterogeneity defect D. (b) Successful shift of the spiral wave from B to A near a curved detector avoiding the defect D. Overlaid dashed lines represent the applied line detector. Scale bar: 1 mm.

assume, the direct way between A and B is blocked by a defect D, as shown in Fig. 9.6(a). A spiral core moving along the straight line between A and B would be captured and attached permanently at this point (see Fig. 9.3b). To control the resulting pinned spiral would be much more complicated as it is very difficult to dislodge a trapped spiral. Thus, in excitable media with defects the straight connection between two sites is not necessary the optimal pathway.

With regard to the described situation we have studied the ability to guide a spiral wave along an arbitrarily shaped line detector. Fig. 9.6(b) shows the successful shift of the spiral wave from point B to A avoiding its anchoring at the defect D. This shift is induced by a feedback control using a curved segment as detector.

Extensive studies revealed the possibility of spiral wave drift near a convex-shaped one-dimensional detector of arbitrary curvature up to the limiting case of the point detector. The drift near a concave arc is more complicated. Indeed, Fig 7(a) illustrates a continuous drift around a cosine-shaped detector. Since the deformation of this detector with respect to a straight line is relatively small, the spiral waves drifts along both concave as well as convex arcs of the detector. However, near a strongly concave arc the drift terminates at some site, as if the spiral wave becomes anchored at some invisible defect, compare Fig. 9.7(b). In fact, the medium is entirely uniform and does not contain any defects.

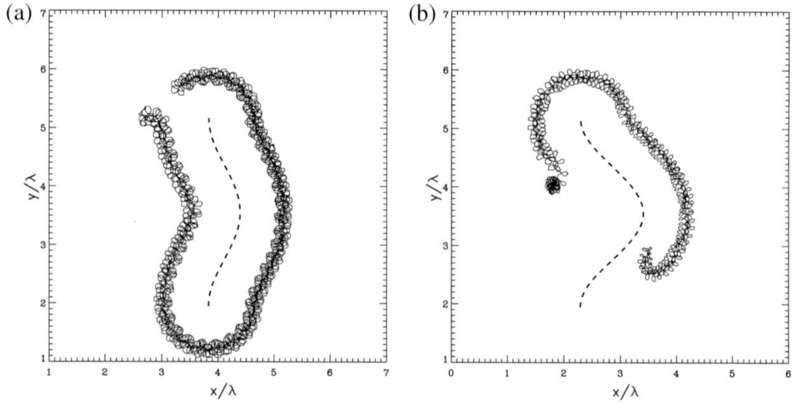

Fig. 9.7. Tip trajectories of a spiral wave observed in a thin layer of the BZ reaction. (a) Drift around a weakly curved one-dimensional detector. (b) Termination of the spiral drift near a strongly curved concave detector. Dashed lines represent the applied detectors.

To explain this amazing result we consider a circular detector of radius R_a centered at the origin of a Cartesian coordinate system. We approximate the spiral wave by an Archimedean spiral located at the point $z = x + iy$. From pure geometrical considerations (compare Fig. 9.8) one finds that the spiral wave will touch the detector at instants t_l satisfying the following equation

$$\bar{\omega} t_l + \Theta_0 - \frac{2\pi r}{\lambda} = \pi + arg(z) - \arctan(\frac{\sigma}{d}) + \arctan(\frac{\sigma}{R_a + d}) + 2\pi l, \quad (9.33)$$

where

$$r(d) = d\sqrt{0.5 + \sqrt{0.25 + (\frac{\lambda}{2\pi d})^2}}, \tag{9.34}$$

$$\sigma(d) = \sqrt{r^2 - d^2} = \frac{\lambda d}{2\pi r(d)}, \tag{9.35}$$

$$x^2 + y^2 = (d + R_a)^2 + r(d)^2 - d^2. \tag{9.36}$$

Here, r denotes the distance between the spiral center and the point of touching, σ is the projection of r on the tangent line, and d is the distance from the spiral center to the tangent line.

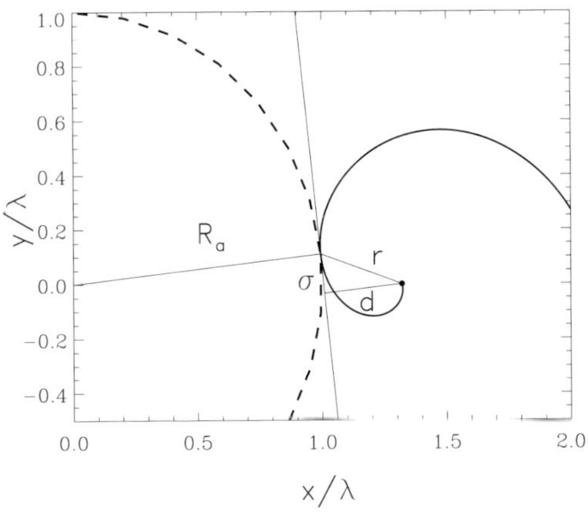

Fig. 9.8. Archimedean spiral (thick solid line) touching a circular detector of radius R_a (dashed line). The thin solid line denotes the common tangent line to both curves. The figure explains the meaning of distances σ, d, and r used in the text.

As before, the stimulating sequence will be generated in a form similar to Eq. (9.15) with the phase shift

$$\phi_{mod} = \pi + arg(z) + \frac{2\pi r}{\lambda} - \Theta_0 + \arctan(\frac{\sigma}{R_a + d}) - \arctan(\frac{\sigma}{d}) + \bar{\omega}\tau. \tag{9.37}$$

This phase shift determines the drift direction in accordance with Eq. (9.16).

An arbitrary curved line detector should be considered as a set of continuously connected arcs with different curvature radii. The determination of the drift velocity field in this case is illustrated in Fig. 9.9 for a detector consisting of three circular arcs. Thin solid curves separate zones of influence of different arcs and both open ends of the detector. In analogy to the case of a line segment discussed before (compare Fig. 9.5(b)), these curves obey Eq. (9.32) written in a local coordinate system with the origin at the end points of the arcs. For instance, within the upper middle region of the considered square-shaped medium in Fig. 9.9 the spiral wave can only touch the upper open end but never can be tangent to the detector. Within the upper right region the spiral wave can either touch the upper open end or be tangent to the middle arc. This region as well as other regions with two sources of influence are shaded in Fig. 9.9. Here, the resulting drift velocity vector is determined as the sum of two particular vectors related to two sources of influence. In the shaded regions the amplitude of the resulting drift velocity vanishes at some sites because the particular vectors have opposite directions. This is the reason for the apparent anchoring of the spiral wave near concave arc shown in Fig. 9.7(b).

To confirm these conclusions we have performed numerical simulations of the feedback mediated drift within the Oregonator model (9.1). The thick solid lines in Fig. 9.9 show the obtained trajectories of a spiral wave center for two different initial locations. The spiral center moves in very good agreement with the predicted drift velocity field and stops near the place where the velocity field vanishes.

9.4. Continuous feedback control

9.4.1. *Superposition principle for feedback-induced spiral drift*

To realize a continuous feedback control of rotating excitation waves, the parameter modulation $I(t)$ is determined to be proportional to the average of a system variable over a certain spatial domain S. For instance, the feedback signal $I(t)$ is computed according to [47, 48, 50–52]:

$$I(t) = k_{fb}[B(t-\tau) - B_0], \tag{9.38}$$

where

$$B(t) = \frac{1}{s}\int_S v(x', y', t)dx'dy'. \tag{9.39}$$

Here, the intensity of the feedback signal is proportional to the integral value B of the variable v over the domain S of area s. Two important control parameters in the feedback loop are the gain k_{fb} and the time delay

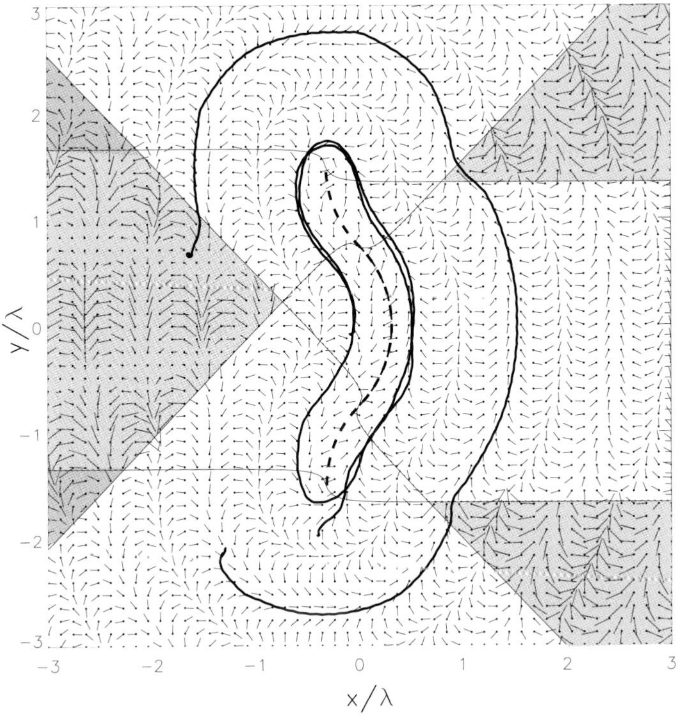

Fig. 9.9. Drift velocity field for a feedback loop based on a strongly curved detector (dashed line) and corresponding trajectories of the spiral center computed for the Oregonator model (9.1). $\tau = 1.4$.

τ. If S is a disk whose center coincides with the center of a rigidly rotating spiral wave, the integral B does not depend on time. The corresponding constant value is denoted B_0 in Eq. (9.38).

Let us assume that the center of an unperturbed spiral wave is located at site $z = x + iy$. Due to the rotation of the spiral wave, the computed integral $B(t|z)$ and, hence, the modulation signal $I(t|z)$ are periodic functions of time with period T_∞. Applying the signal $I(t|z)$ to the medium, the spiral wave is forced to drift in accordance with the general rule (9.14). Direction and magnitude of drift velocity are determined by the first Fourier component of the periodic modulation, which can be expressed as

$$I_1(t|z) = k_{fb} A(z) \cos[\bar{\omega} t - \bar{\omega}\tau - \phi(z)]. \tag{9.40}$$

Here the amplitude $A(z)$ and the phase $\phi(z)$ are defined by the following expression

$$A(z)e^{i\phi(z)} = \frac{2}{T_\infty} \int_0^{T_\infty} B(t|z) \exp(i\bar{\omega}t) dt. \tag{9.41}$$

Higher Fourier components with frequencies $n\bar{\omega}$, where $n \geq 2$, do not induce a long term drift as shown in Sec. 9.2.3

If the induced drift is slow, the actual location of the spiral center z determines the magnitude $V(z) \sim k_{fb}A(z)$ and the direction of the drift velocity with respect to the x-axis

$$\gamma(z) = \varphi + \omega\tau + \phi(z). \tag{9.42}$$

The constant φ specifies the drift direction induced in the case $\tau = 0$ and $\phi = 0$. In the Oregonator model (9.1) one finds with the parameters indicated before $\varphi = -0.5$ [47]. Hence, the drift velocity field can be written as:

$$\dot{z} = V(z) \exp[i\gamma(z)] \tag{9.43}$$

with γ defined by Eq. (9.42).

If the shape of a slightly meandering wave can be approximated by a counterclockwise rotating Archimedean spiral, the first Fourier component of $v(z', t|z)$ reads

$$v_1(z', t|z) = v_m \cos[\omega t - arg(z' - z) - 2\pi|z' - z|/\lambda], \tag{9.44}$$

where v_m is a constant. Substituting Eqs. (9.39) and (9.44) into Eq. (9.41), we get

$$A(z)e^{i\phi(z)} = \frac{v_m}{s} \int_S \exp[i\Phi(z'|z)] dx' dy', \tag{9.45}$$

where

$$\Phi(z'|z) = arg(z' - z) + 2\pi|z' - z|/\lambda. \tag{9.46}$$

It is important to stress that for a domain size much smaller than the spiral wavelength λ, we recover the limiting case of one-point feedback control [47] specified by Eqs. (9.42),(9.43),(9.45) with $(x', y') = (0, 0)$. In this case we have $\phi(z) = \Phi(0|z) = \pi + arg(z) + 2\pi|z|/\lambda$. Hence, the angle γ determining the drift direction under continuous one-point control reads

$$\gamma(z) = \varphi + \pi + arg(z) + 2\pi|z|/\lambda + \bar{\omega}\tau. \tag{9.47}$$

Note, that the only difference between this expression and Eq. (9.20) for discrete one-channel control is the specific value of the constant φ.

To determine the drift velocity field for an arbitrarily shaped spatial domain, we divide it into a set of small subdomains. Each subdomain is treated as a one-point detector generating a feedback signal with a phase shift given by Eq. (9.46). Finally, the resulting drift velocity field is derived according to Eqs. (9.45) and (9.42) as the sum over all drift vectors induced by single subdomains. This superposition principle unifies and simplifies the study of different algorithms for continuous feedback control considerably.

9.4.2. *Two-point feedback control*

The simplest way to illustrate the importance and the efficiency of the superposition principle is to consider a two-point feedback, where the feedback signal is taken from two measuring points at $(z'_+) = (0, a)$ and $(z'_-) = (0, -a)$. Substituting these coordinates into Eq. (9.46), we get two functions $\Phi_+(z)$ and $\Phi_-(z)$ describing the effect of the feedback signal generated from each measuring point separately

$$\Phi_\pm(z) = arg(z'_\pm - z) + \frac{2\pi}{\lambda}|z'_\pm - z|. \tag{9.48}$$

The amplitude $A(z)$ and the phase $\phi(z)$ of the drift velocity field induced by the two points together are determined from Eq. (9.45) as

$$A(z)e^{i\phi(z)} = [\exp(i\Phi_+) + \exp(i\Phi_-)]v_m/2, \tag{9.49}$$

and the drift angle $\gamma(z)$ follows from Eq. (9.42).

It can be easily seen from Eq. (9.48) that the amplitude $A(z)$ vanishes if

$$\Phi_+(z) - \Phi_-(z) = \pi(2m+1), \tag{9.50}$$

where m is an integer. A solution of this equation specifies a smooth line or a discrete set of smooth lines on the z plane.

To analyze the obtained drift velocity field it is suitable to choose the distance between the two measuring points $d_p = 2a$ as control parameter [53]. For $d_p/\lambda \ll 1$, the drift velocity field looks very similar to that induced by one-point feedback (compare section 9.3.1, [47]). It includes a set of circular-shaped attracting manifolds called resonance attractors [21], as shown in Fig. 9.2. This attractor structure still persists for distances $d_p/\lambda < 0.5$. For example, the drift velocity field obtained for $d_p/\lambda = 0.45$ is shown in Fig. 9.10(a). The thick solid line represents the drift

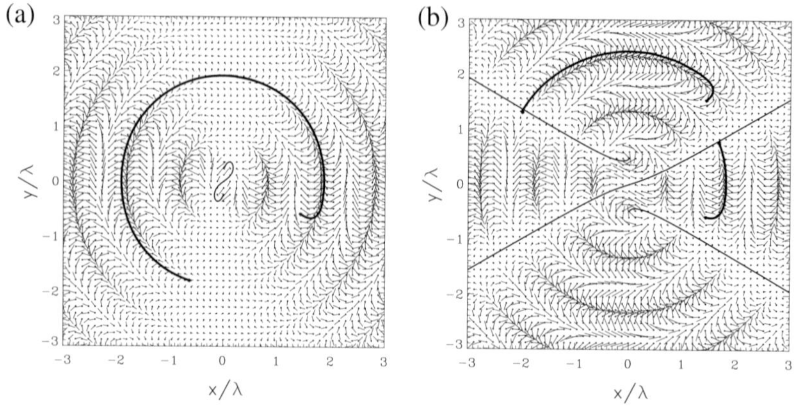

Fig. 9.10. Drift velocity field determined from Eqs.(9.43), (9.42), (9.48) for (a) $d_p/\lambda = 0.45$, (b) $d_p/\lambda = 1.0$. Thin solid lines represent lines of fixed points that satisfy Eq. (9.49) (compare text). Thick solid lines depict trajectories of the spiral center computed for the Oregonator model (9.1) with $k_{fb} = 0.02$ and $\tau = 0$ [53].

of the spiral center obtained numerically for the Oregonator model (9.1) and illustrates the existence of a circular-shaped resonance attractor in quantitative agreement with the drift velocity field predicted analytically. However, in contrast to one-point feedback, the magnitude of the drift velocity is not a constant. Especially, drift becomes very slow in the upper and lower part of the attractors. Even more important, the drift velocity vanishes at a smooth curve that connects the measuring points and reminds of a question mark (thin line in Fig. 9.10(a). It is natural to refer to such unusual equilibrium manifold as fixed line, in analogy to the well-studied isolated fixed points of vector fields.

For $0.5 \leq d_p/\lambda < 1.5$, the drift velocity field changes dramatically, see Fig. 9.10(b). There are three spatially unbounded fixed lines that destroy the circular-shaped attractors existing as long as $d_p/\lambda < 0.5$. In numerical simulations with the Oregonator model (9.1), the spiral wave center follows an approximately circular trajectory until it stops practically at a fixed line in complete agreement with the predictions from the drift velocity field.

These results have been verified experimentally. Figure 9.11 shows measured tip trajectories of spiral waves in the light-sensitive BZ medium starting from six different initial locations of the spiral center. The feedback signal is determined as the sum of the intensities of the transmitted light measured at two detector points located at the distance $d_p = \lambda$. As pre-

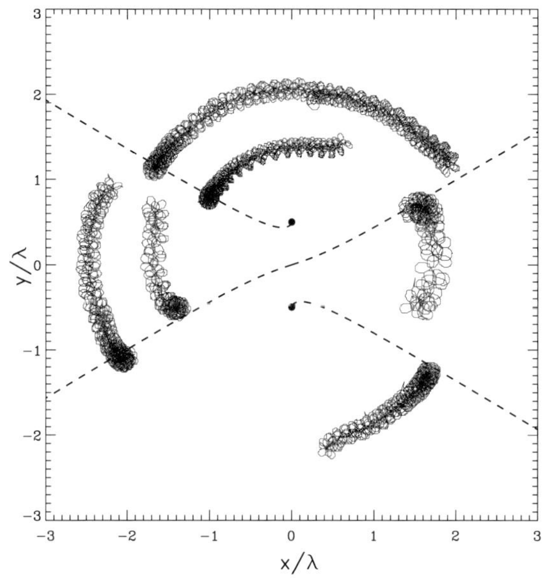

Fig. 9.11. Trajectories of spiral wave tips observed in the experiments with the light-sensitive BZ reaction. Two measuring points are shown by black dots. Dashed lines depict fixed lines determined numerically from Eq.(9.49) [53].

dicted by the corresponding drift velocity field shown in Fig. 9.10(b), the spiral drift stops at the theoretically predetermined fixed lines.

9.4.3. *Global feedback in circular and elliptical domains*

An increase in the radius R_d of the integration domain S from zero to the radius of the gel layer describes the transition from one-point to global feedback control [48–50]. As this transition preserves the rotational symmetry existing for one-point control, one can expect that the drift direction for global feedback within a circular domain should be defined by an expression similar to Eq. (9.47). Indeed, it was shown [47, 50] that Eq. (9.47) has to be generalized and reads

$$\gamma(z) = \varphi + arg(z) + \phi(|z|, R_d) + \bar{\omega}\tau. \qquad (9.51)$$

In contrast to one-point feedback, under global feedback control phase $\phi(|z|, R_d)$ and amplitude $A(|z|)$ of the first Fourier component in the feedback signal are nonlinear functions of $|z|$. An example of these functions obtained numerically for $R_d = 1.5\lambda$ is plotted in Fig. 9.12(a).

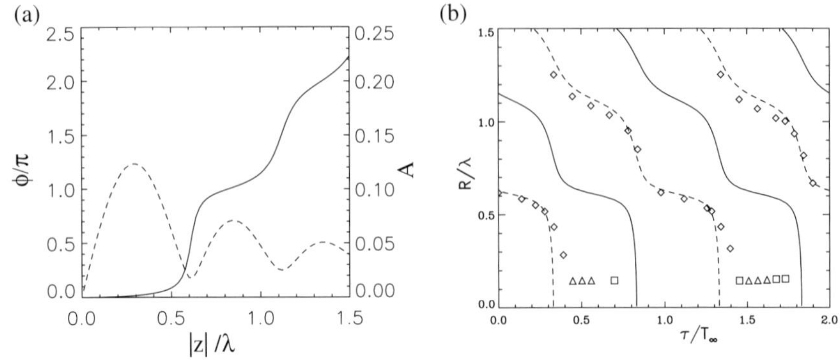

Fig. 9.12. (a) Phase ϕ (solid line) and amplitude A (dashed line) of the feedback signal determined under global control in a circular domain of radius $R_d = 1.5\lambda$. (b) Radius R of the resonance attractor in the domain versus time delay τ in the feedback loop. Radii of observed entrainment and asynchronous attractors are shown by triangles and squares, correspondingly [47].

Since condition (9.22) is valid for a circular orbit in the drift velocity field as well, the radii of stable attracting trajectories and the basins of attraction can be determined from Eqs. (9.51) and (9.22). As a result, the radius of the resonant attractor becomes a nonlinear function of the time delay τ in the feedback loop, as shown in Fig. 9.12(b). As under one-point feedback (compare Fig. 9.3), the theoretical predictions are violated for a small attractor radius $R < r_A \approx 0.2\lambda$. In this case the amplitude of the first Fourier component of the feedback signal practically vanishes, and entrainment or asynchronous attractors are observed instead of the resonance attractor [47].

It is very important to stress that changes in the geometrical shape of the integration domain can induce bifurcations in the drift velocity field [31, 47, 50, 52]. Let us consider, for example, the drift velocity field computed for an elliptical domain with major axis $a = 3\lambda$ and minor axis $b = a/1.1$. As shown in Fig. 9.13(a), instead of the stable limit cycle of the resonance attractor in the circular domain of radius $R_d = 1.5\lambda$ we have two pairs of fixed points where the drift velocity vanishes. In each pair, one fixed point is a saddle and the other one is a stable node. Depending on the initial conditions, the spiral wave approaches one of the two stable nodes. Trajectories of the spiral center obtained by numerical integration of the Oregonator model (9.1) are in perfect agreement with the predicted drift

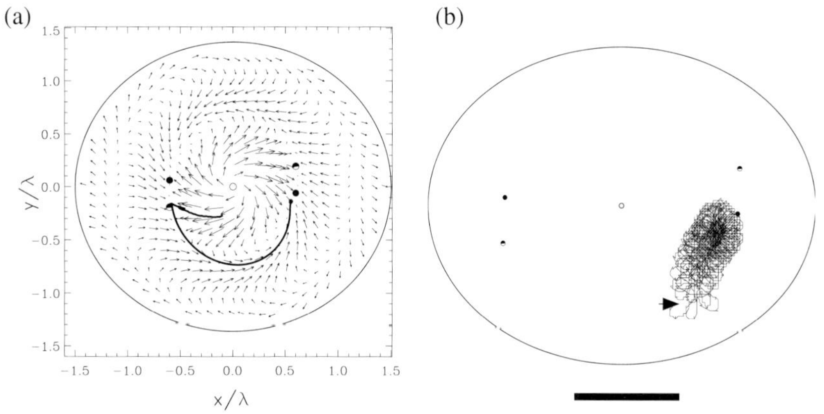

Fig. 9.13. (a) Drift velocity field obtained for global feedback in an elliptical domain ($a/b = 1.1$). (b) Resonant drift induced in experiments with the BZ reaction ($a/b = 1.25$). Open, filled and semi-filled circles denote unstable nodes, stable nodes, and saddle points, respectively. Scale bar: 1 mm [47].

velocity field.

The theoretically predicted destruction of the resonance attractor in response to deviations from the circular shape of the integration domain has been confirmed experimentally within the light-sensitive BZ medium. A spiral wave was exposed to uniform illumination proportional to the total gray level obtained in an elliptical integration domain. Fig. 9.13(b) shows the resonant drift mediated during global feedback control. The spiral wave drifts towards a stable node of the drift velocity field. Close to this fixed point the drift velocity becomes very slow. Thus, the experimentally observed termination of the spiral drift at certain positions in a uniform medium is explained in the framework of the developed theory of feedback-mediated resonant drift.

9.5. Discussion

We have shown that all existing methods of spiral wave control can be considered in the framework of a unified theoretical approach. This approach is based on the well established phenomenon of resonant drift induced by a periodic parametric modulation at the rotation frequency of a spiral wave [14–20]. The direction of resonant drift depends on the initial orientation of the spiral and on the phase of the first Fourier component of the periodic modulation. To specify the spiral orientation we propose to use an Archimedean spiral approximation for a rigidly rotating or a slightly meandering spiral wave [47, 50, 53].

It was extremely important to demonstrate that because the phase of the feedback signal depends on the spiral orientation, the direction of feedback-induced resonant drift is determined only by the spiral location and does not depend on its initial orientation. Thus, under feedback control the dynamics of spiral waves can be described by a drift velocity field [47, 53].

If a feedback can be considered as a sum of signals taken from different sources, the resulting drift direction is found to be the sum of particular drift vectors induced by each separate source. This superposition principle essentially simplifies the analysis of the drift velocity fields [47, 53].

The drift velocity fields can display quite non-trivial properties very interesting from the viewpoint of the general theory of dynamical systems. Depending on the particular feedback algorithm, besides isolated fixed points and limit cycles, unrestricted fixed lines with zero drift velocity resulting from global bifurcations have been observed [53]. The unusual behavior near this new type of equilibrium manifolds is currently investigated [54]. Hopefully, the experimental discovery of fixed lines in velocity fields of feedback-mediated control of spiral waves will motivate further research in this interesting direction.

The deeper understanding of the dynamics of feedback-induced drift allows us to propose robust and efficient methods to move a spiral center from a given initial location along an arbitrary path to a desired final position. In this way, we can avoid eventually existing defects in the medium that could pin the spiral wave. This result might be important for such possible application as low-voltage defibrillation of cardiac tissue.

A very interesting observation is the termination of spiral drift in spatially uniform media. Pinning of spiral waves is nowadays the subject of very general mathematical considerations [55]. A spiral wave can stay fixed at a certain site as if it is anchored at an invisible defect. In fact, this behavior is a consequence of the superposition principle for feedback-induced drift discussed in section 9.4.1. There is a deep analogy between this phenomenon and the destructive interference of linear waves.

The experimental system employed in this paper is the light-sensitive Belousov-Zhabotinsky reaction. For numerical simulations only the underlying Oregonator model was used. However, our theoretical approach is based on a very general description of excitable media and does not rely on specific features of any experimental or model system. Therefore, our results are of general character, and can be applied to control spiral wave dynamics in quite different excitable media.

References

[1] G. Gerisch, Naturwissenschaften **58**, 430 (1971).
[2] A.T. Winfree, Science **175**, 634 (1972).
[3] A.M. Zhabotinsky and A.N. Zaikin, J. Theor. Biol. **40**, 45 (1973).
[4] M.A. Allessie, F.I.M. Bonke, and F.J.G. Schopman, Circ. Res. **33**, 54 (1973).
[5] J.M. Davidenko, A.V. Pertsov, R. Salomonsz, W. Baxter, and J. Jalife, Nature **355**, 349 (1992).
[6] N.A. Gorelova and J. Bures, J. Neurobiol. **14**, 353 (1983).
[7] S. Jakubith, H.H. Rotermund, W. Engel, A. von Oertzen, and G. Ertl, Phys. Rev. Lett. **65**, 3013 (1990).
[8] J. Lechleiter, S. Girard, E. Peralta, and D. Clapham, Science **252**, 123 (1991).
[9] T. Mair and S.C. Müller, J. Biol. Chem. **271**, 627 (1996).
[10] A.T. Winfree: *The Geometry of Biological Time* (Springer, Berlin, Heidelberg, 2000).
[11] Y. Kuramoto: *Chemical Oscillations, Waves and Turbulence* (Springer, Berlin, Heidelberg, 1984).
[12] V. Krinsky and H. Swinney (eds): *Wave and patterns in biological and chemical excitable media* (North-Holland, Amsterdam, 1991).
[13] R. Kapral and K. Showalter (eds.): *Chemical Waves and Patterns* (Kluwer, Dordrecht, 1995).
[14] V.A. Davydov, V.S. Zykov, and A.S. Mikhailov, Usp. Fiz. Nauk **161**, 45 (1991) [Sov. Phys. Usp. **34**, 665 (1991)].
[15] R.M. Mantel and D. Barkley, Phys. Rev. E **54**, 4791 (1996).
[16] V. Hakim and A. Karma, Phys. Rev. E **60**, 5073 (1999).
[17] B. Sandstede, A. Scheel, and C. Wulff, J. Nonlin. Sci. **9**, 439 (1999).
[18] K.I. Agladze, V.A. Davydov, and A.S. Mikhailov, JETP Lett. **45**, 767 (1987).
[19] S. Grill, V.S. Zykov, and S.C. Müller, J. Phys. Chem. **100**, 19082 (1996).
[20] V.N. Biktashev and A. Holden, J. Theor. Biol. **169**, 101 (1994).
[21] S. Grill, V.S. Zykov, and S.C. Müller, Phys. Rev. Lett. **75**, 3368 (1995).
[22] Y.E. Volodin, V.V. Barelko, and A.G. Merzhanov, Sov. J. Chem. Phys. **5**, 1146 (1982).
[23] C. Radehaus, K. Kardell, H. Baumann, P. Jäger, and H.-G. Purwins, Z. Phys. B - Condensed Matter **65**, 515 (1987).
[24] S.L. Lane and D. Luss, Phys. Rev. Lett. **70**, 830 (1993).
[25] E. Schöll, Nonlinear spatio-temporal dynamics and chaos in semiconductors, (Cambridge Univ. Press, Cambridge, 2001).
[26] V.K. Vanag, L. Yang, M. Dolnik, A.M. Zhabotinsky, and I. Epstein, Nature **406**, 389 (2000).
[27] E. Mihaliuk, T. Sakurai, F. Chirila, and K. Showalter, Phys. Rev. E **65**, 065602 (2002).

[28] M. Kim, M. Bertram, M. Pollman, A. von Oertzen, A.S. Mikhailov, H.H. Rotermund, and G. Ertl, Science **292**, 1357 (2001).
[29] H. Brandtstädter, M. Braune, I. Schebesch, and H. Engel, Chem. Phys. Lett. **323**, 145 (2000).
[30] V.S. Zykov, G. Bordiougov, H. Brandtstädter, I. Gerdes, and H. Engel, Phys. Rev. E **68**, 016214 (2003).
[31] V.S. Zykov, G. Bordiougov, H. Brandtstädter, I. Gerdes, and H. Engel, Phys. Rev. Lett. **92**, 18304 (2004).
[32] H.-J. Krug, L. Pohlmann, and L. Kuhnert, J. Phys. Chem. **94**, 4862 (1990).
[33] N. Wiener and A. Rosenblueth, Arch.Inst. Cardiol. Mex. **16**, 205 (1946).
[34] V.S. Zykov, Simulation of Wave Processes in Excitable Media (Manchester Univ. Press, Manchester, 1987).
[35] V.S. Zykov and O.L. Morozova, Biophysics **24**, 739 (1980).
[36] J.J. Tyson, and J.P. Keener, Physica **D 32**, 327 (1988).
[37] J. Gomatam, and D.A. Hodson, Physica **D 49**, 82 (1991).
[38] F.B. Gul'ko and A.A. Petrov, Biofizika **17**, 261 (1972).
[39] V.S. Zykov, Biophysics **25**, 329 (1980).
[40] A. Karma and V.S. Zykov, Phys. Rev. Lett. **83**, 2453 (1999).
[41] P.S. Hagan, SIAM J. Appl. Math. **42**, 762 (1982).
[42] I.S. Aranson and L. Kramer, Rev. Mod. Phys. **74**, 99 (2002).
[43] V.S. Zykov, O. Kheowan, O. Rangsiman, and S.C. Müller, Phys. Rev. E **65**, 026206 (2002).
[44] D. Barkley, M. Kness, and L.S. Tuckerman, Phys. Rev. A **42**, 2489 (1990).
[45] S. Nettesheim, A. von Oertzen, H.H. Rotermund, and G. Ertl, J. Chem. Phys. **98**, 9977 (1993).
[46] M. Braune and H. Engel, Phys. Rev. E **62**, 5986 (2000).
[47] V.S. Zykov and H. Engel, Physica D **199**, 243 (2004).
[48] O. Kheowan, C.K. Chan, V.S. Zykov, O. Rangsiman, and S.C. Müller, Phys. Rew. E **64**, 035201(R), (2001).
[49] O. Kheowan, V.S. Zykov, and S.C. Müller, Phys.Chem.Chem.Phys. **4**, 1334 (2002).
[50] V.S. Zykov and H. Engel, Phys. Rev. E **66**, 16206 (2002).
[51] V.S. Zykov, A.S. Mikhailov, and S.C. Müller, Phys. Rev. Lett., **78**, 3398 (1997).
[52] O. Kheowan and S.C. Müller, Appl. Math. Comput. **164**, 373 (2005).
[53] V.S. Zykov, H. Brandtstädter, G. Bordiougov, and H. Engel, Phys. Rev. E **72**, 065201(R) (2005).
[54] B. Fiedler and S. Liebscher, in: Proc. of the International Congress of Mathematicians, ICM 2002, Vol. III, Higer Education Press, Beijing, pp. 305-316 (2002).
[55] B. Fiedler, M. Georgi, and N. Jangle, Chapter 3 in this volume.

Chapter 10

Radiative driven instabilities

Michael Hegmann* and Erwin Sedlmayr
Zentrum für Astronomie und Astrophysik, Technische Universität Berlin, Hardenbergstr.36, D-10623, Berlin, Germany
**hegmann@astro.physik.TU-Berlin.de*

Contents

10.1. Introduction

Astrophysical media very often exhibit a remarkable degree of inhomogeneity. An example are different types of dust forming objects like the envelopes of old and developed stars or planetary nebulae. High angular resolution observations of such objects reveal a large variety of different spatial-temporal dust structures. In Figure 10.1a, a HST image of the Eskimo nebula, a so called planetary nebula, is shown. The filaments visible here consist of material ejected by the central star. Figure 10.1b shows a speckle interferometry image of the carbon rich star IRC 10216 at 1.125 μm. The structure of the dust shell can be clearly seen. Another example are

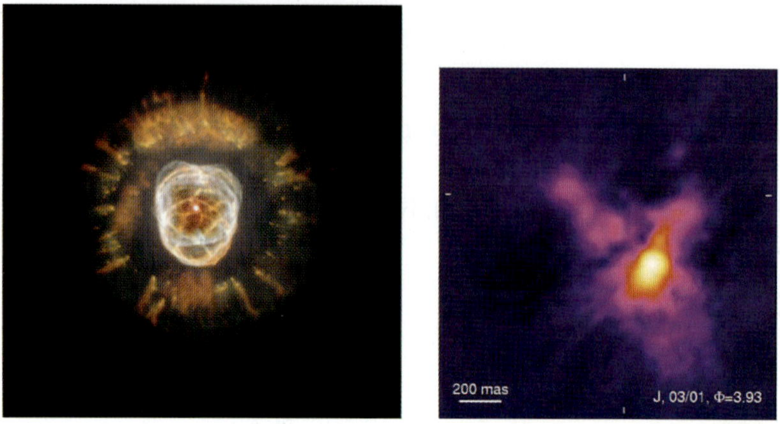

Fig. 10.1. Left: Hubble Space Telescope image of the Eskimo nebula. The filaments visible above consist of material ejected by the central star (now a white dwarf) during its AGB phase. Right: Speckle interferometry of the carbon star IRC 10216 at 1.125 μm. The structure of the dust shell around the star can be clearly seen. Credit: J. Hester, P. Scowen (ASU), HST, NASA

star forming regions, where the usually thin interstellar medium is highly compressed and exposed to the intense FUV radiation of the newly born stars. Since the physical and chemical state inside such a medium is dominated by photons, these regions are also called photon dominated regions. In Figure 10.2, the famous and highly structured Eagle Nebula, a stellar nursery at a distance of approximately 7000 light years, is shown.

For both types of astrophysical objects, the envelopes of evolved stars and star forming regions, the radiation field plays a crucial role in the development of the observed structure. For the formation of stars out of the on average thin interstellar medium great compression of the material has to occur. An important question is under which conditions structures formed in the interstellar medium by for example turbulent compression or by the passage of a shock wave can be become gravitationally and/or thermally unstable and collapse. For star formation to take place, i.e. for large compression to occur, the gas must cool down. This means, that the released compressional and gravitational energy has to be removed from the system. The by far most important cooling mechanism is the transport of energy by radiation. In the case of stellar envelopes, not only the transport of energy by radiation is important, but also the transport of momentum by the photons is crucial. Since in the circumstellar envelopes of evolved

Fig. 10.2. Hubble Space Telescope image of the Eagle nebula. Credit: Andrew Fruchter (STScI) et al., WFPC2, HST, NASA

stars (Red Giants) dust can be effectively produced, the radiative force on dust grains plays a major role in the structuration of these envelopes. In the following we want to examine two different types of instabilities which can trigger the formation of the observed structures. At first, we study the conditions for the appearance of combined thermal/gravitational instability in section10.2. Afterwards, we discuss the mechanisms which lead to a radiative instability in the dust forming envelopes of AGB stars in section 10.4.

10.2. Gravitational/thermal instability

10.2.1. *The Jeans instability*

When a region of gas is compressed , the excess pressure tries to smooth out the compression, which excites acoustic waves. In the compressed region, however, gravitation is enhanced and therefore more gas is attracted by it. The latter effect is negligible for typical sound waves in the earth's atmosphere. For perturbations of gaseous bodies of astronomical size, however, the self-gravity is quite important and the enhanced gravitation in the compressed region may overpower the expansive tendency of the excess pressure, pulling more material there and triggering an instability.

Jeans, who at first demonstrated the nature of this so called Jeans instability, considered perturbations in a uniform infinite gas. Considering the continuity equation, the equation of motion and the Poisson equation:

$$\frac{\partial \rho}{\partial t} + \nabla(\rho \mathbf{v}) = 0 \tag{10.1}$$

$$\rho \left(\frac{\partial \mathbf{v}}{\partial t} + (\mathbf{v}\nabla)\,\mathbf{v} \right) = -\nabla p - \rho \nabla \Phi \tag{10.2}$$

and

$$\Delta \Phi = 4\pi G \rho \tag{10.3}$$

he obtained the following equation which characterizes the growth of the density perturbation $\delta\rho$ owing to the self-gravity in a linear approximation

$$\frac{\partial^2 \delta\rho}{\partial t^2} = c_\mathrm{s}^2 \Delta \delta\rho + 4\pi G \rho \delta\rho \,. \tag{10.4}$$

Here, G is the gravitational constant and c_s is the isothermal sound speed. From eq. (10.4) the following dispersion relation can be found

$$\omega^2 = c_\mathrm{s}^2 k^2 - 4\pi G \rho \,. \tag{10.5}$$

For $k < k_\mathrm{J}$ with

$$c_\mathrm{s}^2 k_\mathrm{J}^2 = 4\pi G \rho \,, \tag{10.6}$$

ω is imaginary, which means that perturbations would increase exponentially. The system is unstable. In other words, if the size of the perturbations is larger than some critical wavelength, the enhanced self-gravity can outweigh the excess pressure so that the perturbations grow. The corresponding critical mass

$$M_\mathrm{J} = \frac{4}{3}\pi \lambda_\mathrm{J}^3 \rho \quad \text{with} \quad \lambda_\mathrm{J} = \frac{2\pi}{k_\mathrm{J}} \tag{10.7}$$

is usually referred to as the Jeans mass.

Under average interstellar conditions, this means a particle density of approximately 1 $\mathrm{cm}^{-3}$and a temperature of 100 K, only masses larger than approximately 10^4 solar masses can become unstable according to eqs. (10.6) and (10.7). As a consequence, stars form usually in clusters. The initial clouds breaks finally up into tens, hundreds, or thousands of fragments. The fragments are much denser (approximately $10^3 - 10^4 \mathrm{cm}^{-3}$). so that now masses in the range of a few solar masses can become unstable.

10.2.2. *Thermal instability*

In deriving eq. (10.6), an isothermal gas and only thermal pressure were assumed. This means that the local pressure and density are connected through

$$p = \alpha\rho\,. \tag{10.8}$$

Magnetic or turbulent pressure were neglected. The situation becomes much more complex if the transport of energy is included. Under interstellar conditions, the thermal pressure is given by the ideal gas equation

$$p = \frac{k_{\rm B}}{\mu}\rho T\,. \tag{10.9}$$

Here, μ is the mean molecular mass and $k_{\rm B}$ is the Boltzmann constant. Eq. (10.9) is a very good approximation, because even in the densest regions of the interstellar medium the particle number density only reaches values as high as $10^5 \rm cm^{-3}$. To obtain the local gas temperature T, the thermal balance equation must be solved:

$$\Gamma(J_\lambda, \rho, T) = \Lambda(J_\lambda, \rho, T)\,, \tag{10.10}$$

where J_λ is the angle averaged spectral intensity of the local radiation field, Γ is the local heating and Λ is the local cooling rate. The by far most important energy transport mechanism in interstellar molecular clouds is the transport of energy by radiation. Hence, the cooling and the heating rate are both sensitive functions of the intensity of the radiation field. The gas is mainly heated by ionizing FUV photons and deeper in the cloud by Cosmic rays. Cooling comes from low-density, sub-thermal collisions. For an optically thin cloud[a] of neutral hydrogen gas, the heating rate is proportional to the density of the gas, whereas the cooling rate is roughly proportional to the square of the density. As a consequence, the cloud cannot only become thermally unstable (the pressure drops with increasing density), but also interstellar gas in thermal balance can coexist at the same pressure with two different temperature/density combinations.

Under the assumption of an optically thin cloud (every emitted photon can leave the cloud without being reabsorbed) a critical wave number similar to eq. (10.6) can be obtained:

$$c_{\rm s}^2 k_{\rm c}^2 = 4\pi G\rho \Big/ \left[1 + \frac{{\rm d}\ln T}{{\rm d}\ln\rho} - \frac{{\rm d}\ln\mu}{{\rm d}\ln\rho}\right] \tag{10.11}$$

[a]Every emitted photon can leave the cloud without being reabsorbed.

with T being the kinetic gas temperature and μ the mean molecular mass of the gas. Again, only thermal pressure is accounted for. According to eq. (10.11), the critical wavelength $2\pi/k_c$ is always smaller than the classical Jeans length λ_J under interstellar conditions. If the term in brackets is negative, all perturbations are instable. This means, that an interstellar cloud can be destabilized by thermal effects as well as by the formation of molecules (increasing μ). However, since most of the energy is transported in a few molecular lines only, astrophysical clouds cannot be assumed to be optically thin in general. Hence, a detailed study of the radiative transfer within the cloud to determine the local heating and cooling rates of the gas is needed. In the next section, we want to discuss some basic radiative transfer effects on the thermal/gravitational stability of an interstellar molecular cloud.

10.2.3. *Stochastic radiative transfer*

10.2.3.1. *Stochastic description of a turbulent velocity field*

There are two basic problems involved with the quantitative discussion of radiative transfer effects in molecular clouds. One is that in general the molecular gas is not in local thermal equilibrium (NLTE). This means that the molecular occupation numbers are not given by a Boltzmann distribution. The other is that the line broadening mechanism has to be known in detail. Due to the low density in the interstellar medium, Doppler broadening is by far the most important broadening mechanism. In most cases observed line widths are considerably larger than the thermal width. This indicates that hydrodynamic motions are present in the cloud. In principle, the basic magnetohydrodynamic equations have to be solved self-consistently with the radiative transport equation and the rate equation which determine the molecular occupation numbers. However, the numerical effort would be immense. Furthermore, the proper initial conditions are not known. Motivated by this, we use an approach which describes distinct physical parameters of the cloud in a statistical sense only. This approach has many applications. To this end, we focused our efforts on the problem of radiative transfer in a medium with an inhomogeneous stochastic density and velocity distribution [7–9]). In the following we assume a turbulent velocity field which can be described in terms of its stochastic properties. This means that we must specify the functional form of the multipoint probability distributions. Following Gail et al. [2], we assume a Markovian type structure for the velocity field along each line of sight. In this approximation only two-point correlations are taken into account, and the stochastic velocity field is completely defined by specifying the one-point distribution of the velocity and the corresponding two-point correlation function. We consider the one-point distribution of v (the turbulent velocity component along the line of sight) to be Gaussian

$$W(v) = \frac{1}{\sqrt{2\pi}\sigma} \exp\left(-\frac{v^2}{2\sigma^2}\right) \tag{10.12}$$

and the two-point correlation function to be exponential

$$f(\Delta s) = \frac{\langle v(s)v(s+\Delta s)\rangle}{\sigma^2} = \exp\left(-\frac{|\Delta s|}{\ell}\right) . \tag{10.13}$$

Here, s is the spatial coordinate along the line of sight, and σ^2 and ℓ are the mean quadratic turbulent velocity and the correlation length of the velocity field, respectively. The correlation length ℓ defines the length scale of the stochastic velocity variation. Since the spatial power spectrum of the velocity field is the Fourier transform of the correlation function, eq. (10.13) implies a one-dimensional power spectrum

$$P(k) \propto \frac{\ell}{1+\ell^2 k^2} \tag{10.14}$$

which is well approximated by a power law for $\ell^2 k^2 \gg 1$.

This stochastic process can also be described by the following Langevin equation:

$$\frac{\mathrm{d}v}{\mathrm{d}s} = -\frac{v}{\ell} + \frac{\sigma}{\sqrt{\ell}}\Gamma(s) , \tag{10.15}$$

where $\Gamma(s)$ is the random noise or Langevin force.

10.2.3.2. *The generalized radiative transfer equation*

If the physical state of the medium is known, the spectral intensity of the radiation field can be determined by the ordinary radiative transfer equation:

$$\frac{dI_\nu}{ds} = -\kappa_\nu I_\nu + \varepsilon_\nu = -\kappa_\nu \left(I_\nu - S_\nu\right) . \tag{10.16}$$

Here, I_ν is the specific intensity at frequency ν, κ_ν the absorption coefficient, ε_ν the emissivity and S_ν the source function. In the case of a local thermal equilibrium S_ν is given by the Planck function B_ν. Both, the emission and the absorption coefficient are a sensitive function of the density, the temperature, the chemical abundances, the velocity and even of the angle averaged intensity J_ν. However, if we describe the velocity by a stochastic ansatz, eq. (10.16) has to be replaced. Due to the stochastic nature of the underlying velocity field, the intensity of the radiation field becomes a stochastic variable, too. It can be shown that the Langevin equation (10.15) together with the ordinary radiative transport equation (10.16) lead to a generalized radiative transfer equation of Fokker-Planck type:

$$\frac{\partial q_\nu}{\partial s} = \frac{1}{\ell}\left(-v\,\frac{\partial q_\nu}{\partial v} + \sigma^2\,\frac{\partial^2 q_\nu}{\partial v^2}\right) - \kappa_\nu\,(q_\nu - S_\nu)\;. \qquad (10.17)$$

Here, $q_\nu(v,s)$ is the conditional intensity, that is the expectation value of the intensity for a given velocity v at point s. By multiplying q_ν with the one-point probability distribution W and integration over all velocities an overall expectation value for the intensity can be obtained:

$$\langle I_\nu\rangle = \int_{-\infty}^{+\infty} q_\nu(v)W(v)\,\mathrm{d}v\;. \qquad (10.18)$$

From eq. (10.17) it becomes clear that the relevant parameter for the radiative transfer is not the correlation length itself but the ratio of correlation length to the mean free path of the photons

$$\tau_\ell = \frac{\ell}{\langle l_{\mathrm{mfp}}\rangle} = \ell\,\langle\kappa_\nu\rangle\;. \qquad (10.19)$$

For a small τ_ℓ, the generalized radiative transfer equation is dominated by the stochastic first term on the right hand side of eq. (10.17), whereas the radiative transfer becomes deterministic for $\tau_\ell \gg 1$.

10.2.3.3. *Non-LTE line formation*

The source function $S_\nu(v,s)$ and the absorption coefficient $\kappa_\nu(v,s)$ are given in the usual way by the occupation numbers n_i and n_j corresponding to the transition considered. Since LTE cannot be assumed in general, the occupation numbers are not given by a Boltzmann distribution. They have to be determined self-consistently with the line intensities from the rate equations. These summarize all processes populating or depopulating the different quantum states:

$$\sum_{j\neq i}\left(n_j(v)\,(A_{ij}+C_{ij}) - n_i(v)\,(A_{ji}+C_{ji})\right) +$$
$$\sum_{j\neq i}\left(\frac{4\,\pi}{c}\,\langle\Phi_{ij}J(v)\rangle\,(n_j(v)\,B_{ij} - n_i(v)\,B_{ji})\right) = 0\,. \qquad (10.20)$$

Here, $\langle \Phi_{ij} J(v) \rangle$ is the mean and angle averaged value of the local radiation field, weighted with the profile function of the local absorption coefficient. The A_{ij} and B_{ij} are the Einstein coefficients for spontaneous and induced transitions, while C_{ij} denotes the probability for a collisional transition from state $j \rightarrow i$. Accordingly, the first row in eq. (10.20) accounts for spontaneous emission and collision of the molecule considered with H_2, whereas in the second row induced emission processes are described. This system of rate equations has to be solved simultaneously with the generalized radiative transfer equation for every point in physical and velocity space.

10.2.4. *On the cooling by CO*

In the cold and dense interstellar clouds, cooling is dominated by the rotational transitions of the abundant CO molecule and its isotopomers. CO possesses a spectrum of lines due to the rotation of the molecule. The permitted energy levels are approximately those allowed by quantum mechanics for a rigid rotator, i.e. a linear molecule that does not change shape as it rotates

$$E = BJ(J+1), \qquad J = 0, 1, 2, \ldots \tag{10.21}$$

with the selection rule

$$\Delta J = \pm 1 .$$

The constant B is given by

$$B = \frac{\hbar}{2\Theta}, \tag{10.22}$$

where Θ is the moment of inertia of the molecule. Thus the massive molecules with large moments of intertia have closely spaced energy levels. For CO, the constant B has the value 3.83×10^{-16}erg. Since this energy is equivalent to a temperature of only 2.77 K, the low rotational transitions of CO are easily excited even for very low gas temperatures.

Of course H_2 is by far more abundant, but its rotational transitions cannot be excited in the cold parts of the interstellar medium. For H_2, electric dipole transitions are forbidden, because no dipole moment exists in the molecule in this state. Transitions occur via electric quadrupole interaction, and only $\Delta J = \pm 2$ transitions occur. The least energetic transition in H_2, the excitation $J = 0 \rightarrow J = 1$, occurs at an energy equivalent to 510 K. We want now to study the influence of radiative transfer effects on the radiative cooling by CO.

10.2.4.1. *The model*

As underlying cloud model, we use the model of a hydrostatic isothermal sphere which is stabilized against gravitational collapse by thermal and turbulent pressure. Although the concept of turbulent pressure is widely used, one should keep in mind that the assumption of a turbulent pressure of the form $\sigma^2 = (\mathrm{d}p_{\mathrm{turb}}/\mathrm{d}\rho)$ implicitly assumes a small-scale and isotropic turbulence. The existence of large scale modes, on the other hand, would be more akin to ram pressure (locally anisotropic, with a well-defined direction) than to isotropic, thermodynamic pressure. Under the assumptions made, the fundamental equations determining the cloud structure are the hydrostatic equation, the Poisson-equation in its integral representation and an equation of state:

$$\frac{\partial p}{\partial r} = -\frac{G\,M(r)\,\rho(r)}{r^2} \tag{10.23}$$

$$\frac{\partial M(r)}{\partial r} = 4\,\pi\,r^2\,\rho(r)\,, \tag{10.24}$$

$$p(r) = (\frac{k\,T}{m} + \sigma^2)\,\rho(r)\,, \tag{10.25}$$

using the following abbreviations:

r	distance from the cloud center
$M(r)$	mass internal to the radius r
$\rho(r)$	local mass density
$p(r)$	local gas pressure
T	kinetic gas temperature
σ	mean turbulent velocity
m	mean molecular mass
G	gravitational constant.

Since an isothermal cloud can only be stable at non-zero external pressure, we define the cloud radius R_{H_2} to be the radius for which the local cloud pressure equals the pressure of the mean interstellar medium p_{ISM}.

Of course, the isothermal equation of state (10.25) is a simplification. In a more realistic cloud model, the thermal pressure is given by the ideal gas equation, where the temperature follows from a thermal balance equation (10.10). First results for such an improved model are given later.

10.2.4.2. *Results*

Figure 10.3 shows the calculated cooling rates (expectation value) normalized to the hydrogen number density as a function of the distance from the cloud center for two different correlation lengths of the turbulent velocity field and for the different rotational transitions of CO. It can be clearly seen that the cooling by the low rotational transitions $J = 1 \to 0$ and $J = 2 \to 1$ rises steeply towards the boundary of the cloud, whereas the cooling rate of the $J = 4 \to 3$ transition becomes smaller. This can be explained as follows: The model of an isothermal sphere implies that the density of the gas drops strongly with increasing distance from the cloud center. On the one hand, this decrease of the density and the spherical geometry lead to an increased escape probability for the photons at the edge of the cloud. As a consequence, the cooling rate increases. On the other hand, the higher rotational levels are less effectively populated by collisions of CO with H_2 for low densities. This leads to a decrease of the cooling rates which becomes more pronounced for the high rotational levels. The effects of the correlation length on the process of line formation and therefore on the cooling rates can be best seen for the $J = 1 \to 0$ and $J = 2 \to 1$ transitions. It can be shown [1] that an effective absorption coefficient of the turbulent medium becomes maximal for $\ell \to 0$. Consequently, the outer layers of the cloud become optically more opaque with decreasing correlation length ℓ. This can be seen in Figure 10.3 where the cooling rates are larger for the smaller correlation length at the cloud's boundary. In contrast, the cooling rates drop with decreasing correlation length near the cloud center. Here, photons escape hardly from the cloud for small ℓ. Along every line of sight there is almost certainly a turbulence element with the appropriate radial velocity to reabsorb the emitted photon.

Figure 10.4 gives the total energy loss added over all lines normalized to the total number density of hydrogen molecules in the cloud as a function of the central density of the model cloud. Evidently, the cooling rates of the higher rotational transitions become dominant with increasing density, whereas the cooling rates of the $= 1 \to 0$ and $J = 2 \to 1$ transitions decrease.

10.3. The structure of photon dominated regions

Up to now we have concentrated on the dense and cool inner regions of molecular clouds which are shielded against the interstellar UV radiation

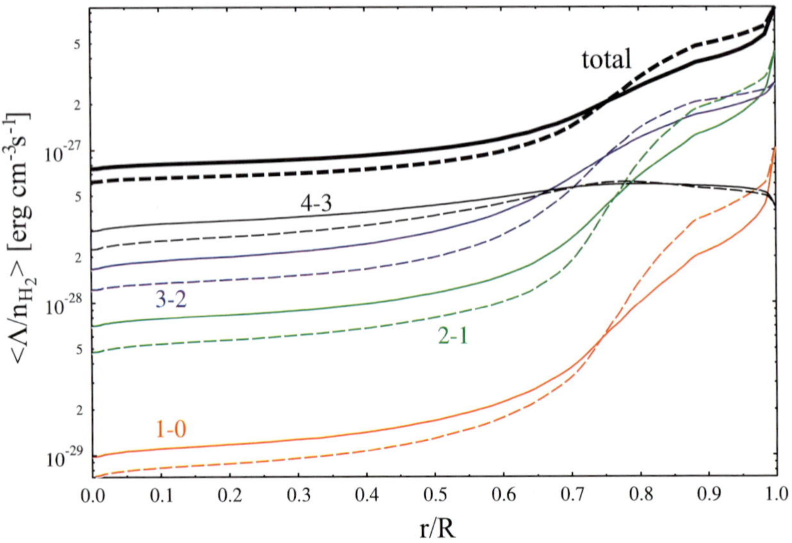

Fig. 10.3. Expectation value of the cooling rate as function of the distance from the cloud center: $n_{\rm cen} = 10^4\,{\rm cm}^{-3}$, $T = 10$ K, $\sigma/v_{\rm therm} = 5$, $\ell = 10^{16}$cm (dashed line) and $\ell = 10^{17}$cm (solid line).

by surrounding gas and dust. On the surface of the interstellar clouds even the most simple molecules like CO or H_2 are not stable, because they are rapidly dissociated by the FUV photons ($6\,{\rm eV} < h\nu < 13.6\,{\rm eV}$) of the interstellar radiation field. Such a region, the physical and chemical structure of which is dominated by the FUV radiation field is called photon dominated region (PDR). In general, PDRs include all interstellar regions where the gas is predominantly neutral but where FUV photons play a significant role in the chemistry and/or heating. Therefore, an important key in understanding the structure and evolution of PDRs is in understanding the transport of radiation inside the medium. It has to be noted, that the radiative transfer leads to a non-local coupling of the physical states within the interstellar medium. Observations of PDRs with a high angular and spectral resolution indicate a highly inhomogeneous structure down to the smallest accessible scales. Such inhomogeneities have a profound impact on the radiative transfer, and, as a consequence, on the chemical abundances and the local heating and cooling rates. However, a self-consistent description of a PDR, including all the small scale hydrodynamics, chemistry, and radiative transfer is far beyond reach of todays super computers. There-

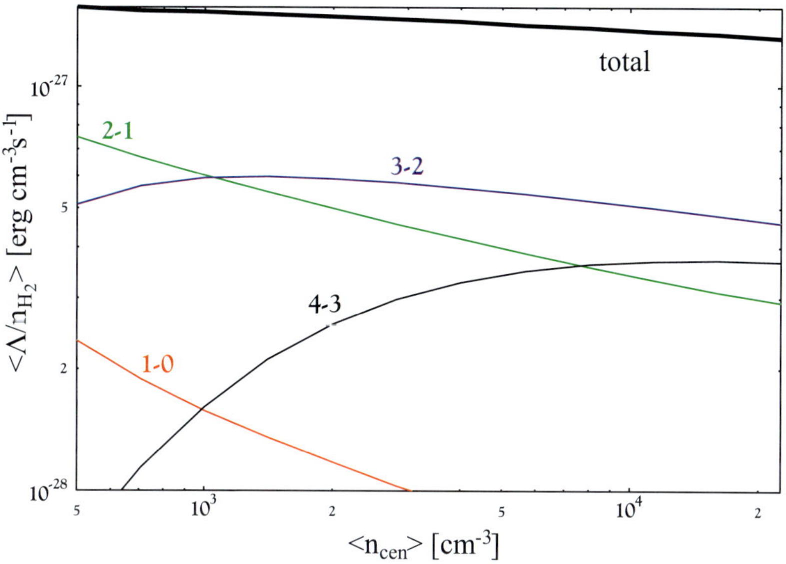

Fig. 10.4. Total cooling rate for different isothermal clouds as function of the central hydrogen density. The gas temperature, the mean quadratic turbulent velocity and the correlation length were kept constant: $T = 10$ K, $\sigma/v_{\rm therm} = 10$ and $\ell = 10^{17}$cm. The cooling rates are normalized to the total number of hydrogen moleculs.

fore, the aim of our current research project is to analyze the influence of stochastic fluctuations of the density and/or the FUV radiation field on the physical and chemical processes inside a PDR. This requires a detailed study of the complex, non-linear interaction between the radiative transfer, the chemical abundances and the local thermal balance.

10.3.1. *Numerical model*

A numerical model of a PDR has at least to combine radiative transfer, chemical kinetics and thermal balance. This means that the generalized radiative transfer equation (10.17) and the thermal balance equation (10.10) have to be solved self-consistently with the rate equations describing the chemical reactions. Due to the very low densities prevailing in the interstellar medium only two-body interactions have to be accounted for. So the chemical rate system can be described by the following set of equations:

$$\frac{dn_i}{dt} = \sum_r \sum_s k_{rs} n_r n_s + \sum_t \zeta_{it} n_t - n_i \left(\zeta_i + \sum_q n_q k_{qi} \right) , \qquad (10.26)$$

where

- k_{rs} rate coefficient for the reaction $R + S \rightarrow I + \ldots$
- k_{qi} rate coefficient for the desctructive reaction $I + Q \rightarrow \ldots$
- ζ photodissociation+photoionization rate
- n_i number density of species i.

Since three-body collisions are negligible under interstellar conditions the formation of even the simple molecules like CO becomes very complex. A direct formation via two-body collisions of two neutral atoms like $C + O \rightarrow CO$ is very unlikely, because the excess energy cannot by simply removed by a third part and the timescale for a radiative transition is too large. At best, only one in 10^5 collisions of C and O produces the molecule CO. So, molecules can be build best via ion-molecule and neutral exchange reactions like:

$$O^+ + H_2 \rightarrow OH + H$$
$$CH + O \rightarrow CO + H\,.$$

However, to allow for reaction chains of the above type which build up the most common molecules in the end, one molecule already existing is needed. This molecule is H_2 which is not generated in the gas phase but on the surface of dust particles.

Figure 10.5 illustrates the chemical structure of a plane parallel PDR by giving the relative abundances of C^+, C and CO as a function of the penetration depth into the model cloud [11]. We assumed a kinetic equilibrium[b] and determined the relative abundances from a chemical network consisting of 38 different species formed and destroyed in 434 reactions. The PDR is illuminated from the left by the mean interstellar radiation field and extends from the predominantly atomic surface region to the point where almost all carbon is bound into CO. One of the difficulties in calculating the chemical and thermal structure of a PDR arises from the effect of self-shielding. Molecules already formed absorb UV photons which are able to dissociate the respective molecule. In other words: they cast a "shadow" into the cloud which enhances the further formation of the respective molecule. This effects becomes especially important for the formation of key molecules like O_2, H_2 and CO. Our current research addresses the question under which physical conditions an instability due to shadowing effects could occur. Another difficulty concerns the effects of small-scale fluctuations of the UV radiation field on the chemical network.

[b]For a kinetice equilibrium, $dn_i/dt = 0$ is assumed.

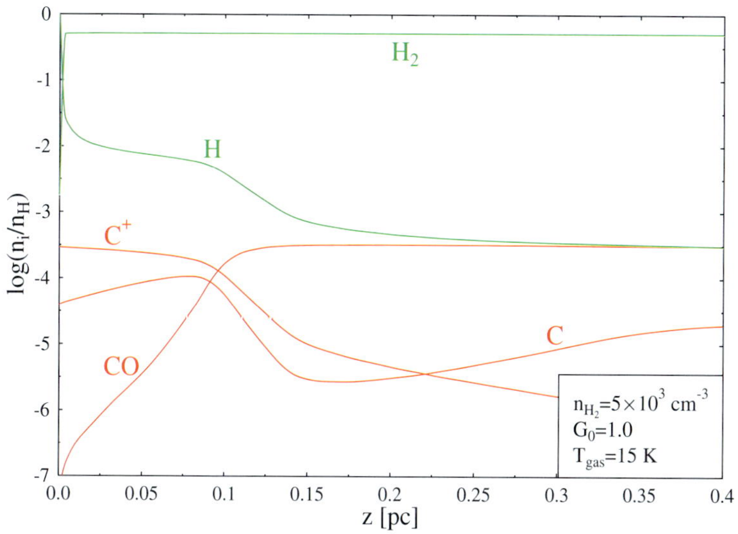

Fig. 10.5. Calculated structure of a PDR as a function of the penetration depth into the cloud. The illuminating source, the mean interstellar radiation field, is to the left. The abundances are given relative to the total hydrogen number density.

As mentioned above, observations of PDRs reveal inhomogeneities down to the smallest accessible scales. For a fixed point in space, the observed spatial fluctuations lead due to hydrodynamic motions to temporal fluctuations in the radiation field. If the time scale for the variation of the radiation field becomes comparable or even smaller than the timescale for the chemical reactions, the assumption of a kinetic equilibrium does not hold. Currently, we perform time dependent calculations for a chemical network consisting of 38 different species in order to study the influence of small scale fluctuations in the UV radiation field.

10.4. Radiative instability of dust formation

10.4.1. *The instability*

Woitke and collaborators [13] identified an instability which leads to a self-organized formation of structure inside a dust forming medium. This instability is based on the strong non-local feedback of newly formed dust on the temperature of the medium due to radiative transfer effects. This feedback is especially strong, because both, the formation and growth of

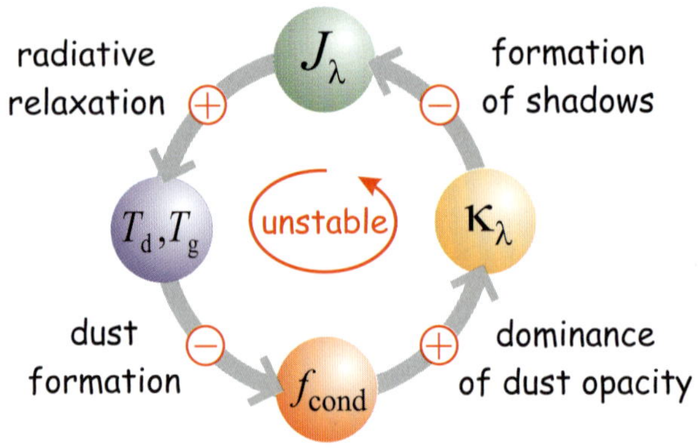

Fig. 10.6. Radiative instability of dust formation. J_λ=mean spectral intensity, T_g=gas temperature, T_d=dust temperature, f_{cond}=degree of condensation, κ_λ=spectral extinction coefficient.

dust are a sensitive function of the temperature. The whole feedback loop is sketched in Figure 10.6: Let us start with a given configuration of the radiation field and with a given degree of condensation f_{cond}, that is with a given amount of gas condensed into dust. A decrease of the mean spectral intensity J_λ of the radiation field is accompanied by a respective decrease of the gas and dust temperature. As a result the formation of dust is favored and the degree of condensation increases. A larger degree of condensation causes a larger opacity, which influences again the radiation field. It is quit obvious that such a local enhancement of the opacity lead to the formation of shadows. Inside the shadows the mean intensity decreases. So the sign of the overall feedback loop becomes finally positive.

10.4.2. *The model*

In order to simulate the radiative instability and to study the related structure formation process around a carbon rich AGB star, an axisymmetric numerical model was developed [14]. It combines equilibrium chemistry and modified classical nucleation theory and time- dependent dust growth [3, 4] with frequency-dependent radiative transfer. The absorption and scattering coefficients of the forming dust grains are calculated by applying Mie-theory to spherical solid particles with the optical data of amorphous carbon according to the resultant dust grain size distribution function. All three components which enter the simulation, the radiation field, the gas phase and the dust are coupled. Both, the gas and the dust are assumed to be in radiative equilibrium, hence the temperature of the dust and the temperature of the gas are a direct result of the radiative transfer part (non-locally coupling). On the other hand, the dust component influences by absorption and scattering the radiation field. The density distribution of the gas is assumed to be constant in time (no hydrodynamics) and exponentially decreasing in the radial direction

$$n_{\mathrm{H}}(\mathbf{r}) \propto \exp\left(-\frac{r}{H_\rho} + \delta(\mathbf{r})\right) , \qquad (10.27)$$

where H_ρ is the density scale height and $\delta(\mathbf{r})$ denotes a small spatial density perturbation of the order of 1.3% and 4.3%, respectively.

10.4.3. *Results*

The development of the forming dust shell in a sample model is presented in Figure 10.7, showing the degree of condensation $f_{\mathrm{cond}}(\mathbf{r}, t)$ and the dust temperature $T(\mathbf{r}, t)$. The initial state of the circumstellar envelope at $t = 0$ was chosen to be dust-free. The denser regions close to the star begin to condense first, because of the strong density dependence of the process of dust formation. Consequently, the spatial dust distribution at first resembles the slightly inhomogeneous gas distribution in the circumstellar shell with a cutoff at the inner edge as a consequence of the temperatures being too high for nucleation. According to the different choices of the density inhomogeneities above and below mid-plan, the resulting spatial variations in the degree of condensation are larger above than below the mid-plane at early phases of the model. The process of dust formation continues for a while in this way, until the first cells close to the star become optically thick. Each optically thick cell casts a shadow into the circumstellar envelope in which the temperatures decrease by several 100 K, which improves the conditions for subsequent dust formation in these shadows. At the same time, scattering and re-emission from the cells that have already condensed intensifies the radiation field in between. The stellar flux finally escapes preferentially through the segments which are still optically thin, thereby heating them up and worsening the conditions for further dust formation there. These two opposing effects amplify the initial spatial contrast of the degree of condensation introduced by the assumed density inhomogeneities. In the end, radially aligned, cool, linear dust structures, henceforth called dust fingers have developed, which point towards the star and are surrounded by warmer, almost dust-free regions at the inner edge of the forming dust shell.

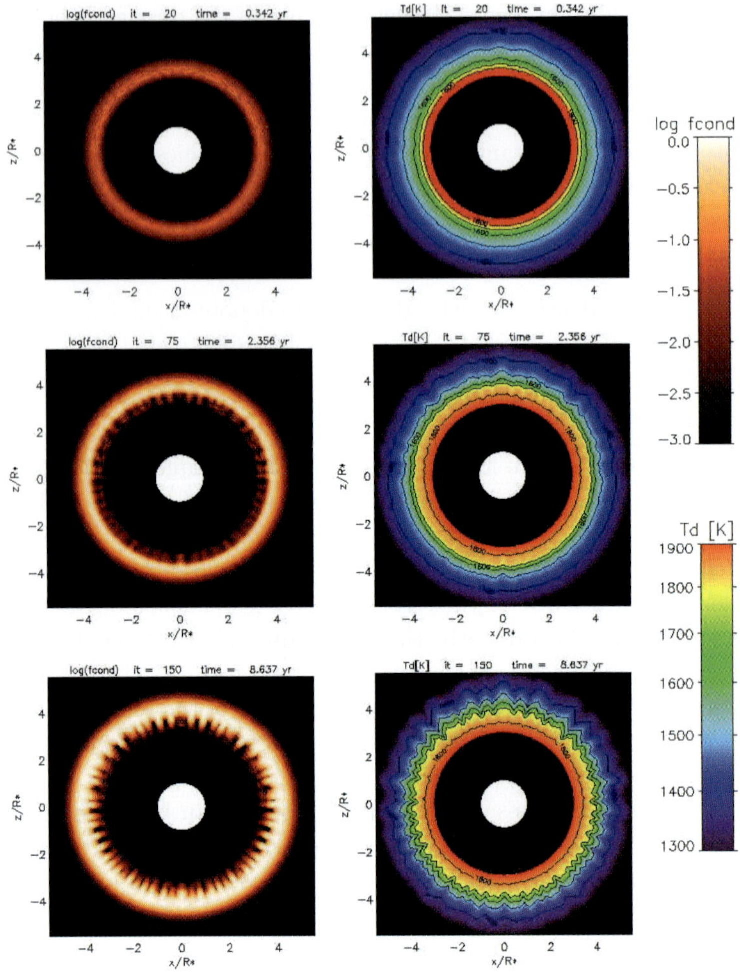

Fig. 10.7. Self-organisation and spatio-temporal structure formation. The figures show contour plots of the degree of condensation (left plot) and the dust temperature (right plot) in the x-z-plane at three selected time steps. The white circle in the center of each figure mark the star.

Apart from the self-organization of the dust in the angular direction, the model shows the formation and evolution of a radial dust shell. In Figure 10.8, the temporal evolution of the angle-averaged temperature and the degree of condensation are shown. The effective formation of dust generally requires a suitable combination of gas density and temperature, called the dust formation window [5]. Initially, favorable temperature conditions for efficient nucleation are only present in a restricted radial zone close to the star. However, as time passes, the dust shell becomes optically thick which dams the outflowing radiation and leads to an increase of the tempera-

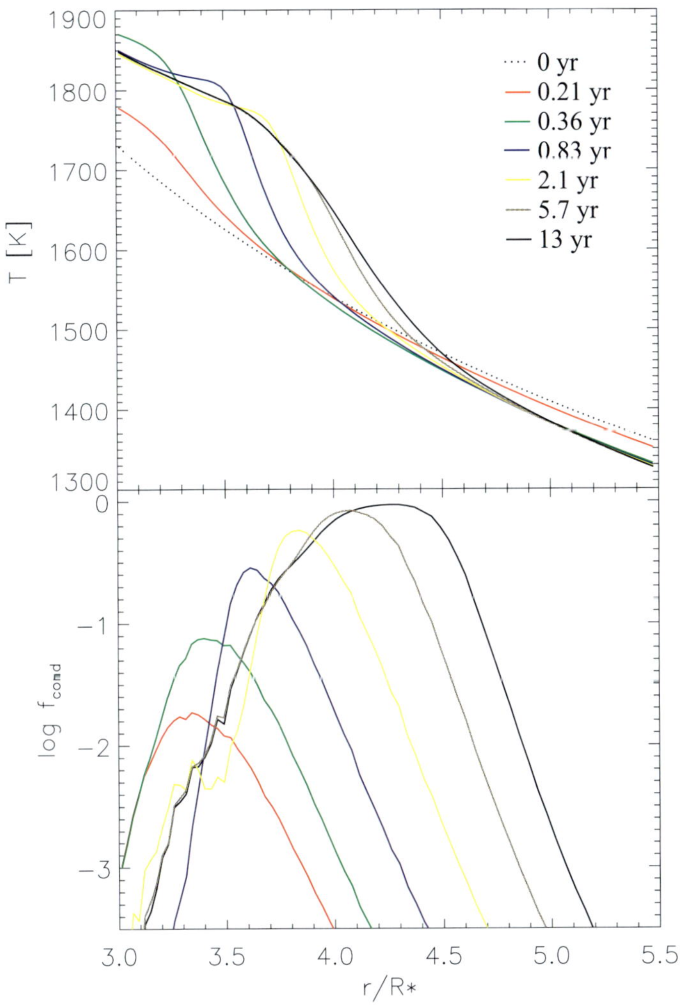

Fig. 10.8. The chemical wave: formation and propagation of a dust shell via dust formation at the outer edge and dust evaporation at the inner edge. The figure shows the time evolution of the angle-averaged dust temperature (upper plot) and degree of condensation (lower plot). The quantities are shown at seven time-steps.

tures inside the shell (radiative backwarming). Consequently, the zone of effective dust formation shifts outward with increasing time. Moreover, the temperatures at the inner edge of the shell temporarily exceed the sublimation temperature, and the dust shell begins to re-evaporate from the inside. The upper plot of Figure 10.8 shows that the temperature temporarily exceeds its equilibrium value which is reached at the later stages. These two effects result in an apparent motion of the dust shell, driven by dust formation at the outer edge and dust evaporation at the inner edge of the shell. This wave-like propagation of the dust-formation front can be denoted as chemical wave, because it is solely based on chemical and radiative processes without bulk velocity fields. Chemical waves are a well-known phenomenon in the laboratory, for example in reaction-diffusion systems, where front-like solutions of the chemical concentrations may exist, sometimes radiatively controlled [12].

References

[1] M. A. Albrecht, ,W. H. Kegel, A&A, **176**, 317 (1987).
[2] H.-P. Gail, E. Hundt, W. H. Kegel, J. Schmid-Burgk, G. Traving, A&A **32**, 65 (1974).
[3] H.-P. Gail, R. Keller, R., E. Sedlmayr, A&A, **133**, 320 (1984).
[4] H. P. Gail, E. Sedlmayr, A&A **206**, 153 (1988).
[5] H. P. Gail,E. Sedlmayr, in *Chemistry and physics of molecules and grains in space*, Faraday Discussion **109**, 303 (1998).
[6] H. P. Gail, E. Sedlmayr,G. Traving, A&A, **44**, 421 (1975).
[7] M. Hegmann,W. H. Kegel, A&A, **359**, 405 (2000).
[8] M. Hegmann, W. H. Kegel, W.H. MNRAS **342**, 453 (2003).
[9] M. Hegmann, Ch. Hengel, M. Röllig, W.H. Kegel, accepted for publication A&A, (2006).
[10] G. Piehler, W. H. Kegel, A&A **297**, 841 (1995).
[11] M. Röllig, M. Hegmann, M., W. H. Kegel,, A&A, **392**, 1081 (2002).
[12] I. Schebesch, H. Engel, Phys. Rev. E, **60**, 6429 (1999).
[13] P. Woitke, E. Sedlmayr, B. Lopez, B. A&A, **358**, 665 (2000).
[14] P. Woitke, G. Niccolini, A&A, **433**, 1101 (2005).

Chapter 11

Building oscillations bottom up: Elemental time scales of intracellular calcium dynamics

Rüdiger Thul* and Martin Falcke+

Hahn-Meitner Institut, Abteilung Theorie, Glienicker Str. 100, D-14109 Berlin, Germany

**thul@hmi.de and +falcke@hmi.de*

Contents

11.1. Introduction

The unraveling of molecular interactions has been providing intriguing insights into cellular organization during the last decades. On the one hand, qualitative results showed connections between formerly unassociated reactions. For instance, one of the molecular mechanisms that underlies cancer was identified to play a central role in Alzheimer's disease [2, 22]. On the other hand, the ongoing improvement in experimental techniques allowed quantitative investigations. They revealed that cells use a plethora of different amounts of molecules for a variety of tasks. Some pathways involve macroscopically large numbers of chemical agents, whereas others are controlled by only some tens of molecules. This discrepancy in the number of reactants attributes a twin role to fluctuations. They can be neglected in the former reactions, but play a pivotal role in the latter. Experiments on gene expression demonstrated that fluctuations can decide upon the phenotype leading to population heterogeneity [19, 23]. For a general review on intracellular noise, see [26].

Modeling intracellular phenomena therefore demands an a priori choice of methods. As long as fluctuations are negligible, deterministic equations correctly capture the dynamics [10, 17]. However, these approaches break down in the presence of noise. Comparisons between stochastic models and their deterministic counterparts have revealed that noise can induce a dynamical behavior that is not present in the absence of fluctuations. For instance, the MinCDE system only oscillates if the experimentally observed small number of interacting molecules is respected [16]. The deterministic equations decay to a fixed point.

The above examples all have in common that the total number of interacting molecules in the entire cell is small. However, fluctuations can also arise due to cellular heterogeneity. Steep concentration gradients may create very heterogeneous conditions within a single cell such that the conditions for a reaction to occur are given in a small part of the cell only. Since these gradients are typically transient we call this phenomenon dynamic compartmentalization. The number of interacting molecules in such a dynamic compartment may only be a tiny fraction of the molecules present in the whole cell. Then, fluctuations remain large and cannot be neglected, although the total number of molecules in the whole cell would allow for mean field behavior. That is the case with intracellular Ca^{2+} dynamics. The dynamic compartment is a single Ca^{2+} releasing channel cluster as is explained below.

The dynamics of the Ca^{2+} concentration in the cytosol of a cell is determined to a large degree by release and uptake of Ca^{2+} by intracellular storage compartments, in particular the endoplasmic reticulum (ER). Release is controlled by inositol-1,4,5-trisphosphate (IP_3) receptor channels (IP_3R). They are arranged in clusters with a diameter of less than 100nm that comprise between 1 and 40 channels and that are randomly distributed on the membrane of the ER with distances between 1-7 μm [21, 30]. IP_3Rs have the important property that their open probability depends on the Ca^{2+} concentration in the cytosol. The details of this dependency will be discussed in section 11.2. A moderate increase in the cytosolic concentration - i.e. on the outside of the storage compartment - increases the opening probability.

The notion of an open probability goes along with the generally stochastic character of ion channels, i.e. a single IP_3R channel opens and closes randomly [14, 35]. Since the number of Ca^{2+} release channels per cluster is small, fluctuations still prevail in a cluster [7]. They lead to random release events called Ca^{2+} puffs. A puff is the spontaneous opening of channels of a single cluster and represents the elemental event of Ca^{2+} liberation. Puffs last from a few tens of milliseconds to a few hundred milliseconds and they cause a huge but strongly localized concentration rise. Many puffs can cooperate to build global phenomena covering the whole cell like waves of release or oscillations [6, 21], i.e. puffs constitute the fundamental building block of Ca^{2+} signals. We illustrate this concept with the initiation and the spreading of a Ca^{2+} wave. Assume that all channels are closed and the cell is in its resting state. Fluctuations due to the random association and dissociation of Ca^{2+} and IP_3 to the IP_3Rs eventually lead to a Ca^{2+} puff. Ca^{2+} is liberated from the ER and diffuses to the surrounding clusters. There, it causes an increase of the open probability and may therefore induce channels of these clusters to open. That gives rise to another Ca^{2+} puff and release has propagated by one cluster distance. In that way a Ca^{2+} wave travels through the cell. However, there is no guarantee that IP_3R channels at neighboring clusters open, because channel opening is a truly random event.

Indeed, fluctuations turned out to be necessary to observe any temporal or spatial structures in intracellular Ca^{2+} dynamics. That role of fluctuations has been established by two complementary approaches. On the one hand, oscillations that agree with experimental findings in stochastic simulations disappear in the deterministic limit of the simulated system [6]. On the other hand, a bifurcation analysis of a deterministic model for a single IP_3R cluster has proofed that the local dynamics is non-oscillatory when realistic Ca^{2+} fluxes and gradients are incorporated [31–33]. The loss of oscillations results from the high Ca^{2+} concentrations at an open cluster. They lead to a saturation of all Ca^{2+} regulating processes and hence do not permit Ca^{2+} oscillations in deterministic models. Fluctuations drive the channel dynamics out of the saturated state and eventually reinstall oscillations. The strong localization of the Ca^{2+} liberation and the entailing large gradients around an open cluster create the dynamic compartmentalization mentioned above. Since the volume of the elevated Ca^{2+} concentration as well as the number of IP_3Rs that experience this highly increased Ca^{2+} concentration is small, fluctuations remain important.

Given the vital part of Ca^{2+} puffs in intracellular Ca^{2+} dynamics and the importance of fluctuations, a stochastic description of a single cluster is the focus of the present report. We will apply it to the initiation of Ca^{2+} puffs, which represents the first step for any Ca^{2+} pattern. Our findings suggest that puff initiation can be mapped to an escape process from the resting state towards the first open channel. The mean first passage time corresponds to the stochastic part of the puff frequency. The mathematical description has to account for the integer number of open channels per cluster. A continuous Ca^{2+} model (using non-integer fractions of open channels) that incorporates realistic fluxes does not permit Ca^{2+} puffs for parameter values that agree with experimental data. The mean first passage times can be represented as an infinite series of exponentials. However, already the first terms in the expansion yield excellent convergence. That hints at a Poissonian character of puff initiation. Although noise is intrinsically multiplicative for intracellular Ca^{2+} dynamics, we provide evidence that additive noise may serve as a reasonable approximation.

We will introduce a Ca^{2+} model for an IP_3 receptor channel cluster in the next section. It serves as input for a master equation in section 11.3, from which we will derive two Fokker-Planck equations in section 11.4. Finally, we will employ these equations to characterize the initiation of Ca^{2+} puffs.

11.2. Ca^{2+} model

The IP_3 receptor channel is a tetramer the subunits of which have binding sites for Ca^{2+} and IP_3. We implement a model for a single subunit that is based on ideas of De Young's and Keizer's [5]. They assume a subunit to possess three binding sites: an activating Ca^{2+} binding site, an inhibiting Ca^{2+} binding site and an activating IP_3 binding site. The occupation of the binding sites controls the state of the subunit. When IP_3 and Ca^{2+} are bound to their activating binding sites, a subunit is in the activated state. As soon as Ca^{2+} binds to the inhibiting binding site, a subunit is inhibited, independent of the state of the other binding sites. It can only be activated again upon dissociation of Ca^{2+} from the inhibiting binding site. Experiments have indicated that an IP_3R channel is conducting when at least 3 subunits are activated [4, 35]. Random binding and unbinding of Ca^{2+} and IP_3 and therefore random state changes of the receptor are the source of stochasticity of intracellular Ca^{2+} dynamics.

The number of open IP_3R channels determines the Ca^{2+} flux from the

ER to the cytosol. Since the release channels are tightly packed within a cluster, a relation between the number of channels in the cluster and cluster size exists. Consequently, we can map the number of open channels to the size of a conducting area (or volume) equal to the area occupied by all open channels. A change in the number of releasing IP_3R channels corresponds to a modulation of the conducting area of a cluster. This region is usually not connected. However, Swillens et al. showed that the spatial arrangement of IP_3R channels does not influence the Ca^{2+} dynamics at an open cluster [30]. Therefore, we map the area of all conducting release channels to an area of the same size concentric to the cluster area. Let a denote the radius of this region, N the total number of channels per cluster and n_o the number of open channels, then $a = a_0 \sqrt[3]{n_o/N}$. That reflects the above notion that the volume of the conducting sphere corresponds to the volume that is occupied by the fraction n_o/N of open channels. If $n_o = 0$ then $a = 0$, and a takes the maximal value a_0 if all N channels are open.

The deterministic dynamics of this cluster model has been investigated in [32, 33]. In addition to IP_3 mediated Ca^{2+} liberation, we considered sarco-endoplasmic reticulum calcium ATPase (SERCA) pumps, which transport Ca^{2+} from the cytosol to the ER, and a leak flux. The stationary Ca^{2+} concentration profile that results from these three fluxes is

$$c(r) = \left[A(a)\frac{\sinh(k_1 r)}{r} + e_1\right]\Theta(a-r) + \left[B(a)\frac{\exp(-k_2 r)}{r} + e_2\right]\Theta(r-a)\,, \tag{11.1}$$

where

$$A(a) = \frac{l(k_2 a + 1)}{\cosh(k_1 a)k_1 + \sinh(k_1 a)k_2}\,, \tag{11.2a}$$

$$B(a) = \frac{l(\sinh(k_1 a) - \cosh(k_1 a)k_1 a)}{\exp(-k_2 a)(\cosh(k_1 a)k_1 + \sinh(k_1 a)k_2)}\,, \tag{11.2b}$$

with

$$l = \frac{-k_c k_p E}{(k_l + k_p + k_c)(k_l + k_p)}\,, \quad k_1 = \sqrt{\frac{k_l + k_p + k_c}{D}}\,, \quad k_2 = \sqrt{\frac{k_l + k_p}{D}}\,, \tag{11.3}$$

and $e_1 := (k_l + k_c)E/(k_l + k_p + k_c)$, $e_2 := k_l E/(k_l + k_p)$. The constants k_l, k_p and k_c denote the leak flux coefficient, the strength of the SERCA pumps and the channel flux coefficient, respectively. The diffusion coefficient is given by D. E denotes the concentration of free Ca^{2+} in the ER.

Simulations have demonstrated that the Ca^{2+} concentration rapidly equilibrates upon a change in the number of open channels [31]. Hence, we will approximate the Ca^{2+} dynamics by its stationary value in the remainder of this work. The number of open channels n_o uniquely determines the Ca^{2+} concentration according to equation (11.1) and $a = a_0 \sqrt[3]{n_o/N}$. The focus of the two subsequent sections is the calculation of n_o.

11.3. Master equation

The number of open channels n_o depends on the state of the subunits of the IP_3Rs. A state of a subunit is determined by the occupation of its binding sites. The De Young Keizer (DK) model has three binding sites per subunit and hence eight subunit states. We reduce these eight states in two steps to three states. Firstly, we eliminate the IP_3 dynamics adiabatically since IP_3 binding and unbinding are much faster than the Ca^{2+} dynamics in the framework of this model. The resulting four states are labeled by a binary pair ij, where the first index represents the Ca^{2+} activating binding site and the second the Ca^{2+} inhibiting binding site [6]. An index equals 1 when the binding site is occupied and 0 otherwise, e.g. 10 corresponds to the activatable state of a subunit. The second approximation uses the fact that we are interested in activation starting from a stationary state.

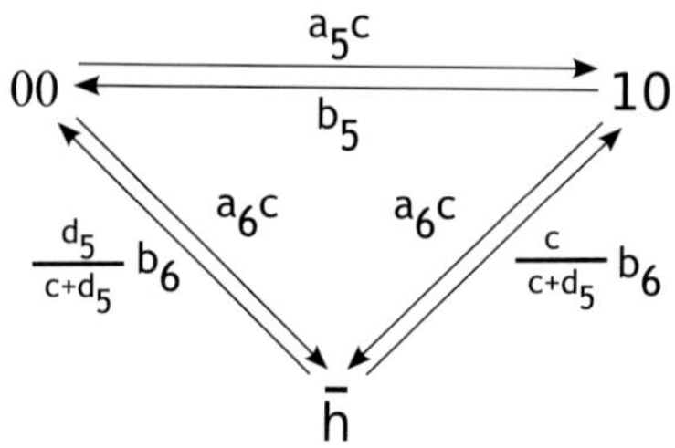

Fig. 11.1. Transition scheme for the three state model of the IP_3 receptor. $d_5 = b_5/a_5$ is the dissociation constant for Ca^{2+} activation, b_6 the dissociation rate of Ca^{2+} from the inhibiting site averaged over both IP_3 binding states. We denote the number of subunits in one of the three states by n_{10}, n_{00} and $n_{\bar{h}}$

Transitions among the inhibited states 11 and 01 have little impact on that activation process. Moreover, these states are rarely populated during puff initiation. Consequently we lump the two inhibited state into one state $\bar{h}$. Figure 11.1 depicts the transition scheme for this 3 state model. The transition rates follow from [5] and [6].

Modeling the dynamics of an IP_3R on the basis of its subunits leads to various consequences for a cluster of N IP_3Rs. As long as every IP_3R is treated individually and subunits are assigned to individual channels - as has been done in stochastic simulations [6] - the state of the cluster is uniquely determined by the states of its subunits. However, an approach based on a population of subunits not grouped into individual channels is more suitable for the derivation of master equations and Fokker-Planck equations which we would like to use. That requires to determine the number of open channels from the total number of activatable subunits in the subunit population. We assume that the activatable subunits are randomly scattered across the channels. The distribution of the n_{10} activatable subunits on the $4N$ subunits of a cluster decides upon the value of n_o and hence the Ca^{2+} concentration. We show in the appendix that this distribution is sharply peaked around its mean value. Therefore, we set $n_o = \langle n_o \rangle = n_a$. n_a is defined in equation (11.52).

The stochastic nature of Ca^{2+} release through IP_3Rs entails that the exact number of subunits in either of the three states 10, 00 or $\bar{h}$ at a given time t, i.e. the triplet $(n_{10}(t), n_{\bar{h}}(t), n_{00}(t))$, cannot be specified exactly any more. On the contrary, only the probability $P(n_{10}, n_{\bar{h}}, n_{00}; t)$ to find a certain realization of $(n_{10}, n_{\bar{h}}, n_{00})$ at time t is accessible. Since the total number of subunits is fixed, the values of n_{10} and $n_{\bar{h}}$ suffice to specify the triplet $(n_{10}, n_{\bar{h}}, n_{00}; t)$, so that $P(n_{10}, n_{\bar{h}}, n_{00}; t) = P(n_{10}, n_{\bar{h}}; t)$.

The probability $P(n_{10}, n_{\bar{h}}; t)$ changes in the time interval $[t, t+dt]$ due to two opposing processes: Being in $(n_{10}, n_{\bar{h}})$ at time t, binding or unbinding of Ca^{2+} alters n_{10} or $n_{\bar{h}}$ during dt and hence reduces $P(n_{10}, n_{\bar{h}}; t)$. On the other hand, transition from states as $(n_{10}+1, n_{\bar{h}})$ or $(n_{10}-1, n_{\bar{h}})$ into $(n_{10}, n_{\bar{h}})$ increases $P(n_{10}, n_{\bar{h}}; t)$. Taking all possible reactions according to the figure 11.1 into account, the time evolution of $P(n_{10}, n_{\bar{h}}; t)$ is captured by the master equation [34]

$$
\begin{aligned}
&\dot{P}(n_{10}, n_{\bar{h}}; t) = \\
&- [n_{10}[b_5 + a_6 c(n_{10})] + n_{\bar{h}} b_6]\, P(n_{10}, n_{\bar{h}}; t) + [n_{10}+1] b_5 P(n_{10}+1, n_{\bar{h}}; t) \\
&- n_{\bar{h}} b_6 P(n_{10}, n_{\bar{h}}; t) + [hN - n_{\bar{h}} - n_{10} + 1] a_6 c(n_{10}) P(n_{10}, n_{\bar{h}} - 1; t) \\
&- [hN - n_{10} - n_{\bar{h}}]\, [a_6 c(n_{10}) + a_5 c(n_{10})]\, P(n_{10}, n_{\bar{h}}; t) \\
&+ \frac{b_6 c(n_{10}-1)[n_{\bar{h}}+1]}{c(n_{10}-1) + d_5} P(n_{10}-1, n_{\bar{h}}+1; t) + \frac{b_6 d_5 [n_{\bar{h}}+1]}{c(n_{10}) + d_5} P(n_{10}, n_{\bar{h}}+1; t) \\
&+ [hN - n_{\bar{h}} - n_{10} + 1] a_5 c(n_{10}-1) P(n_{10}-1, n_{\bar{h}}; t) \\
&+ [n_{10}+1] a_6 c(n_{10}+1) P(n_{10}+1, n_{\bar{h}} - 1; t)\,.
\end{aligned}
\tag{11.4}
$$

For instance being in $(n_{10}, n_{\bar{h}})$, the term proportional to $a_6 c$ in the first line denotes a transition from 10 to $\bar{h}$, so that the final state is $(n_{10} - 1, n_{\bar{h}} + 1)$. The Ca^{2+} concentration is given by equation (11.1). The Ca^{2+} concentration in the master equation depends on n_{10}, which is indicated by the notation $c(n_{10})$. The radius a in equation (11.1) follows from the number of activatable subunits as $a = a_0 \sqrt[3]{n_a/N}$ according to the preceding discussions.

The adiabatic elimination of the IP_3 dynamics leads to non-integer values for the number of open channels. That demands a careful interpretation of the size of the conducting membrane patch, which was assumed to take only discrete values due to the discreteness of n_o. One approach is to truncate the rational values of n_o as $[n_o]_+$, where $[n_o]_+$ denotes the largest integer that is less or equal n_o. It entails $c = c_b$ as long as $n_0 < 1$, where c_b denotes the base level of the Ca^{2+} concentration. This approach favors the closed configuration during puff initiation. In another approach we will keep the non-integer value of n_o and consider a as a quasi continuous function. We will discuss the effects of both approaches with respect to puff initiation.

Equation (11.4) is an accurate description of the stochastic dynamics represented by the scheme in Figure 11.1. We will derive approximations like Fokker-Planck equations to calculate escape time characteristics from this master equation.

11.4. Fokker-Planck equations

The discrete nature of master equations often impedes an analytic treatment. That holds in particular for master equations with nonlinearities or artificial boundary conditions. In these cases, several approximations have been put forward [11, 12, 18, 25, 34]. Despite the plethora of methods, there is still no consensus which approximation is best [9]. Each of them possesses advantages and drawbacks, so that the problem at hand finally decides which procedure to use. We will concentrate on van Kampen's Ω expansion and a method that is similar to a Kramers-Moyal expansion. The latter keeps the nonlinearities of the master equation in the fluctuations, whereas the former approximates them in a linear fashion. Moreover, van Kampens's expansion is only valid when the macroscopic equation displays a single stable fixed point.

The Ω expansion requires a small parameter $1/\Omega$ in the master equation, which for our purposes is the inverse number of subunits, i.e. $\Omega = 4N$. The systematic expansion of equation (11.4) in powers of Ω is based on the transformations $n_{10} = \Omega\phi(t) + \Omega^{1/2}\xi$ and $n_{\bar{h}} = \Omega\psi(t) + \Omega^{1/2}\eta$. They decompose the variables of the master equation into macroscopic parts (ϕ, ψ) and fluctuations (ξ, η). Inserting this ansatz into equation (11.4), the first non vanishing order of Ω yields the macroscopic equations

$$\frac{\partial \phi}{\partial t} = -\,\phi(a_5 c + a_6 c + b_5) + \psi\left(\frac{b_6 c}{c + d_5} - a_5 c\right) + a_5 c\,, \tag{11.5a}$$

$$\frac{\partial \psi}{\partial t} = -\,(a_6 c + b_6)\psi + a_6 c\,, \tag{11.5b}$$

with $c = c(a_0 \sqrt[3]{\phi_a})$ and $\phi_a := r^3\phi^3(4 - 3r\phi)$. r denotes the fraction of subunits in the state 10 that are activated: $r := I/(I + d_1)$. Equations (11.5) correspond to the rate equations that are associated with the transition scheme in figure 11.1, when the conservation condition $n_{10} + n_{\bar{h}} + n_{00} = 4N$ is applied. Note that ϕ_a is the continuous limit ($N \to \infty$) of equation (11.52). Therefore, ϕ_a is the probability that at least 3 of the 4 subunits of an IP_3R are activated. The solutions of equation (11.5) represent the deterministic part of the above transformation of variables. They have the stationary values

$$\bar{\phi} = \frac{d_6 c}{(c + d_5)(c + d_6)}\,, \qquad \bar{\psi} = \frac{c}{c + d_6}\,, \tag{11.6}$$

which agree with results in [33]. $d_6 = d_2(I + d_1)/(I + d_3)$ is an effective dissociation constant. d_2 denotes the dissociation constant for Ca^{2+} inhibition when the IP_3 binding site is ligated, d_1 and d_3 represent the dissociation constants of IP_3 binding [5].

The next order in Ω determines the fluctuations through the probability $P(n_{10}, n_{\bar{h}}; t)$

$= P(\Omega\phi + \Omega^{1/2}\xi, \Omega\psi + \Omega^{1/2}\eta; t) =: \Pi(\xi, \eta; t)$ according to

$$\frac{\partial \Pi}{\partial t} = -\left[g_{11}\frac{\partial}{\partial \xi} + g_{21}\frac{\partial}{\partial \eta}\right](\xi\Pi) - \left[g_{12}\frac{\partial}{\partial \xi} + g_{22}\frac{\partial}{\partial \eta}\right](\eta\Pi) + \frac{1}{2}\left[h_{11}\frac{\partial^2}{\partial \xi^2} + 2h_{12}\frac{\partial^2}{\partial \eta \partial \xi} + h_{22}\frac{\partial^2}{\partial \eta^2}\right]\Pi . \tag{11.7}$$

The matrices (g_{ij}) and (h_{ij}) with $h_{12} = h_{21}$ are defined as

$$g_{11} := b_6 d_5 \psi c^1/(c+d_5)^2 - a_6\left(c + \phi c^1\right) - b_5 - a_5\left(c - (1-\phi-\psi)c^1\right), \tag{11.8a}$$

$$g_{21} := a_6 c^1 - a_6 \psi c^1 , \tag{11.8b}$$

$$g_{12} := b_6 c/(c + d_5) - a_5 c , \tag{11.8c}$$

$$g_{22} := -\left(a_6 c + b_6\right) , \tag{11.8d}$$

and

$$h_{11} := a_5(1-\psi-\phi)c + b_6\psi c/(c+d_5) + a_6\phi c + b_5\phi , \tag{11.9a}$$

$$h_{21} := -b_6\psi c/(c+d_5) - a_6\phi c , \tag{11.9b}$$

$$h_{22} := a_6(1-\psi)c + b_6\psi . \tag{11.9c}$$

with

$$c^1 := \frac{dc}{da}(a_0\sqrt[3]{\phi_a})\frac{a_0}{3}\sqrt[3]{\phi_a^{-2}}\left[18r^4\phi^3 - 12r^3\phi^3 - 12r^3\phi^2\right] . \tag{11.10}$$

Equation (11.10) arises from inserting equation (11.52) into $a = a_o\sqrt[3]{n_a/N}$ and then expanding $c(a)$ in powers of Ω. The matrix (g_{ij}) coincides with the matrix of the linearized macroscopic equations (11.5). The fluctuations enter through the matrix (h_{ij}). The Hurwitz criterion [15] assures that this matrix is positive semi definite, which means that equation (11.7) is a linear multivariate Fokker-Planck equation.

The linear treatment of the noise in equation (11.7) has cast some doubt on the validity of the Ω expansion. Therefore, a different class of Fokker-Planck equations have been proposed that keep the nonlinearities of the master equation. Kramers and Moyal have treated the shifts $n_{10} \pm 1, n_{\bar{h}} \pm 1$ of n_{10} and $n_{\bar{h}}$ in equation (11.4) by means of a Taylor expansion [18, 25]. Following this procedure and defining the new variables $\phi := n_{10}/\Omega$ and $\psi := n_{\bar{h}}/\Omega$, we obtain a Fokker-Planck equation for the probability $p =$

$p(\phi, \psi, t)$:

$$\begin{aligned}\frac{\partial p}{\partial t} =& \frac{\partial}{\partial \phi}\left[\phi a_6 c + \phi b_5 - (1-\psi-\phi)a_5 c - \frac{b_6 c}{c+d_5}\psi\right] p \\ &+ \frac{\partial^2}{2\Omega\partial\phi^2}\left[\phi a_6 c + \phi b_5 + (1-\psi-\phi)a_5 c + \frac{b_6 c}{c+d_5}\psi\right] p \\ &+ \frac{\partial}{\partial\psi}\left[b_6 - (1-\psi)a_6 c\right] p + \frac{\partial^2}{2\Omega\partial\psi^2}\left[b_6 + (1-\psi)a_6 c\right] p \\ &- \frac{\partial}{\Omega\partial\psi\partial\phi}\left[\phi a_6 c + \frac{b_6 c}{c+d_5}\right] p\,. \end{aligned} \tag{11.11}$$

The nonlinearities are introduced through $c = c(a_0\sqrt[3]{\phi_a})$ with ϕ_a defined as after equation (11.5).

Equations (11.4), (11.7) and (11.11) constitute the starting point for a systematic study of puff frequencies. Given a configuration $(n_{10}^0, n_{\bar{h}}^0)$ at time $t = 0$, they all yield the probability for a configuration $(n_{10}^t, n_{\bar{h}}^t)$ at time $t > 0$. If we identify $(n_{10}^0, n_{\bar{h}}^0)$ with the resting state of a cluster and $(n_{10}^t, n_{\bar{h}}^t)$ with the first channel opening, such a transition in the configuration space gives the probability for a Ca^{2+} puff. Consequently, we interpret a puff as an escape process from the state $(n_{10}^0, n_{\bar{h}}^0)$ to the state $(n_{10}^t, n_{\bar{h}}^t)$. Although the above equations allow the calculation of this escape probability, no general solutions are known for two dimensional escape processes (see [13] for a recent result). However, the time scale separation between Ca^{2+} activation and Ca^{2+} inhibition leads to a reduction of the two dimensional equations to one dimension. Since the inhibiting processes are much slower than binding and unbinding of Ca^{2+} to the activating binding site, we assume that $n_{\bar{h}}$ remains unchanged during the initiation of a puff. That is identical to setting $n_{\bar{h}} = \text{const}$, and the master equation simplifies to

$$\begin{aligned}\dot{P}(n_{10}) =& -\frac{b_6 c(n_{10})}{c(n_{10})+d_5} n_{\bar{h}} P(n_{10}) + \frac{b_6 c(n_{10}-1)}{c(n_{10}-1)+d_5} n_{\bar{h}} P(n_{10}-1) \\ &- a_6 c(n_{10}) n_{10} P(n_{10}) + a_6 (n_{10}+1) c(n_{10}+1) P(n_{10}+1) \\ &- (4N - n_{10} - n_{\bar{h}}) a_5 c(n_{10}) P(n_{10}) - b_5 n_{10} P(n_{10}) \\ &+ (4N - n_{10} - n_{\bar{h}} + 1) a_5 c(n_{10}-1) P(n_{10}-1) \\ &+ b_5 (n_{10}+1) P(n_{10}+1)\,. \end{aligned} \tag{11.12}$$

For a later analysis, it is convenient to rewrite equation (11.12) in the form

$$\dot{P}(n_{10}) = -\left(g_{n_{10}} + r_{n_{10}}\right) P(n_{10}) + g_{n_{10}-1} P(n_{10}-1) + r_{n_{10}+1} P(n_{10}+1) \tag{11.13}$$

with

$$g_{n_{10}} = \frac{b_6 c(n_{10})}{c(n_{10}) + d_5}[\Omega\bar{\psi}] + (4N - [\Omega\bar{\psi}] - n_{10})a_5 c\,, \tag{11.14a}$$

$$r_{n_{10}} = b_5 n_{10} + a_6 c(n_{10}) n_{10}\,, \tag{11.14b}$$

and $\bar{\psi}$ defined as in equation (11.6). From equation (11.12), we could again derive Fokker-Planck equations in the same manner as before. However, a more direct approach for the one dimensional Ω expansion is setting η equal to zero in equation (11.7) due to $n_{\bar{h}}$ = const. Keeping only the derivatives with respect to ϕ in equation (11.11) gives the nonlinear Fokker-Planck equation. Note that these one dimensional equations are only valid during the initiation phase of a puff, whereas equations (11.4), (11.7) and (11.11) capture the full time evolution. Nevertheless, we will concentrate on equation (11.12) and the entailing Fokker-Planck equations in the remainder of this work, because they admit analytic solutions and provide far reaching insights into puff frequencies. The existence of analytic solutions is one of the most prominent features of van Kampen's expansion, so that we will treat the corresponding Fokker-Planck equation most generally in the next section.

11.5. Escape times

The initiation of a Ca^{2+} puff corresponds to an escape from the stationary state to the first channel opening. That requires the definition of the boundaries of the phase space area from which the escape occurs. Since we restrict the discussion to one dimension in phase space, the boundary consists of two points. We see from equation (11.12) that the lower boundary d is at $n_{10} = 0$ and that it is reflecting. That agrees with the interpretation of n_{10} as the number of activatable subunits, which is always positive. The value of the upper boundary b is chosen such that the number of open channels $n_o = 1$. The upper boundary corresponds to the escape site, so that the boundary condition is of absorbing type [8].

The time t to reach the absorbing boundary is a stochastic variable. It is described by the probability density $\rho(t)$ i.e. $\rho(t)dt$ is the probability that the stochastic process reaches b between t and $t+dt$. ρ is most conveniently computed from $G(x,t) = 1 - \int_0^t \rho(x,\tau)d\tau$, which represents the probability that $d \leq n_{10} < b$ at time t when it started at $x = n_{10}^0$ at $t = 0$. The time evolution of G is governed by $\tilde{\mathcal{L}}$, which is the adjoint of the Fokker-Planck operator $\mathcal{L}$ [8]. Up to now, no general solution has been obtained for arbitrary $\mathcal{L}$. Yet, an analytic expression exists for G in the case of a linear Fokker-Planck operator, e.g. van Kampen's Ω expansion. Since the following derivation always holds and is not restricted to the current problem, we introduce new constants v and w. They are given by $v = -g_{11}$ and $w = h_{11}$ defined as in equations (11.8a) and (11.9a), respectively, in the present study. G obeys the linear backward Fokker-Planck equation [8]:

$$\frac{\partial G(x,t)}{\partial t} = -vx\frac{\partial G(x,t)}{\partial x} + \frac{w}{2}\frac{\partial^2 G(x,t)}{\partial x^2}, \qquad v, w > 0\,, \tag{11.15}$$

with initial and boundary conditions

$$G(x,0) = \begin{cases} 1, & d \le x \le b \\ 0, & else \end{cases}, \quad \frac{\partial G(d,t)}{\partial x} = 0 \quad \forall t\,, \quad G(b,t) = 0 \quad \forall t\,. \tag{11.16}$$

The initial condition states that $d \le x < b$ at $t = 0$ with probability one. The reflecting boundary condition at $x = d$ in the adjoint Fokker-Planck equation is expressed by a no-flux boundary condition. Setting $G \equiv 0$ at the right boundary corresponds to an absorbing boundary. We solve equation (11.15) with the ansatz $G(x,t) = \exp(-\lambda t)u(x)$, $\lambda \ge 0$ so that it reduces to the ordinary differential equation

$$\frac{d^2u}{dx^2} - \frac{2vx}{w}\frac{du}{dx} + \frac{2\lambda}{w}u = 0\,. \tag{11.17}$$

Applying the transformation $z := x^2/4$ we find for $\bar{u}(z) := u(x)$

$$z\frac{d^2\bar{u}}{dz^2} + \left(\frac{1}{2} - \frac{4vz}{w}\right)\frac{d\bar{u}}{dz} + \frac{2\lambda}{w}\bar{u} = 0\,. \tag{11.18}$$

It equals Kummer's equation for $\tilde{u}(\tilde{z}) := \bar{u}(z)$ with $\tilde{z} := 4vz/w$

$$\tilde{z}\frac{d^2\tilde{u}}{d\tilde{z}^2} + \left(\frac{1}{2} - \tilde{z}\right)\frac{d\tilde{u}}{d\tilde{z}} + \frac{\lambda}{2v}\tilde{u} = 0\,. \tag{11.19}$$

Two independent solutions of equation (11.17) are [1]

$$u_1(x) := M\left(-\frac{\lambda}{2v}, \frac{1}{2}, \frac{vx^2}{w}\right), \quad u_2(x) := xM\left(\frac{1}{2} - \frac{\lambda}{2v}, \frac{3}{2}, \frac{vx^2}{w}\right). \tag{11.20}$$

M designates the confluent hypergeometric function

$$M(a,b,x) := \sum_{k=0}^{\infty} \frac{(a)_k}{(b)_k}\frac{x^k}{k!}\,, \tag{11.21}$$

where $(a)_0 := 1$ and $(a)_k := a(a+1)\dots(a+k-1)$. The boundary condition at $n_{10} = b$ entails that a solution of equation (11.17) is

$$v(x) := C_1 \left[u_1(x) - \frac{u_1(b)}{u_2(b)} u_2(x) \right] = u_1(x) - \frac{u_1(b)}{u_2(b)} u_2(x) . \tag{11.22}$$

Without loss of generality we set $C_1 = 1$ because it merely serves as normalization. The second boundary condition fixes the still unknown eigenvalues λ. They constitute an infinite countable set $\{\lambda_n\}$ due to the finiteness of d and b. Therefore, the general solution of equation (11.15) can be expressed as

$$G(x,t) = \sum_{n=0}^{\infty} a_n \exp(-\lambda_n t) v_n(x) . \tag{11.23}$$

The subscript of $v_n(x)$ indicates that equation (11.22) has to be evaluated at $\lambda = \lambda_n$ (see equation (11.20)). The coefficients a_n are determined by the initial condition $G(x,0)$ which results in

$$a_n = \int_d^b r(x) v_n(x) dx \Big/ \int_d^b r(x) v_n^2(x) dx , \quad r(x) := \frac{2}{w} \exp\left(-\frac{v}{w} x^2\right) . \tag{11.24}$$

Here we used the orthogonality relation of the eigenfunctions $v_n(x)$:

$$\int_d^b v_n(x) v_m(x) r(x) dx = \delta_{m,n} \int_d^b v_n^2(x) r(x) dx . \tag{11.25}$$

The probability $\rho(x,t)$ that the absorbing state is reached between t and $t + dt$ is readily computed from $G(x,t)$ as $\rho = -\partial_t G(x,t)$. Note that ρ is already normalized due to the initial condition $G(x,0)$. Hence, the mean first passage time $T(x)$ equals

$$T(x) := \langle t(x) \rangle = \int_0^{\infty} t\rho(x,t) dt = -\int_0^{\infty} t \partial_t G(x,t) dt = \sum_n \frac{a_n v_n(x)}{\lambda_n} . \tag{11.26}$$

Equation (11.26) includes an infinite number of eigenvalues. We found that the first three terms of the sum over n were sufficient to achieve results indistinguishable from the exact results of equation (11.30).

An alternative approach to the mean first passage time follows from the differential equation [8]

$$-vx \frac{dT(x)}{dx} + \frac{w}{2} \frac{d^2 T(x)}{dx^2} = -1 , \tag{11.27}$$

with the solution

$$T(x) = \frac{2}{w}\int_x^b \frac{dy}{h(y)}\int_d^y h(z)dz\,, \quad h(x) := \exp\left\{-\frac{v}{w}(x^2 - d^2)\right\}. \tag{11.28}$$

Performing the z integration we find

$$\begin{aligned} T(x) =& \sqrt{\frac{\pi}{vw}}\int_x^b dy \exp\left(\frac{v}{w}y^2\right)\operatorname{erf}\left(\sqrt{\frac{v}{w}}y\right) \\ &+ \frac{\pi}{2v}\operatorname{erf}\left(\sqrt{\frac{v}{w}}d\right)\left\{\operatorname{erfi}\left(\sqrt{\frac{v}{w}}x\right) - \operatorname{erfi}\left(\sqrt{\frac{v}{w}}b\right)\right\}. \end{aligned} \tag{11.29}$$

The functions $\operatorname{erf}(x)$ and $\operatorname{erfi}(x) = \operatorname{erf}(ix)/i$ denote the Gaussian error function and the imaginary Gaussian error function, respectively. The remaining integral can be solved by series expansion so that the final expression for the mean first passage time takes the form

$$\begin{aligned} T(x) =& \frac{b^2}{w}F_{2;2}\left(1,1;\frac{3}{2},2;\frac{w}{v}b^2\right) - \frac{x^2}{w}F_{2;2}\left(1,1;\frac{3}{2},2;\frac{w}{v}x^2\right) \\ &+ \frac{\pi}{2v}\operatorname{erf}\left(\sqrt{\frac{v}{w}}d\right)\left\{\operatorname{erfi}\left(\sqrt{\frac{v}{w}}x\right) - \operatorname{erfi}\left(\sqrt{\frac{v}{w}}b\right)\right\}. \end{aligned} \tag{11.30}$$

We employed the generalized hypergeometric function

$$F_{p;q}(a_1,\ldots,a_p;b_1,\ldots,b_q;x) = \sum_{l=0}^{\infty}\frac{(a_1)_l\cdots(a_p)_l}{(b_1)_l\cdots(b_q)_l}\frac{x^l}{l!}\,, \tag{11.31}$$

and used the identiy

$$\frac{j!}{2j+2}\sum_{l=0}^{j}\frac{(-1)^l}{(2l+1)(j-l)!l!} = \frac{1}{2}\frac{(1)_j(1)_j}{\left(\frac{3}{2}\right)_j(2)_j}. \tag{11.32}$$

We defer the proof to the appendix. The reason for presenting two methods for evaluating the mean first passage time is based on their different scopes of applicability. If we were only interested in T, then equation (11.30) would be preferable because it requires less computation. However, we are limited to the first moment [8]. The advantage of the first approach is that we obtain any moment by one integration. Moreover, we have access to the time evolution of the escape process which allows for a more detailed analysis.

The above results could only be obtained analytically because the corresponding Fokker-Planck equation was linear. In the case of a nonlinear Fokker-Planck equation, all quantities have to be computed numerically.

The mean first passage time is evaluated best from a generalization of equation (11.28). For $\mathcal{L} = -\partial_x A(x) + \partial_x^2 B(x)/2$ we find [8]

$$T(x) = 2\int_x^b \frac{dy}{h(y)} \int_d^y \frac{h(z)}{B(z)} dz\,, \quad h(x) := \exp\left\{ \int_d^x \frac{2A(y)}{B(y)} dy \right\}. \tag{11.33}$$

The study of Fokker-Planck equations instead of master equations is often motivated by easier treatment. That holds in particular in higher dimensions, because a broader spectrum of tools is available for Fokker-Planck equations than for master equations [27] and even analytical calculations may be possible as in the case of equation (11.12). That constitutes one of the reasons for the derivations in section 11.4. However, Fokker-Planck equations always represent approximations. The only way to test their quality is a comparison with results obtained from a master equation.

To this end, we consider a general one step process, to which class equation (11.12) belongs. We assume that this jump process starts at a site m at $t = 0$. Being at site n the particle hops to the right with a rate g_n and to the left with a rate r_n, respectively. When it reaches the left boundary L, it is reflected. Then, the mean first passage time to arrive at a site $R > m$ reads [34]

$$T_{R,m} = \sum_{i=m}^{R-1} \left(\frac{1}{g_i} + \sum_{l=L+1}^{i} \frac{r_i r_{i-1} \cdots r_l}{g_i g_{i-1} \cdots g_l} \frac{1}{g_{l-1}} \right). \tag{11.34}$$

That allows us to estimate the validity of the preceding approximations. The transition rates r_i and g_i follow from equation (11.14) for the current investigation.

11.6. Results

11.6.1. *Mean first passage time*

The calculation of the mean first passage times according to equations (11.26), (11.33) and (11.34) necessitates a further specification of the boundaries. Since we consider a cluster with N channels, the upper boundary ϕ_b for the nonlinear Fokker-Planck equation is given by the solution of $(r\phi_b)^3(4 - 3r\phi_b) = 1/N$. The left hand side corresponds to the fraction of open channels as discussed after equation (11.5). For the lower boundary, we have $\phi = 0$. This value holds for the master equation, too. The upper boundary for the master equation is obtained by rounding off $\Omega\phi_b$ to its nearest integer value $[\Omega\phi_b]$. Before specifying the boundary conditions for van Kampen's expansion we note that it describes the strength of the fluctuations ξ around the fixed point $\bar{\phi}$. The left boundary is imposed by $n_{10} > 0$, whereas the right boundary has to satisfy $\phi_b = \bar{\phi} + \Omega^{-1/2}\xi$. Consequently, the boundaries of ξ are $-\Omega^{1/2}\bar{\phi}$ and $(\phi_b - \bar{\phi})\Omega^{1/2}$, respectively, with $\bar{\phi}$ given by equation (11.6).

The mean first passage time depends strongly on the Ca^{2+} concentration (see e.g. equations (11.14) and (11.34)). The results presented throughout the sections 11.6.1 - 11.6.3 are calculated with a constant base level concentration. The number of open channels is an integer variable and there is no Ca^{2+} channel flux before the first channel opens. The Ca^{2+} concentration remains at a steady value until a Ca^{2+} puff occurs. That leads to $c^1 \equiv 0$ in equation (11.10) and to coefficients linear in ϕ and ψ in equation (11.11).

Figure 11.2 depicts the mean first passage time as a function of the IP_3 concentration for two different values of the basal Ca^{2+} concentration.

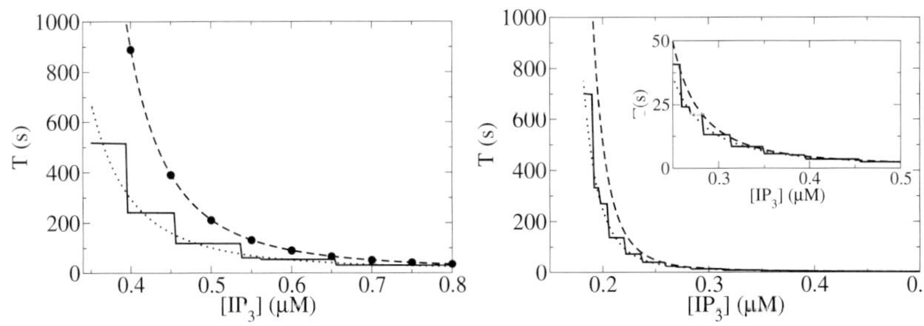

Fig. 11.2. Mean first passage time for $c_b = 50nM$ (left) and $c_b = 80nM$ (right) computed from the master equation (solid), the Ω expansion (dashed) and the nonlinear Fokker-Planck equation (dotted) for $d_1 = 0.13\mu M, d_2 = 3\mu M, d_3 = 0.9434\mu M, d_4 = 0.4133\mu M, d_5 = 0.24\mu M, a_2 = a_4 = 0.2\,(\mu Ms)^{-1}, a_5 = 5\,(\mu Ms)^{-1}, N = 25$. The dots in the left panel represent the variance of the Ω expansion. The inset in the right panel shows a blow up of the plot for large IP_3 concentration.

The master equation and the two Fokker-Planck equations exhibit an increase of the mean first passage time with decreasing IP_3 concentration. This increment diverges for lower values of the IP_3 concentration.

The nonlinear Fokker-Planck equation interpolates the master equation very efficiently. The results agree well with experimental findings for puff periods, although the mean first passage time only constitutes its mean stochastic fraction [21]. The discreteness of the master equation leads to discontinuities in the mean first passage time. The plateaus correspond to ranges of ϕ_b that are mapped to a single integer for the absorbing boundary of the master equation. Whenever that integer increases by 1, a jump occurs in the mean first passage time. Van Kampen's expansion yields good results for higher IP_3 concentrations, but overestimates the escape times otherwise (figure 11.2).

Figure 11.3 depicts the influence of the base level on the mean first passage time. The higher the basal concentration in this regime, the faster the first channel opens. Van Kampen's expansion improves with increasing base level as a comparison between the two panels in figure 11.2 and the right panel of figure 11.3 shows. The zigzag behavior of the relative difference $\tau := (T_{vK} - T_{ME})/T_{ME}$ results from the discontinuities of T_{ME}, see figure 11.2. Additionally, this quantifies the finding that the difference of the mean first passage time between the master equation and the Ω expansion diminishes with increasing IP_3 concentration.

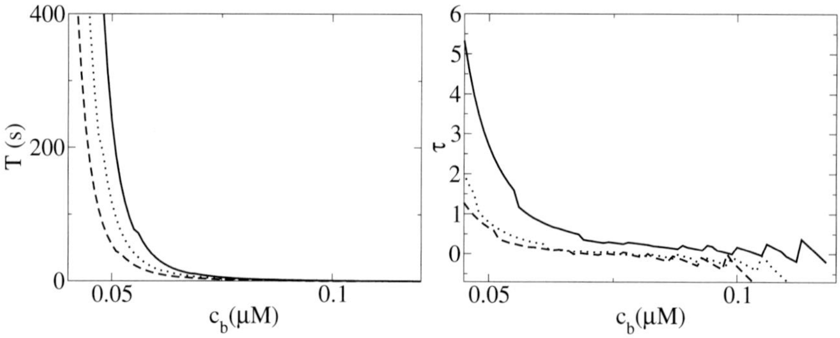

Fig. 11.3. Mean first passage time for the master equation (left) and the relative difference $\tau := (T_{vK} - T_{ME})/T_{ME}$ of the mean first passage time between van Kampen's method T_{vK} and the master equation T_{ME} (right) in dependence on the base level c_b for different values of the IP_3 concentration: $I = 0.4\mu$M (solid), 0.5 μM (dotted), 0.6 μM (dashed). Parameter values as in figure 11.2 and $a_5 = 1\,(\mu\text{Ms})^{-1}$.

11.6.2. *Role of fluctuations*

The most important difference between the nonlinear Fokker-Planck equation (11.11) and van Kampen's expansion (11.7) is in the diffusion term. It is constant in van Kampen's expansion - describing additive noise - and linear in ϕ and ψ in the nonlinear Fokker-Planck equation thus describing multiplicative noise. As expected intuitively, the results in figure 11.2 show a better agreement between the nonlinear Fokker-Planck equation and the master equation than between van Kampen's expansion and the master equation. However, van Kampen's expansion approximates the master equation results rather well for high IP_3 and high base level of Ca^{2+}. That is quantified in figure 11.3. Consequently, additive noise is probably a good approximation in these parameter areas.

11.6.3. *Distribution of first passage time*

Van Kampen's expansion allows a direct computation of the probability density of the first passage time $\rho(0,t)$. $\rho(0,t)dt$ is the probability that the absorbing boundary is reached between t and $t+dt$. The starting point of the escape process in the Ω expansion is $\xi = 0$. The IP_3R cluster is exactly in the macroscopic state $\bar{\phi}$ at $t=0$, so that the noise vanishes at $t=0$. The results for ρ are depicted in figure 11.4. A convergence of the probability density according to equation (11.23) requires less than 10 eigenvalues. The curves show the well known rising phase of ρ and the exponential decay. We find a maximal probability that shifts toward shorter times for higher IP_3 concentrations. The two graphs in figure 11.4 illustrate again the influence of the base level. Lowering c_b from 60nM to 40nM leads to an extreme broadening of the probability distribution and hence to an increase of the mean first passage time (see figure 11.3).

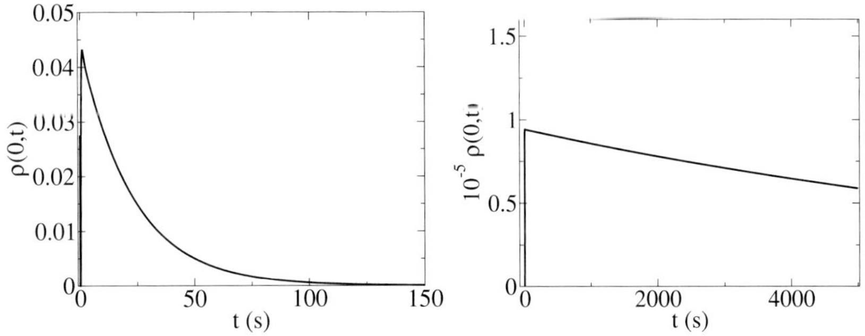

Fig. 11.4. Probability density $\rho(0,t)$ for van Kampen's expansion. Parameter values as in figure 11.2 and $I = 0.5\mu$M. Left panel: $c_b = 0.06\mu$M. Right panel: $c_b = 0.04\mu$M.

The probability density ρ permits an efficient computation of all moments of t for the escape process. Since the eigenvalues λ_n and the coefficients a_n are known, we immediately arrive at $\langle t^m \rangle = \sum_n m! a_n \lambda_n^{-m}$ in analogy to equation (11.26) due to $v_n(0) = 1$ for all n. The dots in figure

11.2 depict the results for the variance. The first six eigenvalues suffice for an excellent convergence. That is a direct consequence of the spectrum of the backward Fokker-Planck operator in equation (11.15).

Figure 11.5 shows the ratio of the first two eigenvalues λ_1/λ_0. λ_1 is only a few times larger than λ_0 for large IP_3 and d_5. However, the ratio increases with decreasing IP_3 concentrations and spans more than one order of magnitude for IP_3 concentrations smaller than 1μM. Hence, already the second term in the expansion (11.23) is considerably damped in the parameter range in which we are interested ($IP_3 < 1\mu$M). Since the eigenvalues constitute a strictly increasing series, i.e. $\lambda_i < \lambda_j$ for $i < j$, the subsequent terms in the expansion decay even more rapidly. The prominent role of the first term is additionally supported by the expansion coefficients a_i. The ratio a_1/a_0 is depicted in the right panel of figure 11.5. It decreases upon lowering the IP_3 concentration and tends to zero for very little concentrations. a_1 is much smaller than a_0 in parameter ranges where $\lambda_1/\lambda_0 \gg 1$ holds, i.e. where the second term of the series in equation (11.23) decays much faster than the first one. Consequently, higher terms only contribute marginally in this parameter regime. A detailed analysis of the spectrum and further implications will be provided in an upcoming report.

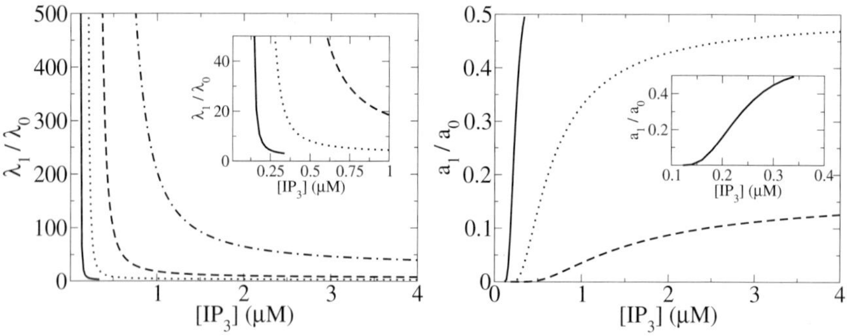

Fig. 11.5. Ratio of the first two eigenvalues (left panel) and ratio of the first two expansion coefficients (right panel) of equation (11.23) in dependence on the IP_3 concentration. Parameter values as in figure 11.3 and $d_5 = 0.08234\mu$M (solid), 0.13234μM (dotted), 0.183234μM (dashed) and 0.23234μM (chain-dotted). Insets show a blow-up for small IP_3 concentrations.

11.6.4. *Continuous* Ca^{2+} *model*

The results presented so far have been based on a discrete description of the number of open channels. The most important consequence is that the Ca^{2+} concentration remains constant as long as no channel opens. In the past, investigations on stochastic properties of IP_3R clusters often employed a continuous model of the ratio of open channels [24, 28, 29]. In these models, the Ca^{2+} concentration changes even for fractions of open channels corresponding to less than one channel. Therefore, we have analyzed the impact of a continuously modulated number of open channels on the mean first passage time. The nullclines of the deterministic dynamics for such an ansatz with the same parameter values as before are displayed in the left panel of figure 11.6. There is only one stationary state, which is linearly stable for all IP_3 concentrations. A prerequisite for a puff is that a sufficient number of subunits can be activated during the escape process from this fixed point. The value of ψ indicates that a large fraction of subunits is inhibited at already moderate IP_3 concentrations. It turns out that the remaining fraction of subunits is too low to induce a Ca^{2+} puff. The high degree of inhibition results from the large Ca^{2+} fluxes that occur at an open cluster [31]. These fluxes entail Ca^{2+} concentrations already in the μM range for sizes of the conducting area that are much smaller than that of a single channel. Since these concentrations exceed the dissociation constants for inhibition, most of the subunits are inhibited. Lowering the IP_3 concentration does not invoke puffs, either. Although the fraction of inhibited subunits diminishes, the number of subunits that can be activated decreases as well.

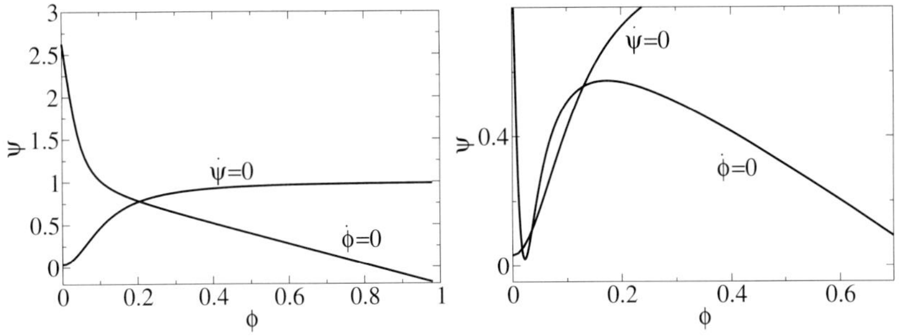

Fig. 11.6. Nullclines of equation (11.5). Parameter values as in figure 11.3 and $k_l = 0.002\text{s}^{-1}, k_c = 34500\text{s}^{-1}, k_p = 80\text{s}^{-1}, E = 750\mu\text{M}, a_0 = 0.03\mu\text{m}, D = 40\mu\text{m}^2\text{s}^{-1}$, $d_5 = 0.1646\mu\text{M}, I = 0.06\mu\text{M}$ (left panel), $d_5 = 1.6468\mu\text{M}, I = 0.053\mu\text{M}$ (right panel).

We compensate for the elevated Ca^{2+} concentrations with an increase in the dissociation constant for Ca^{2+} activation, d_5. The ensuing nullclines are depicted in the right panel of figure 11.6. The left stationary state is linearly stable and corresponds to a low degree of inhibition. The motion of ϕ in phase space proceeds along an almost horizontal line through this fixed point during puff initiation. These dynamics are bistable as the potential in figure 11.7 highlights. A Ca^{2+} puff parallels an escape process from the left well over the barrier to the first channel opening.

The time for such an escape process depends on the position of the absorbing boundary with respect to the barrier of the potential. Figure 11.7 shows the mean first passage time in dependence on the location ϕ of this boundary. ϕ varies from the value of the potential maximum (see inset) to the value of the first channel opening ϕ_b (see section 11.6.1). The steep increase of T for small ϕ reflects the influence of the left well. As long as the absorbing boundary is close to the maximum of the potential, reentrance in the left well is possible. That becomes less dominant with increasing ϕ, so that the mean first passage time reaches the plateau. For the upper range of IP_3 concentrations in figure 11.7, the value of the plateau equals the mean first passage time. Consequently, the time scale of the puff is set by the properties of the left well. The strong increase of the mean first passage time for smaller IP_3 concentrations is due to two reasons. On the one hand, the left well of the bistable potential becomes broader and deeper with lower values of I. On the other hand, the absorbing boundary increases in a disproportionally high manner and moves higher on the right branch of the potential.

We exclude van Kampen's expansion in the above analysis, because its validity requires a single stationary state throughout the stochastic motion [34]. In contrast to a constant Ca^{2+} concentration, the nonlinear Fokker-Planck equation underestimates the results of the master equation. Nevertheless, the results in figure 11.7, which correspond to the stochastic fraction of the puff frequency, are in the same range as experimentally determined puff periods [21].

11.7. Discussion

We have derived a master equation and two Fokker-Planck equations for channel cluster behavior in IP_3 mediated Ca^{2+} dynamics. Among the different approaches to approximate a master equation by a Fokker-Planck equation we have chosen van Kampen's Ω expansion and an ansatz based

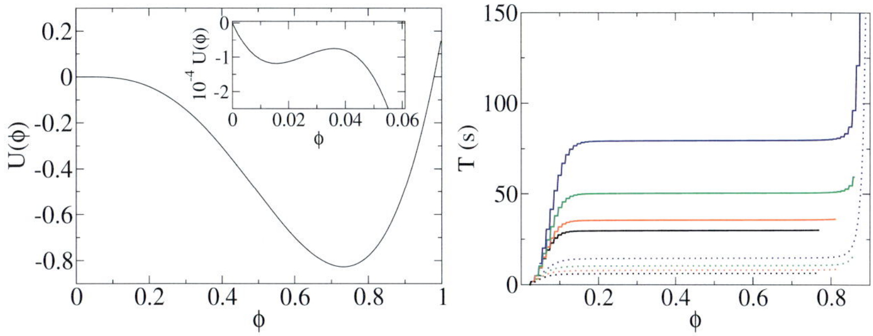

Fig. 11.7. Left: Potential $U(\phi)$ for $I = 0.0483\mu$M. The inset depicts an enlarged view for $\phi \approx 0$. Note the difference in scale for the axis. Right (Color): Mean first passage time computed from the master equation (solid lines) and the nonlinear Fokker-Planck equation (dotted lines) for $I = 0.0553\mu$M (black), 0.0513μM (red), 0.0473μM (green) and 0.0433μM (blue) in dependence on the position of the absorbing boundary ϕ. Parameter values as in figure 11.6, right panel.

on the Kramers-Moyal expansion. Master equations and corresponding Fokker-Planck equations for intracellular Ca^{2+} dynamics have been investigated in the past [24, 28, 29], but the study at hand is founded on different ideas. Most of the previous contributions employ the Li-Rinzel model [20] for the dynamics of a single subunit of an IP_3 receptor. It describes the time evolution of the fraction of subunits that are not inhibited yet, taking advantage of the time scale separation between IP_3 activation, Ca^{2+} activation and Ca^{2+} inhibition. We have used a state scheme for one subunit that only eliminates the IP_3 dynamics adiabatically. It focuses on Ca^{2+} activation, which is the driving force behind puff initiation. Therefore, we consider Ca^{2+} activation as the fluctuating variable, whereas Ca^{2+} inhibition is the random variable in the Li-Rinzel model.

The Ca^{2+} concentration plays a pivotal role in the initiation of Ca^{2+} puffs. On the one hand, it fixes the resting state of a cluster, i.e. the starting point of the escape process. On the other hand, it determines the transition rates. The present work has demonstrated that the Ca^{2+} concentration needs to stay at base level until the first channel opens. Theory provides Ca^{2+} puffs that are in agreement with experimental results [21] at physiological parameter values only if the Ca^{2+} concentration remains constant during the entire escape process. These findings underline the discrete character of IP_3R channels in a cluster [32, 33].

We use a realistic value for the channel flux constant k_c in difference to earlier studies [24, 28, 29]. That value is based on detailed simulations [31] and leads to Ca^{2+} concentrations 2-3 orders of magnitude larger than base level at an open channel. That causes models with a continuous number of open channels to fail. The non-vanishing Ca^{2+} flux at fractions of open channels smaller than 1 resulted in highly elevated Ca^{2+} concentrations at a cluster due to the large flux density [31]. In turn, that induced a high degree of inhibition. Decreasing the IP_3 concentration reduced the level of inhibition, but the number of subunits that could be activated decreased, too. The lack of Ca^{2+} puffs was resolved by increasing the dissociation constant for Ca^{2+} activation d_5. The ensuing mean first passage times again complied with experimental results, but at unphysiological values of d_5. These results demonstrate that parameter values may decide upon the underlying mechanisms. The large Ca^{2+} fluxes demand a discrete modeling of the Ca^{2+} release channels. This discrete modelling is one of the aspects of this study setting it apart from previous investigations of stochastic cluster dynamics [24, 28].

At a constant Ca^{2+} concentration, the main difference between van Kampen's expansion and the nonlinear Fokker-Planck equation is in the character of fluctuations. They correspond to additive noise for the Ω expansion and to multiplicative noise in the latter approach. Although the noise is intrinsically multiplicative, van Kampen's expansion provides a reasonable approximation, which improves with increasing base level and growing IP_3 concentration. It opens up the opportunity for further studies since the Ω expansion is the only method that yields analytic expressions for the probability density and all higher moments. That distinguishes it from the master equation and the nonlinear Fokker-Planck equation, for which only the first moment is directly accessible.

The dependencies of the mean first passage time on the Ca^{2+} concentration as well as on the IP_3 concentration comply with physiological findings. An increase of the basal Ca^{2+} concentration enhances the open probability of the IP_3R channel [4]. Consequently, the mean first passage time is to decrease with growing Ca^{2+} concentration. Our results fully agree with this activating role of Ca^{2+} (see figure 11.3). The same tendency was observed when we increased the IP_3 concentration, which agrees with the activating role of IP_3.

The present study has provided a framework for a quantitative determination of Ca^{2+} puff frequencies. The mean first passage times correspond to the stochastic fraction of the inter-puff interval, which is governed by the activation of the IP_3Rs. The second contribution to the inter-puff interval is a deterministic part controlled by puff duration, inhibition and recovery from it. Taking into account that Ca^{2+} puffs represent the fundamental building blocks of global Ca^{2+} patterns, our calculations may serve as a starting point to compute periods of Ca^{2+} waves. Experiments and theoretical studies suggest that the initiation of Ca^{2+} waves occurs by a nucleation process. Therefore, knowledge of the frequency of Ca^{2+} puffs is the first step in the calculation of wave frequencies and leads to a deeper understanding of intracellular Ca^{2+} dynamics.

11.8. Appendix A: Combinatorics for subunits

Measurements on the IP_3 receptor have revealed that a minimum number of subunits h_m needs to be activated for the channel to open [4]. A single IP_3R possesses a non zero open probability only if at least h_m subunits are in the state 10. Activation in the cell occurs of course for a subunit already associated with a certain receptor. With our model, the number of open channels depends on the arrangement of n_{10} activatable subunits on the receptors. Here, we derive the distribution of open channels resulting from such a random scattering of activatable subunits and its properties, whereas the mean was used earlier. To this aim we consider N receptors with h subunits each. Let $n_i, i = 1, \ldots, h$ denote the number of receptors with i activatable subunits, then the number of possible configurations for a given set $\{n_i\} := \{n_1, \ldots, n_h\}$ that satisfies

$$n_0 + \ldots + n_h = N\,, \quad n_1 + 2n_2 + \ldots + hn_h = n_{10} \tag{11.35}$$

is

$$M(\{n_i\}) := \frac{N!}{n_0! \cdots n_h!} \binom{h}{0}^{n_0} \binom{h}{1}^{n_1} \cdots \binom{h}{h}^{n_h}. \tag{11.36}$$

The fraction represents the number of permutations for the set $\{n_i\}$, whereas the binomial coefficients take into account the number of ways how to distribute i activatable subunits on a single receptor. The total number of configurations is given by

$$\Gamma := \sum_{\{n_i\}}^{\star} M(\{n_i\}) \tag{11.37}$$

The asterisk indicates the summation with the restrictions of equation (11.35). To evaluate equation (11.37), we introduce a generating function

$$f_1(z) := \sum_{\{n_i\}}{}' M(\{n_i\}) z^l\,, \qquad l = n_1 + \ldots + hn_h\,. \tag{11.38}$$

The prime refers to the restriction $n_0 + \ldots + n_h = N$. Therefore, the total number of configurations follows from the generating function as

$$\Gamma = \frac{1}{n_{10}!} \frac{d^{n_{10}}}{dz^{n_{10}}} f_1(z)\Big|_{z=0} . \tag{11.39}$$

Due to the identity

$$\begin{aligned} f_1(z) &= \sum_{n_i=0}^{N}{}' \frac{N!}{n_0! \cdots n_h!} \binom{h}{0}^{n_0} \left[\binom{h}{1} z^1\right]^{n_1} \cdots \left[\binom{h}{h} z^h\right]^{n_h} \\ &= \left[\binom{h}{0} + \ldots + \binom{h}{h} z^h\right]^N = (1+z)^{hN} = \sum_{j=0}^{hN} \binom{hN}{j} z^j , \end{aligned} \tag{11.40}$$

we finally arrive at $\Gamma = \binom{hN}{n_{10}}$, which complies with the combinatorics of choosing n_{10} subunits from a total of hN subunits. Consequently, the probability distribution of n_j for a fixed value of $j \in \{0, \ldots, h\}$ is given by

$$\begin{aligned} p(n_j) =& \frac{1}{\Gamma} \sum_{\substack{\{n_i\} \\ i \neq j}}^{\star} \frac{N!}{n_0! \cdots n_h!} \binom{h}{0}^{n_0} \binom{h}{1}^{n_1} \cdots \binom{h}{h}^{n_h} \\ =& \frac{1}{\Gamma} \binom{N}{n_j} \binom{h}{j}^{n_j} \sum_{\substack{\{n_i\} \\ i \neq j}}^{\star} (N-n_j)! \prod_{\substack{l=0 \\ l \neq j}}^{h} \frac{1}{n_l!} \binom{h}{l}^{n_l} . \end{aligned} \tag{11.41}$$

Equation (11.41) is most conveniently computed as

$$p(n_j) = \frac{1}{\Gamma} \binom{N}{n_j} \binom{h}{j}^{n_j} \frac{1}{n_{10}!} \frac{d^{n_{10}}}{dz^{n_{10}}} f_2(z)\Big|_{z=0} , \tag{11.42}$$

where we used the generating function

$$\begin{aligned} f_2(z) :=& \sum_{\substack{\{n_i\} \\ i \neq j}}{}' \tilde{N}! \prod_{\substack{l=0 \\ l \neq j}}^{h} \frac{1}{n_l!} \left[\binom{h}{l} z^l\right]^{n_l} \\ =& \sum_{i=0}^{\tilde{N}} \sum_{l=0}^{hi} \binom{\tilde{N}}{i} \binom{hi}{l} \left[-\binom{h}{j}\right]^{\tilde{N}-i} z^{l+j(\tilde{N}-i)} . \end{aligned} \tag{11.43}$$

Here, the prime denotes the restriction

$$n_0 + \ldots + n_{j-1} + n_{j+1} + \cdots + n_h = N - n_j =: \tilde{N} . \tag{11.44}$$

In the case $j = 0$ the derivatives in equation (11.42) can be performed explicitly, so that

$$p(n_0) = \frac{1}{\Gamma}\binom{N}{n_0}\sum_{j=0}^{\tilde{N}}\binom{\tilde{N}}{j}\binom{jh}{n_{10}}(-1)^{\tilde{N}-j}. \tag{11.45}$$

The above analysis remains valid, when we interchange the number of activatable subunits n_{10} and the number of the remaining $Nh - n_{10}$ subunits. Such a transition corresponds to the exchange of balls and voids in classical combinatorics. In that picture, equation (11.45) would represent the probability distribution of fully occupied receptors, i.e.

$$p(n_h) = \frac{1}{\Gamma}\binom{N}{n_h}\sum_{j=0}^{\tilde{N}}\binom{\tilde{N}}{j}\binom{jh}{Nh-n_{10}}(-1)^{\tilde{N}-j}. \tag{11.46}$$

Equation (11.46) arises from equation (11.45) by substituting n_{10} by $Nh - n_{10}$ and n_0 by n_h.

To gain further insight into the probability distributions we calculate the first two moments. For the average we start with

$$\langle n_j \rangle = \frac{1}{\Gamma}\sum_{\{n_i\}}^{\star} n_j M(\{n_i\}), \tag{11.47}$$

because a closed expression for the probability distribution is only available for the two cases presented above. Defining the corresponding generating function

$$f_3(z) := \frac{1}{\Gamma}\sum_{\{n_i\}}{}' n_j M(\{n_i\})\, z^l, \quad l = n_1 + \ldots + hn_h, \tag{11.48}$$

we find

$$\langle n_j \rangle = \frac{1}{n_{10}!}\frac{d^{n_{10}}}{dz^{n_{10}}} f_3(z)\Big|_{z=0} = \frac{N}{\Gamma}\binom{h}{j}\binom{h(N-1)}{n_{10}-j}. \tag{11.49}$$

In the limit $N \to \infty, n_{10} \to \infty$ we recover the result from [3]. Analogously evaluation of the second moments results in

$$\langle n_l n_k \rangle = \frac{N(N-1)}{\Gamma}\binom{h}{l}\binom{h}{k}\binom{h(N-2)}{n_{10}-l-k} + \delta_{k,l}\frac{N}{\Gamma}\binom{h}{l}\binom{h(N-1)}{n_{10}-l}. \tag{11.50}$$

Applying these general expressions to IP_3Rs requires values for h, h_m and N. The tetrameric structure of the receptor ensues $h = 4$. However, previous results by different groups are based on $h = 3$. We therefore compute the statistics for both cases. Experiments on a single channel have shown four conductance levels, each a multiple of 20pS, with a predominance of opening to the third level [4, 35]. Thus, we set $h_m = 3$. The number of receptors in a cluster has not been measured yet. We employ $N = 25$ following recent estimates by Swillens and Dupont [30].

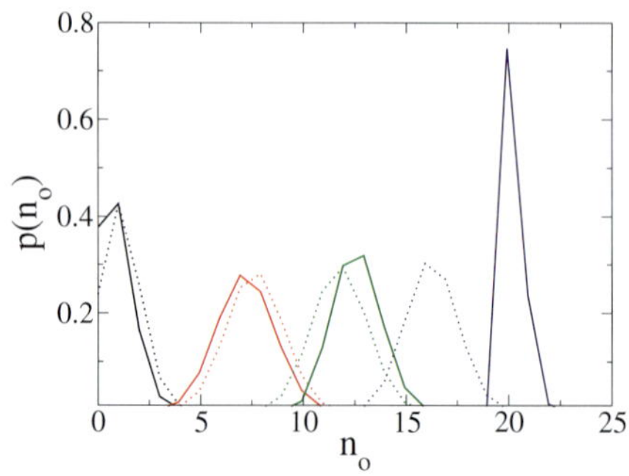

Fig. 11.8. (Color) Probability distribution $p(n_o)$ for $n_o = n_3, h = 3$ (solid) and $n_o = n_3 + n_4, h = 4$ (dotted) for $N = 25$ and different n_{10}. Values of n_{10} are 25 (black), 50 (red), 60 (green) and 70 (blue).

The probability distributions $p(n_3 + n_4)$ with $h = 4$ and $p(n_3)$ with $h = 3$ are depicted in figure 11.8. They both agree very well. This is also supported by their mean and variance as shown in figure 11.9. In the left panel we also include the position of the maxima of the distributions indicated by dots. They closely follow the average. Due to the narrowness of the distributions demonstrated by the small variance as well as the accordance between the mean and the maximum we calculate the number of open channels n_c from the average for a given value of n_{10}:

$$n_a^{(3)} = Nr^3 \frac{n_{10}}{3N} \frac{n_{10} - 1}{3N - 1} \frac{n_{10} - 2}{3N - 2}, \tag{11.51}$$

$$n_a^{(3,4)} = Nr^3 \frac{n_{10}}{4N} \frac{n_{10} - 1}{4N - 1} \frac{n_{10} - 2}{4N - 2} \left[\frac{n_{10} - 3}{4N - 3}(4 - 3r) + 4\left(1 - \frac{n_{10}}{4N}\right) \right] \tag{11.52}$$

Here $r := I/(I + d_1)$ denotes the fraction of subunits in the activatable state 10 that are activated. The subscripts (3) and (3,4) indicate that we used $p(n_3), h = 3$ and $p(n_3 + n_4), h = 4$ for averaging, respectively. Note that in the limit $N \to \infty, n_{10} \to \infty$ equations (11.51), (11.52) reduce to the well known expressions of the deterministic description. All results in section 11.6 are based on equation (11.52), which can be further simplified by approximating all denominators by $4N$ due to $4N \gg 1$.

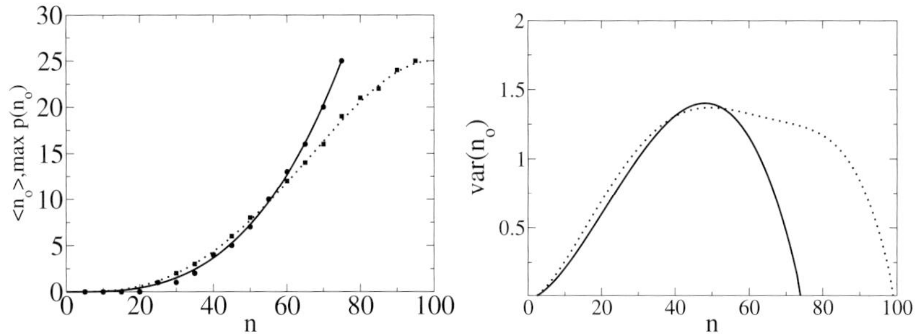

Fig. 11.9. Mean (left) and variance (right) of n_o for $n_o = n_3, h = 3$ (solid) and $n_o = n_3 + n_4, h = 4$ (dotted). The left panel shows the position of $\max p(n_o)$ as dots and squares, respectively.

11.9. Appendix B: Proof of equation (32)

This section deals with the proof of equation (11.32). It is based on the identity

$$\sum_{k=0}^{j} \binom{j}{k} \frac{(-1)^k}{2k+1} = \frac{2^{2j}\,(j!)^2}{(2j+1)!}, \tag{11.53}$$

which we now proof. We transform the left hand side of equation (11.53) according to

$$\sum_{k=0}^{j} \binom{j}{k} (-1)^k \int_0^1 t^{2k} dt = \int_0^1 \sum_{k=0}^{j} \binom{j}{k} (-t^2)^k dt = \int_0^1 \left(1-t^2\right)^j dt. \tag{11.54}$$

It can be simplified with Euler's Beta function $B(z, w)$. From its definition

$$B(z,w) := \int_0^1 t^{z-1}(1-t)^{w-1} dt \tag{11.55}$$

follows

$$\int_a^b (t-a)^{z-1}(b-t)^{w-1} dt = (b-a)^{z+w-1} B(z,w). \tag{11.56}$$

Hence we express the integral in equation (11.54) through

$$\int_0^1 \left(1-t^2\right)^j dt = \frac{1}{2}\int_{-1}^1 (t+1)^j (1-t)^j dt = 2^{2j} B(j+1, j+1). \tag{11.57}$$

According to [1] the Beta function is related to the Gamma function $\Gamma(z)$ via $B(z,w) = \Gamma(z)\Gamma(w)/\Gamma(z+w)$, so that we find

$$\sum_{l=0}^{j} \binom{j}{l} \frac{(-1)^l}{2l+1} = 2^{2j} \frac{\Gamma(j+1)^2}{\Gamma(2j+2)} = \frac{2^{2j}\,(j!)^2}{(2j+1)!} \tag{11.58}$$

due to $n! = \Gamma(n+1)$, which proofs equation (11.53). Expanding the right hand side yields

$$\frac{2^{2j}\,(j!)^2}{(2j+1)!} = \frac{2\cdot 1}{2}\cdot\frac{2}{3}\cdot\frac{2\cdot 2}{4}\cdot\frac{2}{5}\cdot\frac{2\cdot 3}{6}\cdots\frac{2\cdot j}{2j}\cdot\frac{2}{2j+1}j! = \frac{j!}{\left(\frac{3}{2}\right)_j}. \tag{11.59}$$

This proofs equation (11.32) when we use $j! = (1)_j$.

References

[1] M. Abramowitz and I. Stegun, editors. *Handbook of mathematical functions.* Dover Publications, New York, 1974.

[2] Jerry M. Adams and Suzanne Cory. The bcl-2 protein family: Arbiters of cell survival. *Science*, 281(5381):1322–1326, 1998.

[3] M. Bär, M. Falcke, H. Levine, and L. S. Tsimring. Discrete stochastic modeling of calcium channel dynamics. *Phys. Rev. Lett.*, 84:5664–5667, 2000.

[4] I. Bezprozvanny, J. Watras, and B.E. Ehrlich. Bell-shaped calcium-response curves of Ins(1,4,5)P_3- and calcium-gated channels form endoplasmic reticulum of cerebellum. *Nature*, 351:751–754, 1991.

[5] G.W. De Young and J. Keizer. A single inositol 1,4,5-triphosphate-receptor-based model for agonist-stimulated oscillations in Ca^{2+} concentration. *Proc. Natl. Acad. Sci. USA*, 89:9895–9899, 1992.

[6] M. Falcke. On the role of stochastic channel behavior in intracellular Ca^{2+} dynamics. *Biophys. J.*, 84:42–56, 2003.

[7] M. Falcke. Reading the pattern in living cells - the physics of Ca^{2+} signaling. *Advances in Physics*, 53:255–440, 2004.

[8] C.W Gardiner. *Handbook of stochastic methods.* Springer, Berlin, 3 edition, 2004.

[9] M. Gitterman and G.H. Weiss. Some comments on approximations to the master equation. *Physica A*, 170:503–510, 1991.

[10] A. Goldbeter. *Biochemical Oscillations and Cellular Rhythms.* Cambridge University Press, Cambridge, 1996.

[11] H. Grabert, P. Hänggi, and I.Oppenheim. Fluctuations in reversible chemical reactions. *Physica*, 117A:300–316, 1983.

[12] P. Hänggi, H. Grabert, P. Talkner, and H. Thomas. Bistable systems: Maser equation versus Fokker-Planck modelling. *Phys. Rev. A*, 29:371–378, 1984.

[13] H. Henry and H. Levine. Wave nucleation in excitable systems in the low noise limit. *Phys. Rev. E*, 68:031914–1–5, 2003.

[14] B. Hille. *Ion channels of excitable membranes.* Sinauer Assoicates, Inc. Publishers, Sunderland, MA USA, 3rd edition, 2001.

[15] Roger A. Horn and Charles R. Johnson. *Matrix Analysis.* Cambridge University Press, Cambridge, 1999.

[16] M. Howard and A. D. Rutenberg. Pattern formation inside bacteria: Fluctuations due to low copy number of proteins. *Phys. Rev. Let.*, 90:128102–1–4, 2003.

[17] J. Keener and J. Sneyd. *Mathematical physiology.* Springer, New York, 1998.

[18] H.A. Kramers. Brownian motion in a field of force and the diffusion model of chemical reactions. *Physica*, 7:284–304, 1940.

[19] MD Levin, CJ Morton-Firth, WN Abouhamad, RB Bourret, and D Bray. Origins of individual swimming behavior in bacteria. *Biophysical Journal*, 74(1):175–181, 1998.

[20] Y. Li and J. Rinzel. Equations for $InsP_3$ receptor-mediated $[Ca^{2+}]_i$ oscillations derived from a detailed kinetic model: A hodgkin-huxley like formalism. *J. theor. Biol.*, 166:461–473, 1994.

[21] J.S. Marchant and I. Parker. Role of elementary Ca^{2+} puffs in generating repetitive Ca^{2+} oscillations. *EMBO J.*, 20:65–76, 2001.

[22] Mark P. Mattson. Pathways towards and away from alzheimer's disease. *Nature*, 430(7000):631–639, 2004.

[23] Harley H. McAdams and Adam Arkin. Stochastic mechanisms in gene expression. *Proceedings of the National Academy of Sciences*, 94(3):814–819, 1997.

[24] L. Meinhold and L. Schimansky-Geier. Analytical description of stochastic calcium-signal periodicity. *Phys. Rev. E*, 66:050901(R)–1–4, 2002.

[25] J.E. Moyal. Stochastic processes and statistical physics. *J. Roy. Stat. Soc. B*, 11:150–210, 1949.

[26] Christopher V. Rao, Denise M. Wolf, and Adam P. Arkin. Control, exploitation and tolerance of intracellular noise. *Nature*, 420:231–237, 2002.

[27] H. Risken. *The Fokker-Planck equation.* Springer, Berlin, 1984.

[28] J. W. Shuai and P. Jung. Optimal intracellular calcium signalling. *Phys. Rev. Lett.*, 88(6):068102–1–4, 2002.

[29] J. W. Shuai and P. Jung. Stochastic properties of Ca^{2+} release of inositol 1,4,5-trisphosphate receptor clusters. *Biophys. J.*, 83(1):87–97, 2002.

[30] S. Swillens, G. Dupont, L. Combettes, and P. Champeil. From calcium blips to calcium puffs: Theoretical analysis of the requirements for interchannel communication. *Proc. Natl. Acad. Sci. USA*, 96:13750–13755, 1999.

[31] R. Thul and M. Falcke. Release currents of IP_3 receptor channel clusters and concentration profiles. *Biophys. J.*, 86:2660–2673, 2004.

[32] R. Thul and M. Falcke. Stability of membrane bound reactions. *Phys. Rev. Lett.*, 93:188103–1–4, 2004.

[33] R. Thul and M. Falcke. Reactive clusters on a membrane. *Phys. Biol.*, 2:51–59, 2005.

[34] N.G. van Kampen. *Stochastic processes in physics and chemistry.* North-Holland, Amsterdam, 2001.

[35] J. Watras, I. Bezprozvanny, and B. Ehrlich. Inositol 1,4,5-trisphosphate-

gated channels in cerebellum: presence of multiple conductance states. *J. Neurosci.*, 11:3239–3245, 1991.

Chapter 12

Continuous wavelet spectral analysis of climate dynamics

Douglas Maraun[1,*], Jürgen Kurths[1,+] and Matthias Holschneider[2,!]

[1] *Nonlinear Dynamics Group, Institute of Physics, University of Potsdam, D-14415 Potsdam, Germany*

[2] *Mathematics Institute, University of Potsdam, D-14415 Potsdam,Germany*

[*] *maraun@agnld.uni-potsdam.de*, [+] *jkurths@agnld.uni-potsdam.de*, [!] *hols@math.uni-potsdam.de*

Contents

12.1. Introduction

Climate is a fascinating complex system. It spans a wide range of spatial and temporal scales, reaching from the behaviour of a single cloud up to ocean wide phenomena, from 60 day rythms of tropical rainfall to the thousands of years lasting glacial cycles. These processes are subject to instationarities on a variety of scales: Although the equations describing the El Niño/Southern Oscillation (ENSO) phenomenon are deterministic, changing background conditions lead to a decadally changing behaviour. Although the glacial cycles itself are triggered by mere orbital oscillations, the response of the climate system induces complex nonlinear regime shifts on millenial scales.

This vast spectral bandwidth illustrates the necessity of a reliable scale and time resolved decomposition of available observations to separate and describe single processes as individual parts of the whole system. Often, the comlex interplay between climate subsystems plays an essential role and the understanding of coupling mechanisms is of crucial importance for the study and prediction of at first sight independent phenomena. Continuous wavelet transformation (CWT) is the prototypic instrument to address these tasks: As an important application, it transforms time series to the time/scale domain for estimating the linear non-stationary spectral properties of the underlying process.

The continuous wavelet spectra of paradigmatic processes as Gaussian white noise [8] or fractional Gaussian noise [10] have been studied. The method has been applied to various real world problems of physics, climatology [6], life sciences [5] and other fields of research. Hudgins et. al. [9] defined the wavelet cross spectrum to investigate scale and time dependent linear relations between different processes. This measure found its application e.g. in atmospheric turbulence [9], the analysis of time varying relations between El Niño/Southern Oscillation and the Indian monsoon [20] as well as interrelations of business cycles from different national economies [3].

It was the merit of Torrence and Compo [19] to place continuous wavelet spectral analysis into the framework of statistical data analysis by formulating pointwise significance tests against reasonable background spectra. However, given a realization of white noise, large patches of spurious significance are detected, making it - without further insight - impossible to judge which features of an estimated wavelet spectrum differ from a background noise and which are just artefacts. Furthermore, given realizations of two independent processes, regions of spurious interrelations show up in the estimated wavelet cross spectrum.

In this contribution we describe a framework of non-stationary Gaussian processes defined in the wavelet domain. These processes are characterized by their time dependent spectral properties and thus can be utilized to investigate the behavior of wavelet spectral estimators. To overcome the problems arising from pointwise significance testing, we present an areawise significance test that takes advantage of basic properties of continuous wavelet transform. Also, we discuss the pitfalls of cross wavelet analysis and advocate for coherency analysis as an alternative.

12.2. Continuous wavelet transformation

CWT might be interpreted as a generalization of the Gabor transformation, which in turn is the time resolved extension of Fourier transformation. For a detailed discussion of CWT basics please refer to the comprehensive literature [2, 4, 8].

Given a time series $s(t)$, its wavelet transformation $W_g s(b, a)$ at time b and scale a (scale refers to 1/frequency) with respect to the chosen wavelet $g(t)$ is given as

$$W_g s(b, a) = \int dt \, \frac{1}{\sqrt{a}} \, \bar{g}\left(\frac{t-b}{a}\right) s(t) \tag{12.1}$$

The L^2-normalization $1/a^{\frac{1}{2}}$ is chosen to ensure that the spectra defined in Sec. 12.3.2 exhibit the same scale dependency as the corresponding estimated spectra (Sec. 12.4.1). Being important for the further discussion, we recall the covariances of CWT:

$$W_g s[t - b'](b, a) = W_g s[t](b - b', a) \tag{12.2}$$

$$W_g s[t/a'](b, a) = W_g s[t](b/a', a/a') \tag{12.3}$$

Here, the brackets [.] denote dependencies of a variable, whereas (.) denote dependencies of the resulting transformation.

For every wavelet in a strict sense, $g(t)$, a reconstruction wavelet $h(t)$ fulfilling certain properties can be found [8]. Utilizing this, one can define an inverse transformation from the positive half plane $\mathbb{H}$ to the real axis, i.e. the time domain.

$$M_h r(t) = \int_{\mathbb{H}} \frac{db\, da}{a^2} \, r(b, a) \, \frac{1}{\sqrt{a}} \, h\left(\frac{t-b}{a}\right) \tag{12.4}$$

We point out that this operation from two dimensions to one dimension is not the unique possible inverse transformation. Two properties of CWT will appear to be important for our study: Not every function on the positive half-plane is a wavelet transformation. Thus the successive transformation of an arbitrary function in time and scale to the time domain and back to the time/scale domain projects this function onto the subspace of all wavelet transformations.

As a consequence, $r(b,a)$ is a wavelet transformation, if and only if

$$r(b,a) = \int_0^\infty \frac{da'}{a'} \int_0^\infty db' \frac{1}{a'} K_{g,h}\left(\frac{b-b'}{a'}, \frac{a}{a'}\right) r(b',a') \tag{12.5}$$

and $K_{g,h}((b-b')/a', a/a') = W_g h((b-b')/a')$ is called the reproducing kernel [8]. The reproducing kernel of the Morlet wavelet is plotted in Fig. 12.1.

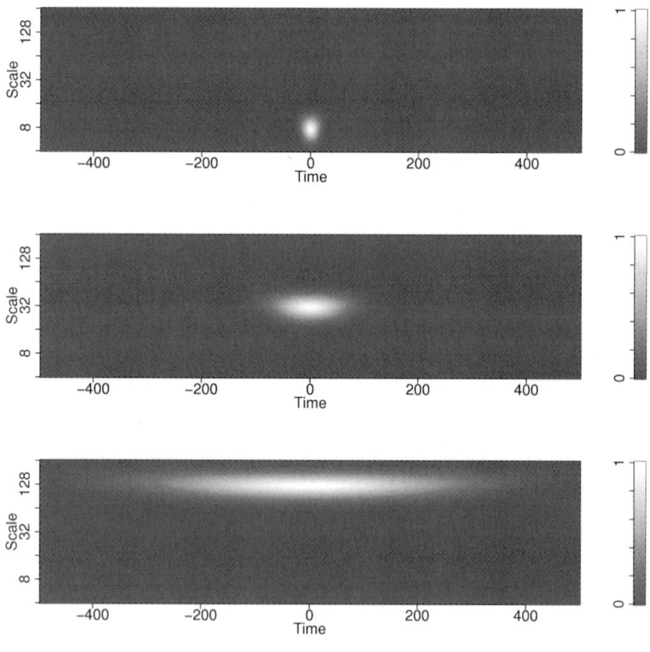

Fig. 12.1. (Taken from [17]) Reproducing kernel of the Morlet wavelet for three different scales: (a) s=8, (b) s=32 and (c) s=128. The width in time and in scale direction increases linearly with scale (i.e. in scale direction it appears constant on a logarithmic scale axis)

The redundancies of CWT are reflected in internal correlations of any wavelet transformation $r(b,a)$. For Gaussian white noise $\eta(t)$, these correlations are given by the reproducing kernel [2, 8]:

$$\begin{aligned} C(b,a;b',a') &= \langle W_g\eta[t](b,a), W_g\eta[t](b',a')\rangle \\ &\sim K_{g,h}\left(\frac{b-b'}{a'},\frac{a}{a'}\right) \end{aligned} \quad (12.6)$$

These correlations constitute a fundamental difference of any time-frequency (or scale) resolved analysis to time independent Fourier analysis, where neighboring frequencies are asymptotically uncorrelated.

12.3. Gaussian processes defined in the wavelet domain

Stationary Gaussian processes are completely defined by their Fourier spectrum $S(\omega)$. A realization of any such process can be simulated by transforming Gaussian white noise to the Fourier domain, multiplying it with a function $f(\omega)$ and transforming it back to the time domain (see e.g. [18]). The spectrum of this process is then given by $|f(\omega)|^2$ where $f(\omega)$ is called a Fourier multiplier. We suggest to extend this concept to non-stationary Gaussian processes using wavelet multipliers $m(b,a)$ as a function of time and scale.

12.3.1. *Definitions*

We define an equivalence class of non-stationary Gaussian processes in the wavelet domain by the wavelet multipliers $m(b,a)$. Realizations are given as

$$s(t) = M_h m[b,a] W_g \eta(t) \quad (12.7)$$

with driving Gaussian white noise $\eta(t) \sim \mathcal{N}(0,1)$ and $\langle \eta(t_1)\eta(t_2)\rangle = \delta(t_1 - t_2)$.

Following the projection property, the realization $m[b,a]W_g\eta(t)$ in the time/scale domain is in general not a wavelet transformation and thus realizations $s(t)$ in the time domain depend - usually weakly - on the chosen wavelet $g(t)$ and the reconstruction wavelet $h(t)$ respectively. To ensure at least asymptotic independence of the (reconstruction) wavelet, we define the processes such that for smaller and smaller scales more and more reproducing kernels fit into local structures of the process:

$$\begin{aligned} \partial_a m(b,a) &< O(a^{-1+\epsilon}) \\ \partial_b m(b,a) &< O(a^{-1+\epsilon}) \end{aligned} \quad (12.8)$$

$m(b,a) \equiv m(a)$ defines a stationary Gaussian process in the wavelet domain. In this special case, the stationary Fourier spectrum exists with

$$f(\omega) \approx m\left(\frac{1}{\omega}\right) C_1 + \frac{1}{\omega} m'\left(\frac{1}{\omega}\right) C_2, \tag{12.9}$$

C_1 and C_2 being constants. As expected, the Fourier spectrum is given by the wavelet spectrum plus a correction term. The latter depends on the localization of the used wavelets and on the slope of the wavelet spectrum. For the asymptotic behavior defined in Eq. (12.8), it vanishes for high frequencies. For details and derivations refer to [17].

12.3.2. *Spectral measures*

For non-stationary processes, continuous wavelet spectral measures have been defined as the expectation value of the corresponding estimator [11]. This brings along certain difficulties: In reality, one usually observes only one realization of a certain process, i.e. one has no access to the expectation value as an ensemble average. Furthermore, the analysis is local in time and hence replacing the ensemble average by the time average is not valid. Using wavelet multipliers one can define time dependent spectral measures, that elegantly overcome these difficulties.

Given wavelet multipliers $m(b,a)$, one can define a wavelet independent spectrum

$$S_g(b,a) = |m(b,a)|^2 \tag{12.10}$$

The spectrum denotes the variance of the process at a certain time b and scale a. With the chosen normalization of the wavelet transformation (Eq. (12.1)), white noise is given by $S_g(b,a) = |m(b,a)|^2 = const.$.

Given two covarying processes given by wavelet multipliers $m_{1c}(b,a)$ and $m_{2c}(b,a)$, one can define a cross spectrum

$$CS_g(b,a) = m_{1c}(b,a)m^*_{2c}(b,a) \tag{12.11}$$

In general, this is a complex function and may be decomposed into amplitude and phase:

$$CS_g(b,a) = |CS_g(b,a)| \exp(\arg(CS_g(b,a))) \tag{12.12}$$

The cross spectrum denotes the covarying power of two processes. Realizations of such two processes are constructed by the same driving noise realization (plus some possible superimposed independent noise): $s_1(t) = M_h m_{1c}[b,a] W_g \eta(t)$ and $s_2(t) = M_h m_{2c}[b,a] W_g \eta(t)$ respectively.

The coherency is defined as the normalized cross spectrum. Exhibiting values between zero and one, it denotes the linear relationship between two processes. The squared coherency reads

$$COH_g^2(b,a) = \frac{|m_{1c}(b,a)m_{2c}^*(b,a)|}{|m_{1c}(b,a)+m_{1i}(b,a)|^2\,|m_{2c}(b,a)+m_{2i}(b,a)|^2}, \tag{12.13}$$

where $m_{1c}(b,a)$ and $m_{2c}(b,a)$ denote the covarying part and a superimposed independent contribution is given by m_{1i} and m_{2i}.

12.3.3. *Example*

We illustrate the concept with the following example: We define a linear stochastic chirp given by

$$m(b,a) = \exp\left(-\frac{(b-b_0(a))^2}{2\sigma^2(a)}\right) \tag{12.14}$$

with $b_0(a) = \beta_0 + c\log(a)$ and $\sigma(a) = \sigma_0 a^{1-\epsilon}$, i.e. every voice (stripe of constant scale) is given by a Gaussian with time position and width varying with scale. The power of $1-\epsilon$ ensures the process exhibiting the desired asymptotical behavior. Figures 12.2(a) and (b) show the spectrum $|m(b,a)|$ and a typical realization respectively.

12.4. Estimating wavelet spectra

12.4.1. *Spectral estimators*

When estimating any continuous wavelet spectral measure, one has to restrict the analysis to a finite number of scales. The amount of independent scale information is limited by the reproducing kernel (see Eq. (12.6)). Thus, a reasonable choice of scales is $a_j = a_0 2^{\frac{i-1}{N_{voice}}}$ with $i = 1 \ldots N_{voice} \cdot N_{octave}$. Here, a_0 corresponds to a frequency lower than or equal to the Nyquist frequency, N_{octave} denotes the number of octaves (i.e. powers of two), and N_{voice} the number of voices (i.e. calculated scales) per octave. The matrix of calculated wavelet spectral coefficients is called wavelet spectral matrix.

Given a realization $s(t)$ of a non-stationary process, one can estimate its spectrum (i.e. calculate the wavelet sample spectrum) using a wavelet $g(t)$ by

$$\widehat{S}_g(b,a) = A(|W_g s(t)|^2), \tag{12.15}$$

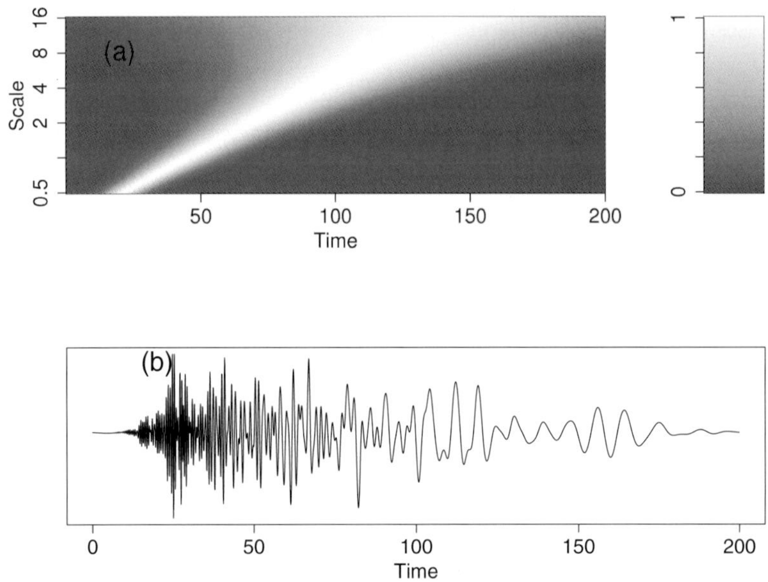

Fig. 12.2. (Taken from [17]) Stochastic chirp with $\epsilon = 0.3$. (a) The spectrum $|m(b,a)|^2$. (b) A typical realization in the time domain, calculated with a Morlet Wavelet, $\omega_0 = 6$.

where $A(.)$ denotes an averaging operator defined in Sec. 12.4.2. Following the notation of Fourier analysis, without averaging, the wavelet spectral estimate is either called scalogram or wavelet periodogram. Estimators of a process property are marked by a hat $\widehat{.}$.

Given realizations $s_1(t)$ and $s_2(t)$ of two processes, the cross spectrum can be estimated as

$$\widehat{CS}_g(b,a) = A(W_g s_1(t) W_g^* s_2(t)), \tag{12.16}$$

or decomposed into amplitude and phase,

$$\widehat{CS}_g(b,a) = |\widehat{CS}_g(b,a)| \exp(\arg(\widehat{CS}_g(b,a)), \tag{12.17}$$

whereas the squared coherency is estimated as

$$\widehat{COH}_g^2(b,a) = \frac{|\widehat{CS}_g(b,a)|}{(\widehat{S}_{g,1}(b,a)\widehat{S}_{g,2}(b,a))^{1/2}}. \tag{12.18}$$

For coherency, averaging is essential. Otherwise, one attempts to infer covarying oscillations without observing the oscillations and consequently obtains a trivial value of one for any two processes.

12.4.2. *Variance of the wavelet sample spectrum*

As discussed in Sec. 12.3.1, we constructed the class of nonstationary Gaussian processes such, that they become locally stationary for small scales (see Eq. (12.8)). Hence, when the averaging kernel $A(.)$ is adapted to the process variability on each scale, it includes more and more reproducing kernels, such that the variance of the spectral estimate vanishes in the limit of small scales. Then, the following relation for the variance $Var_A(a)$ of the averaged wavelet spectral estimate holds:

$$Var_A(a) \sim Var(a)a^{1-\alpha}, \qquad (12.19)$$

where $Var(a)$ denotes the variance of the wavelet scalogram at each scale and α describes the scaling of the averaging window. The factor a^1 results from the width of the reproducing kernel in smoothing direction.

The prior discussion appears to be rather academic: In general, one does not a priori know the asymptotical behavior for small scales of an observed process. Also, finite sampling restrains the investigation of small scales. Finally, one often is interested in nonstationary behavior on scales large compared to the sampling time.

In fact, retaining a scale independent variance appears to be a reasonable choice. This might be accomplished by averaging the same amount of independent information on every scale, i.e. by choosing the length of the averaging kernel according to the reproducing kernel [11]:

- Averaging in scale direction should be done with a window exhibiting constant length (for logarithmic scales), see Fig. 12.3(a). Half window length is w_a in the same units as N_{voice}.
- Averaging in time direction should be done with a window exhibiting a length proportional to scale, see Fig. 12.3(b). Half window length is $w_b \cdot a$.

The (scale independent) variance as a function of the width of a square averaging window is shown in Fig. 12.4. The graphs for averaging in scale as well as in time direction resemble the shape of the reproducing kernel. An averaging window short compared to the effective width of the reproducing kernel includes only a minor part of independent information and thus fails to notably reduce the variance.

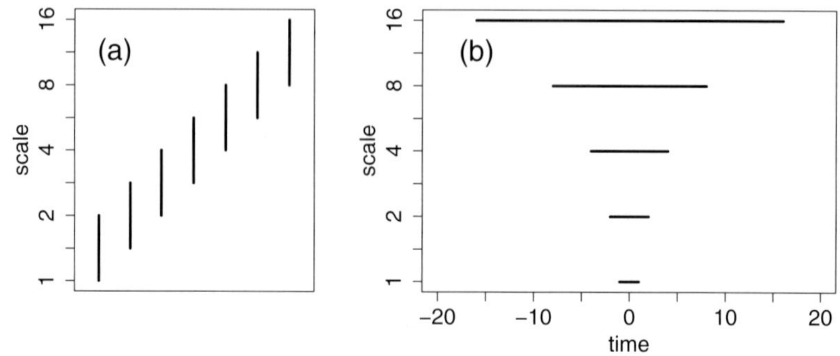

Fig. 12.3. (Taken from [17]) Smoothing according to the reproducing kernel to provide a constant variance for all scales. (a) In scale direction, the length of the smoothing window stays constant (for a logarithmic scale axis). (b) In time direction, the length of the smoothing window increases linearly with scale.

12.4.3. *Bias of the wavelet sample spectrum*

The bias of the estimated wavelet spectrum reads

$$\mathrm{Bias}(\widehat{S}_g(b,a)) = \\ \langle\, A(\,|\, \underbrace{W_k M_h}_{P_{h\to k}}\, m(b,a)\, W_g \eta(t)\,|^2\,)\,\rangle - |m(b,a)|^2, \tag{12.20}$$

where W_g and M_h denote the pair of wavelets used to simulate the process and W_k the wavelet used for the estimation. For processes following Eq. (12.8), this term vanishes for small scales [17]. For averaging on finite scales, one has to consider the trade off between bias and variance.

12.4.4. *Example*

We recall the example from Sec. 12.3.3. Figure 12.5(a) depicts the wavelet scalogram of the realization shown in Fig. 12.2(b). It is easy to see, that a single realization without averaging yields a rather insufficient estimation of the real spectrum. Averaging, shown in (b) and (c), reduces the variance, but produces a bias. The estimation based on 1000 realizations (d) yields a pretty accurate result of the underlying process.

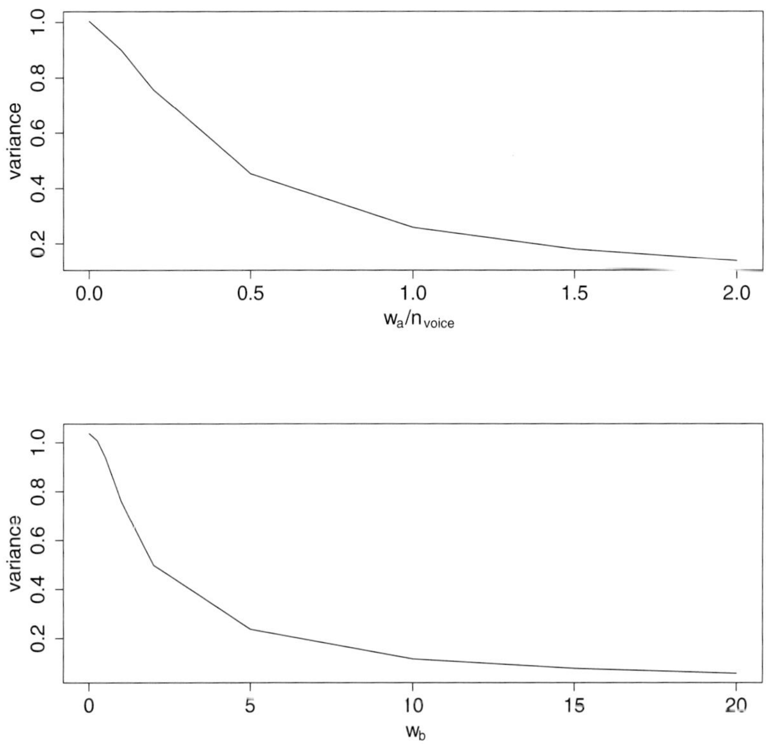

Fig. 12.4. (Taken from [17]) Variance of the wavelet sample spectrum as a function of the length of an averaging square window in scale (a) and time (b) direction. The graphs resemble the shape of the reproducing kernel.

12.5. Significance testing

12.5.1. *Sensitivity vs. Specificity*

Accounting for its purpose, every significance test has to be designed individually: A medical test should always warn the patient in case of a severe disease; on the other hand, establishing a result by means of statistics, a test should preferably reject false positive results. These antithetic demands are represented by the terms of *sensitivity* and *specificity*:

$$Sensitivity = \frac{N_{TP}}{N_{TP} + N_{FN}} \tag{12.21}$$

$$Specificity = \frac{N_{TN}}{N_{FP} + N_{TN}} \tag{12.22}$$

Here, N_{TN}, N_{TP}, N_{FN} and N_{FP} refer to the numbers of true negative/positive and false negative/positive results, respectively. Given a null hypothesis H_0 and N applications of the test, where H_0 is right in N_R cases and wrong in $N_W = N - N_R$ cases. Then the sensitivity relates the number N_{TP} of true rejections of H_0 to the total number of wrong H_0, $N_W = N_{TP} + N_{FN}$. On the other hand, the specificity measures the number N_{TN} of true acceptances of H_0 in relation to the total number of right H_0, $N_R = N_{FP} + N_{TN}$. A sensitive test rejects H_0 in preferably every case it is wrong (low β-error), whereas a specific test preferably only rejects H_0 when it is definitely wrong (low α-error). For finite data, no test can be perfectly sensitive and specific simultaneously.

12.5.2. *Pointwise testing of the wavelet spectrum*

To our knowledge, Torrence and Compo [19] were the first to establish significance tests for wavelet spectral measures. They assumed a reasonable background spectrum for the null hypothesis and tested for every point in the time/scale plane separately (i.e. pointwise) whether the power exceeded a certain critical value corresponding to the chosen significance level. Since the critical values of the background model are difficult to be accessed analytically [11], they need to be estimated based on a parametric bootstrap:

- Choose a significance level $1 - \alpha$.
- Choose a reasonable model (e.g. an AR[1] process in case of climate data following Hasselmann [7]) as null hypothesis H_0 and fit it to the data.
- Estimate the $(1 - \alpha)$-quantile S_{crit} (i.e. the critical value) of the corresponding background spectrum by Monte Carlo simulations. Depending on the chosen background model and the chosen normalization of the spectral estimator, the critical value in general depends on scale.
- Check for every point in the wavelet domain, whether the estimated spectrum exceeds the corresponding critical value. The set of all pointwise significant wavelet spectral coefficients is given as

$$P_{pw} = \{ (b, a) \mid \widehat{S}_g(b, a) > S_{crit} \} \tag{12.23}$$

12.5.3. *Areawise testing of the wavelet spectrum*

The concept of pointwise significance testing always leads to multiple testing: Given a significance level $1-\alpha$, a repetition of the test for N wavelet spectral coefficients leads to, on an average, $\alpha \cdot N$ false positive results. For any time/scale resolved analysis, a second problem comes into play: According to the reproducing kernel Eq. (12.6), neighboring times and scales of a wavelet transformation are correlated. Correspondingly, false positive results always occur as contiguous patches. These spurious patches reflect oscillations, which are randomly stable [a] for a short time.

For the interpretation of data from a process with unknown spectrum, these effects mark a problem: Which of the patches detected in a pointwise manner remain significant when considering multiple testing effects and the intrinsic correlations of the wavelet transformation? Fig. 12.6 illustrates that a mere visual judgment based on a sample spectrum will presumably be misleading: Even in the case of white noise, the test described in Sec. 12.5.2 yields a large number of - by construction spuriously - significant patches.

We develop an areawise test, which utilizes information about the size and geometry of a detected patch to decide whether it is significant or not. The main idea is simple: If the intrinsic correlations are given by the reproducing kernel, Eq. (12.6), then also the typical patch area for random fluctuations is given by the reproducing kernel. Following the dilation covariance Eq. (12.3) and as illustrated in Fig. 12.1, the typical patch width in time and scale direction should grow linearly with scale.

However, investigating the wavelet spectral matrices Fig. 12.6 reveals that many spurious patches are not formed "typical" but rather arbitrary and complex. Patches might exhibit a large extend in one direction, but be very localized in the other direction (Patch A in Fig. 12.6). Other patches might consist of rather small patches connected by thin "bridges" (Patch B in Fig. 12.6). These patches are spurious even though their area might be large compared to the corresponding reproducing kernel. Thus, not only the area but also the geometry has to be taken into account: Given a pointwise significant patch, a point inside this patch is areawise significant, if a reproducing kernel (dilated according to the investigated scale) containing this point totally fits into the patch. Consequently, small as well as long but thin patches or bridges are sorted out as being insignificant.

[a] As physicists, we would rather say coherent. We prefer to use the term stable to avoid confusion with the term coherence, which here refers to the interrelation between two processes.

We call the particular area given by the dilated and translated reproducing kernel at time b and scale a critical area $P_{crit}(b,a)$, as it corresponds to a critical value in a pointwise significance test. It is given as the subset of the time/scale domain, where the dilated reproducing kernel exceeds the threshold of a certain critical level K_{crit}:

$$P_{crit}(b,a) = \{ (b',a') \,|\, (K(b,a;b',a') > K_{crit} \} \qquad (12.24)$$

The set of all patches P_{pw} is given by Eq. (12.23). Then the subset of additionally areawise significant wavelet spectral coefficients is given as

$$P_{aw} = \bigcup_{P_{crit}(b,a) \subset P_{pw}} P_{crit}(b,a) \qquad (12.25)$$

The determination of the critical area as a function of a desired significance level appeared to be rather non-trivial.

The characteristic functions of the pointwise and areawise patches P_{pw} and P_{aw} are defined as

$$\begin{aligned} \chi_{P_{pw}}(a,b) &= \begin{cases} 1 & \text{if} \quad (a,b) \in P_{pw}, \\ 0 & \text{otherwise} \end{cases} \\ \chi_{P_{aw}}(a,b) &= \begin{cases} 1 & \text{if} \quad (a,b) \in P_{aw}, \\ 0 & \text{otherwise} \end{cases} \end{aligned} \qquad (12.26)$$

Then the corresponding areas A_{pw} and A_{aw} result as

$$\begin{aligned} A_{pw} &= \int_{(a,b)} \frac{db\,da}{a^2} \chi_{P_{pw}}(a,b) \\ A_{aw} &= \int_{(a,b)} \frac{db\,da}{a^2} \chi_{P_{aw}}(a,b) \end{aligned} \qquad (12.27)$$

Note, that on every scale a, the area is related to the corresponding measure a^2. We now define the significance level of the areawise test as

$$1 - \alpha_{aw} = 1 - \langle \frac{A_{aw}}{A_{pw}} \rangle, \qquad (12.28)$$

i.e. one minus the average ratio between the areas of areawise significant patches and pointwise significant patches.

As a matter of fact, we had to estimate the corresponding critical area P_{crit} as a function of a desired significance level $1 - \alpha_{aw}$ by a root finding algorithm individually for every pair (ω_0, w_s). The idea of this algorithm is outlined in [17]. It turned out that the critical area does not depend on the chosen background model.

The actual areawise test is performed as follows:

- Perform the pointwise test according to Sec. 12.5.2 on the $1 - \alpha$ level.
- Stretch the reproducing kernel for every scale according to Eq. (12.3), choose a significance level $1 - \alpha_{aw}$ for the areawise test and the corresponding critical area $P_{crit}(b, a)$ of the reproducing kernel.
- Slide the critical area $P_{crit}(b, a)$ (for every scale the corresponding dilated version) over the wavelet matrix. Every point inside a patch is defined as areawise significant, if the critical area containing this point totally lays within the patch.

Figure 12.7 illustrates the areawise test based on the result of the pointwise test for a Gaussian white noise realization shown in Fig. 12.6. The areawise test is capable of identifying about 90% of the spuriously significant area from the pointwise test.

12.5.4. *Sensitivity and specificity of the areawise test*

For the time/scale analysis of real data, it is important to have an idea of what time and scale dependent structures in principle can be identified. Especially geophysical processes exhibit power on a wide range of scales, where only a narrow band of time localized oscillations might be interesting. The question arises, how strong the localization in time and scale might be in relation to the background noise to be in principle identifiable. This question addresses the sensitivity of the test. On the other hand, it is relevant to know, how many true features the test detects compared to the number of false positive results. This question addresses the specificity of the test. To investigate these questions, we defined Gaussian bumps

$$m(b, a) = \exp\left(-\frac{(b - b_0)^2}{2\sigma_b^2}\right) \cdot \exp\left(-\frac{(c\log(a) - c\log(a_0))^2}{2\sigma_a^2}\right), \tag{12.29}$$

where b_0 and a_0 denote mean time and scale respectively, whereas σ_b and σ_b define the width in time and scale direction. The logarithm of the scale provides a bump with an elliptic Gaussian amplitude in the typical logarithmic scale axis wavelet matrix.

We performed the following study [17]:

- We simulated Gaussian bumps of different widths σ_b and σ_a, superimposed by background noise. As a simple model we chose Gaussian white noise $\eta \sim \mathcal{N}(0, \sigma)$ with zero mean and variance σ. For each set of values $(\sigma_b, \sigma_a, \sigma)$, we simulated $N = 10000$ realizations.

- To every realization, we applied the pointwise ($\alpha_{pw} = 0.95$) and the areawise test ($\alpha_{aw} = 0.9$).

Based on this experiment, we compared the sensitivity and specificity of the areawise significance test to those of the pointwise test. We define the area of the bump (i.e. the set of points where we assume H_0 as being true) as $P_B = \{(a,b) \,|\, m(a,b) > 1/e^2\}$, the complement as $P_{NB} = \{(a,b) \,|\, m(a,b) \leq 1/e^2\}$, the true positive patches as $P_{TP} = P \cap P_B$ and the true negative patches $P_{TN} = \bar{P} \cap P_{NB}$, the false positive $P_{FP} = P \cap P_{NB}$ and the false negative $P_{FN} = \bar{P} \cap P_B$, where P stands for either P_{pw} or P_{aw} and $\bar{P}$ denotes the complement. Using characteristic functions as in Eq. (12.26), we calculate the corresponding areas A_B, A_{NB}, A_{TP}, A_{TN}, A_{FP} and A_{FN} as in Eq. (12.27). Then we can define *Sensitivity* $= \frac{A_{TP}}{A_B}$ and *Specificity* $= \frac{A_{TN}}{A_{NB}}$.

We summarize the results from [17]: On the one hand, the sensitivity of the pointwise test is higher than that of the areawise test, as the latter one sorts out small patches in the area of the bump. The sensitivity depends strongly on the signal to noise ratio: For low background noise, both tests perform very well, although the part of the bump area not detected by the areawise test is around twice as large than that not detected by the pointwise test, because the areawise test sorts out small patches. As the noise-level increases to the order of the bump's driving noise, the sensitivity decreases rapidly. For a zero signal to noise ratio, the sensitivity of the pointwise test should converge to $\alpha_{pw} = 0.05$, that of the areawise test to $\alpha_{pw} \cdot \alpha_{aw} = 0.005$. However, the ratio between the parts of the area not detected by the two tests converges to $(1 - \alpha_{pw})/(1 - \alpha_{pw} \cdot \alpha_{aw}) \approx 0.95$. In other words, for a very bad signal to noise ratio, the performance of the pointwise test is not really better. The decrease of the sensitivity with noise is much larger for small bumps than for large bumps. That means that small patches get rather invisible as they get superimposed by strong noise.

On the other hand, the specificity of the areawise test is higher than that of the pointwise test, as the latter one detects many more patches outside the area of the bump. Whereas the specificity of the areawise test appears to be almost independently of the signal to noise ratio close to one, that of the pointwise test decreases for high background noise, as more and more spurious patches appear. The ratio of the numbers of false positive results ranges from 1 for a high signal to noise ratio to $1/\alpha_{aw} = 10$ for a zero signal to noise ratio in favor of the areawise test.

For typical geophysical data sets exhibiting a broad spectrum (i.e. a low signal to noise ratio), this study brings along important consequences: Many of the patches resulting from the pointwise patches are likely to be spurious. Applying the areawise test drastically increases the reliability of the interpretation.

12.5.5. *Testing of covarying power*

Compared to testing the single wavelet spectrum, the inference of covarying power is rather non-trivial. Such as for the stationary Fourier cross spectrum and the covariance (its time domain counterpart), no significance test for the wavelet cross spectrum exists. As it is not a normalized measure, it is impossible to decide whether a cross spectral coefficient is large because the one or the other process exhibits strong power or if actually covarying power does exist. Maraun and Kurths [11] illustrated this problem and analyzed a prominent example. To overcome this problems, one normalizes the cross spectrum and tests against zero coherence. Also here, a pointwise and an areawise test might be done. However, the areawise test alone is not capable to infer coherence in a strict sense.

The structure of the pointwise test is similar to that developed for the wavelet spectrum. As the coherence is normalized to the single wavelet spectra, the critical value becomes independent of the scale as long as the smoothing is done properly according to Sec. 12.4.2, i.e. when the geometry of the reproducing kernel is accounted for.

In the case of Fourier analysis, the coherence critical value is independent of the processes to be compared, if they sufficiently well follow a linear description [1, 15]. This independency, however, holds exactly only in the limit of long time series. As wavelet analysis is a localized measure, this condition is not fullfilled. Hence, for different AR[1] processes (from white noise to almost nonstationary processes), we found a marginal dependency on the process parameters.

Also here, an areawise test can be performed to sort out false positive patches being artefacts from time/frequency resolved analysis. The procedure is exactly the same as for the wavelet spectrum, only the critical patchsize $P_{crit}(b, a)$ has to be reestimated. Areawise significant patches denote significant common oscillations of two processes. Here, common means that two processes exhibit a rather stable phase relation on a certain scale for a certain time intervall.

If one is also interested in deriving significant coherence in the sense of coupling between the processes, the areawise test has to be succeeded by another step: It has to be tested, if the time intervall of the common oscillations is significantly long compared to typical randomly common oscillations of independent processes. The length distribution of such intervals is related to the decorrelation times of the individual processes and might be estimated by a bootstrap ansatz.

If the covariance is restricted to the dominant oscillations of the underlying processes and only the time resolved phase relation but no amplitude and scale information is desired, another method might be superior to wavelet coherency analysis: An instantanious phase relation might be derived by Hilbert phase analyis. This method originated from synchronization analysis [14, 16]. We combined this approach with a geometrically motivated filtering [13] and successfully utilized it for the analysis of the coupling between El Niño/Southern Oscillation (ENSO) and the Indian monsoon [12].

12.6. Conclusions

We outlined a framework to define nonstationary Gaussian processes in the wavelet domain: A driving Gaussian noise is transformed to the wavelet domain, multiplied with wavelet multipliers defining the desired nonstationary spectrum and finally transformed back to the time domain with a suitable reconstruction wavelet. These processes are completely defined by its wavelet multipliers and the wavelets used for the construction. We defined the process behavior in such a way, that the dependency on the wavelets vanishes for small scales. This concept elegantly allows to define wavelet spectra and wavelet cross spectra. For the stationary case, these spectra are closely related to Fourier spectra. Based on this framework, we investigated the variance and bias of continuous wavelet spectral estimation and developed an areawise significance test to overcome multiple testing effects.

For kernel averaging to reduce the variance of a wavelet sample spectrum, the extension of the averaging kernel has to be chosen corresponding to the reproducing kernel on every scale. Otherwise, the variance and bias will change with scale. The reproducing kernel also gives a reasonable minimum length of the averaging kernel for effectively reducing the variance.

Due to multiple testing, the wavelet sample spectrum of uncorrelated noise exhibits typical spurious patches with a size distribution related to the reproducing kernel. Consequently, a pointwise significance test disregarding information about the size of detected patches yields many false positive results. We developed an areawise significance test utilizing information about the area of detected patches to decide whether they contain significant structure. The minimum area of a patch to be areawise significant is determined by the reproducing kernel, smaller patches are in principle indistinguishable from noise. As it sorts out patches small in relation to the reproducing kernel, the areawise test is less sensitive but more specific. Given a typical geophysical data set with a broad spectrum, the conventional test mimics a misleading structure that is successfully uncovered by the areawise test.

However, even though the effect of multiple testing has been dramatically reduced by the areawise test, the outcome is still of a mere statistical nature. As for any statistical result, it is up to the researcher to provide a reasonable interpretation. A wavelet analysis in principle can barely be the end of a study but rather the starting point for a deeper physical understanding.

The presented framework is paradigmatically suitable for nonparametric bootstrapping in the wavelet domain. Aside from the construction of nonstationary surrogate data, this approach allows to perform significance testing with a more complex nonstationary background spectrum. Amongst others, this is important for the analysis of processes with a trend in the variance. The concept might also easily be extended to non-Gaussian noise. These ideas, however, will be subject of future research.

To estimate wavelet spectra and to perform areawise significance tests, we developed a free R-package based on the package Rwave by Carmona et. al. [2]. All wavelet plots in this paper have been realized with this software. It can be downloaded from http://www.agnld.uni-potsdam.de/~maraun/wavelets

References

[1] P. J. Brockwell and R. A. Davis. *Time Series: Theory and Models.* Heidelberg: Springer, 1991.

[2] R. Carmona, W.-L. Hwang, and B Torresani. *Practical Time-Frequency Analysis. Gabor and Wavelet Transforms with an Implementation in S.* Academic Press, 1998.

[3] P.M. Crowley, D. Maraun, and D. Mayes. How hard is the euro area core? A wavelet analysis of growth cycles in Germany, France and Italy. *submitted to Econ. J.*, 2006.

[4] I. Daubechies. *Ten Lectures on Wavelets.* Society for Industrial and Applied Mathematics, 1992.

[5] B.T. Grenfell, O.N. Bjœrnstad, and J. Kappey. Travelling waves and spatial hierarchies in measles epidemics. *Nature*, 414:716–723, 2001.

[6] D. Gu and S.G.H. Philander. Secular changes of annual and interannual variability in the tropics during the past century. *J. Climate*, 8:864–876, 1995.

[7] K. Hasselmann. Stochastic climate models: Part I. Theory. *Tellus*, 28(6):473–485, 1976.

[8] M. Holschneider. *Wavelets: an analysis tool.* Oxford University Press, 1998.

[9] L. Hudgins, C.A. Friehe, and M.E. Mayer. Wavelet Transform and Atmospheric Turbulence. *Phys. Rev. Lett.*, 71(20):3279–3282, 1993.

[10] Ta-Hsin Li. Wavelet Spectrum and Its Characterization Property for Random Processes. *IEEE Trans. Inf. Th.*, 48(11):2922–2937, 2002.

[11] D. Maraun and J. Kurths. Cross wavelet analysis. Significance testing and pitfalls. *Nonlin. Proc. Geoph.*, 11(4):505–514, 2004.

[12] D. Maraun and J. Kurths. Epochs of phase coherence between El Niño/ Southern Oscillation and Indian monsoon. *Geophys. Res. Lett.*, 32:L15709, doi:10.1029/2005GL023225, 2005.

[13] G.V. Osipov, B. Hu, C.S. Zhou, M.V. Ivanchenko, and J. Kurths. Three types of transitions to phase synchronization in chaotic oscillators. *Phys. Rev. Lett.*, 91:024101, 2003.

[14] A.S. Pikovsky, M.G. Rosenblum, and J. Kurths. *Synchronization: A universal concept in nonlinear sciences.* Cambridge Univ. Press, 2001.

[15] M. B. Priestley. *Spectral Analysis and Time Series.* Academic Press, 1992.

[16] M.G. Rosenblum, A.S. Pikovsky, and J. Kurths. Phase synchronization of chaotic oscillators. *Phys. Rev. Lett*, 76(11):1804–1807, 1996.

[17] J. Saynisch, J. Kurths, and D. Maraun. A Conceptual ENSO Modell under Realistic Noise Forcing. *accepted for publication in Nonlin. Proc. Geophys.*, 2006.

[18] J. Timmer and M. König. On generating power law noise. *A&A*, 300(3):707–710, 1995.

[19] C. Torrence and G. Compo. A practical guide to wavelet analysis. *Bull. Amer. Meteor. Soc.*, 79:61–78, 1998.

[20] C. Torrence and P.J. Webster. The annual cycle of persistence in the El Niño Southern Oscillation. *Quart. J. Roy. Met. Soc.*, 124:1985–2004, 1998.

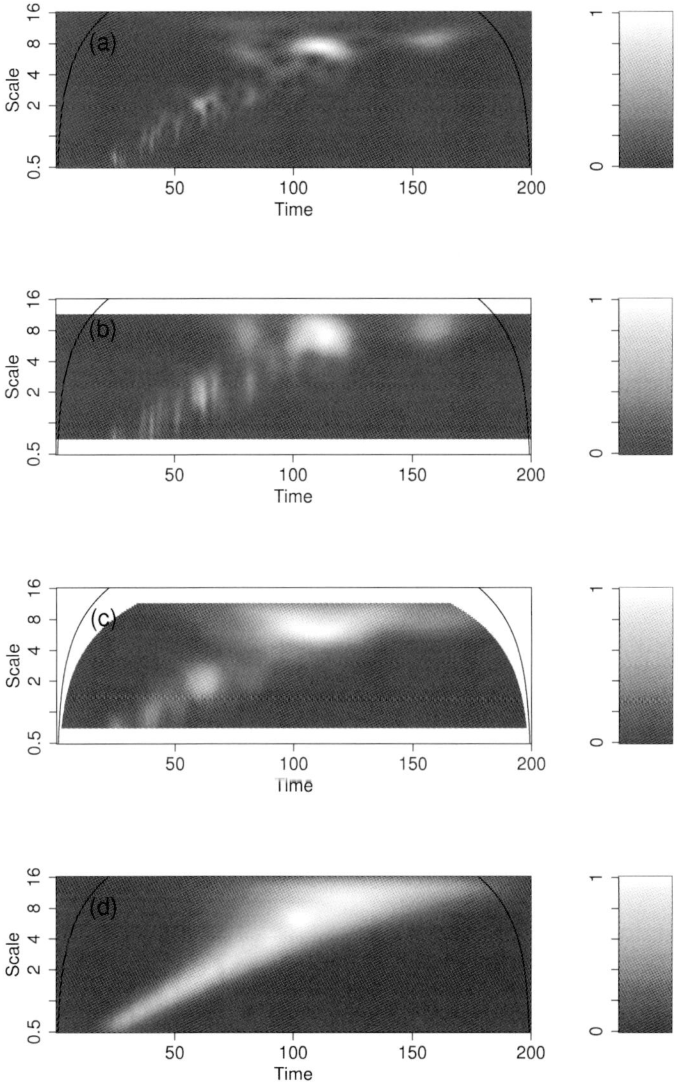

Fig. 12.5. (Taken from [17]) Estimation of the stochastic chirp based on the realization in Fig. 12.2(b). (a) The wavelet scalogram, i.e. the sample spectrum without averaging. (b) Averaged sample spectrum with $w_a/N_{voice} = 0.5$. (c) Averaged sample spectrum with $w_a/N_{voice} = 0.5$ and $w_b = 3$. (d) The spectrum estimated as the mean of 1000 realizations.

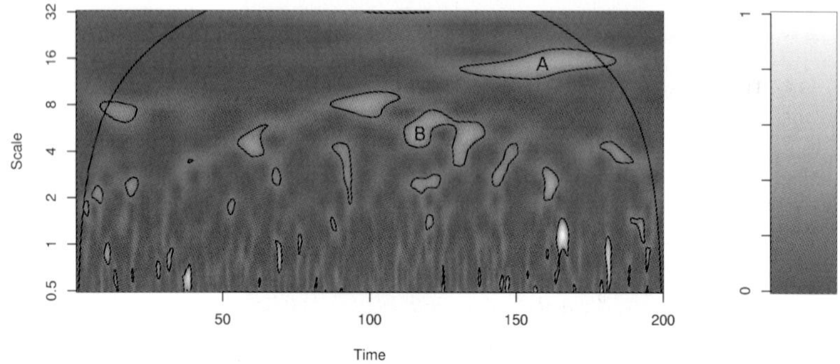

Fig. 12.6. (Taken from [17]) Pointwise significance test of the wavelet sample spectrum of Gaussian white noise (Morlet wavelet, $\omega_0 = 6$, $w_s = 0$) against a white noise background spectrum of equal variance. Spuriously significant patches appear.

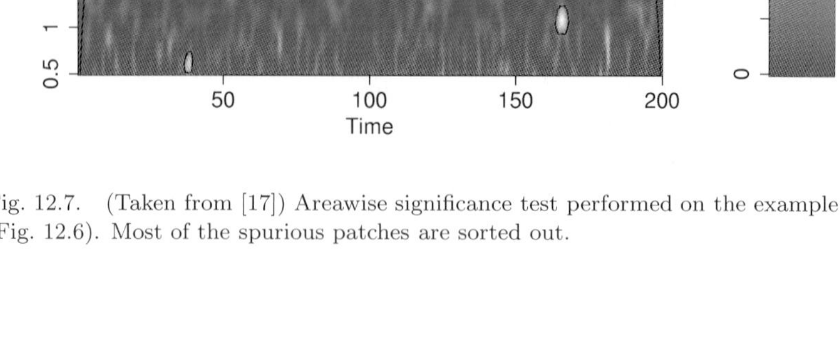

Fig. 12.7. (Taken from [17]) Areawise significance test performed on the example from (Fig. 12.6). Most of the spurious patches are sorted out.

Chapter 13

Synchronization of complex systems: Analysis and control

Michael Rosenblum* and Arkady Pikovsky+

Institute of Physics, University of Potsdam, D-14415 Potsdam, Germany

**mros@agnld.uni-potsdam.de and +pikovsky@stat.physik.uni-potsdam.de*

Contents

13.1. Introduction

Synchronization is a fundamental nonlinear phenomenon. Its essence can be explained quite simply: weakly coupled oscillating objects can adjust their rhythms and start oscillating with common frequencies and tightly related phases. This qualitative description goes back to famous Christiaan Huygens who discovered and experimentally investigated this phenomenon at the beginning of the era of the modern science [23]. Being the oldest scientifically described nonlinear effect, synchronization continues to attract attention of researches from various fields of natural science and engineering. Though the mechanisms leading to synchrony in relatively simple systems like a periodically forced limit cycle oscillator or two interacting periodic oscillators are well understood theoretically and are widely used in practice, the theory of synchronization is still far from being completed.

The foundations of the classical theory has been laid by B. van der Pol and E. Appleton [2, 51] some eighty years ago. Further developments [3, 28] provide a quite complete description of frequency and phase locking of periodical noise-free self-sustained oscillators and rotators. An important contribution to the theory has been done by R. Stratonovich [46] who extended the theory to account for the presence of fluctuations.

In spite of well-developed classical theory, the interest to investigation of synchronization phenomena essentially increased within last two decades and this discipline still remains a field of active research, due to several reasons. First, a discovery and analysis of chaotic dynamics in low-dimensional deterministic systems posed a problem of extension of the theory to cover the case of chaotic oscillators as well. Second, a rapid development of computer technologies made a numerical analysis of complex systems, which still cannot be treated analytically, possible. Finally, a further development of synchronization theory is stimulated by new fields of application in physics (e.g., systems of coupled lasers and Josephson junctions), chemistry (oscillatory reactions), and in biology, where synchronization phenomena play an important role on all levels of organization, from cells to physiological subsystems and even organisms.

One can outline the following main directions of contemporary studies of synchronization phenomena.

(1) **Chaotic systems.** Here the mere notion of synchrony is non-trivial, and several concepts have been developed. The effect of phase synchronization is a direct extension of the classical theory to the case of a subclass of self-sustained continuous time chaotic oscillators which admit a description in terms of phase. Synchronization of these systems can be described as a phase and frequency locking, in analogy to the theory of synchronization of noisy systems. An alternative approach considers a synchronization of arbitrary chaotic systems as a coincidence of their state variables (complete synchronization) or as an onset of a functional relationship between state variables of two unidirectionally coupled systems (generalized synchronization). Although physical mechanisms behind the two latter phenomena essentially differ from the mechanisms of phase and frequency locking, all these effects constitute the field of application of the modern synchronization theory.

(2) **Noise-induces oscillations and excitable systems.** These systems represent a separate class exhibiting synchronization. Without noise, these systems are stable and do not oscillate. In the presence of noise the dynamics is similar to the dynamics of noisy self-

sustained oscillators, and therefore synchronization-like effects are possible. Analysis of these effects and further references can be found, e.g., in [7, 16, 17, 36, 50].

(3) **Oscillators with multiple time scales.** A typical example of multiscale dynamics is neural bursting (see discussion and plots below). These systems have attracted the interest quite recently [29] and their theoretical description is still under development.

(4) **Large systems.** If two oscillators can adjust their rhythms due to an interaction, then one can expect similar behavior in large populations of units. Among different models under investigation we outline regular oscillator lattices (which in the limit case provide a description of extended systems) [9], ensembles of globally (all-to-all) coupled elements, the main topic of the present contribution, random oscillator networks (small world networks, scale-free networks, etc), and spatially-extended systems [10, 44, 55].

(5) **Systems with delays.** Oscillators with internal delays [18], delays in coupling [53], and delayed-feedback control of chaotic and noise-induced oscillations [6, 45] represent another field of actual research.

Different aspects of synchronization theory can be found in a number of monographs [1, 11, 28, 30, 34, 38, 47]. It is important to emphasize that synchronization is an essentially nonlinear effect. In contrast to many classical physical problems, where consideration of nonlinearity gives a correction to a linear theory, here the account of nonlinearity is crucial: the phenomenon occurs only in intrinsically nonlinear, so-called self-sustained systems.

13.2. Simplest case: Periodically forced self-sustained oscillator

13.2.1. *Self-sustained oscillators*

Self-sustained oscillators are mathematical models behind natural oscillating objects, and these models are essentially nonlinear. To be not too abstract, we consider a classical device, that gave birth to synchronization theory, the pendulum clock. Let us discuss how it works. Its mechanism transforms the potential energy of the lifted weight (or compressed spring, or electrical battery) into the oscillatory motion of the pendulum. In its turn, this oscillation is transferred into the rotation of the hands on the clock's face. We are not interested in the particular design of the mechanism; important is only that it takes energy from the source in order to compensate the loss of energy due to dissipation, and in this way maintains a steady oscillation of the pendulum, which continues without any change until the supply of energy expires. The next important property is that the exact form of the oscillatory motion is entirely determined by the internal parameters of the clock and does not depend on how the pendulum was put into motion. Moreover, after being slightly perturbed, following some transient process, the pendulum restores its previous internal rhythm.

In physics such oscillatory objects are denoted as self-sustained oscillators. Mathematically, such an oscillator is described by an autonomous (i.e., without an explicit time dependence) nonlinear dynamical system. It differs both from linear oscillators (which, if a damping is present, can oscillate only due to external forcing) and from nonlinear energy conserving systems, whose dynamics essentially depends on initial state. Dynamics of oscillators is typically described in the phase (state) space. Periodic oscillations, like those of the clock, correspond to a closed attractive curve in the phase space, called the *limit cycle*. The limit cycle is a simple attractor, in contrast to a *strange (chaotic) attractor*. The latter is a geometrical image of *chaotic* self-sustained oscillations.

Examples of self-sustained oscillatory systems are electronic circuits used for the generation of radio-frequency power, lasers, Belousov–Zhabotinsky and other oscillatory chemical reactions, pacemakers (sinoatrial nodes) of human hearts or artificial pacemakers that are used in cardiac pathologies, and many other natural and artificial systems. An outstanding common feature of such systems is their ability to be synchronized.

This ability of periodic self-sustained oscillators is based on the existence of a special variable, phase ϕ. Mathematically, ϕ can be introduced as the variable parameterizing the motion along the stable limit cycle in the state space of an autonomous continuous-time dynamical system. One can always choose phase proportional to the fraction of the period, i.e. in a way that it grows uniformly in time,

$$\frac{d\phi}{dt} = \omega_0 \,, \tag{13.1}$$

where ω_0 is the natural frequency of oscillations. The phase is neutrally stable: it's perturbations neither grow no decay. (In terms of nonlinear dynamics, the neutral stability means that the phase is a variable that corresponds to the zero Lyapunov exponent of the dynamical system.) Thus, already an infinitesimal perturbation (e.g. external periodic forcing or coupling to another system) can cause large deviations of the phase – contrary to the amplitude, which is only slightly perturbed due to the transversal stability of the cycle. The main consequence of this fact is that *the phase can be very easily adjusted by an external action, and as a result the oscillator can be synchronized!*

13.2.2. *Entrainment by external force: An example*

We begin our discussion of synchronization phenomena by considering the simplest case, an entrainment of a self-sustained oscillator by an external periodic force. Before we describe this effect in mathematical terms, we illustrate it by an example. We will again speak about clocks, but this time about biological clocks that regulate daily and seasonal rhythms of living systems – from bacteria to humans.

In 1729 Jean-Jacques Dortous de Mairan, the French astronomer and mathematician, who was later the Secretary of the Académie Royale des Sciences in Paris, reported on his experiments with a haricot bean. He noticed that the leaves of this plant moved up and down in accordance with the change of day into night. Having made this observation, de Mairan put the plant in a dark room and found that the motion of the leaves continued even without variations in the illuminance of the environment. Since that time these and much more complicated experiments have been replicated in different laboratories, and now it is well-known that all biological systems, from rather simple to highly organized ones, have internal biological clocks that provide their "owners" with information on the change between day and night. The origin of these clocks is still a challenging problem, but it is well established that they can adjust their circadian rhythms (from *circa* = about and *dies* = day) to external signals: if the system is completely isolated from the environment and is kept under controlled constant conditions (constant illuminance, temperature, pressure, parameters of electromagnetic fields, etc.), its internal cycle can essentially differ from a 24-hour cycle. Under natural conditions, biological clocks tune their rhythms in accordance with the 24-hour period of the Earth's daily cycle.

Experiments show that for most people the internal period of biological clocks differs from 24 h, but it is entrained by environmental signals, e.g., illuminance, having the period of the Earth's rotation (Fig. 13.1). Obviously, the action here is unidirectional: the revolution of a planet cannot be influenced by mankind (yet); thus, this case constitutes an example of synchronization by an external force. In usual circumstances this force is

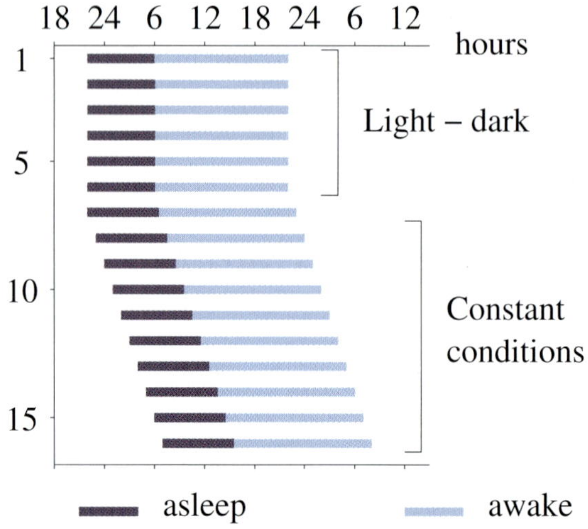

Fig. 13.1. Schematic diagram of the behavioral sleep–wake rhythm. This cycle (termed circadian rhythm) represents the fundamental adaptation of organisms to an environmental stimulus, the daily cycle of light and dark. Here the circadian rhythm is shown entrained for five days by the environmental light–dark cycle and autonomous for the rest of the experiment when the subject is placed under constant light conditions. The intrinsic period of the circadian oscillator is in this particular case greater than 24 hours. Correspondingly, the phase difference between the sleep–wake cycle and daily cycle increases: the internal "day" begins later and later. Such plots are typically observed in experiments with both animals and humans, see, e.g., [4, 15, 33].

strong enough to ensure perfect entrainment; in order to desynchronize a biological clock one can either travel to polar regions or go caving. It is interesting that although normally the period of one's activity is exactly locked to that of the Earth's rotation, the phase shift between the internal clock and the external force varies from person to person: some people say that they are "early birds" whereas others call themselves "owls". Perturbation of the phase shift strongly violates normal activity. Every day many people perform such an experiment by rapidly changing their longitude (e.g., crossing the Atlantic) and experiencing jet lag. It can take up to several days to re-establish a proper phase relation to the force; in the language of nonlinear dynamics one can speak of different lengths of transients leading to the stable synchronous state. As other commonly known examples of synchronization by external force we mention radio-controlled clocks and cardiac pacemakers.

13.2.3. *Phase dynamics of a forced oscillator*

For a mathematical treatment of synchronization we recall that the phase of an oscillator is neutrally stable and can be adjusted by a small action, whereas the amplitude is stable. This property allows a description of the effect of small forcing/coupling within the framework of the phase approximation. Considering the simplest case of a limit cycle oscillator, driven by a periodic force with frequency ω and amplitude ε, we can write the equation for the perturbed phase dynamics in the form

$$\frac{d\phi}{dt} = \omega_0 + \varepsilon Q(\psi, \omega t)\,, \tag{13.2}$$

where the coupling function Q depends on the form of the limit cycle and of the forcing (see [28, 38] for derivation of the phase equation (13.2) from coupled equations for state variables). As the states with the phases ϕ and $\phi + 2\pi$ are physically equivalent, the function Q is 2π-periodic in its both arguments, and therefore can be represented as a double Fourier series. If the frequency of the external force is close to the natural frequency of the oscillator, $\omega \approx \omega_0$, then the series contains fast oscillating and slow varying terms, the latter can be written as $q(\phi - \omega t)$. Introducing the difference between the phases of the oscillation and of the forcing $\psi = \phi - \omega t$ and performing an averaging over the oscillation period we get rid of the oscillating terms and obtain the following basic equation for the phase dynamics:

$$\frac{d\psi}{dt} = -(\omega - \omega_0) + \varepsilon q(\psi)\,. \tag{13.3}$$

Function q is 2π-periodic, and in the simplest case $q(\cdot) = \sin(\cdot)$ Eq. (13.3) is called the Adler equation. One can easily see that on the plane of the parameters of the external forcing (ω, ε) there exist a region $\varepsilon q_{min} < \omega - \omega_0 < \varepsilon q_{max}$, where Eq. (13.3) has a stable stationary solution. This solution corresponds to the conditions of phase locking (the phase ϕ just follows the phase of the force, i.e. $\phi = \omega t + \text{constant}$) and frequency entrainment (the observed frequency of the oscillator $\Omega = \langle\dot{\phi}\rangle$ exactly coincides with the forcing frequency ω; brackets $\langle\rangle$ denote time averaging).

Generally, synchronization is observed for high-order resonances $n\omega \approx m\omega_0$ as well. In this case the dynamics of the generalized phase difference $\psi = m\phi - n\omega t$, where n, m are integers, is described by the equation similar to Eq. (13.3), namely by $d(\psi)/dt = -(n\omega - m\omega_0) + \varepsilon\tilde{q}(\psi)$. Synchronous regime then means perfect entrainment of the oscillator frequency at the rational multiple of the forcing frequency, $\Omega = \frac{n}{m}\omega$, as well as phase locking

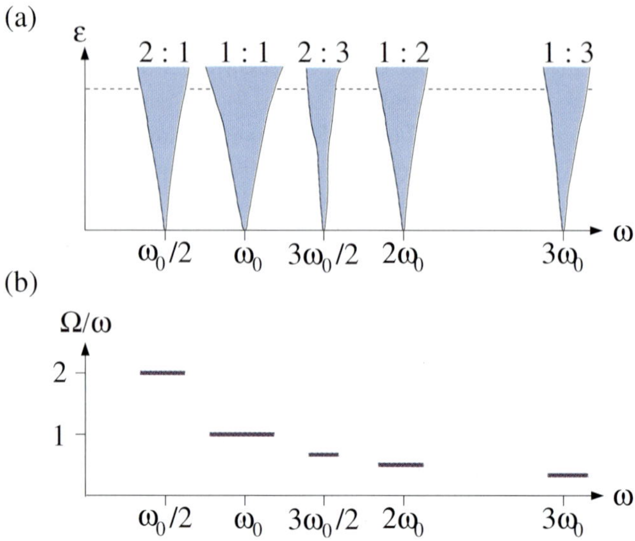

Fig. 13.2. Family of synchronization regions, or Arnold tongues (schematically). The numbers on top of each tongue indicate the order of locking; e.g., 2 : 3 means that the relation $2\omega = 3\Omega$ is fulfilled. (b) The Ω/ω vs. ω plot for a fixed amplitude of the force (shown by the dashed line in (a)) has a characteristic shape, known as the *devil's staircase*. (In this scheme the variation of the frequency ratio between the main plateaus of the staircase is not shown).

$m\phi = n\omega t +$ constant. The overall picture can be shown on the (ω,ε) plane: there exists a family of triangular-shaped synchronization regions touching the ω-axis at the rational multiples of the natural frequency $\frac{m}{n}\omega_0$, these regions are usually called Arnold tongues (Fig. 13.2(a)). This picture is preserved for moderate forcing, although now the shape of the tongues generally differs from being exactly triangular. For a fixed amplitude of the forcing ε and varied driving frequency ω one observes different phase locking intervals where the motion is periodic, whereas in between them it is quasiperiodic. The curve Ω vs. ω thus consists of horizontal plateaus at all possible rational frequency ratios; this fractal curve is called Devil's staircase (Fig. 13.2(b)). A famous example of such a curve is the voltage–current plot for a Josephson junction in an ac electromagnetic field; in this context synchronization plateaus are called Shapiro steps. Note that a junction can be considered as a rotator (rotations are maintained by a dc current); this example demonstrates that synchronization properties of rotators are very close to those of oscillators.

Finally, we note that phase difference in the synchronous state is not necessarily constant, but may oscillate around a constant value. Indeed, a solution $m\phi - n\omega t =$ constant was obtained from Eq. (13.2) by means of averaging, i.e. by neglecting the fast oscillating terms. If we take this terms into account, then we have to reformulate the condition of phase locking as $|m\phi - n\omega t| <$ constant. Thus, in the synchronous regime the phase difference is bounded, otherwise it grows infinitely.

13.3. Globally coupled oscillators

Having considered synchronization of a single oscillator driven by external force, we proceed with a much more complicated case. Now we study synchronization phenomena in large ensembles of oscillators, where each element interacts with all others. This is usually denoted as *global*, or all-to-all coupling. To illustrate this by a representative example, we cite the Dutch physician Engelbert Kaempfer [25][a] who, after his voyage to Siam in 1680 wrote:

> "The glowworms ... represent another shew, which settle on some Trees, like a fiery cloud, with this surprising circumstance, that a whole swarm of these insects, having taken possession of one Tree, and spread themselves over its branches, sometimes hide their Light all at once, and a moment after make it appear again with the utmost regularity and exactness ... ."

This very early observation reports exactly on synchronization in a large population of oscillating systems. The same physical mechanism that makes the insects to keep in synchrony is responsible for an onset of rhythms in neuronal populations or for an emergence of synchronous clapping in a large audience. The latter phenomenon, a self-organization in a large applauding audience, has probably been experienced by everyone, e.g., in a theater. Indeed, if the audience is large enough, then one can often hear a rather fast (several oscillatory periods) transition from noise to a rhythmic, nearly periodic, applause. This happens when the majority of the public applaud in unison, or synchronously.

The each-to-each interaction is also denoted as a *mean field* coupling. Indeed, each firefly is influenced by the light field that is created by the whole population (see Fig. 13.3). Similarly, each applauding person hears the sound that is produced by all other people in the hall. Thus, we can say that all elements are exposed to a common force. This force results from the summation of outputs of all elements. Let us denote these outputs by $x_k(t)$, where $k = 1, \ldots, N$ is the index of an oscillator, and N is

[a]Citation taken from [12].

the number of elements in the ensemble; x can be variation of light intensity or of the acoustic field around some average value, or, generally, any other oscillating quantity. Then the force that drives each oscillator is proportional to $\sum_k x_k(t)$. It is conventional to write this proportionality as $\varepsilon N^{-1} \sum_k x_k(t)$, so that it includes the normalization by the number of oscillators N. The term $N^{-1} \sum_k x_k(t)$ is just an arithmetic mean of all oscillations, what explains the origin of the term "mean field coupling".

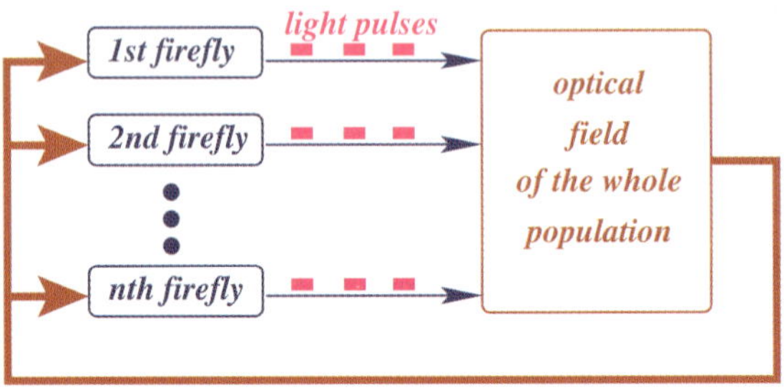

Fig. 13.3. Qualitative explanation of the emergence of collective synchrony in a mean field coupled ensemble.

Thus, the oscillators in a globally coupled ensemble are driven by a common force. Clearly, this force can entrain many oscillators if their frequencies are close. The problem is that this force (the mean field) is not predetermined, but arises from interaction within the ensemble. This force determines whether the systems synchronize, but it itself depends on their oscillation – it is a typical example of self-organization [20]. To explain qualitatively the appearance of this force (or to compute it, as is done in [28, 38]) one should consider this problem self-consistently.

First, assume for the moment that the mean field is zero. Then all the elements in the population oscillate independently, and their contributions to the mean field nearly cancel each other. Even if the frequencies of these oscillations are identical, but their phases are independent, the average of the outputs of all elements of the ensemble is small if compared with the amplitude of a single oscillator. (According to the law of large numbers, it tends to zero when the number of interacting oscillators tends to infinity; the fluctuations of the mean field are of the order $N^{-1/2}$.) Thus, the asynchronous, zero mean field state obeys the self-consistency condition.

Next, to demonstrate that synchronization in the population is also possible, we suppose that the mean field is non-vanishing. Then, naturally, it entrains at least some part of the population, the outputs of these entrained elements sum up coherently, and the mean field is indeed nonzero, as assumed. Which of these two states – synchronous or asynchronous – is realized, or, in other words, which one is stable, depends on the strength of interaction between each pair and on how different the elements are. The interplay between these two factors, the coupling strength and the distribution of the natural frequencies, also determines how many oscillators are synchronized, and, hence, how strong the mean field is.

We discuss now how the synchronization transition occurs, taking the applause in an audience as an example (experimental study of synchronous clapping is reported in [35]). Initially, each person claps with an individual frequency, and the sound they all produce is noisy.[b] As long as this sound is weak, and contains no characteristic frequency, it does not essentially affect the ensemble. Each oscillator has its own frequency ω_k, each person applauds and each firefly flashes with its individual rate, but there always exists some value of it that is preferred by the majority. Definitely, some elements behave in a very individualistic manner, but the main part of the population tends to be "like the neighbor". So, the frequencies ω_k are distributed over some range, and this distribution has a maximum around the most probable frequency. Therefore, there are always at least two oscillators that have very close frequencies and, hence, easily synchronize. As a result, the contribution to the mean field at the frequency of these synchronous oscillations increases. This increased component of the driving force naturally entrains other elements that have close frequencies, this leads to the growth of the synchronized cluster and to a further increase of the component of the mean field at a certain frequency. This process develops (quickly for relaxation oscillators, relatively slow for quasilinear ones), and eventually almost all elements join the majority and oscillate in synchrony, and their common output – the mean field – is not noisy any more, but rhythmic.

The physical mechanism we described is known as the Kuramoto self-synchronization transition [27]. The scenario of this transition does not depend on the origin of the oscillators (biological, electronic, etc.) or on the origin of interaction. In the above presented examples the coupling occurred via an optical or acoustic field. Global coupling of electronic systems

[b]Naturally, the common (mean) acoustic field is nonzero, because each individual oscillation is always positive; the intensity of the sound cannot be negative, it oscillates between zero and some maximal value. Correspondingly, the sum of these oscillations contains some rather large constant component, and it is the deviation from this constant that we consider as the oscillation of the mean field and that is small. Therefore, the applause is perceived as some noise of almost constant intensity.

can be implemented via a common load; in this case the voltage applied to individual systems depends on the sum of the currents of all elements. (As an example we mention an array of the Josephson junctions.) Chemical oscillators can be coupled via a common medium, where concentration of a reagent depends on the reaction in each oscillator and, on the other hand, influences these reactions. The Kuramoto transition can be treated as a nonequilibrium phase transition, the mean oscillating field serving as an order parameter (Fig. 13.4 below).

The scenarios of the Kuramoto transition may be also more complicated, e.g., if the distribution of the individual frequencies ω_k has several maxima. Then several synchronous clusters can be formed; they can eventually merge or coexist. Clustering can also happen if, say, the strength of interaction of an element of the population with its nearest (in space) neighbors is larger than with those that are far away.

In summary, investigation of synchronization in large populations of interacting oscillatory elements is an intensively developing branch of nonlinear science [20, 26, 28, 38, 48, 52, 54], relevant to many problems of physics, chemistry, and life sciences, in particular, to neuroscience. For example, synchronization can occur in arrays of lasers and Josephson junctions, where this phenomenon may play a constructive role for generation of a strong coherent field. In other cases synchronization may be harmful; an illustrative example is the excitation of the left-to-right swaying motion of the London's Millennium Bridge observed at its opening day[c]. This motion appeared due to mutual synchronization of steps of hundreds of pedestrians. To prevent the onset of such synchronization, the bridge has been reconstructed in a way that the damping of its corresponding oscillatory mode was essentially increased. In many cases, when such a direct intervention into the system is not possible, it is nevertheless desirable to control the synchronous motion, in particular, to suppress it, when it appears.

13.3.1. *Globally coupled ensemble as a model of neural synchrony*

An important issue is the collective dynamics of neuronal populations. Indeed, synchronization of individual neurons is believed to play the crucial role in the emergence of pathological rhythmic brain activity in Parkinson's disease, essential tremor and in epilepsies; the detailed discussion of

[c]The corresponding video can be found on the WEB-page of the firm that constructed the bridge: `www.arup.com/MillenniumBridge/`.

this topic and numerous citations can be found in [19, 31, 48]. Obviously, the development of the techniques for suppression of the undesired neural synchrony constitutes an important clinical problem. Technically, this problem can be solved by means of implantation of the micro-electrodes into the impaired part of the brain with subsequent electric stimulation through these electrodes [8, 13]. However, in spite of successful experimental studies followed by clinical applications, the physiological mechanisms of such stimulation remain unclear and the development of effective stimulation techniques is a challenging problem of neuroscience and biological physics. In particular, it is important to minimize the intervention caused by stimulation, e.g. by designing the techniques which allow suppression of the pathological rhythm by weak though precisely timed pulses [48].

The model of globally coupled oscillators is commonly used as a simplest model of neural synchrony. We illustrate this using a computationally efficient neuronal model, proposed by Rulkov [42, 43]. In this model a neuron is described by a 2D map. In spite of its simplicity, this model reproduces most regimes exhibited by the full Hodgkin-Huxley model, but at essentially lower computational costs, thus allowing detailed analysis of the dynamics of large ensembles. The model reads

$$\begin{aligned} x_i(n+1) &= \frac{4.3}{1+x_i^2(n)} + y_i(n) + \varepsilon X(n) + \mathcal{C} \,, \\ y_i(n+1) &= y_i(n) - 0.01(x_i(n)+1) \,, \end{aligned} \tag{13.4}$$

where n is the discrete time, $i = 1, \ldots N$ is the index of a neuron in the population, and

$$X(n) = \frac{1}{N}\sum_{1}^{N} x_i(n)$$

is the mean field. The term $\mathcal{C}$, describing an additional control of the ensemble, will be specified below.

Global coupling in an ensemble is a theoretically convenient approximation. Here we numerically explore how good this approximation works, if the neurons are in fact coupled in a more complex, random, way. There are many possibilities to model a randomly coupled network. We consider that each of $N = 10000$ neurons has N_l links, i.e., it is coupled to N_l randomly chosen elements of the population. The coupling strength ε within each pair is taken to be the same. In Fig. 13.4 we plot the dependence of the variance of the mean field var(x) on ε for different number of links N_l.

The synchronization transition is similar to the case of the global coupling, though it takes place for higher values of ε. We see that the bifurcation curve for $N_l = 50$ practically coincide with that for the global coupling case. Thus, the dynamics of an ensemble with random coupling to N_l neighbors is qualitatively close to the dynamics of the globally coupled population.

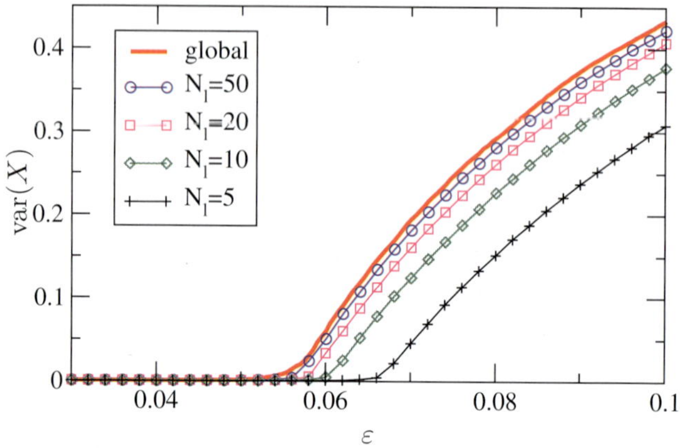

Fig. 13.4. Comparison of synchronization transition in ensembles with all-to-all coupling and random coupling, when each unit is coupled to N_l units. Bold line shows the transition for the case of all-to-all coupling. It is seen that the dynamics of the ensemble with global coupling is very close to the case of an ensemble with $N_l = 50$. For smaller number of links, the dynamics is qualitatively similar, though the synchronization transition happens for larger coupling ε.

13.4. Controlling neural synchrony

A particular problem of high practical importance is to develop techniques for *control* of collective neuronal activity. The importance of this task is related to the hypothesis that pathological brain rhythms, observed e.g. in patients with Parkinson's disease and epilepsies, appear due to synchrony in many thousands of neurons. Correspondingly, suppression of undesirable collective synchrony in a population of neurons is a challenging problem of neuroscience. Its solution may essentially improve the so-called *Deep Brain Stimulation (DBS)* technique, currently used in medical practice to suppress Parkinsonian and essential tremor [8]. This technique implies implantation of microelectrodes deep into the brain of a patient, either to subthalamic nucleus or to globus pallidus, and continuous stimulation of this target by a high frequency (about 100 or 120 Hz) periodic pulse train; the stimuli are delivered by a controller implanted into the chest. Important, that in spite of rather broad usage of DBS[d] the neurophysiological mechanisms of such stimulation are poorly (if at all) understood, and therefore its parameters are chosen by trial and error. Most likely, high frequency DBS mimics lesion of the tissue by quenching the firing of neurons. This fact, as well as the necessity to apply the (rather strong) stimulation continuously, calls for development of more efficient suppression techniques.

Development of model based DBS techniques exploiting the methods of nonlinear dynamics and statistical physics was pioneered by P. A. Tass, who proposed a number of approaches. The main idea of these approaches is that suppression of the pathological rhythm should be achieved in such a way that (i) activity of individual units is not suppressed, but only their firing becomes asynchronous, and (ii) the stimulation should be minimized, e.g., it is desirable to switch it off as soon as the synchrony is suppressed (see [48, 49] and references therein). Following these ideas we suggested in our previous publications [40, 41] a delayed feedback suppression control scheme (Fig. 13.5), cf. delayed and non-delayed techniques for stabilization of low-dimensional systems [5, 22, 39] and for control of noise-induced motion [24]. In our approach it is assumed that the collective activity of many neurons is reflected in the local field potential (LFP) which can be registered by an extracellular microelectrode. Delayed and amplified LFP signal can be fed back into the systems via the second or same electrode (see [37] and references therein for a description of one electrode measurement – stimulation setup.) Numerical simulation as well as analytical analysis of the delayed feedback control demonstrate that it indeed can be exploited for suppression of the collective synchrony.

For an introduction of the delayed feedback control of collective synchrony we consider suppression of the mean field in an ensemble of $N = 10000$ identical Hindmarsh-Rose neurons (Eqs. (13.5)) in the regime of chaotic bursting. The dynamics of the ensemble is described by the following set of equations,

$$\begin{aligned} \dot{x}_i &= y_i - x_i^3 + 3x_i^2 - z_i + 3 + KX + K_f(X(t-T) - X(t)) \,, \\ \dot{y}_i &= 1 - 5x_i^2 - y_i \,, \\ \dot{z}_i &= 0.006 \cdot (4(x_i + 1.56) - z_i) \,, \end{aligned} \tag{13.5}$$

where $X = N^{-1} \sum_{i=1}^{N} x_i$ is the mean field, and the terms KX and $K_f(X(t-T) - X(t)) = \mathcal{C}$ describe the global coupling and the feedback control,

[d]For example, one of the producers of DBS controllers, the Medtronic Inc, reports on over 20 thousands of patients using their devices.

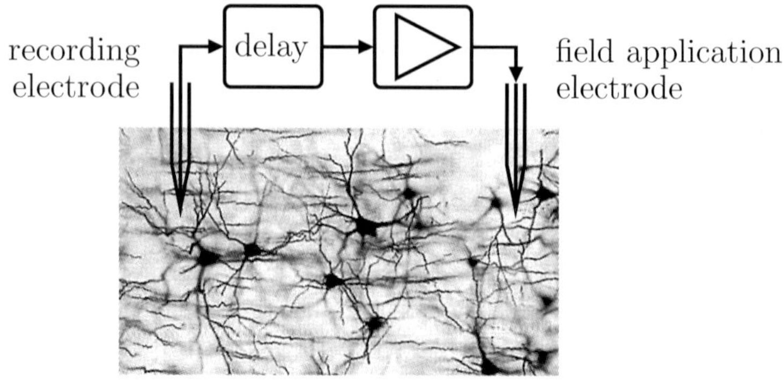

Fig. 13.5. Scheme of the suggested approach to deep brain stimulation [40, 41]. The local electrical field in a neuronal population should be measured by the recording electrode and fed back via the field application electrode. The device should contain a delay line and an amplifier.

respectively.

The results are presented in the Fig. 13.6 for the strength of the internal coupling $K = 0.08$. The feedback control was switched on at $t_0 = 5000$, i.e. $K_f = 0$ for $t < t_0$ and $K_f = 0.036$ for $t \geq t_0$; the delay time is $\mathcal{T} = 72.5$. Here the panels (a) and (b) show the mean field and the control signal, respectively. It is clearly seen that switching on of the feedback results in a quick suppression of the mean field in the ensemble, so that only small noise-like fluctuations remain (we remind, that the mean field models here the pathological brain activity). Important is that the control signal decays rapidly and then the asynchronous state of the ensemble is maintained by feeding back a very weak signal. Another important feature of the technique is illustrated in the panels (c) and (d) where we show the bursting dynamics of two neurons before (c) and after (d) the feedback was switched on. One can see that the dynamics of individual units barely change, however they burst incoherently and therefore produce no macroscopic oscillation. Thus, the feedback control suppresses the collective synchrony in the ensemble without suppressing the firing of individual neurons and maintains such state with a weak intervention. Simulations with the Rulkov models (13.4) [40] demonstrate that the variance of the field fluctuations in the suppressed state scale decrease with the ensemble size as $1/\sqrt{N}$. This means that for very large ensembles the control signal tends to zero, and therefore the control is *non-invasive*.

A theoretical analysis of the delay controlled ensemble dynamics is based

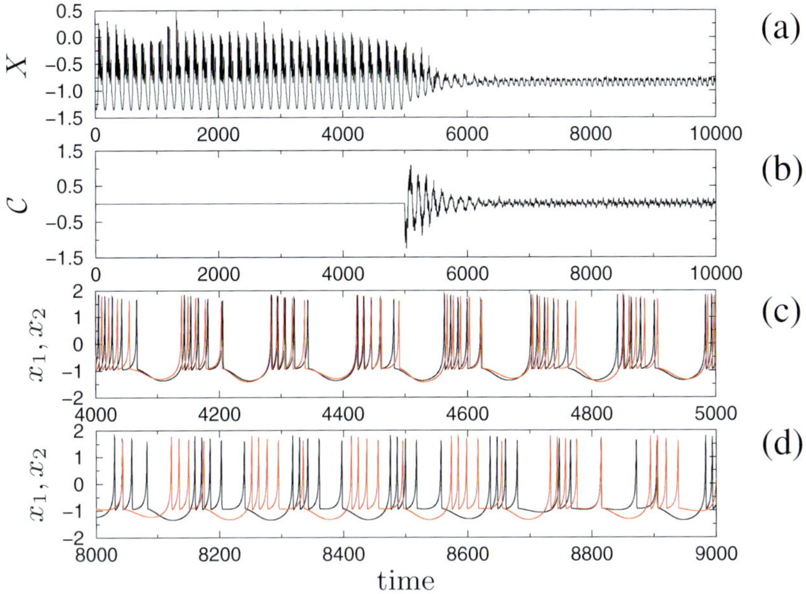

Fig. 13.6. (a) Suppression of the mean field X in the ensemble of 10000 Hindmarsh-Rose neurons (Eqs. (13.5)). The delayed feedback is switched on at $t = 5000$. (b) The control signal $\mathcal{C} = \varepsilon_f(X(t - \mathcal{T}) - X(t))$ quickly decays to the noise level and the desired, asynchronous, state of the system is maintained with a minimal intervention. (c,d) Synchronous and asynchronous bursting of two neurons in the absence and in the presence of the feedback, respectively.

on the assumption that emergence of collective synchrony can be understood as the Kuramoto transition [27, 28, 32, 52]. The analytical treatment of the problem can be performed only for an idealized model under assumption of (i) global (each-to-each) interaction, (ii) infinitely large ensemble size, and (iii) weak coupling (i.e., in the phase approximation) [40]. Another approach is based on a quite general consideration of the Kuramoto synchronization transition in an ensemble as the Hopf bifurcation for the mean field, and on the analysis of the corresponding model amplitude equation (normal form) for the dynamics of the complex mean field A [14, 28]. With the account of a delayed feedback loop, this equation takes the form [41]

$$\dot{A} = (\varepsilon - \varepsilon_{cr} + i\omega)A + \varepsilon_f e^{-i\alpha}\mathcal{C} - \zeta|A|^2 A \,. \tag{13.6}$$

Here the factor ε describes the internal global coupling in the ensemble; if it exceeds the critical value ε_{cr} then synchrony sets in in the population and macroscopic mean field appears. ε_f describes the strength of the feedback and τ is the delay; the phase shift α depends on the properties of individual oscillators and on how the feedback term appears in their equations. The term $\mathcal{C}$ has different form for different feedback schemes, discussed and compared below. Qualitatively, the effect of the delayed feedback can be understood in the following way: for proper feedback parameters ε_f, τ the delayed term compensates the instability due to internal coupling (described by the increment $\varepsilon - \varepsilon_{cr}$) and thus changes the linear stability of the system.

A theoretical analysis of Eq. (13.6) provides the domains of control, i.e. the ranges of delay time and amplification in the feedback loop for which the control is effective. These results are in a good correspondence with the numerical simulation of the ensemble dynamics with different neuron models used for the description of individual units (Bonhoefer - van der Pol or Hindmarsh-Rose equations [21], Rulkov map model [42]).

We have used two types of feedback: direct control $\mathcal{C} = X(t-\tau)$ and differential control $\mathcal{C} = (X(t-\tau) - X(t))$. The parameters in Eqs. (13.4) are chosen in such a way that individual units are in the regime of chaotic bursting. The efficiency of suppression is quantified by the suppression factor

$$S = (\text{var}(X)/\text{var}(X_f))^{1/2} . \tag{13.7}$$

High values of $S = S(\tau, \varepsilon_f)$ (Fig. 13.7) provide the domains of control. Certainly, the suppression in the system (13.4) is not perfect: the suppression factor remains finite due to finite-size effects, discussed below. In other words, the suppressed mean field is not zero, but exhibits some irregular fluctuations due to a finite number of the elements in the ensemble. Correspondingly, the borders of the control domains can be determined only approximately, by setting some cut-off level. However, the shape and position of domains of control are in a good correspondence with the results of the theoretical analysis of Eq. (13.6), which describes the idealized case of infinitely large population size.

Figure 13.7 demonstrates that suppression of the synchrony can be achieved both by the direct and differential control schemes. This example also shows that differential control provides less number of the domains of control, but these domains are generally larger and the suppression factor is higher. Another advantage of the differential control scheme is that it provides noninvasive suppression, in the sense that $\lim_{N\to 0} \mathcal{C} = 0$, whereas for the direct control suppression is generally invasive, $\lim_{N\to 0} \mathcal{C} = \text{const}$.

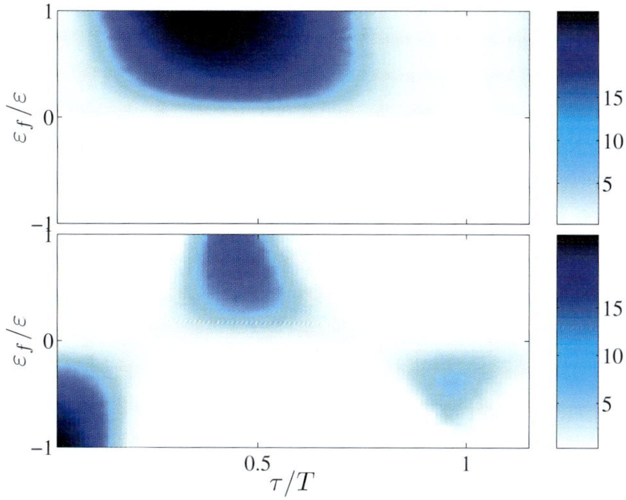

Fig. 13.7. Domains of control for the system (13.4) in the parameter plane delay – feedback strength, for differential (top panel) and direct (bottom panel) schemes. Color codes the coefficient of suppression (see text). Internal coupling is $\varepsilon = 0.06$.

Now we discuss a possibility of control of randomly coupled oscillators, Fig. 13.4. The delayed feedback control (direct scheme with $\tau = 30$ was used here) shifts the bifurcation curves (Fig. 13.8), thus allowing one to control the transition, as in case of global coupling.[e] Indeed, a shift of the bifurcation curve to the right means decrease of the mean field variance for given value of the internal coupling in the ensemble, thus providing suppression up to the level of noise fluctuations in the system if the bifurcation value becomes larger than the internal coupling. This is supported by the computation of the control domains (the results are similar to those shown Fig. 13.7 and therefore are not given here).

Black curve in Fig. 13.8 illustrates the case when the feedback signal is not the mean field of the whole population but the local field, acting on one neuron. It is, $\mathcal{C} = \varepsilon_f N_l^{-1} \sum_l x_l$, where index $l = 1, \ldots, N_l$ numerates the neurons linked to the chosen one. This example demonstrates that such an imperfect measurement however provides suppression, though it is not so effective as the control via mean field X (magenta curve). Summarizing the results presented in Fig. 13.8, we conclude that mean field approximation works very well even for moderate connectivity of the network (50 links in

[e]Note that sufficiently strong feedback, $\varepsilon_f = 0.04$, changes the type of the bifurcation; now the transition exhibits hysteresis, see blue curve in Fig. 13.8.

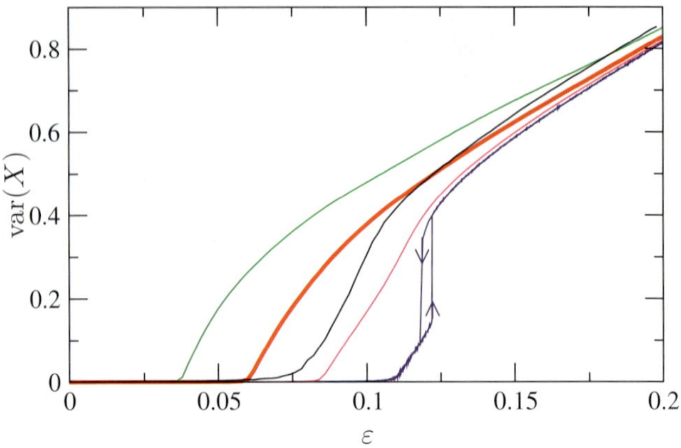

Fig. 13.8. Control of a randomly coupled ensemble of Rulkov neurons, $N_l = 10$. Red curve illustrates the transition in the uncontrolled system. Green, magenta, and blue curves correspond to the case when the mean field of the whole population has been used for the control, with the feedback strength $\varepsilon_f = -0.02$, $\varepsilon_f = 0.02$, and $\varepsilon_f = 0.04$, respectively. The black curve corresponds to the feedback with the local field measured at a randomly chosen cite, $\varepsilon_f = 0.02$; this field is computed as an average over the $N_l = 10$ units coupled to this cite.

an network of 10000 elements in our example).

13.5. Conclusion

In this chapter we described general properties of synchronization of coupled oscillators. Remarkably, many properties of coupled complex systems of different nature can be described in a framework of a simple model of coupled phase oscillators. Moreover, it appears to be possible to control synchrony in coupled ensembles. Different control schemes, as well as experimental implementations of the control, certainly deserve an thorough investigations in the future.

References

[1] V. Anishchenko, A. Neiman, V. Astakhov, T. Vadiavasova, and L. Schimansky-Geier. *Chaotic and Stochastic Processes in Dynamic Systems.* Springer Verlag, Berlin, 2002.
[2] E. V. Appleton. The automatic synchronization of triode oscillator. *Proc. Cambridge Phil. Soc. (Math. and Phys. Sci.)*, 21:231–248, 1922.
[3] V. I. Arnold. Small denominators. I. Mappings of the circumference onto

itself. *Izv. Akad. Nauk Ser. Mat.*, 25(1):21–86, 1961. (In Russian); English Translation: AMS Transl. Ser. 2, v. 46, 213-284.

[4] J. Aschoff, S. Daan, and G. A. Groos. *Vertebrate Circadian Systems. Structure and Physiology.* Springer, Berlin, 1982.

[5] F. M. Atay. Delayed-feedback control of oscillations in non-linear planar systems. *Int. J. Control*, 75:297–304, 2002.

[6] A. G. Balanov, N. B. Janson, and E. Schöll. Period delayed feedback control of chaos: Bifurcation analysis. *Phys. Rev. E*, 71:016222 (2005).

[7] V. Beato, I. Sendina-Nadal, I. Gerdes, and H. Engel. Noise-induced wave nucleations in an excitable chemical reaction. *Phys. Rev. E*, 71:035204 (2005).

[8] A. L. Benabid, P. Pollak, C. Gervason, D. Hoffmann, D. M. Gao, M. Hommel, J.E. Perret, and J. De Rougemont. Long-term suppression of tremor by chronic stimulation of the ventral intermediate thalamic nucleus. *Lancet*, 337:403–406, 1991.

[9] B. Blasius, A. Huppert, and L. Stone. Complex dynamics and phase synchronization in spatially extended ecological systems. *Nature*, 399:354 (1999).

[10] B. Blasius and R. Tönjes. Quasiregular concentric waves in heterogeneous lattices of coupled oscillators. *Phys. Rev. Lett.*, 95:084101 (2005).

[11] I. I. Blekhman. *Synchronization in Science and Technology.* Nauka, Moscow, 1981. (In Russian); English translation: 1988, ASME Press, New York.

[12] J. Buck and E. Buck. Mechanism of rhythmic synchronous flashing of fireflies. *Science*, 159:1319–1327, 1968.

[13] S. A. Chkhenkeli. Direct deep brain stimulation: First steps towards the feedback control of seizures. In J. Milton and P. Jung, editors, *Epilepsy as a Dynamic Disease*, pages 249–261. Springer, Berlin, 2003.

[14] J. D. Crawford. Amplitude expansions for instabilities in populations of globally-coupled oscillators. *J. Stat. Phys.*, 74(5/6):1047–1084, 1994.

[15] C. A. Czeisler, J. S. Allan, S. H. Strogatz, J. M. Ronda, R. Sánches, C. D. Ríos, W. O. Freitag, G. S. Richardson, and R. E. Kronauer. Bright light resets the human circadian pacemaker independent of the timing of the sleep-wake cycle. *Science*, 233:667–671, 1986.

[16] U. Erdmann, W. Ebeling, and A. S. Mikhailov. Noise-induced transition from translational to rotational motion of swarms. *Phys. Rev. E*, 71:051904 (2005).

[17] A. Ganopolski and S. Rahmstorf. Abrupt Glacial Climate Changes due to Stochastic Resonance. *Phys. Rev. Lett.*, 88:038501 (2002).

[18] D. Goldobin, M. Rosenblum, and A. Pikovsky. Controlling oscillator coherence by delayed feedback. *Phys. Rev. E*, 67(6):061119, 2003.

[19] D. Golomb, D. Hansel, and G. Mato. Mechanisms of synchrony of neural activity in large networks. In F. Moss and S. Gielen, editors, *Neuro-informatics and Neural Modeling*, volume 4 of *Handbook of Biological Physics*, pages 887–968. Elsevier, Amsterdam, 2001.

[20] H. Haken. *Advanced Synergetics: Instability Hierarchies of Self-Organizing Systems.* Springer, Berlin, 1993.

[21] J. L. Hindmarsh and R. M. Rose. A model for neuronal bursting using three coupled first order differential equations. *Proc. Roy. Soc. London Ser. B*,

221:87, 1984.

[22] P. Hövel and E. Schöll. Control of unstable steady states by time-delayed feedback methods. *Phys. Rev. E*, 72:046203, 2005.

[23] Ch. Huygens (Hugenii). *Horologium Oscillatorium.* Apud F. Muguet, Parisiis, France, 1673. English translation: *The Pendulum Clock*, Iowa State University Press, Ames, 1986.

[24] N. B. Janson, A. G. Balanov, and E. Schöll. Delayed Feedback as a Means of Control of Noise-Induced Motion. *Phys. Rev. Lett.*, 93:010601 (2004).

[25] E. Kaempfer. *The History of Japan (With a Description of the Kingdom of Siam).* Sloane, London, 1727. Posthumous translation; or reprint by McLehose, Glasgow, 1906.

[26] H. Kori and A. S. Mikhailov. Entrainment of Randomly Coupled Oscillator Networks by a Pacemaker. *Phys. Rev. Lett.*, 93:254101 (2004).

[27] Y. Kuramoto. Self-entrainment of a population of coupled nonlinear oscillators. In H. Araki, editor, *International Symposium on Mathematical Problems in Theoretical Physics*, page 420, New York, 1975. Springer Lecture Notes Phys., v. 39.

[28] Y. Kuramoto. *Chemical Oscillations, Waves and Turbulence.* Springer, Berlin, 1984.

[29] P. S. Landa, M. Rosenblum, and A. Pikovsky. Synchronization of multi-scale oscillatory systems. in preparation, 2006.

[30] S. C. Manrubia, A. S. Mikhailov, and D. H. Zanette. *Emergence of Dynamical Order: Synchronization Phenomena in Complex Systems*, volume 2 of *Lecture Notes in Complex Systems.* World Scientific, Singapore, 2004.

[31] J. Milton and P. Jung, editors. *Epilepsy as a Dynamic Disease.* Springer, Berlin, 2003.

[32] R. Mirollo and S. Strogatz. Synchronization of pulse-coupled biological oscillators. *SIAM J. Appl. Math.*, 50:1645–1662, 1990.

[33] R. Y. Moore. A clock for the ages. *Science*, 284:2102–2103, 1999.

[34] E. Mosekilde, Yu. Maistrenko, and D. Postnov. *Chaotic Synchronization. Applications To Living Systems.* World Scientific, Singapore, 2002.

[35] Z. Néda, E. Ravasz, Y. Brechet, T. Vicsek, and A.-L. Barabási. Tumultuous applause can transform itself into waves of synchronized clapping. *Nature*, 403(6772):849–850, 2000.

[36] A. B. Neiman and D. F. Russell. Synchronization of noise-induced bursts in noncoupled sensory neurons. *Phys. Rev. Lett.*, 88(6):138103, 2002.

[37] I. Ozden, S. Venkataramani, M. A. Long, B. W. Connors, and A. V. Nurmikko. Strong coupling of nonlinear electronic and biological oscillators: Reaching the "amplitude death" regime. *Phys. Rev. Lett.*, 93:158102, 2004.

[38] A. Pikovsky, M. Rosenblum, and J. Kurths. *Synchronization. A Universal Concept in Nonlinear Sciences.* Cambridge University Press, Cambridge, 2001.

[39] K. Pyragas, V. Pyragas, I. Z. Kiss, and J. L. Hudson. Adaptive control of unknown unstable steady states of dynamical systems. *Phys. Rev. E.*, 92:026215, 2004.

[40] M. G. Rosenblum and A. S. Pikovsky. Controlling synchrony in ensemble of

globally coupled oscillators. *Phys. Rev. Lett.*, 92:114102, 2004.

[41] M. G. Rosenblum and A. S. Pikovsky. Delayed feedback control of collective synchrony: An approach to suppression of pathological brain rhythms. *Phys. Rev. E.*, 70:041904, 2004.

[42] N. F. Rulkov. Regularization of synchronized chaotic bursts. *Phys. Rev. Lett.*, 86(1):183–186, 2001.

[43] N. F. Rulkov. Modeling of spiking-bursting neural behavior using two-dimensional map. *Phys. Rev. E.*, 65:041922, 2002.

[44] O. Rudzick and A. S. Mikhailov. Front Reversals, Wave Traps, and Twisted Spirals in Periodically Forced Oscillatory Media. *Phys. Rev. Lett.*, 96:018302 (2006).

[45] G. Stegemann, A. G. Balanov, and E. Schöll Delayed feedback control of stochastic spatiotemporal dynamics in a resonant tunneling diode. *Phys. Rev. E*, 73:016203 (2006).

[46] R. L. Stratonovich. *Topics in the Theory of Random Noise.* Gordon and Breach, New York, 1963.

[47] S. H. Strogatz. *Sync: The Emerging Science of Spontaneous Order.* Hyperion, NY, 2003.

[48] P. A. Tass. *Phase Resetting in Medicine and Biology. Stochastic Modelling and Data Analysis.* Springer-Verlag, Berlin, 1999.

[49] P. A. Tass. A model of desynchronizing deep brain stimulation with a demand-controlled coordinated reset of neural subpopulations. *Biological Cybernetics*, 89:81–88, 2003.

[50] O. V. Ushakov, H.-J. Wünsche, F. Henneberger, I. A. Khovanov, L. Schimansky-Geier, and M. A. Zaks. Coherence Resonance Near a Hopf Bifurcation *Phys. Rev. Lett.*, 95:123903 (2005).

[51] B. van der Pol. Forced oscillators in a circuit with non-linear resistance. (Reception with reactive triode). *Phil. Mag.*, 3:64–80, 1927.

[52] A. T. Winfree. *The Geometry of Biological Time.* Springer, Berlin, 1980.

[53] H.-J. Wünsche. Synchronization of delay-coupled oscillators: A study of semiconductor lasers. *Phys. Rev. Lett.*, 94:163901, 2005.

[54] M. A. Zaks, A. B. Neiman, S. Feistel, and L. Schimansky-Geier. Noise-controlled oscillations and their bifurcations in coupled phase oscillators *Phys. Rev. E*, 68:066206 (2003).

[55] V. S. Zykov, G. Bordiougov, H. Brandtstädter, I. Gerdes, and H. Engel. Global Control of Spiral Wave Dynamics in an Excitable Domain of Circular and Elliptical Shape. *Phys. Rev. Lett.*, 92:018304 (2004).

Chapter 14

Critical states of seismicity – Implications from a physical model for the seismic cycle

Gert Zöller[1,*], Matthias Holschneider[2,+] and Jürgen Kurths[1,!]
Nonlinear Dynamics Group, Institute of Physics, University of Potsdam, D-14415 Potsdam, Germany
[2] *Mathematics Institute, University of Potsdam, D-14415 Potsdam, Germany*
*gert.zoeller@uni-potsdam.de, +hols@math.uni-potsdam.de and !jkurths@agnld.uni-potsdam.de

Contents

14.1. Introduction

In 2003, the geophysicist Vladimir Keilis-Borok, director of the *International Institute of Earthquake Prediction Theory and Mathematical Geophysics* in Moscow issued an alarm for an upcoming earthquake of magnitude 6.4 or greater within a 12.440 square miles area of southern California that includes portions of the eastern Mojave Desert, Coachella Valley, Imperial Valley (San Bernardino, Riverside and Imperial Counties) and eastern San Diego County, during a time interval of nine months (January 5 - September 5, 2004). This prediction was based on previous observations of microearthquake patterns forming chains. Keilis-Borok and co-workers claimed to have predicted two earthquakes correctly by means of such chains – one in Hokkaido, Japan in September 2003 and the second in San Simeon, California in December 2003. However, the deadline of the recent forecast passed and no earthquake fitting the alarm occurred.

Apart from the social and the economic dimension, this failed prediction raises also basic scientific questions in earth sciences: Is a prediction of earthquakes solely based on the emergence of seismicity patterns reliable? In other words, is there a "magic parameter", which becomes anomalous prior to a large earthquake? Is it necessary that such a parameter is based on a physical model? Are pure observational methods without specific physical understanding, like the pattern recognition approach of Keilis-Borok, also sufficient? Taking into account that earthquakes are monitored continuously only since about 100 years and the best available data sets ("earthquake catalogs") cover only a few decades, it seems questionable to forecast earthquakes solely on the basis of observed seismicity patterns, because large earthquakes have recurrence periods of decades to centuries; consequently, data sets for a certain region include not more than ten large events making a reliable statistical testing questionable.

The relation between frequency and magnitude of earthquakes in a large seismically active region is given by the empirical Gutenberg-Richter law [20]

$$\log N = bM - a, \tag{14.1}$$

where N is the frequency of earthquakes with magnitude equal to or greater than M; a is a measure of the overall seismicity level in a region and the slope b is the Richter b value, which determines the relation between large and small earthquakes.

A key problem is the evaluation of the relevance of observed seismicity patterns. First, it is important to decide whether an observed pattern has a physical origin or is an artifact, arising for example from inhomogeneous reporting or from man-made seismicity, like quarry blasts or explosions. Second, the non-artifical events have to be analyzed with respect to their underlying mechanisms. This leads to an inverse problem with a non-unique solution, which can be illustrated for the most pronounced observed seismicity pattern, the occurrence of aftershocks. It is empirically known that the earthquake rate $\dot{N}$ after a large event at time t_M follows the modified Omori law [37, 50]

$$\dot{N} = \frac{c_1}{(c_2 + t - t_M)^p}, \tag{14.2}$$

where t is the time, c_1 and c_2 are constants, and the Omori exponent p is close to unity. In particular, aftershocks are an almost universal phenomenon; that is, they are observed nearly after each mainshock. The underlying mechanisms leading to aftershocks are, however, unknown. Various physical models have been designed in order to explain aftershock occurrence following Eq. (14.2). These models assume physical mechanisms including viscoelasticity [22], pore fluid flow [34], damage rheology [7, 44], and special friction laws [18]. The question, which mechanism is realistic in a certain fault zone, remains open. Detailed comparisons of observed and modeled seismicity with respect to the aftershock rate, the duration of aftershock sequences, the dependence on the mainshock size, and other features are necessary to address this problem. Additionally, the results from lab experiments on rupture dynamics, and satellite observations in fault zones, provide important constraints for the evaluation of such models.

Apart from aftershock activity, other seismicity patterns are well-known from observations, e.g. foreshocks [28], seismic quiescence [24, 52, 59], and accelerating seismic moment release [12, 27]. These patterns have been documented in several cases before large earthquakes. They occur, however, less frequent than aftershocks. For example, foreshocks are known to preceed only 20-30% of large earthquakes [51]. Therefore, their predictive power is questionable. Moreover, it is not clear whether or not these findings can be attributed to physical processes or to random fluctuations in the highly noisy earthquake catalogs. This problem can be addressed by using conceptual fault models which allow to simulate long earthquake sequences over at least 1000 years. If the models are to some extent physical, the occurrence of seismicity patterns can be studied with reasonable statistics. The main ingredients of such models are the geometry of a fault region, empirically known friction laws, quenched spatial heterogeneities, and stress and displacement functions in accordance to dislocation theory [15, 35]. In order to allow for detailed studies of the relations between the imposed mechanisms and the observed seismicity functions, it is important that the number of adjustable parameters is limited. It is emphasized that these models do not aim to reproduce an observed earthquake catalog in detail. Instead, the main goal is to address questions like: Why is the Parkfield segment of the San Andreas fault characterized by relatively regular occurrence of earthquakes with magnitude $M \approx 6$, while on the San Jacinto fault in California the properties of earthquake occurrence are more irregular?

Conceptual models for seismicity are mainly based on one or more solid blocks, which are driven by a plate over a rough surface. The plate and the blocks are connected with springs. This model can produce stick-slip motion of the blocks, where a slip event is considered to simulate an earthquake. The model setup allows to govern a wide range of complexity, beginning with a single-block model which produces periodic occurrence of earthquakes of uniform size, ending with a network of connected blocks leading to complex sequences of earthquakes with variable size. The latter model has been proposed by Burridge and Knopoff [13]. In order to reduce the computational effort, i.e. solving coupled differential equations, the use of cellular automata became popular [30, 36]. Mathematically, these models include maps instead of differential equations; physically, this corresponds to instantaneously occurring slip events, neglecting inertia effects. The main ingredients of such models are (1) external driving (plate motion), and (2) sudden change of system parameters (stress), when a critical value (material strength) is reached, followed by an avalanche of block slips (stress drop and co-seismic stress transfer during an earthquake). While the first process lasts for years to several decades, the second occurs on a time scale of a few seconds. The simplest model including these feature has been formulated by Reid [39] and is known as *Reid's elastic rebound theory*; in terms of spring-block models, this corresponds to a single-block model with constant plate velocity. Accounting for spatial heterogeneity and fault segmentation, many interacting blocks, or fault segments, have to be considered. This leads to a spatiotemporal stress field instead of a single stress value. In general, the material strength will also become space-dependent. Such a model framework can be treated with the methodology of statistical physics similar to the Ising model or percolation models [32]. In this context, large earthquakes are associated with second-order phase transitions [45]. The view of earthquakes as phase transitions in a system with many degrees of freedom and an underlying critical point, is hereinafter referred to as the "critical point concept". It is interesting to note that the period before such a phase transition is characterized by a preparation process, or a "critical state", e.g. in terms of growing spatial correlation length following a power law [9]. However, depending on the parameters of a model, different scenarios are conceivable: the system trajectory can enter the critical state and the critical point frequently ("supercritical") or it becomes never critical ("subcritical"). A case of special interest is the class of models showing *self-organized criticality* (SOC) [2], which have their origin in a simple cellular automaton model for a sandpile [3]. Here, the system drives itself permanently in the vicinity of the critical point with almost scale-free characteristics. Consequently, each small event can grow into a large earthquake with some probability [19].

Long simulations of earthquake activity allow to calculate statistical features like the frequency-size distribution with high precision. Despite the scaling behavior (Eq. (14.1)) for small and intermediate earthquakes, which is observed for all sets of model parameters, clear deviations become visible for large magnitudes. Such deviations are known from real catalogs, but their statistical significance is not clear in all cases. The model simulations suggest that deviations from scaling for strong earthquakes can be attributed to physical properties. One important property is the spatial disorder of brittle parameters of the fault. The presence of strong heterogeneities suppresses system-wide events with some probability, whereas such events can evolve more easily on smooth faults. The degree of quenched (time-independent) spatial heterogeneity turns out to be a key parameter for statistical and dynamical properties of seismicity. This includes the temporal regularity of mainshock occurrence, various aspects of stress and displacement field, and a spontaneous mode-switching between different dynamical regimes without changing parameters. It is interesting that the degree of heterogeneity can act as a tuning parameter that allows for a continuous change of the model dynamics between the end-member cases of supercritical and subcritical behavior. Such a dependence, which is observed also for other parameters, can be visualized in phase diagrams similar to the phase diagram for the different aggregate states of water. For increasing complexity of a model, the number of axis of the phase diagram, representing the relevant model parameters, will increase. The above mentioned question of distinguishing different faults like the Parkfield segment and the San Jacinto fault can be rephrased to the problem of assigning the faults to different points in such a diagram. An important step in this direction is the physical modeling of observed seismicity patterns like aftershocks (Eq. (14.2)), foreshocks, and the acceleration of seismic energy release before large earthquakes. The latter phenomenon which is known to occur over large regions including more than one fault, can be interpreted in terms of the approach towards a critical point. This view is supported by an observational study of the growth of the spatial correlation length which is a different aspect of the same underlying physics.

In Section 14.2, the physical fault model is developed. Results from numerical simulations are presented and discussed in Section 14.3. A summary is given in Section 14.4.

14.2. Modeling seismicity in real fault regions

Numerous conceptual models have been used to simulate seismic cycles (see e.g. [8, 13, 22, 26, 36] and references therein). These include spring-block models and continuous cellular automata, which are so-called "inherently discrete" models; that is, they are not obtained by discretizing the differential equations from a continuous model – the discreteness is an inherent feature of the imposed physics. In this section, we focus on the question, how the framework of conceptual models can be adjusted in order to simulate seismicity of a real fault region, e.g. the Parkfield segment of the San Andreas fault in California.

14.2.1. *Fault geometry and model framework*

A first constraint for a specific model is to include the geometry of the fault segment. As shown in Fig. 14.2, the region of Parkfield is characterized by a distribution of fault segments, which have in good approximation the same orientation. It is therefore reasonable to map these segments in the model on a straight line from SE to NW. Using a similar procedure in depth leads to a rectangular fault plane. The dimensions of the fault segment are chosen to be $70km$ in length and $17.5km$ in depth. As discussed in [8], this geometry corresponds approximately to the San Andreas fault near Parkfield. It is emphasized that the plate boundary is assumed to have infinite length, but the brittle processes are calculated on the above defined segment of finite length. The discretization of the plane is imposed by a computational grid with 128×32 computational cells of uniform size, where stress and slip are calculated. The size of the computational cells is not determined by observational findings, rather it depends on the magnitude range under consideration and the computational effort; a single cell would correspond to a single magnitude. A higher resolution of the grid increases the magnitude range, because the magnitude is calculated from the slip of all cells during an earthquake. The degree of complexity as a function of the model geometry is determined by the resolution of the computational grid. A change of the physical dimensions leads only to a rescaling of time and magnitude axis. Following [8], the material surrounding the fault is assumed to be a homogeneous elastic half space of infinite size, which is characterized by two quantities:

(1) The elastic properties are expressed by the Lamé constants λ and μ, which connect stress and strain in Hook's law. For many rocks, these constants are almost equal; therefore we use $\lambda = \mu$, denoting μ as the rigidity. An elastic solid with this property is called a *Poisson solid.* Because the strain is dimensionless, μ has the same dimension as the stress. In the present study, we use $\mu = 30GPa$.

(2) The (static) Green's function $G(\mathbf{y_1}, \mathbf{y_2})$ defines the static response of the half space at a position $\mathbf{y_1}$ on a displacement at $\mathbf{y_2}$, which may arise from (coseismic) slip or (aseismic) creep motion. Due to the discretization of the fault plane into computational cells, we use the Green's function for static dislocations on rectangular fault patches of width dx and height dz, which is given in [15] and [35]. For a model including a single vertical fault, the Green's function is calculated only on a plane: $G(x_1, z_1; x_2, z_2)$, where x and z denote the coordinates along strike and in depth (for a sketch see Fig. 14.1(a)). Further reduction of the computational effort is given by the symmetry along strike $G(x_1, z_1; x_2, z_2) = G(|x_1 - x_2|, z_1, z_2)$. Therefore, $x_2 = 0$ can be used without loss of generality and the Green's function has the following form:

$$G(x_1, z_1; 0, z_2) = \frac{\mu}{2\pi} \cdot \sum_{i=1}^{4} \sigma_i \cdot \left[\frac{2}{3}(\varepsilon_A^i + \varepsilon_B^i) + \frac{1}{2}\varepsilon_C^i \right]. \tag{14.3}$$

The sign σ_i is defined by

$$\begin{aligned} \sigma_1 &= 1 \quad \sigma_2 = -1 \\ \sigma_3 &= -1 \; \sigma_4 = 1. \end{aligned} \tag{14.4}$$

With the notation

$$\begin{aligned} t_1 &= dx/2 - x_1 \quad t_2 = dx/2 - x_1 \\ t_3 &= -dx/2 - x_1 \; t_4 = -dx/2 - x_1 \end{aligned} \tag{14.5}$$

$$\begin{aligned} q_1 &= z_2 + dz/2 - z_1 \; q_2 = z_2 - dz/2 - z_1 \\ q_3 &= z_2 + dz/2 - z_1 \; q_4 = z_2 - dz/2 - z_1 \end{aligned} \tag{14.6}$$

$$\begin{aligned} p_1 &= z_2 + dz/2 + z_1 \; p_2 = z_2 - dz/2 + z_1 \\ p_3 &= z_2 + dz/2 + z_1 \; p_4 = z_2 - dz/2 + z_1, \end{aligned} \tag{14.7}$$

the three parts in Eq. (14.3) can be written as follows:

$$\varepsilon_A^i = t_i \cdot \left(\frac{1}{\sqrt{t_i^2 - q_i^2} \cdot (q_i + \sqrt{t_i^2 - q_i^2})} + \frac{1}{\sqrt{t_i^2 - p_i^2} \cdot (p_i + \sqrt{t_i^2 - p_i^2})} \right),$$
$$\varepsilon_B^i = t_i \cdot \left(\frac{\frac{1}{4}\sqrt{t_i^2 - p_i^2} + q_i}{\sqrt{t_i^2 - p_i^2} \cdot \left(p_i + \sqrt{t_i^2 - p_i^2}\right)^2} - \frac{(p_i^2 - q_i^2) \cdot \left(2\sqrt{t_i^2 - p_i^2} + p_i\right)}{2(t_i^2 - p_i^2)^{3/2} \left(p_i + \sqrt{t_i^2 - p_i^2}\right)^2} \right), \quad (14.8)$$
$$\varepsilon_C^i = \frac{q_i}{\sqrt{t_i^2 - q_i^2} \cdot \left(\sqrt{t_i^2 - q_i^2} + t_i\right)} + \frac{p_i}{\sqrt{t_i^2 - p_i^2} \cdot \left(\sqrt{t_i^2 - p_i^2} + t_i\right)}.$$

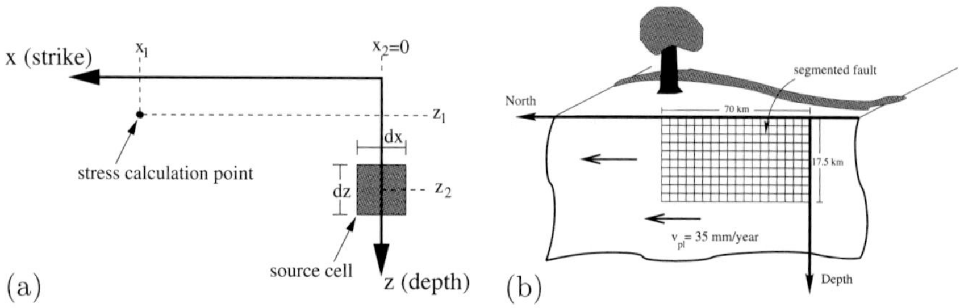

Fig. 14.1. (a) Sketch to illustrate the terms in Eq. (14.3)-(14.8); (b) Sketch of the fault model framework.

The main difference of this Green's function to the nearest-neighbor interaction of spring-block models is the infinite-range interaction following a decay according to $1/r^3$, where r is the distance between source cell and receiver point.

A sketch of the fault model framework is given in Fig. 14.1(b).

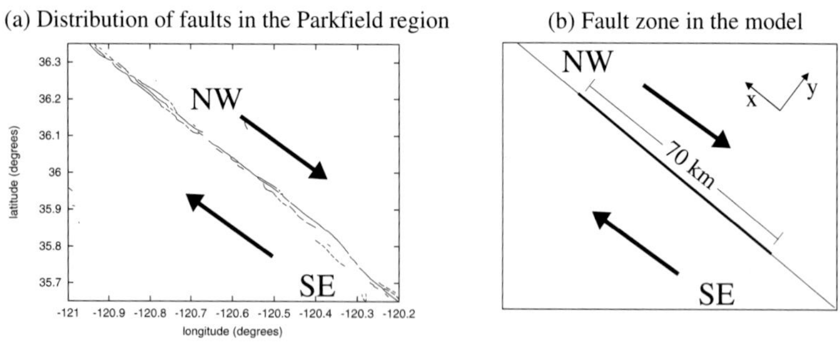

Fig. 14.2. (a) Distribution of faults in the Parkfield (California) region; (b) fault region in the model.

14.2.2. *Plate motion*

The motion of the tectonic plates, indicated in Fig. 14.2, is responsible for the build-up of stress in the fault zone. Satellite-based measurements of surface displacements allow to estimate the velocity of the plates. For the San Andreas fault, a value of $v_{pl} = 35mm/year$ as a long-term average is widely accepted and is therefore adopted for the model [42]. The displacement $\Delta u(i,j)$ of a cell with coordinate i along strike and j in depth during a time period Δt is simply $\Delta u(i,j) = v_{pl} \cdot \Delta t$. While the average slip rate $\dot{u}$ is independent of the location of the cell, the stress rate $\dot{\tau}$ depends on space. The assumption that the fault zone is embedded in a medium which performs constant creep, suggests that cells at the boundaries of the grid are in general higher loaded than cells in the center of the grid. The properties of the elastic medium are determined by the Green's function $G(i,j;k,l)$, which defines the interaction of points (i,j) and (k,l) in the medium. In particular, the stress response at a position (i,j) on a static change of the displacement field $\Delta u(k,l)$ is given by

$$\Delta\tau(i,j) = -\sum_{(k,l)\in halfspace} G(i,j;k,l) \cdot \Delta u(k,l), \tag{14.9}$$

where the minus sign stems from the fact that forward (right-lateral) slip of regions around a locked fault segment is equivalent to back (left-lateral) slip of the locked fault segment. Taking into account that

$$\sum_{(k,l)\in halfspace} G(i,j;k,l) = 0, \tag{14.10}$$

Eq. (14.9) can be written as

$$\tau(i,j;t) = -\sum_{(k,l)\in halfspace} G(i,j;k,l) \cdot [u(k,l;t) - v_{pl}t], \tag{14.11}$$

where $u(k,l;t)$ is the total displacement at position (k,l) and time t since the begin of the simulation. Because the surrounding medium performs stable sliding, $u(k,l;t) = v_{pl}t$ for $(k,l) \notin grid$, the slip deficit outside the fault region vanishes and it is sufficient to perform the summation on the computational grid:

$$\tau(i,j;t) = \sum_{(k,l)\in grid} G(i,j;k,l) \cdot [v_{pl}t - u(k,l;t)]. \tag{14.12}$$

Equation (14.12) can be decomposed in a part for the tectonic loading and a residual part for other processes, especially coseismic slip. The tectonic loading follows the formula

$$\tau_{load}(i,j;t) = \gamma(i,j) \cdot t \tag{14.13}$$

with the space-dependent, but time-independent loading rate

$$\gamma(i,j) = v_{pl} \cdot \sum_{(k,l)\in grid} G(i,j;k,l). \tag{14.14}$$

14.2.3. *Friction and coseismic stress transfer; quasidynamic approach*

It is widely accepted that most earthquakes are due to frictional processes on pre-existing faults. The friction is therefore an important empirical ingredient of a fault model [43]. Numerous laboratory experiments have been carried out to characterize frictional behavior of different materials (see e.g. [14]). An important finding is that the friction defined as the ratio of shear stress τ_{shear} and normal stress τ_{normal}, $\mu_f = \tau_{shear}/\tau_{normal}$ at the initiation of slip, is approximately constant for many materials; the value of μ_f lies between 0.6 and 0.85. This observation, known as *Byerlee's law*, is related to the Coulomb failure criterion [11] for the Coulomb stress CS,

$$CS = \tau_{shear} - \mu_f \tau_{normal}. \tag{14.15}$$

The Coulomb stress depends on a plane, where shear stress and normal stress are calculated. The Coulomb criterion for brittle failure is

$$CS \geq 0, \tag{14.16}$$

which is for $CS = 0$ Byerlee's law.

The North-American plate and the Pacific plate move in opposite direction along the fault plane performing strike-slip motion. The absence of normal and thrust faulting reduces the problem to a one-dimensional motion: all parts of the fault move along the fault direction. The stress state of the fault is fully determined by the shear stress τ_{xy} in the coordinates given in Fig. 14.2(b). Slip is initiated, if τ_{xy} exceeds $\mu_f \tau_{yy}$. This quantity, which is called the material strength or static strength τ_s, is constant in time, if μ_f is assumed to be constant. Note that the normal stress on a strike-slip fault does not change [1]. The shear stress τ_{xy} will be denoted simply by τ. In this notation, the failure criterion Eq. (14.16) reduces to

$$\tau \geq \tau_s. \tag{14.17}$$

When a cell (k, l) fails, the stress drops in this cell to the arrest stress τ_a:

$$\tau(k, l) \to \tau_a, \tag{14.18}$$

with a constant value τ_a, which may become space-dependent later. In terms of slip, this corresponds to a displacement

$$\Delta u(k, l) = \frac{\tau(k, l) - \tau_a}{G(k, l; k, l)} \tag{14.19}$$

with the self-stiffness $G(k, l; k, l)$ of cell (k, l).

The observational effect of dynamic weakening includes also a strength drop from the static strength to a lower dynamic strength:

$$\tau_s \to \tau_d. \tag{14.20}$$

In particular, slipping material becomes weaker during rupture and recovers to the static level at the end of the rupture. This behavior of the strength is known as the static-kinetic friction law.

The values τ_s, τ_d, and τ_a are connected by the dynamic overshoot coefficient D:

$$D = \frac{\tau_s - \tau_a}{\tau_s - \tau_d}. \tag{14.21}$$

Motivated by elastodynamic model simulations, Madariaga [31] proposes $D = 1.25$, which is used to constrain the choice of these values in our model.

The redistribution of the stress release $\Delta\tau(k, l) = \tau(k, l) - \tau_a$ from cell (k, l) to a point (i, j) at time t is

$$\tau(i, j; t) = G(i, j; k, l) \cdot \delta\left(t - \frac{r(i, j; k, l)}{v_s}\right) \cdot \frac{\tau(k, l) - \tau_a}{G(k, l; k, l)}, \tag{14.22}$$

where $\delta(x, y)$ denotes the δ-function, which is 1 for $x = y$ and 0 else; v_s is the constant shear-wave velocity, and $r(i, j; k, l)$ is the distance between source cell (k, l) and receiver position (i, j). That is, cells far from the slipping cell receive their stress portion later than cells close to the slipping cell. The value of v_s is assumed to be constant. Each "stress transfer event" denotes an instantaneous transfer of a stress $\Delta\tau$ from a source cell (k, l) to a receiver cell (i, j) at time t. This time-dependent stress transfer is called the *quasidynamic* approach in contrast to the *quasistatic* approach used in most of the similar models.

The evolution of stress and strength in a cell, where an earthquake is initiated (hypocenter cell), is shown in a sketch in Fig. 14.3. When the earthquake is initiated, the stress and the strength drop. Due to coseismic stress transfer during the event, the cell may slip several times, before the earthquake is terminated and instantaneous healing takes place in all cells. The piecewise constant failure envelope (dashed line) indicates static-kinetic friction.

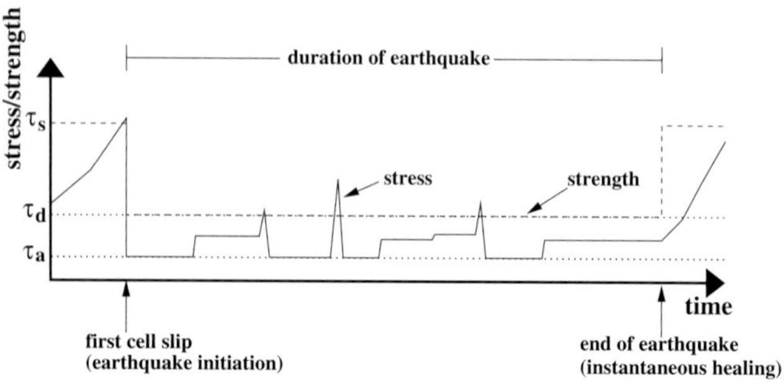

Fig. 14.3. Pictorial evolution of stress (solid line) and strength (dashed line) of a hypocenter cell in the quasidynamic approach.

We note that the Green's function leads to an infinite interaction range. Using open boundary conditions with respect to the computational grid, the stress release from a slipping cell is not conserved on the grid, but on the (infinite) fault plane.

14.2.4. *Model algorithm*

Equation (14.19) shows the correspondence of stress and slip in our model. The model can be formulated either by maintaining stress or by maintaining slip. The algorithm is given for both formulations.

Stress formulation:

(1) Load the fault according to Eq. (14.13), until the first cell is *critical*; that is, the cell fulfills the failure criterion Eq. (14.17). Initiate earthquake.

(2) Reduce stress in critical cell to τ_a and strength to τ_d. Schedule stress transfer events according to Eq. (14.22) on the intra-event timescale.

(3) Perform stress transfer event with the smallest time. Check whether receiver cell is critical.

   (a) If no, remove current stress transfer event from the scheduler: (i) if last stress transfer event, terminate earthquake and go to point 4; (ii) else go back to the beginning of point 3.

(b) If yes, go to point 2.

(4) Set strength to τ_s for all cells (instantaneous healing).

Slip formulation:

(1) Load the fault according to Eq. (14.13), until the first cell is *critical*; that is, the cell fulfills the failure criterion Eq. (14.17). Initiate earthquake.
(2) Reduce stress in critical cell to τ_a and strength to τ_d; update coseismic slip $u \to u + \Delta u$ in this cell with Δu from Eq. (14.19). Schedule stress transfer events according to Eq. (14.22) on the intra-event timescale.
(3) Perform stress transfer event with the smallest time. Check whether receiver cell is critical.
   (a) If no, remove current stress transfer event from the scheduler: (i) if last stress transfer event, terminate earthquake and go to point 4; (ii) else go back to the beginning of point 3.
   (b) If yes, go to point 2.
(4) Set strength to τ_s for all cells (instantaneous healing).
(5) Calculate stresses $\tau(i,j)$ of all cells after the earthquake from the initial stresses $\tau(i,j;0)$ and the new positions $u(k,l;t)$:

$$\tau(i,j;t) = \tau(i,j;0) + \sum_{(k,l)\in grid} G(i,j;k,l) \cdot [v_{pl}t - u(k,l;t)]. \tag{14.23}$$

14.2.5. *Data*

The model produces two types of data, earthquake catalogs and histories of stress and displacement fields. As demonstrated below, all parameters of the model have physical dimensions and can therefore be compared directly with real data, where they are available. This is in contrast to most of the slider-block and cellular automaton models.

Earthquake catalogs include time, hypocenter coordinates, and earthquake size. The time of an earthquake is the time of the first slip; the hypocenter is determined by the position of the corresponding cell along strike and depth. The size of an event can be described by different measures: The rupture area A is the total area, which slipped during an earthquake. The potency

$$P = \int_{grid} \Delta u(x,z)dxdz \tag{14.24}$$

measures the total slip during the event and is related to the seismic moment m_0 by the rigidity: $m_0 = \mu P$. The (moment) magnitude M can be calculated from the potency using

$$M = (2/3)\log_{10}(P) + 3.6, \tag{14.25}$$

where P is given in $cm \cdot km^2$.

14.3. Results

Numerous simulations of the model described in the previous section have been performed. The first catalogs [60] produced by simulations have been studied with respect to the influence of the quasidynamic approach and the discretization of the intra-event time scale in comparison with the quasistatic model of [8]. Then, a large fraction of the parameter space has been analyzed to find relationships between input parameters and observed seismicity features. In this section, a selection of key results is presented and discussed in the light of critical states of seismicity.

14.3.1. *Frequency-size distributions*

The frequency-size (FS) distribution is one of the most important characteristics of observed seismicity. For worldwide seismicity as well as for large faults systems, this distribution is given by the Gutenberg-Richter law (Eq. (14.1)). Figure 14.4 shows the FS distribution of California from 1970 to 2004. Here we use the non-cumulative version of Eq. (14.1), where N is the number of earthquakes with magnitude between M and $M + dM$ with a time bin dM.

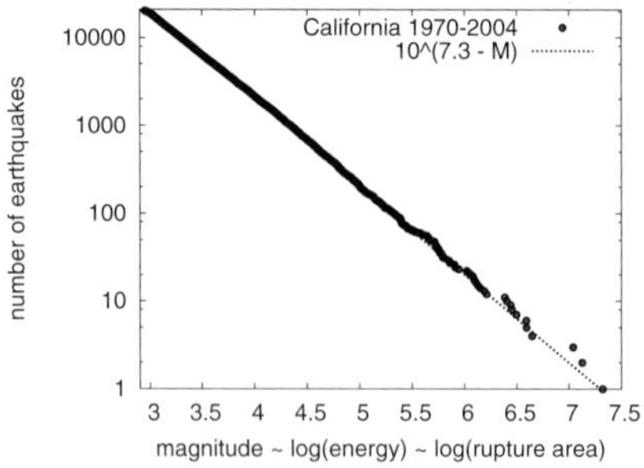

Fig. 14.4. Frequency-size distribution for California from 1970 to 2004; the dashed line denotes a power-law fit to the data.

For individual faults or small fault systems, the FS distribution can deviate from Eq. (14.1), especially for high magnitudes. An example is

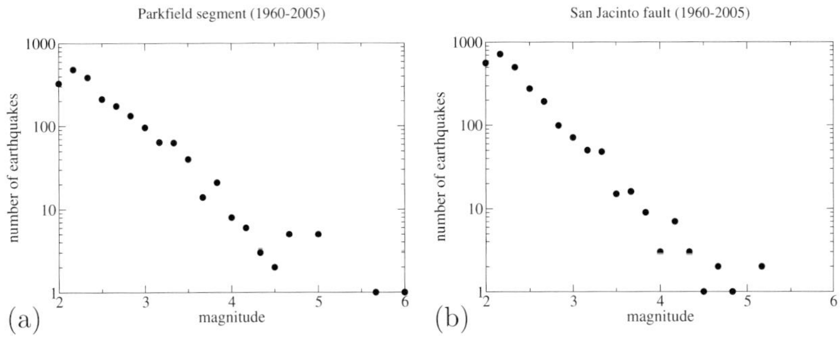

Fig. 14.5. Frequency-size distribution for two faults in California: (a) the Parkfield segment, and (b) the San Jacinto fault calculated over 45 years.

given in Fig. 14.5, which shows the FS distribution of the Parkfield segment (Fig. 14.5(a)) and for the San Jacinto fault (Fig. 14.5(b)) in California calculated for a time span of 45 years. The distribution of the Parkfield segment consists of two parts: A scaling regime for $2.2 \leq M \leq 4.5$ and a significant "bump" for $4.5 < M \leq 6.0$. For the San Jacinto fault, the scaling range is observed for almost all events ($2.2 \leq M \leq 5.0$). The slight decrease for $M \approx 2$ in both plots is due to lacking catalog completeness.

A FS distribution as shown in Fig. 14.5(a) is called a characteristic earthquake distribution, because of the increased probability for the occurrence of a large ("characteristic") event. In contrast, distributions with a broad scaling regime following power-law behavior according to the Gutenberg-Richter law, are denoted as "scale-free", because a power law distribution indicates the absence of a characteristic scale of the earthquake size [48]. In terms of critical point processes, the absence of a characteristic length scale indicates that the system is close to the critical point. In this state, earthquakes of all magnitudes can occur, or each small rupture can grow into a large one. Therefore, the frequency-size distribution can serve as a proxy for the current state of a system in relation to a critical point.

In a model, the easiest way to tune the FS distribution is a variation of the mean stress $\langle\tau\rangle$ on the fault, where $\langle\rangle$ denotes the spatial average. This can be achieved, for instance, by varying brittle properties, e.g. in terms of the dynamic overshoot coefficient D (Eq. (14.21)), or by introducing dissipation [21, 60]. Figure 14.6 shows FS distributions for two different values of D, first $D = 1.25$ (Fig. 14.6(a)) from [31], and second a higher value $D = 1.67$ (Fig. 14.6(b)). While Fig. 14.6(a) follows a characteristic earthquake behavior similar to the Parkfield case (Fig. 14.5(a)), Fig. 14.6(b) resembles the shape of the FS distribution of the San Jacinto fault (Fig. 14.5(b)).

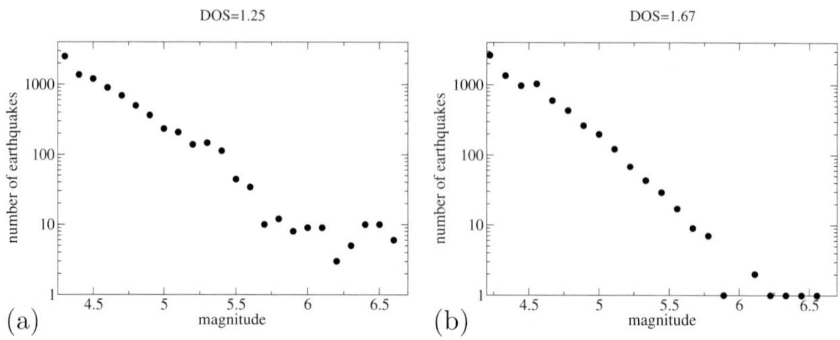

Fig. 14.6. Frequency-size distribution for model realizations with different dynamic overshoot coefficients (Eq. (14.21)): panel (a) $D = 1.25$, panel (b) $D = 1.67$.

As an outcome, three cases can be distinguished by means of a critical mean stress τ_{crit}:

(1) subcritical fault ($\langle\tau\rangle < \tau_{crit}$): the mean stress on the fault is too small to produce large events. The system is always far from the critical point. The FS distribution is a truncated Gutenberg-Richter law.
(2) supercritical fault ($\langle\tau\rangle > \tau_{crit}$): the mean stress is high and produces frequently large events. After a large earthquake (critical point), the stress level is low (system is far from the critical point) and recovers slowly (approaches the critical point). The FS distribution is a characteristic earthquake distribution.
(3) critical fault ($\langle\tau\rangle \approx \tau_{crit}$): the system is always close to the critical point with scale-free characteristics. The FS distribution is a Gutenberg-Richter law with a scaling range over all magnitudes.

This result demonstrates that, for isolated faults, the Gutenberg-Richter law is not the rule, but the exception. If the FS distribution is plotted as a function of the parameters controlling $\langle\tau\rangle$, the result can be visualized by a phase diagram [16, 21, 60, 61].

While it is obvious that the mean stress drop $\langle\Delta\tau\rangle \approx \langle\tau_s - \tau_a\rangle$ controls the mean stress $\langle\tau\rangle$ on the fault, it has also been found that the spatial distribution of the stress drop has significant influence on the FS distribution. In [61], it is demonstrated that the degree of spatial disorder of the stress drop acts as a tuning parameter for the FS distribution and can thus be added as an independent axis in a phase diagram.

14.3.2. *Temporal occurrence of large earthquakes*

The finding that quenched spatial heterogeneities have influence on the rupture propagation suggests that also temporal clustering properties will be affected. While the end member case of a smooth fault produces regular occurrence of events, it can be assumed that a strongly disordered fault will show irregular earthquake occurrence. In Fig. 14.7, we focus on the regularity of the largest earthquakes in a simulation, e.g. earthquakes with $M \geq 5.7$, for two different degrees of spatial disorder. Each plots is based on an earthquake sequence covering 1000 years. Figure 14.7(a) corresponds to a smooth fault and Fig. 14.7(b) to a rough (disordered) fault; the ordinate has no meaning. Although no strictly periodic mainshock occurrence is observed, the sequence for the smooth fault (Fig. 14.7(a)) is characterized by relatively regular mainshock sequences. In contrast, the sequence for the highly disordered fault Fig. 14.7(b) shows a more disordered and clustered behavior. It is interesting to note that sequences similar to the Parkfield sequence can be found in Fig. 14.7(a), e.g. between $t \approx 600 years$ and $t \approx 720 years$, where an almost periodic mainshock sequence is followed by a gap before the next large event occurs. This resembles the most recent Parkfield event on September 28, 2004, which occurred 16 years after it was predicted based on the approximate period of 22 years for $M6$ events on this fault segment.

A more quantitative measure for temporal clustering properties of earthquake sequences is the coefficient of variation [57]

$$CV = \sigma_{\Delta t}/\langle \Delta t \rangle \tag{14.26}$$

calculated for the interevent-time distribution, where $\sigma_{\Delta t}$ is the standard deviation and $\langle \Delta t \rangle$ the mean value of the interevent-time distribution. High values of CV denote clustered activity, while low values represent quasiperiodic occurrence of events. The limit case $CV = 1$ corresponds to a random Poisson process [17]. Figure 14.8 shows CV as a function of the earthquake magnitude for a relatively smooth fault. The figure demonstrates that the degree of temporal regularity increases for growing earthquake sizes. In [61], the coefficient of variation is calculated for different degrees of spatial disorder. The results support the relationship between disorder and CV suggested by Fig 14.7.

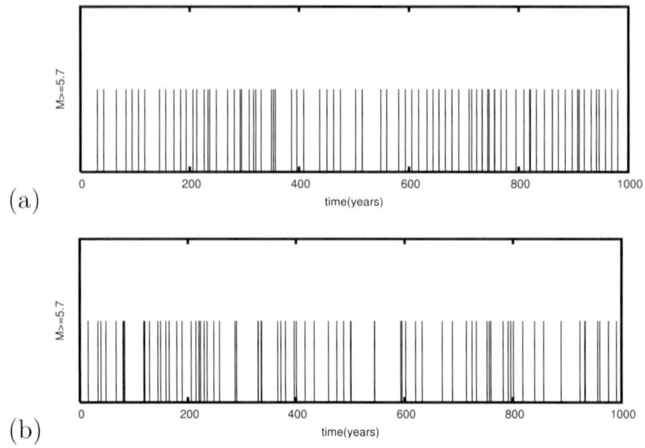

Fig. 14.7. Mainshock sequence from model simulations of a smooth fault (panel (a)) and a rough fault (panel (b)). The plots show earthquakes with $M \geq 5.7$; the ordinate has no meaning.

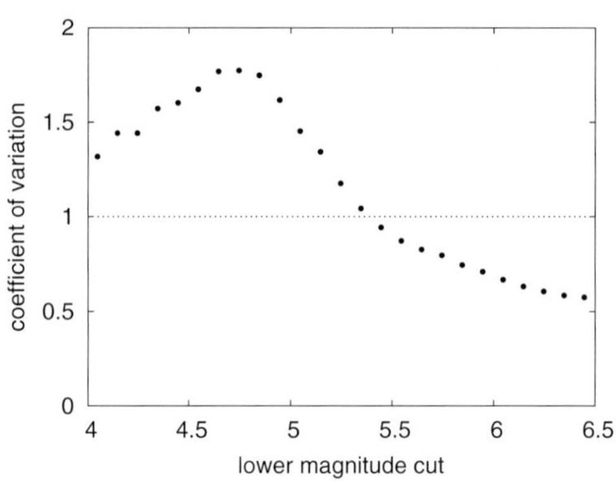

Fig. 14.8. Temporal earthquake occurrence quantified by the coefficient of variation (Eq. (14.26)) as a function of the lower magnitude cutoff.

14.3.3. *Aftershocks and foreshocks*

The most pronounced pattern in observed seismicity is the emergence of strongly clustered aftershock activity following a large earthquake. Apart from the Omori law (Eq. (14.2)), it is widely accepted that aftershocks are characterized by the following properties:

(1) The aftershock rate scales with the mainshock size [38].
(2) Aftershocks occur predominantly at the edges of the ruptured fault

segments [49].

(3) Båth's law [4]: The magnitude of the largest aftershock is $M_m - D_1$, where M_m is the mainshock magnitude and $D_1 \approx 1.2$.

Deviations from the Omori law, especially for rough faults, are discussed in [33]. While aftershocks are observed after almost all large earthquakes, foreshocks occur less frequent [51]. As a consequence, much less is known about the properties of these events. Kagan and Knopoff [29] and Jones and Molnar [28] propose a power law increase of activity according to an

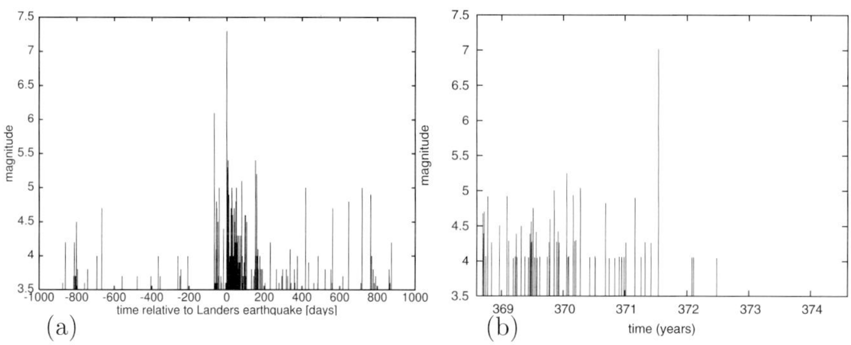

Fig. 14.9. Earthquakes before and after a mainshock: (a) the $M7$ Landers (California) earthquake; (b) $M7$ earthquake in the basic version of the model.

Figure 14.9(a) shows an example for the aftershock sequence following the $M7$ Landers earthquake in California on June 28, 1992. An earthquake of similar size generated by the model is given in Fig. 14.9(b). The absence of aftershocks in the simulation is clearly visible. The reason for the lack of aftershocks is the unloading of the fault resulting from the mainshock: When a large fraction of the fault has ruptured, the stress in this region will be close to the arrest stress after the event. Consequently, the seismic rate will be almost zero until the stress field has recovered to a moderate level.

It is not surprising that a model which imposes only tectonic loading and coseismic stress redistribution, produces no aftershocks, because it is likely that aftershocks are due to additional mechanisms triggered by the mainshock. A discussion on candidates for such mechanisms is given in [58]. A common feature is the presence of postseismic stress which generates aftershock activity. In [22] , for instance, postseismic stress has been attributed to a viscoelastic relaxation process following the mainshock. In the present work, continuous creep displacement following the constitutive law $\dot{u}_{creep}(i,j;t) = c(i,j) \cdot \tau^3(i,j;t)$ with space dependent, but time-independent creep coefficients $c(i,j)$ is assumed [5]. Additionally, the computational grid is divided by aseismic barriers from the free surface to depth into a couple of seismically active fault segments. As discussed in [58], this modification results in a concentration of stress in the aseismic regions during rupture and, subsequently in a release of stress after the event according to the coupled creep process. This stress release triggers aftershock sequences obeying the Omori law (Eq. (14.2)). A typical aftershock sequence after a $M6.8$ event is shown in Fig. 14.10(a). According to Båth's law, the strongest aftershock has the magnitude $M = 5.5$. The sequence shows also the effect of secondary aftershocks, namely aftershocks of aftershocks [47]. The stacked earthquake rate as a function of the time after the mainshock is given in Fig. 14.10(b). In this case, where the barriers are characterized by creep coefficients, which are by a factor of 10^5 higher than the creep coefficients in the seismic patches, a realistic Omori exponent of $p = 1$ is found.

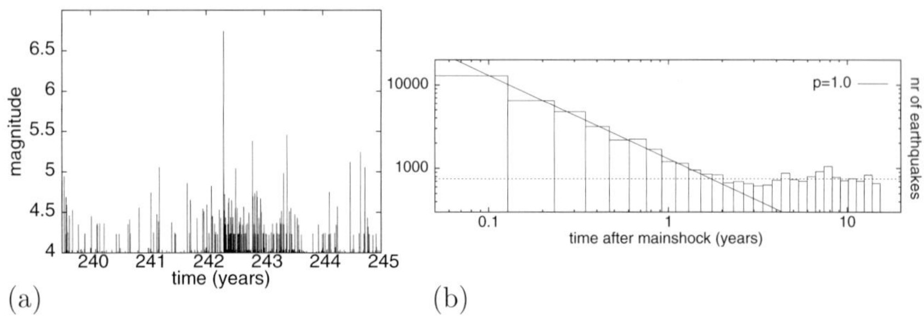

Fig. 14.10. (a) Earthquakes before and after a mainshock with $M = 6.8$ in the modified model; (b) Earthquake rate as a function of time for the model with seismic and aseismic regions. The calculation is based on a simulation with 20,000 earthquakes covering about 500 years; the earthquake rates are averaged over about 30 mainshocks. A fit of the Omori law (Eq. (14.2)) with $p = 1$ is denoted as a solid line. The dashed line gives the estimated background level of seismicity.

Aftershock sequences like in in Fig. 14.10(a) emerge after all large events in the extended model. In contrast, there is no clear foreshock signal visible in single sequences. However, stacking many sequences together, unveils a slight increase of the earthquake rate prior to a mainshock supporting the observation of rarely occurring foreshock activity. An explanation of these events can be given in the following way: Between two mainshocks, the stress field organizes itself towards a critical state, where the next large earthquake can occur. This critical state is characterized by a disordered stress field and the absence of a typical length scale, where earthquakes of all sizes can occur. The mainshock may occur immediately or after some small to moderate events. The latter case can be considered as a single earthquake, which is interrupted in the beginning. This phenomenon of delayed rupture propagation has already provided a successful explanation of foreshocks and aftershocks in a cellular automaton model [23, 25].

The hypothesis that foreshocks occur in the critical point and belong, in principle, to the mainshock, can be verified by means of the findings from Section 14.3.1. In particular, the frequency-size distribution in the critical point (or close to the critical point) is expected to show scale-free statistics. If an overall smooth model fault following characteristic earthquake statistics is studied over a long time period, the approach of the critical point can be calculated precisely in terms of a change of the frequency-size distribution towards Gutenberg-Richter behavior. This change of frequency-size statistics is observed in the model (Fig. 14.11) and supports thus the validity of the critical point concept [57].

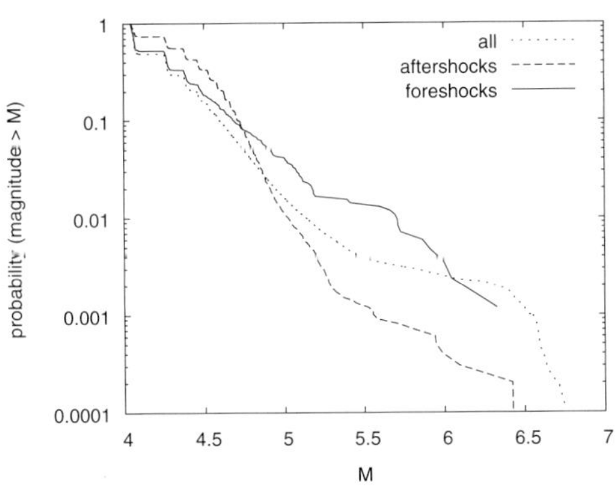

Fig. 14.11. Frequency-magnitude distribution of all earthquakes, foreshocks and aftershocks, respectively. Foreshocks and aftershocks are defined as earthquakes occurring within one month before and after an earthquake with $M \geq 6$.

14.3.4. *Accelerating moment release*

In the previous section, it has been argued that large earthquakes are associated with a critical point and the preparation process is characterized by increasing disorder of the stress field and increasing tendency to scale-free characteristics in the frequency-size distribution. Further support for critical point dynamics has been provided by the observational finding of [12] that the cumulative Benioff strain $\Sigma\Omega(t)$ follows a power law time-to-failure relation prior to the $M7$ Loma Prieta earthquake on October 17, 1989:

$$\Sigma\Omega(t) = \sum_{i=1}^{N(t)} \sqrt{E_i} = A + B(t_f - t)^m \tag{14.27}$$

Here, E_i is the energy release of earthquake i, and $N(t)$ is the number of earthquakes before time t; t_f is the failure time and A, B and $m > 0$ are constants. The systematic study of [10] verifies this behavior also for other large earthquakes in California. Similar studies for numerous seismically active regions followed (see [56] and references therein).

The time-to-failure relation Eq. (14.27) has been proposed by [41] and [46] from the viewpoint of renormalization theory. Moreover, they demonstrated that a complex exponent $m = \alpha + i\beta$ results in an additional term of log-periodic oscillations decorating the power-law increase of $\Sigma\Omega(t)$. This law has been fitted by [46] to the data of [12]. Although the fit shows good agreement with the data, there is no evidence that this concept is feasible for the prediction of earthquakes so far. In particular, the fit operates with a large number of free parameters including amplitude and phase of the fluctuations and a cutoff time. Therefore, Eq. (14.27) with real m will be used to describe accelerating moment release in this study.

Similar to the findings about foreshocks, this pattern is not universal. Therefore, a stacking procedure is adopted in order to obtain robust results on the validity of Eq. (14.27) in the model. This is not straightforward, since the interval of accelerating moment release is not known a priori and the duration of a whole seismic cycle, as an upper limit, is not constant. To normalize the time interval for the stacking, the potency release (Eq. (14.24)) is computed as a function of the (normalized) stress level (Fig. 14.12). Taking into account that the stress level increases almost linearly during a large fraction of the seismic cycle, the stress level axis in Fig. 14.12 can effectively be replaced by the time axis leading to a power-law dependence of the potency release on time. The best fit is provided with an exponent $s = -1.5$. Transforming the potency release to the cumulative Benioff strain (Eq. (14.27)), results in an exponent $m = 0.25$ in Eq. (14.27). This finding is based on a simulation over about 5000 years; the exponent is in good agreement with the theoretical work of Rundle et al. [40], who also derive $m = 0.25$ from a spinodal model, and the analytical result of

$m = 0.3$ in the damage mechanics model of Ben-Zion and Lyakhovsky [6]. An observational study of California seismicity finds m between 0.1 and 0.55 [10].

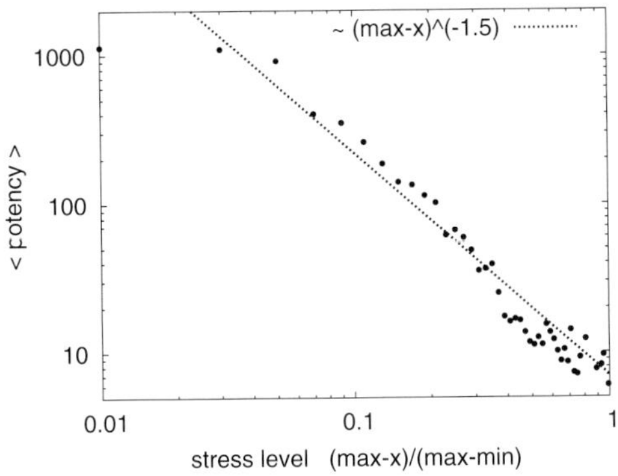

Fig. 14.12. Mean potency release (Eq. (14.24)) as a function of the stress level. The stress level is normalized to the maximum (max) and minimum (min) observed stress.

14.4. Summary and conclusions

The present work deals with the analysis, the understanding and the interpretation of seismicity patterns with a special focus on the critical point concept for large earthquakes. Both, physical modeling and data analysis are employed. This study aims at practical applications to model data and earthquake catalogs from real fault systems. A point of particular interest is the detection of phenomena prior to large earthquakes and their relevance for a possible prediction of these events. While aftershocks are an almost universal phenomenon, there is no precursor obeying a similar degree of universality. It is, therefore, interesting to study less frequent precursory phenomena by means of long model simulations.

For this aim, we use a numerical model which is on the one hand to some degree physical, and on the other hand simple enough that it allows to perform long simulations. The basic version of the model consists of a segmented two-dimensional strike-slip fault in a three-dimensional elastic half space and is inherently discrete, because it does not arise from discretizing a continuous model.

The results of the simulations indicate an overall good agreement of the synthetic seismicity with real earthquake activity, with respect to frequency-size distributions and various features of earthquake sequences. An important role for the characteristics of a simulated catalog seems to play the degree of spatial heterogeneity on the fault, which is implemented by means of space-dependent brittle parameters. Smooth faults are governed by characteristic earthquake statistics, regular occurrence of mainshocks and overall smooth stress fields. On the other hand, rough faults obey scale-free Gutenberg-Richter statistics, irregular mainshock occurrence, and overall rough stress fields. A closer look at the disorder of the stress field shows, however, that even on a smooth fault a gradual roughening takes place when the next large earthquake is approached. This is reflected in the frequency-size distribution which evolves towards the Gutenberg-Richter law. Such a state is denoted as a critical state of seismic dynamics. This finding allows to establish a relation between the closeness to the underlying critical point, the (unobservable) stress field, and the (observable) frequency-size distribution. Moreover, the concepts of "self-organized criticality" and the "critical earthquake concept" can be interpreted as special cases of a generalized concept, which is further supported by the observation of accelerating moment release and growing spatial correlation length prior to large earthquakes [53–56].

We conclude that a multiparameter approach including parameters related to different seismicity patterns is promising for the improvement of seismic hazard assessment.

References

[1] K. Aki, P.G. Richards: *Quantitative seismology* (University Science Books., 2002)
[2] P. Bak: *How Nature Works. The Science of Self-Organised Criticality* (Oxford University Press., 1996)
[3] P. Bak, C. Tang: J. Geophys. Res. **94**, 15,635 (1989)
[4] M. Båth: Tectonophysics **2**, 483 (1965)
[5] Y. Ben-Zion: J. Geophys. Res. **101**, 5677 (1996)
[6] Y. Ben-Zion, V. Lyakhovsky: Pure Appl. Geophys. **159**, 2385 (2002)
[7] Y. Ben-Zion, V. Lyakhovsky: Eos Trans. AGU **84**(46), Fall Meet. Suppl., Abstract NG12C-02 (2003)
[8] Y. Ben-Zion, J.R. Rice: J. Geophys. Res. **98**, 14109 (1993)
[9] J.J. Binney, N.J. Dowrick, A.J. Fisher, M.E.J. Newman: *The theory of critical phenomena* (Oxford University Press, 1993)
[10] D.D. Bowman, G. Oullion, C.G. Sammis, A. Sornette, D. Sornette: J. Geophys. Res. **103**, 24359 (1998)
[11] W.F. Brace: J. Geophys. Res. **65**, 3477 (1960)

[12] C.G. Bufe, D.J. Varnes: J. Geophys. Res. **98**, 9871 (1993)
[13] R. Burridge, L. Knopoff: Bull. Seim. Soc. Am. **57**, 341 (1967)
[14] J.D. Byerlee: Pure Appl. Geophys **116**, 615 (1978)
[15] M. Chinnery: Bull. Seim. Soc. Am. **53**, 921 (1963)
[16] K. Dahmen, D. Ertas, Y. Ben-Zion: Phys. Rev. E **58**, 1494 (1998)
[17] D.J. Daley, D. Vere-Jones: *An Introduction to the Theory of Point Processes* (Springer, New York, 1988)
[18] J.H. Dieterich: J. Geophys. Res. **99**, 2601 (1994)
[19] R.J. Geller, D.D. Jackson, Y.Y. Kagan, F. Mulargia: Science **275**, 1616 (1997)
[20] B. Gutenberg, C.F. Richter: Bull. Seismol. Soc. Am. **46**, 105 (1956)
[21] S. Hainzl, G. Zöller: Physica A **294**, 67 (2001)
[22] S. Hainzl, G. Zöller, J. Kurths: J. Geophys. Res. **104**, 7243 (1999).
[23] S. Hainzl, G. Zöller, J. Kurths: Nonlin. Proc. Geophys. **7**, 21 (2000)
[24] S. Hainzl, G. Zöller, J. Kurths, J. Zschau: Geophys. Res. Lett. **27**, 597 (2000)
[25] S. Hainzl, G. Zöller, F. Scherbaum: J. Geophys. Res. **108**,2013, doi 10.1029/2001JB000610 (2003)
[26] J. Huang, D.L. Turcotte: Geophys. Res. Lett. **17**, 223 (1990)
[27] S.C. Jaumé, L.R. Sykes: Pure Appl. Geophys **155**, 279 (1999)
[28] L.M. Jones, P. Molnar: J. Geophys. Res. **84**, 3596 (1979)
[29] Y.Y. Kagan, L. Knopoff: Geophys. J. R. Astron. Soc. **55**, 67 (1978)
[30] J. Lomnitz-Adler: J. Geophys. Res. **98**, 17745 (1999)
[31] R. Madariaga: Bull. Seim. Soc. Am. **66**, 639 (1976)
[32] I.G. Main, G. O'Brian, J.R. Henderson: J. Geophys. Res. **105**, 6105 (2000)
[33] C. Narteau, P. Shebalin, S. Hainzl, G. Zöller, M. Holschneider: Geophys. Res. Lett. **30**, 1568, doi 10.1029/2003GL017110 (2003)
[34] A. Nur, J.R. Booker: Science **175**, 885 (1972)
[35] Y. Okada: Bull. Seism. Soc. Am. **82**, 1018 (1992)
[36] Z. Olami, H.S. Feder, K. Christensen: Phys. Rev. Lett. **68**, 1244 (1992)
[37] F. Omori: J. Coll. Sci. Imp. Univ. Tokyo **7**, 111 (1894)
[38] P. Reasenberg: J. Geophys. Res. **90**, 5479 (1985)
[39] H.F. Reid: *The Mechanics of the Earthquake, The California Earthquake of April 18, 1906, Report of the State Investigation Commission, Vol.2* (Carnegie Institution of Washington, Washington, D.C., 1910)
[40] J.B. Rundle, W. Klein, D.L. Turcotte, B.D. Malamud: Pure Appl. Geophys **157**, 2165 (2000)
[41] H. Saleur, C.G. Sammis, D. Sornette: J. Geophys. Res. **101**, 17661 (1996)
[42] J.C. Savage, J.L. Svarc, W.H. Prescott: J. Geophys. Res. **104**, 4995 (1999)
[43] C.H. Scholz: Nature **391**, 37 (1998)
[44] R. Shcherbakov, D.L. Turcotte: Pure and Applied Geophys, **161**, 2379, doi 10.1007/s00024-004-2570-x (2004)
[45] D. Sornette: *Self-organization and Disorder: Concepts & Tools* (Springer Series in Synergetics, Heidelberg, 2004)
[46] D. Sornette, C.G.Sammis: J. Phys. 1 **5**, 607 (1995)
[47] D. Sornette, A. Sornette: Geophys. Res. Lett. **6**, 1981 (1999)

[48] D.L. Turcotte: *Fractals and chaos in geology and geophysics* (Cambridge University Press, New York, 1997)
[49] T. Utsu: Statistical features of seismicity. In *International handbook of earthquake and engineering seismology, Vol. 81A*, ed by the Int'l Assoc. Seismol. & Phys. Earth's Interior (Academic Press, 2002) pp 719–732
[50] T. Utsu, Y. Ogata, R.S. Matsu'ura: J. Phys. Earth **43**, 1 (1995)
[51] M. Wyss: Science **278**, 487 (1997)
[52] M. Wyss, R.E. Habermann: Pure Appl. Geophys **126**, 319 (1988)
[53] I. Zaliapin, Z. Liu, G. Zöller, V. Keilis-Borok, D. Turcotte: Computational Seismology **33**, 141 (2002)
[54] G. Zöller, S. Hainzl: Natural Hazards and Earth System Sciences **1**, 93 (2001)
[55] G. Zöller, S. Hainzl: Geophys. Res. Lett. **29**, 1558, doi 10.1029/2002GL014856 (2002)
[56] G. Zöller, S. Hainzl, J. Kurths: J. Geophys. Res. **106**, 2167 (2001)
[57] G. Zöller, S. Hainzl, Y. Ben-Zion, M. Holschneider: Tectonophysics *in press* (2006)
[58] G. Zöller, S. Hainzl, M. Holschneider, Y. Ben-Zion: Geophys. Res. Lett. **32**, L03308, doi 10.1029/2004GL021871 (2005)
[59] G. Zöller, S. Hainzl, J. Kurths, J. Zschau: Natural Hazards **26**, 245 (2002)
[60] G. Zöller, M. Holschneider, Y. Ben-Zion: Pure Appl. Geophys **161**, 2103 (2004)
[61] G. Zöller, M. Holschneider, Y. Ben-Zion: Pure Appl. Geophys **162**, 1027, doi 10.1007/s00024-004-2660-9 (2005)

Chapter 15

Predator-prey oscillations, synchronization and pattern formation in ecological systems

Bernd Blasius* and Ralf Tönjes

[1] *Institut für Physik, Universität Potsdam, Postfach 601553,D-14415 Potsdam, Germany*

* *bernd@agnld.uni-potsdam.de*

Contents

15.1. Introduction

Ecological systems and their component biological populations exhibit a broad spectrum of non-equilibrium dynamics ranging from characteristic natural cycles to more complex chaotic oscillations [1]. Perhaps the most spectacular example of this dynamic is Ecology's well known hare-lynx cycle . Despite unpredictable population fluctuations from one cycle to the next in the snowshoe hare (*Lepus americanus*) and the Canadian lynx (*Lynx canadensis*), the overall oscillation tends to follow a tight rhythm with a period of ~ 10 years. Curiously, hare-lynx populations from different regions of Canada synchronize in phase to a collective cycle that manifests over millions of square kilometers [2–6]. Similar spatially synchronized fluctuations have been observed across widely separated sites for many other ecological populations and are also prominent in the dynamics of many biological, ecological and epidemiological contexts and may involve disparate animal taxa across widely separated sites [6, 7]. The underlying causes of such population cycles, which persist in time and are synchronised over space, remain a longstanding enigma.

The spontaneous onset of synchronization , as exhibited by the lynx populations over Canada, is one of the most remarkable phenomena found in biological systems and relies on the coordination and interaction among many scattered organisms. Synchronization arises in a large class of systems of various origins, ranging from physics and chemistry to biology and social sciences. Examples include swarms of fireflies that flash in unison, crickets that chirp in synchrony, and synchronous firing cardiac pacemaker cells. The presence, absence or degree of synchronization can be an important part of the function or malfunction of a biological system, as for example epileptic seizures or heart fibrillation. In ecology, the synchronization of populations is often seen as detrimental to spatially structured populations. This is because asynchrony enhances the global persistence of a population through rescue effects, even when there are local extinctions.

Many hypotheses have been advanced to explain the causes of spatially synchronized population oscillations [8, 9]. Moran [5] suggested that extrinsic large-scale climatic forcing may often be responsible for entraining populations over vast geographic distances [10]. Intrinsic predator-prey and consumer-resource relationships within the foodweb, including density dependent and time-delayed effects, may also be responsible for generating population cycling, with local migration enhancing spatial synchronization [4, 6]. The plethora of ecological processes at work underscores the need to develop deterministic conceptional models capable of realistically reproducing these complex population dynamics. Simply formulated robust foodweb models are able to achieve this goal and provide the necessary framework for studying ecological synchronization and pattern formation effects .

15.2. Predator prey systems and oscillations

15.2.1. *The Lotka-Volterra model - does war favour sharks?*

Historically the mathematical description of predator-prey systems goes back to the times of the first World War. The Italian biologist Umberto D'Ancona made the observation that in the years after the war the proportion of predatory fishes, which were caught in the Upper Adriatic Sea,

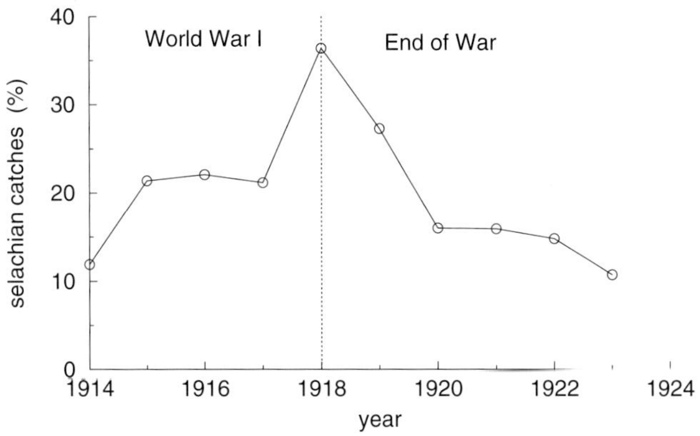

Fig. 15.1. Relative frequency of predator fish *selachian* at the market of Fiume during and after WWI (data from Kot (2001)).

was found to be considerably higher than in the years before the war (see Fig. 15.1). Obviously, during the war fishery in the Adriatic had been strongly reduced. However, it was unclear why this should give benefit to the predatory compared to the prey fish. In other words, why should war favor sharks? D'Ancona (Volterra's future son in law) put the question of this unexpected side-effect of war to Vito Volterra, who set up the following model to describe the dynamics of the predator and prey species

$$\begin{aligned} \dot{R} &= aR - k_1 RN \\ \dot{N} &= -bN + k_2 RN. \end{aligned} \tag{15.1}$$

In these equations, N is the predator and R is the prey species. It is assumed that the dynamics of the prey and predator populations in the absence of the other species is given by exponential growth $\dot{R} = aR$ or decay $\dot{N} = -bN$. Predation is taken into account in the form of a mass action RN term with rates k_1 and k_2. Model (15.1) became famous as the Lotka-Volterra model [11] (some of the work of Volterra was preceded by that of Alfred J. Lotka in a chemical context [12]).

The model contains two fixed points: the trivial equilibrium $(R^*, N^*) = (0, 0)$ where both predator and prey disappear and the feasible equilibrium $(R^*, N^*) = (b/k_2, a/k_1)$. Interestingly, in the feasible equilibrium the abundance of the predator or prey population is independent of their own respective growth or death rates but instead is set by the productivity of the other species. Stability analysis reveals that the trivial equilibrium is a saddle point and the feasible equilibrium is a neutrally stable center with

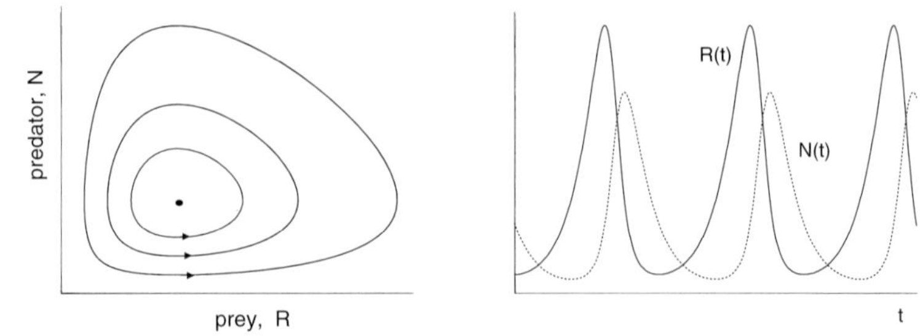

Fig. 15.2. Predator-prey cycles in Lotka-Volterra system (15.1). Left: neutrally stable center in the phase plane. Oscillations are seen as closed loops of orbits around the center. Different initial values lead to different trajectories. Right: typical time series of predator $N(t)$ (solid line) and of prey $R(t)$ (dotted line).

purely imaginary eigenvalues $\lambda_{1,2} = \pm i\sqrt{ab}$. As a consequence, solutions show oscillations around the center point with a frequency that is given by the square root of the growth and death terms $\omega = \sqrt{ab}$ (as shown in Fig. 15.2). The intuitive explanation for the origin of the oscillations are delayed predator-prey interactions. Assume that initially there are not many predators. Then, at first, the prey population is rising. With increasing prey abundance also the growth condition for the predators is improving and after a typical time delay predator numbers begin to grow. However with sufficiently predators to cash in, the prey numbers become decimated, which again is followed by a decrease of the predators since they have over-exploited their own food. Now with small predation pressure prey numbers are on the rise again, and the cycle repeats anew.

Going back to D'Ancona's problem of inhibited fishing in the Adria, we include the effects of fishing into the Lotka-Volterra model (15.1) by introducing a linear loss term with harvesting rate h for both predator and prey species

$$\begin{aligned} \dot{R} &= aR - k_1 RN - hR \\ \dot{N} &= -bN + k_2 RN - hN. \end{aligned} \tag{15.2}$$

These equations correspond exactly to the original model (15.1), however with reduced prey birth rate, $a-h$, and increased predator mortality, $b+h$. Consequently, with the inclusion of fishing the feasible equilibrium reads

$$R^*(h) = \frac{b+h}{k_2}, \quad N^*(h) = \frac{a-h}{k_1}. \tag{15.3}$$

Thus, under fishing pressure the equilibrium level of the prey is increasing whereas that of the predator is reduced. This result is known as Volterra's

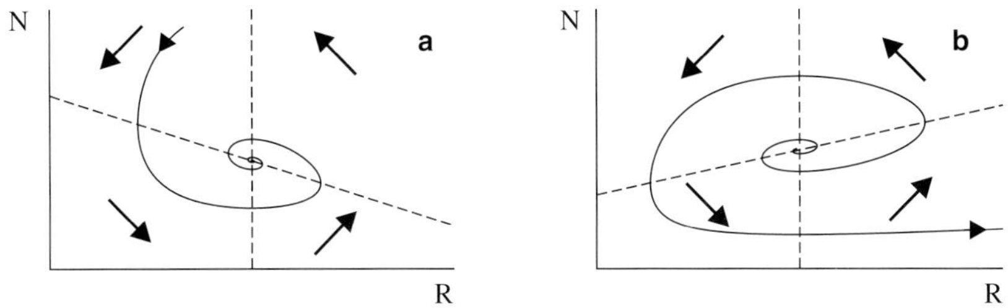

Fig. 15.3. Phase portrait of the modified Lotka-Volterra models in the (R, N)-phase plane (solid lines). Graphical analysis revolves around plotting the isoclines in the prey-predator phase plane that denote zero-growth of the model predator and prey populations [13] (dashed lines). a) Nonlinear density dependence $g(R)$ leads to a decreasing prey isocline and is stabilizing. b) Type-II functional response $f(R)$ gives rise to an increasing prey isocline and is destabilizing.

principle: whenever the population sizes of predator and prey determine each other, a reduction of the growth rates (e.g., by harvesting) leads to an increase of the prey and a decrease of the predator population. Strictly speaking, this principle holds only for the equilibrium values, whereas the typical model dynamics are oscillatory. However, by separation of variables in the equations (15.1) one can easily show that the equilibrium value in the Lotka-Volterra model equals the time average over one period

$$R^* = \frac{1}{T}\int_{\tau}^{\tau+T} R(t)\,dt, \quad N^* = \frac{1}{T}\int_{\tau}^{\tau+T} N(t)\,dt. \tag{15.4}$$

The principle of Volterra provides an answer to the unexpected dynamics of the predator fish numbers in the Adriatic Sea. The predator abundance during the war (low fishing) corresponds to the unperturbed model, whereas after, or before, the war, with the onset of fishing according to Eq. (15.3), predator numbers must be reduced. Volterra's principle has been confirmed by some catastrophic failures in insect pest control. Pesticides usually act not only on the pest species but also, with even stronger impact, on their natural enemies, the predators. Thus often, although the pesticide does indeed hurt the prey, it is simultaneously reducing the predator numbers with an, even larger, positive secondary influence on the prey. As a result of this indirect effect treatment can counterintuitively lead to an effective increase of the pest species.

15.2.2. *The Rosenzweig-MacArthur model*

Being able to provide a theoretical explanation for the commonly observed predator-prey oscillations and also by predicting the effects of harvesting in such systems, the Lotka-Volterra equations constitute one of the first big successes of Theoretical Ecology. However, the same model later was heavily criticized for a number of disturbing drawbacks, all of which can be traced back to the neutral stability of the feasible equilibrium. As shown in Fig.15.2, the amplitude of oscillation depends on the initial values, meaning that in contrast to ecological observation there is no typical amplitude of the solution. Even more disturbing, from a theoretical point of view, is the structural instability of the model due to the nonhyperbolic fixed point, which means that the oscillatory solutions are destroyed by arbitrarily small changes in the structure of the model.

To overcome these shortcomings, modified predator-prey models were sought for which should improve the most important oversimplifications of the Lotka-Volterra model. In general, the dynamics of two populations, which are coupled by predation, can be generalized as follows

$$\begin{aligned} \dot{R} &= g(R) - f(R)N \\ \dot{N} &= -bN + \chi f(R)N. \end{aligned} \tag{15.5}$$

These equations contain two functional relationships which take into account for saturation effects with increasing prey density R. The first term $g(R)$ describes the density dependence of the prey growth due to intraspecific competition. The second term $f(R)$, the so-called functional response , models the saturation of the food uptake with the density of the prey. The parameter χ describes the efficiency of energy conversion and b is the predator mortality.

To analyze the possible dynamic outcomes of model (15.5) we study both possible model modifications separately. At first we only consider a nonlinear density dependence using a logistic growth for the prey $g(R) = aR(1 - \frac{R}{K})$ with carrying capacity K, while the uptake function remains linear $f(R) = kR$ (type-I functional response). The modified dynamics can be understood from the geometry of the iscoclines in phase space. The R-isocline (obtained by setting $\dot{R} = 0$) leads to a straight line with negative slope in phase space $N = \frac{a}{k}(1 - \frac{R}{K})$, whereas the N-isocline ($\dot{N} = 0$) is a vertical line $R = b/(\chi k)$ (see Fig.15.3a). Compared to the basic model (15.1), nonlinear density dependence leads to a change of the R-isocline from a horizontal to a tilted straight line, with a slope that depends on the carrying capacity K. As an effect, the rotating trajectory is 'squeezed' in phase space and finally results in damped rotations, for every value of K. In this way, the introduction of arbitrarily weak nonlinear density dependence destroys the Lotka-Volterra oscillations, i.e., it has a stabilizing effect.

Now we keep the prey growth rate linear $g(R) = aR$ but instead use a more realistic type-II functional response, $f(R) = kR/(K_M + R)$, to describe the food-uptake. As shown in Fig.15.3b again the N-isocline is a vertical line, but the R-isocline $N = \frac{a}{k}(K_M + R)$ is now a straight line with increasing slope. This has the consequence that the rotating trajectory is increasing in amplitude. Thus, the effect of a functional response is to destabilize the Lotka-Volterra oscillations.

As revealed by Fig.15.3 both modification types of the original Lotka-Volterra model (15.1) have a profound impact on the dynamics. Using a more realistic description for the prey growth destroys the oscillations, while a saturation of the food uptake is destabilizing. So the question remains what are the minimal requirements to come up with a model that is able to show robust fluctuating population numbers. It was the insight of Rosenzweig and MacArthur [14] that the solution is to simultaneously include both antagonistic model modifications, i.e., both stabilizing and destabilizing forces, into the model, which leads to the following equations

$$\begin{aligned} \dot{R} &= aR(1 - \frac{R}{K}) - kN\frac{R}{K_M + R} \\ \dot{N} &= -bN + \chi kN\frac{R}{K_M + R}. \end{aligned} \qquad (15.6)$$

As demonstrated in Fig.15.4 the Rosenzweig-MacArthur model (15.6) can give rise to self-sustained oscillations and limit cycle behaviour. This becomes apparent by examination of the models isoclines. In the model (15.6), including both logistic prey growth and Holling type-II functional response, the prey isocline is given by the parabola $N = \frac{a}{k}(1 - \frac{R}{K})(K_M + R)$. Stability depends on whether the intersection between the two isoclines in phase space occurs in the increasing or the decreasing branch of the parabola (see Fig. 15.4). This is evident, for example, by studying the effect of enrichment in the predator-prey model (15.6), i.e. by increasing the carrying capacity K and so allowing the prey to grow to higher densities in the absence of predators [15]. As demonstrated in Fig.15.4 increasing the availability of resources will destabilize the community dynamics from a stable equilibrium to oscillatory dynamics. The phenomenon has been coined as the "paradox of enrichment" [15–17]. Attempts to establish the

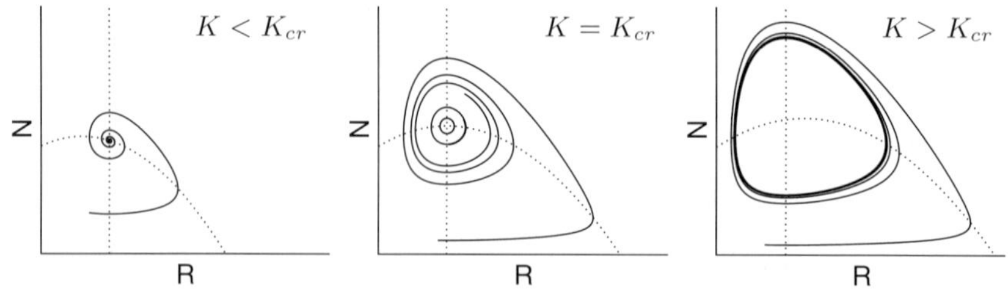

Fig. 15.4. Enrichment in the Rosenzweig-MacArthur model (15.6) with logistic prey growth and Holling type-II functional response. Plotted are typical trajectories (solid lines) in the (R, N)-phase plane for increasing values of the carrying capacity K (from left to right) and corresponding isoclines (dotted lines). Left) For small values of K the intersection occurs in the right, decreasing portion of the prey isocline, and the equilibrium is locally stable. Middle) Critical enrichment gives rise to a neutrally stable center, similar to the Lotka-Volterra model (see Fig. 15.2). Right) With sufficient enrichment $K > K_{cr}$ the intersection occurs in the left, increasing branch of the prey isocline and mark unstable equilibria, leading to limit cycle dynamics.

destabilizing effects of resource enrichment in experiments or in the wild met with only partial success [18]. This indicates that real communities respond to enrichment in a more complicated way than simple models suggest. Environmental conditions and properties of the community (web-like structure, shift to inedible prey) have been offered as explanations for why communities might fail to destabilize as a consequence of enrichment [18]. Here, we provide a different explanation based on a hidden supersensitivity to model structure which is contained in the general predator-prey model (15.5) [19].

A loophole to structural instability The classical Rosenzweig-MacArthur model (15.6) is probably the simplest formulation of a trophic community to produce realistic predator-prey cycles. However, as is the case for most biological processes, the exact analytical form of the functions, which are involved in the model such as the functional response $f(R)$, are not known. In praxis, therefore, one has to use functions which serve as phenomenological descriptor of resource uptake. Minimal requirements for realistic uptake functions $f(R)$ are that the function be zero at zero resource concentration, be monotonically increasing with resource density, and be saturating when resource density goes to infinity. Here, we restrict ourselves to type-II functional response curves with negative curvature over the whole prey range (the case of sigmoidal uptake functions will be discussed below). The most widely used functions that fulfil these requirements are

Holling's type-II function $f_H(R) = k_H R/(1 + \kappa_H R)$ [20] and Ivlev's function $f_I(R) = k_I[1 - \exp(-\kappa_I R)]$ [21]. However also other functions, e.g. trigonometric functions $f_T(R) = k_T \tanh(\kappa_T R)$ may be used.

In [19] it was studied whether the specific analytical form of the functional response affects the community dynamics that the model (15.5) predicts. Given the error with which resource uptake by real organisms is measured, it is usually unjustified to identify a best fitting model. Thus, structurally different analytical forms may be used interchangeably. In Fig. 15.5 we chose a generic parameterization of Ivlev's functional response and used a nonlinear least squares fit to maximize the phenomenological similarity with Holling's type-II and a response curve based on a trigonometric function. Interestingly, the almost identical uptake curves lead to drastic differences in the dynamic stability of the system (Fig. 15.5, right). Although the system with Hollings function is far from the stability boundary at large K, a subtle change in model structure may stabilize the equilibrium. With enrichment (by raising the carrying capacity) all three functions eventually become destabilizing, however not to the same degree. In fact, the enrichment level at which the equilibrium is destabilized varies by a factor of > 20 in the three uptake functions. The vastly differing conditions at which destabilization occurs will be disconcerting to anyone using mathematical models as a predictive tool. This phenomenon of supersensitivity to model structure where analytical functions, which are seemingly identical for all practical purposes, can give completely different outcomes in terms of model dynamics has been described for several ecological models [19, 22–24].

To come up with an explanation, we note that logistic prey growth, $g(R)$ and resource uptake $f(R)$ are structurally very similar functions at small prey numbers R. This has consequences for the prey isocline $\hat{N}(R) = g(R)/f(R)$. Recall that in model (15.5) the stability of the equilibrium depends on the slope of $\hat{N}(R)$ at equilibrium. In the extreme case that the two functions are exactly identical, the isocline is a constant $\hat{N}(R) = 1$, and has slope zero everywhere. Thus, linear stability is not well defined and the system is structurally unstable. But assume that in some range close to the equilibrium point the two functions differ slightly $cf(R) = g(R) + \epsilon(R)$, where $\epsilon(R)$ is a small function. Now the isocline can be approximated as $\hat{N}(R) = c[1 - \epsilon(R)/g(R)]$ and in this range the sign of the slope of $\hat{N}(R)$ entirely depends on the difference $\epsilon(R)$. Therefore, arbitrary small deviations of resource uptake $f(R)$ from the prey growth rate $g(R)$ determine the stability of the equilibrium.

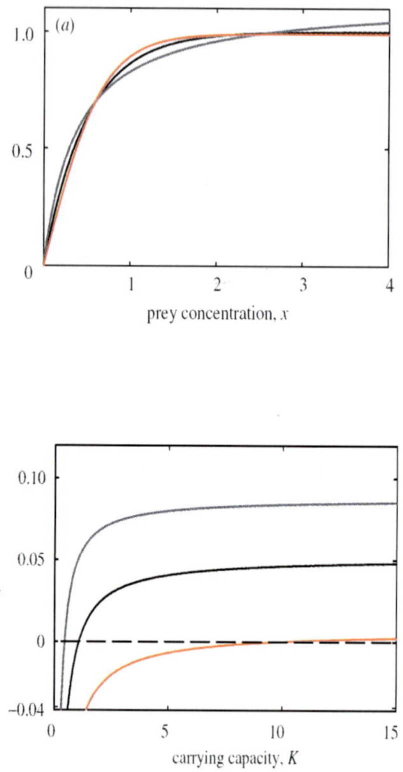

Fig. 15.5. Response to enrichment in the Rosenzweig-MacArthur model. Top: Three nearly congruent resource uptake curves. Black: "Ivlev"; blue: "Holling"; red: "trigonometric". Nonlinear least squares fits to Ivlev's response with $k_I = 1$, $\kappa_I = 2$. (Holling: $k_H = 3.05$, $\kappa_H = 2.68$, trigonometric: $k_T = 0.99$, $\kappa_T = 1.484$). Bottom: stability analysis. Real part, τ, of the largest eigenvalue of the community matrix vs. the carrying capacity K for different uptake curves. Positive values of τ indicate an unstable equilibrium. Other parameters: $a = 1$, $b = 0.1$, $\chi = 1$.

We emphasize that the whole argument relies on the fact that in the relevant interval the prey growth $g(R)$ and resource uptake $f(R)$ are structurally very similar functions. In the model (15.5) this is always the case for small prey numbers because both $g(R)$ and $f(R)$ start from 0 as a negatively curved function (as becomes apparent from a Taylor expansion of $f(R)$ up to second order). Thus, whenever the equilibrium is at small prey levels, e.g. for small mortality m (as in Fig. 15.5), the model is sensitive to

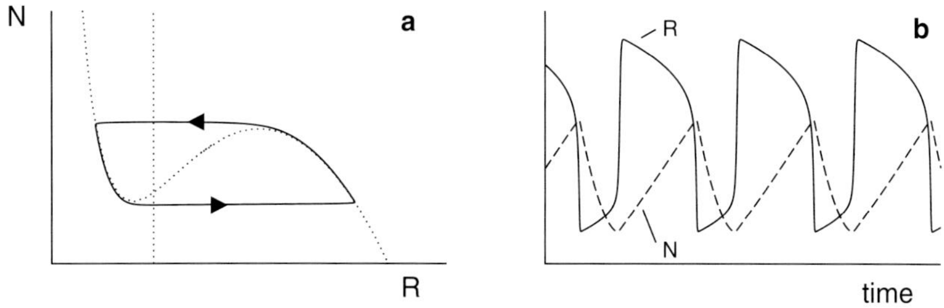

Fig. 15.6. Typical dynamics of model (15.7) in the oscillatory regime. a) Phase portrait in the (R, N)-phase plane showing the limit cycle (solid line) and the isoclines (dotted lines). Note, that the prey isocline is a 'cubic' function. b) Time series of prey (solid line) and predator (dashed line) show recurrent population outbreaks.

minor variations in the form of the functional response curve. In contrast, if the equilibrium is at large value of R, where prey growth and resource uptake are significantly different functions, the effect of sensitivity to model structure is not observed.

The same mechanism lies at the heart of the above mentioned major drawback of the original Lotka-Volterra model (15.1), which can be seen as the special case of the general model (15.5) where the prey growth and resource uptake are linear functions, i.e. $g(R) = aR$ and $f(R) = bR$. In this model per definition $g(R)/f(R) = \text{const}$, which leads to structural instability in the whole parameter range. With the introduction of nonlinear logistic prey growth and a saturating functional response Rosenzweig and MacArthur tried to circumvent these problems. While this works out for most parameter ranges the same difficulties of sensitivity to infinitesimal variations in the model structure are still inherent in the general model (15.5).

15.2.3. *Excitable systems and population outbreaks*

So far, we have only considered type-II functional responses, where the function $g(R)$ is saturating with negative curvature over the whole range of prey densities. In many cases, however, it has been observed that the food-uptake (or predation rate) is switched on only when the prey numbers exceed a certain, critical density. In such cases the functional response becomes a sigmoidal function (type-III functional response). This can be realized for example with Holling type-III function $g(R) = bN\, R^\alpha/(\kappa^\alpha + R^\alpha)$. If $\alpha = 1$ this reduces to a function of type-II, however, for $\alpha > 1$ the functional response is sigmoidal. Typically values for α are even integer

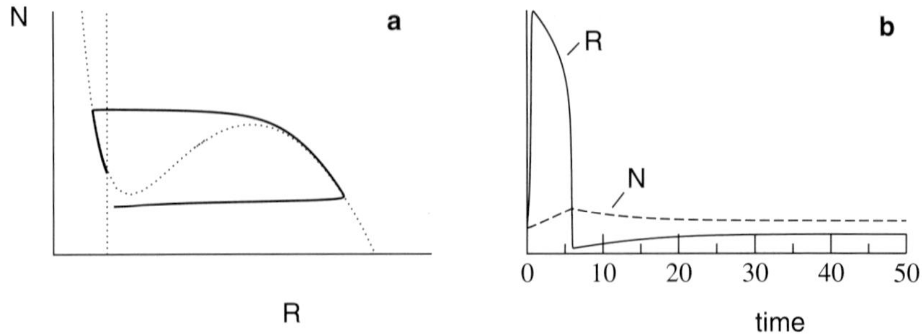

Fig. 15.7. Model (15.7) in the excitatory regime. a) Compared to Fig. 15.6 the predator isocline is shifted to the left so that the intersection is in the decreasing portion of the prey isocline. b) Time series of prey (solid line) and predator (dashed line) show a single population outbreak after a small perturbation from the stable fixed point.

numbers such as $\alpha = 2$ or $\alpha = 4$.

The major new property, when including a type-III food uptake in the predator prey model (15.5) is that the prey isocline becomes a "cubic-like" function. Again, the dynamics depend on the exact location of the intersections of predator- and prey-isoclines. If the intersection is in one of the two decreasing branches of the prey isocline (see Fig. 15.7), the model exhibits a stable fixed point. Otherwise, the fixed point becomes unstable giving rise to limit cycle oscillations (see Fig. 15.6).

The phase portraits in Figs. 15.6 and 15.7 are reminiscent to those of other excitable systems. This is even more so if the characteristic time scales of the prey are fast compared to that of the predators (as it is typically the case, since prey species usually have smaller biomass than their predators). The time scale separation can be made explicit in the model with the help of a new dimensionless parameter ϵ. This leads to the following excitatory predator-prey model

$$\dot{R} = \frac{1}{\epsilon}\left[aR(1-\frac{R}{K}) - kN\frac{R^\alpha}{\kappa^\alpha + R^\alpha}\right]$$
$$\dot{N} = -bN + \chi kN\frac{R^\alpha}{\kappa^\alpha + R^\alpha}. \qquad (15.7)$$

Depending on the intersection of the isoclines in phase space, the model can be in the oscillatory or in the excitatory regime. In the excitatory regime, a perturbation from the fixed point may result in a large excursion in phase space, before the stable fixed point is re-approached (Fig. 15.7). In ecological terms, the prey species undergoes a single outbreak in population numbers, which is triggered by a small perturbation. In contrast, in the oscillatory regime, the systems shows robust relaxation oscillations and perturbations have nearly no influence on the dynamics. This is shown in (Fig. 15.6) where the prey undergoes recurrent population outbreaks.

Model model (15.7) has originally been introduced to model insect pest outbreaks of the spruce budworm *Choristoneura fumiferana* [25]. This forest defoliator lives in the spruce and fir forests of north US and Canada. The budworm population can be either in small numbers (rest state) which are under predatory control by birds. Around approximately every 40 years it comes to an outbreak in the budworm numbers (excited state). Then the insects develop to a pest that can defoliate the trees with enormous environmental and economic damage.

The requirements to describe population outbreaks with a predator-prey model (15.5) are the following: (i) sigmoidal (type-III) functional response; (ii) top-down control, i.e., the pest population is the prey species of the model; (iii) time scale separation, where the changing rate of the prey is much faster, than that of the predator. As emphasized above, if these three conditions are met, model (15.7) shows all the generic properties that are well known also from other excitable systems [26]. In particular, under the influence of stochastic fluctuations the model will exhibit noise induced transitions [26] such as coherence resonance [27].

15.2.4. *Three-trophic foodchains and chaos*

Many population cycles have the unusual property that their period length remains remarkably constant while their abundance levels are highly erratic. Fig. 15.8a demonstrates these features for one of the most celebrated time series in Ecology - the Canadian hare-lynx cycle. In [28, 29] is was shown that such more complex oscillations can be achieved in simple predator-prey models by including more trophic levels. To describe the main dynamics of the lynx the following ecological foodweb model was presented

$$\begin{aligned}\dot{u} &= a\,(u-u^\star) \;-k_1\,f_1(u,v)\\ \dot{v} &= -b\,(v-v^\star) \;+k_1\,f_1(u,v) - k_2\,f_2(v,w)\\ \dot{w} &= -c\,(w-w^\star) \qquad\qquad\qquad\quad + k_2\,f_2(v,w).\end{aligned} \tag{15.8}$$

The model describes a three level vertical food-chain, where the resource or vegetation u is consumed by herbivores v, which in turn are prey on by top predators w. The parameters a, b and c represent the respective growth rates of each species in the absence of interspecific interactions ($k_1 = k_2 = 0$). The functions $f_i(x, y)$ describe interactions between

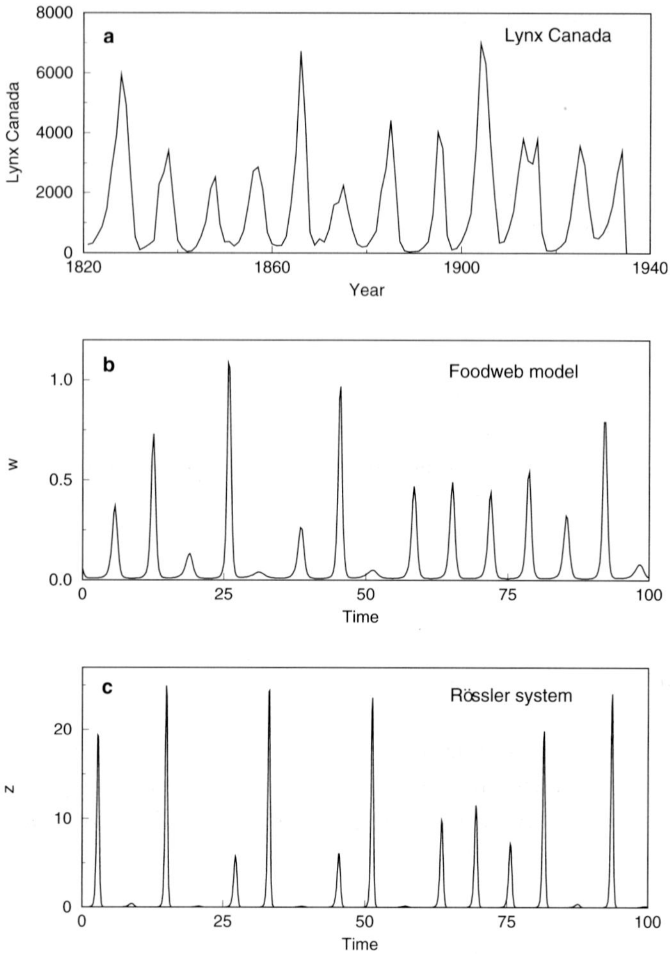

Fig. 15.8. (a) Ten year cycle in the Canadian lynx in the Mackenzie river area [2]. (b) Chaotic oscillations of the top predator w in the foodweb model (15.8). For simulation runs we used the following parameters: $a=1$, $b=1$, $c=10$, $k_1=0.2$, $k_2=1$, $\kappa_1=0.05$, $\kappa_2=0$, $w^\star=0.006$. (c) Time series in the Rössler system (15.9) with $a=0.2$, $b=1$, $c=0.2$ and $\gamma=5.7$.

species with strength k_i. Predator-prey and consumer-resource interactions are incorporated into the equations via either the Lotka-Volterra term $f_i(x,y)=xy$, or the Holling type II term $f_i(x,y)=xy/(1+\kappa_i x)$. We also assume the existence of a (stable or unstable) fixed point (u^*,v^*,w^*) in the absence of species interactions, and expand the system linearly around this steady state. Note, that here the steady state (u^*, v^*, w^*) is not necessarily set at the origin. This, for example, allows for the predator w to maintain a low equilibrium level $w = w^\star$ even during times when its usual prey v is rare, a case which might arise when there are alternative food sources available for the predator.

Despite its minimal structure, the equations capture complex dynamics including equilibrium and limit cycle behaviour, as well as large parameter ranges for which there are well defined chaotic oscillations (Fig. 15.9). Fig. 15.8b provides a time-series of a typical model run in the phase coherent chaotic regime. Observe that the top predator w, and indeed all model populations, oscillate with a nearly constant frequency although the maximum or peak amplitude in each cycle is highly unpredictable. These twin features of Uniform Phase evolution and Chaotic Amplitudes (UPCA) endow the model with a strong resemblance to the Canadian hare-lynx system. The time series of the foodweb model also resembles the UPCA found in the standard phase coherent Rössler system [30] (see Fig. 15.8c):

$$\dot{x} = -b(y+z), \quad \dot{y} = bx + ay, \quad \dot{z} = c + z(x - \gamma). \tag{15.9}$$

In order to study the oscillation frequency of the chaotic model it is important to develop a means for decomposing a chaotic signal into its phase and amplitude components. This is non-trivial for chaotic systems where there is often no unambiguous definition of phase. In our case, the motion always shows phase coherent dynamics, so that a phase can be defined as an angle in (x, y)-phase plane or via the Hilbert-transform [31]. Here, we use an alternative method which is based on counting successive maxima, that allows analysis even if the signal is "spiky". In this scheme we estimate the instantaneous phase $\phi(t)$ by counting successive maxima, e.g. we locate the times t_n of the n'th major local maxima of, say, the top predator population w. We define that the phase increases by 2π between two successive maxima and interpolate linearly in between for intermediate times [32]

$$\phi(t) = 2\pi \frac{t - t_n}{t_{n+1} - t_n} + 2\pi n, \quad t_n < t \le t_{n+1}. \tag{15.10}$$

Given the phase evolution, the rotation frequency is then given as the long time average of phase velocity, $\omega = \langle \dot{\phi}(t) \rangle$. As indicated in Fig. 15.9c the frequency of the foodweb model's cycle is a monotonically increasing function for almost the entire range of the control parameter b. In fact, the frequency of the foodweb model is largely determined by the underlying

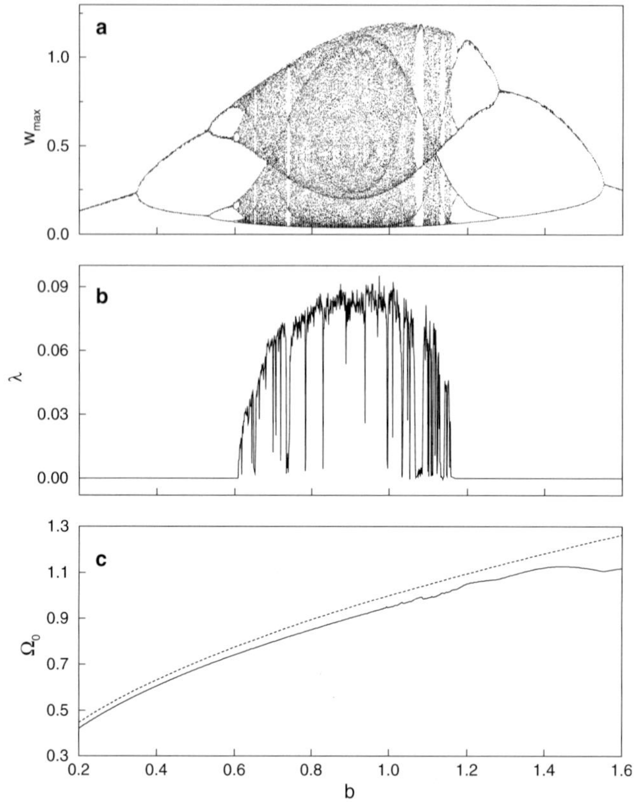

Fig. 15.9. Dynamics of the foodweb model (15.8) in the phase coherent regime as a function of the control parameter b. (a) Bifurcation diagram, plotted are the maxima of w; (b) largest Lyapunov exponent λ; (c) mean frequency ω (solid line). Further indicated is the approximation $\omega_0(b) = \sqrt{b}$ (dotted line).

Lotka-Volterra cycle in the (u, v) plane, whose intrinsic frequency equals $\omega_0 = \sqrt{ab}$. Fig. 15.9c shows that this simple formula gives an excellent estimation for the mean frequency of the chaotic three variable system.

15.3. Spatially coupled systems

15.3.1. *Coupled foodwebs*

The population models which were presented in the previous sections describe a single isolated "patch" or community. We now add spatial structure by analysing a set of patch models which are interconnected by diffusive

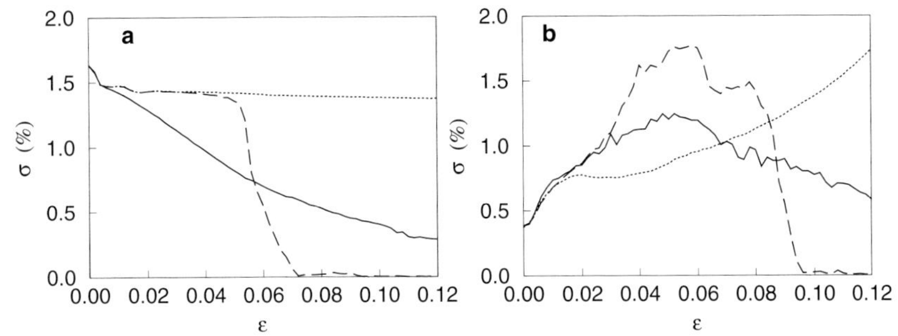

Fig. 15.10. Standard deviation of frequencies $\sigma(\epsilon)$ in a population of 500 coupled oscillators; (a) Rössler system (15.9) with $a = 0.15$, $c = 0.4$, $\gamma = 8.5$, and (b) foodweb model (15.8) with $a = 1$, $u^\star = 1.5$, $v^\star = 0$, $w^\star = 0.01$, $k_1 = 0.1$, $k_2 = 0.6$, $c = 10$ and $\kappa_{1,2} = 0$. Oscillators have been coupled in the y variable, $C = \text{diag}(0, 1, 0)$, with strength ϵ to either next neighbours in a ring with periodic boundaries (solid line), with global coupling (dashed line), or using approximation (15.12) (dotted line). Parameters b_i were taken as uniformly distributed random numbers in the range 0.97 ± 0.025.

migration of strength ϵ to form what might constitute a "metacommunity". This leads us in general to consider a system of N coupled oscillators

$$\dot{\mathbf{x}}_i = \mathbf{F}(\mathbf{x}_i, \chi_i) + \epsilon C \sum_{j \in N_i} (\mathbf{x}_j - \mathbf{x}_i), \quad i = 1 \cdots N. \tag{15.11}$$

In the absence of coupling each autonomous oscillator, $\mathbf{x}_i \in \mathbb{R}^n$, follows its own local (predator-prey) dynamics $\dot{\mathbf{x}}_i = \mathbf{F}(\mathbf{x}_i, \chi_i)$ which we assume to be either a limit cycle or phase coherent chaos. The oscillators are coupled by local dispersal with strength ϵ over a predefined set N_i of next neighbours and using the diagonal coupling matrix $C = \text{diag}(c_1, c_2 \ldots, c_n)$.

It is a natural assumption, that local communities in habitats at different geographic locations should vary in their local growth conditions. Therefore, quenched disorder is imposed onto the system by assigning to each local model i an independent set of control parameters $\chi_i = (a_i, b_i ..)$ (usually taken to be the growth rates), which affect the natural frequency $\omega_i = \omega(\chi_i)$ at each patch. For example, in the previously studied predator-prey models the natural frequency ω_i of each uncoupled community is, to a good approximation, proportional to the square root of the prey's growth rate $\sqrt{b_i}$. This natural disorder in local growth rates leads to a frequency mismatch between the oscillators.

The question arises as to whether or how these local communities might mutually synchronize after coupling. In general, synchronization arises as an interplay of the interaction and the frequency mismatch between the oscillators. Thereby, in general, all frequencies $\Omega_i = \Omega_i(\epsilon)$ will be detuned from the natural frequency, i.e. $\omega_i = \Omega_i(0)$. Here, we denote the observed oscillator frequency in the presence of coupling with a capital $\Omega_i(\epsilon_i)$ in contrast to the natural frequency ω_i of the uncoupled oscillator, i.e. $\omega_i = \Omega_i(0)$. It is convenient to measure the amount of synchronization with the standard deviation of all oscillator frequencies, $\sigma(\epsilon)$. Phase synchronization refers to the fact that with sufficient coupling strength $\epsilon > \epsilon_c$ all oscillators rotate with the same frequency and implies $\sigma(\epsilon) = 0$.

It is long known that phase synchronization arises naturally in two interacting limit cycle systems [32], but it is also possible in two coupled phase coherent chaotic oscillators by maintaining chaotic amplitudes [31]. To study the synchronization in the disordered systems (15.11) we compare the transition to synchronization in ensembles of two phase coherent chaotic oscillators, namely the Rössler system (15.9) and the chaotic predator-prey model (15.8). Both systems have a free parameter b_i which is taken for each oscillator from the same statistical distribution. Despite the fact that both models have a very similar attractor topology, we find fundamental differences in their response to the interaction (see Fig. 15.10). In the ensemble of Rössler systems we observe the usual onset of synchronization where the frequency disorder $\sigma(\epsilon)$ is a monotonically decreasing function of coupling strength. In contrast, the ensemble of foodweb models (15.8) shows a totally different behaviour. Here, with increasing coupling strength $\sigma(\epsilon)$ is first amplified and synchronization sets in only for much larger coupling with a maximal decoherence for intermediate values of ϵ. This counterintuitive increase of disorder with coupling strength has been denoted as anomalous phase synchronization and has been demonstrated to arise naturally in a large class of oscillator types and coupling topologies [33–35].

It is possible to describe the frequency disorder in the regime of weak global coupling by noting that in the absence of coupling the oscillators are rotating independently of each other. With the onset of weak coupling one can safely assume that the oscillators (15.11) remain independent. Thus, for $\epsilon \ll 1$ and in the thermodynamic limit the ensemble average is constant in time, $\langle \mathbf{x}_j \rangle = \xi$, and we can approximate the interacting system as a system of N uncoupled oscillators with modified dynamics (see Fig. 15.10)

$$\dot{\mathbf{x}}_i = \mathbf{F}(\mathbf{x}_i, \chi_i) - \epsilon C(\mathbf{x}_i - \xi). \tag{15.12}$$

Consequently, for $\epsilon \ll 1$ the frequency detuning of each oscillator i depends only on its own parameters, $\Omega_i = \Omega(\chi_i, \epsilon) \approx \omega(\chi_i) + \epsilon\, \kappa(\chi_i) + O(\epsilon^2)$, where $\kappa(\chi_i)$ describes the frequency response of each oscillator to the onset of

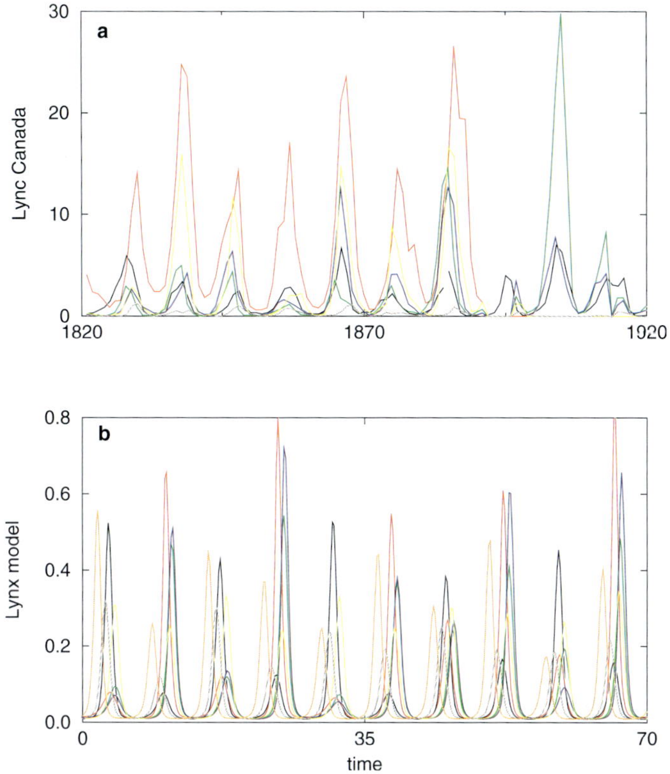

Fig. 15.11. (a) Time series of lynx numbers in six different regions in Canada (after [2]) provide an excellent example of phase synchronization in an ecological data set. (b) Time series of predators w_i in a 20×20 lattice with next neighbor coupling ($\epsilon = 0.035$) and periodic boundary conditions. Plotted are the simulation results for seven patches along the lattice diagonal.

interaction. The crucial fact is that the effective functions $\omega(\chi_i)$ and $\kappa(\chi_i)$, in general, are not functionally independent, with the consequence that the frequency disorder $\sigma(\epsilon)$ is determined by the covariance between the values of ω_i and κ_i over the ensemble, which finally can give rise to anomalous synchronization [33, 34].

As shown in Fig. 15.10 in a spatial lattice of patches, only small levels of local migration are required to induce broad-scale phase synchronization . The result of the simulation in the phase synchronized regime is visualized in Fig. 15.11, which demonstrates that all populations in the lattice are phaselocking to the same collective rhythm. Similar to the synchronized oscillations of the Canadian lynx, also in the lattice simulation despite the

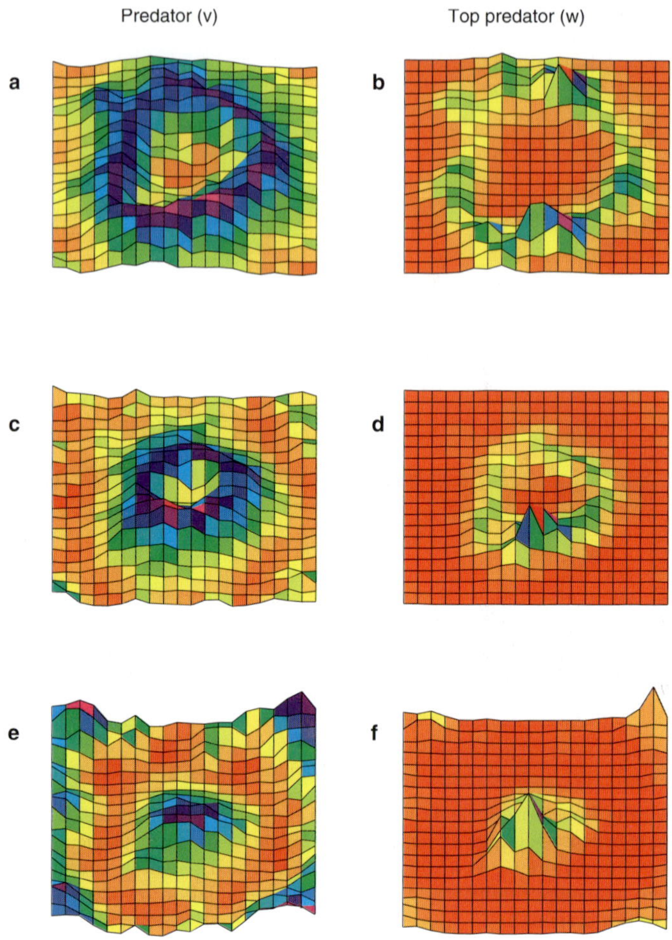

Fig. 15.12. Evolution of a chaotic target wave. Snapshots of predators v_i (left column) and top predators w_i (right column) in a 20×20-lattice (Fig. 15.11b) for three consecutive time instances. Abundance levels are color coded.

strong phase locking the peak population abundances remain chaotic and largely uncorrelated ($r < 0.2$).

Note, that for spatially extended systems full synchronization leads only to trivial spatial patterns, since phase and amplitude dynamics are then identical across the entire lattice. In the region of phase synchronization, however, synchronized patch populations are typically separated by phase lags (as seen in Fig. 15.11), which when summed up over the whole lattice can give rise to complex spatio-temporal patterns. Most remarkably, in the coupled foodchain model we find a coherent regular traveling wave structure where population abundances remain chaotic, but unusual circular waves form and spread in time across the landscape (see Fig. 15.12). The wave pattern repeats in an endless cycle, with patches having chaotic amplitudes, making each cycle different from the rest [28, 29].

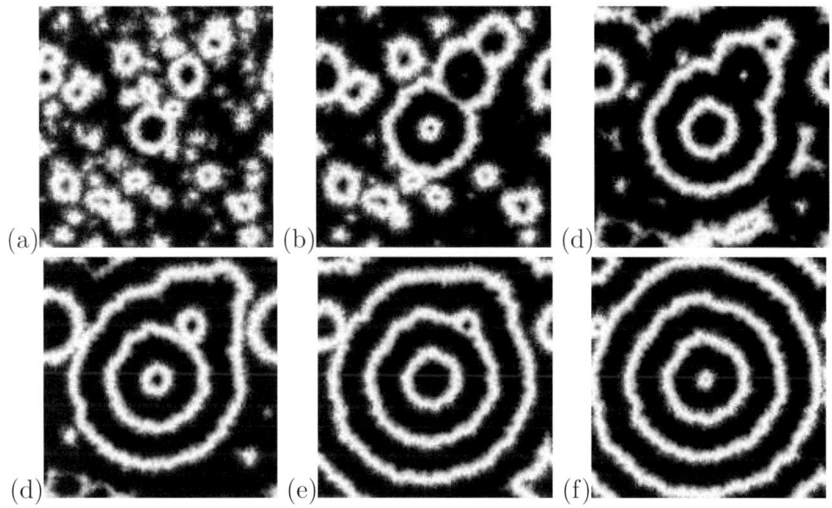

Fig. 15.13. Spontaneous formation of quasi-regular concentric waves in a heterogenous medium. Plotted are the simulation results in a 2-dimensional lattice (15.11) of 150x150 predator-prey oscillators (15.6) with periodic boundary conditions at 6 consecutive time instances (from a to f) starting from homogeneous initial conditions. Plotted is the density of the prey as grey level. Parameters: $K = 3$, $k = 3.5$, $\chi = 1$, $K_M = 1$, $b = 1$, $\epsilon = 0.1$, $C = \mathrm{diag}(1, 1)$. Growth rates a_i are taken from a uniform distribution in the range 5 ± 0.4.

As demonstrated in Fig. 15.13, these concentric target waves do not result from the chaotic dynamics, but are present in a very similar form also in disordered limit-cycle predator-prey models. This observation suggests that the origin of the target waves may be found in the intrinsic heterogeneity of the considered spatial models, which will be discussed in detail below.

15.3.2. *Phase models*

Target waves are one of the most prominent patterns in oscillatory media and are usually associated with the presence of local impurities in the system [36–38]. These pacemakers change the local oscillation frequency and are able to enslave all other oscillators in the medium, which finally results in regular ring waves [37, 39, 46]. However, the assumption of a discrete set of localized pacemaker regions in an otherwise homogeneous medium is somewhat artificial. Especially biological systems are often under the constraint of large heterogeneity. In such a disordered system no point can clearly be distinguished as a pacemaker and it is not obvious whether such a system can sustain highly regular target patterns and where they should originate.

The emergence of target patterns in heterogeneous oscillatory media was first reported and explained in [40] and subsequently observed in [28]. In [41] it was shown that the random nature of the medium itself plays a key role in the formation of the patterns. As the disorder in a rather homogeneous synchronized medium is increased one observes the formation of quasi regular target waves, which result from an intricate interplay between the heterogeneity and a symmetry breaking of the coupling function.

Consider a system of N coupled phase oscillators [37]

$$\dot{\theta}_i = \omega_i + \epsilon \sum_{j \in N_i} \Gamma(\theta_j - \theta_i), \quad i = 1, \ldots, N. \tag{15.13}$$

Here, θ_i represents the phase of oscillator i, which is coupled with strength ϵ to a set of nearest neighbors N_i in a one- or two-dimensional lattice. The natural frequencies ω_i are fixed in time, uncorrelated and taken from a distribution $\rho(\omega)$. A scaling of time and a transformation into a rotating reference frame can always be applied so that $\epsilon = 1$ and the ensemble mean frequency $\overline{\omega}$ is equal to zero. We refer to the variance $\sigma^2 = \mathbf{var}(\omega_i)$ of the random frequencies as the disorder of the medium.

The effects of coupling are represented by an interaction function Γ which, in general, is a 2π-periodic function of the phase difference with $\Gamma(0) = 0$. For weakly coupled, weakly nonlinear oscillators Γ has the universal form [37, 42]

$$\Gamma(\phi) = (\sin(\phi) + \gamma[1 - \cos(\phi)]) \,. \tag{15.14}$$

The symmetry breaking parameter γ describes the nonisochronicity of the oscillations [37].

It is well known that for sufficiently small disorder the oscillators eventually become entrained to a common locking frequency Ω [32, 37, 43]. Since the oscillators are nonidentical, even in this synchronized state they are usually separated by fixed phase differences. These can sum up over the whole lattice to produce spatio-temporal patterns, which are characterized by a stationary phase profile, $\theta_i - \theta_1$. This is demonstrated in Fig. 15.14, where we have simulated system (15.13) with interaction (15.14) in a two-dimensional lattice. Fig. 15.14, left demonstrates the well known formation of a target wave from a central pacemaker region of locally faster natural frequencies. However, in the right panel of Fig. 15.14 it is shown that very similar waves can arise in a fully disordered system. It is a counterintuitive observation that the isotropic medium with random frequency distributions of no spatial correlation can generate and sustain very regular wave patterns.

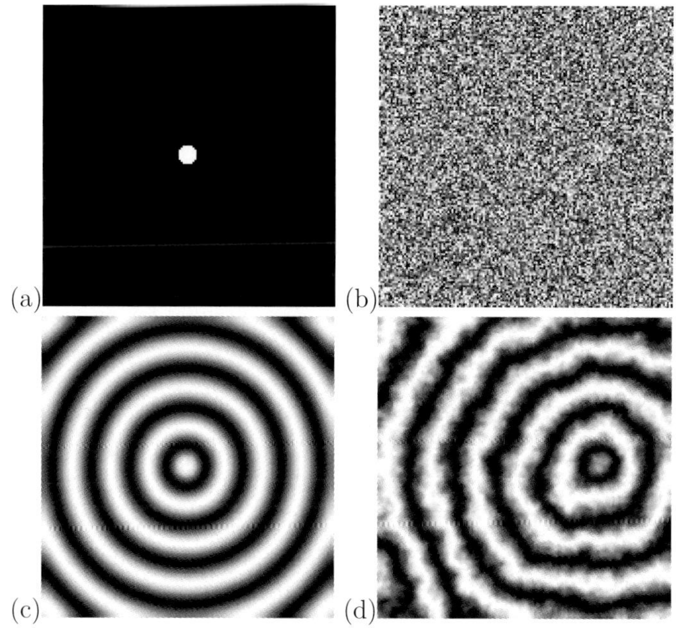

Fig. 15.14. Simulation results in a 2-dimensional lattice of 150x150 phase oscillators (15.13,15.14) with nearest neighbor coupling, periodic boundary conditions, $\epsilon = 1$, $\gamma = 1$ starting from homogeneous initial conditions. Left column: results with a central pacemaker region of local frequencies $\omega_i = 0.3$, while the rest of the system is homogeneous with $\omega_i = 0.1$. Right: heterogenous system, where the random frequencies are taken from a uniform distribution of variance σ^2 with $\sigma = 0.3$. Top row: the corresponding natural frequencies ω_i are indicated in grey levels. Bottom row: sine of the simulated phases θ_i as grey level.

In the synchronized state all oscillators rotate with the constant locking frequency $\dot{\theta}_i = \Omega$, so that system (15.13) becomes a set of N equations, which have to be solved self consistently for the phases θ_i and Ω under some imposed boundary conditions. To determine Ω suppose first that the coupling function Γ is fully antisymmetric $\Gamma(-\phi) = -\Gamma(\phi)$, e.g. $\gamma = 0$ in Eq.(15.14). In this case, by summing up all equations in (15.13) we obtain $\Omega = \overline{\omega} = 0$ in the rotating frame. Thus, nontrivial locking frequencies $\Omega \neq 0$ only arise if $\Gamma(\phi)$ has a symmetric part $\Gamma_S(\phi) = \frac{1}{2}(\Gamma(\phi) + \Gamma(-\phi))$,

$$\Omega = \frac{1}{N} \sum_{i,j \in N_i} \Gamma_S(\theta_j - \theta_i). \tag{15.15}$$

For any coupling function Γ given a realization of the natural frequencies ω_i we ask for the resulting phase profile θ_i. Note, that the inverse problem is easy to solve: for any regular phase profile θ_i we can calculate Ω from Eq. (15.15), which after inserting into Eq.(15.13) yields the frequencies ω_i.

Insights into the pattern formation can be gained from a one-dimensional chain of phase oscillators [40, 44]

$$\Omega = \omega_i + [\Gamma(\phi_i) + \Gamma(-\phi_{i-1})]. \tag{15.16}$$

Here we use $\phi_i = \theta_{i+1} - \theta_i$ for the phase differences between neighboring oscillators, $i = 1 \ldots N-1$. Further, we assume open boundary conditions $\phi_0 = \phi_N = 0$. The self consistency problem is trivial for an antisymmetric Γ where $\Omega = 0$. In this case the $\Gamma(\phi_i)$ simply describe a random walk $\Gamma(\phi_i) = -\sum_{j=1}^{i} \omega_j$. Thus, for small $|\phi_i|$ the phase profile θ_i is essentially given by a double summation, i.e. a smoothening, over the disorder ω_i (see Fig.15.15a,b). Note, that synchronization can only be achieved as long as the random walk stays within the range of Γ. Thus, with increasing system size N synchronization becomes more and more unlikely.

The emergence of target waves is connected to a breaking of the coupling symmetry. To explore this we study a unidirectional coupling with respect to ϕ

$$\Gamma(\phi) = f(\phi)\,\Theta(\phi), \quad \text{for } |\phi| \ll 1, \tag{15.17}$$

with the Heaviside function $\Theta(\phi)$ and $f(\phi > 0) > 0$. Here, the phase of oscillator i is only influenced from neighboring oscillators which are ahead of i. If the solution are small phase differences we are not concerned about the periodicity of $\Gamma(\phi)$ as the coupling is only required to be unidirectional close to zero. For open boundaries $\phi_0 = \phi_N = 0$ the solution to (15.16,15.17) is given by

$$\phi_i = \begin{cases} f^{-1}(\Omega - \omega_i), & i < m \\ -f^{-1}(\Omega - \omega_{i+1}), & i \geq m. \end{cases} \tag{15.18}$$

Here, the index m is the location of the oscillator with the largest natural frequency, which also sets the synchronization frequency

$$\Omega = \omega_m = \mathbf{max}_i(\omega_i). \tag{15.19}$$

The phase differences (15.18) are positive to the left of the fastest oscillator, $i < m$, and negative to the right $i \geq m$. As a consequence, the phase profile has a tent shape with a mean slope that is given by averaging (15.18) with respect to the frequency distribution (Fig.15.15 c,d). We call this solution type a quasi-regular concentric wave. This example illustrates, that the asymmetry of the coupling function increases the influence of faster oscillators and effectively creates pacemakers with the potential to entrain the whole system. Note, that the solution (15.18,15.19) is not possible without disorder, i.e. for $\sigma = 0$. Further, in contrast to the antisymmetric coupling, here synchronization can be achieved for chains of arbitrary length.

In general, the coupling function Γ will interpolate between the two extremes of fully antisymmetry and unidirectional coupling in the vicinity of zero, e.g. Eq. (15.14) with $\gamma \neq 0$. As shown in Fig.2e this also gives rise to quasi regular concentric waves, very similar to the exactly solvable system Fig.2c. To further investigate the origin of these patterns note that for any given Ω system (15.16) implicitly defines two transfer maps, T_Ω : $\{\phi_{i-1}, \omega_i\} \mapsto \phi_i$ and T_Ω^{-1} : $\{\phi_i, \omega_i\} \mapsto \phi_{i-1}$, which describe the evolution of the phase differences into the right or the left direction of the chain, respectively. The random frequencies ω_i can be seen as noise acting on the map (see Fig.2f)

$$\phi_i = T_\Omega(\phi_{i-1}, \omega_i) = \Gamma^{-1}\left[\Omega - \omega_i - \Gamma(-\phi_{i-1})\right]. \tag{15.20}$$

It is easy to see that the breaking of symmetry leads to a pair of fixed points, ϕ^* and $-\phi^*$, in the noisy maps

$$\phi^* = T_\Omega(\phi^*, \overline{\omega}) = \Gamma_S^{-1}\left(\tfrac{\Omega}{2}\right). \tag{15.21}$$

The transfer map can be linearized at the fixed points so that $T_\Omega(\phi^* + \zeta, \omega) \approx \phi^* + a\zeta - b\omega$ with $a = \frac{\Gamma'(-\phi^*)}{\Gamma'(\phi^*)}$ and $b = \frac{1}{\Gamma'(\phi^*)}$. While one fixed point, ϕ^* in the case (15.14) with $\gamma > 0$, is linearly stable ($|a| \leq 1$) the other fixed point is necessarily unstable. These stability properties are inverted for T_Ω^{-1}. Thus, when iterating to the right of the chain the ϕ_i are concentrated around ϕ^* and around $-\phi^*$ when iterating to the left. As a consequence, the general solution of the selfconsistency problem (15.16) is build up from two branches around the two fixed points $\pm\phi^*$, superimposed

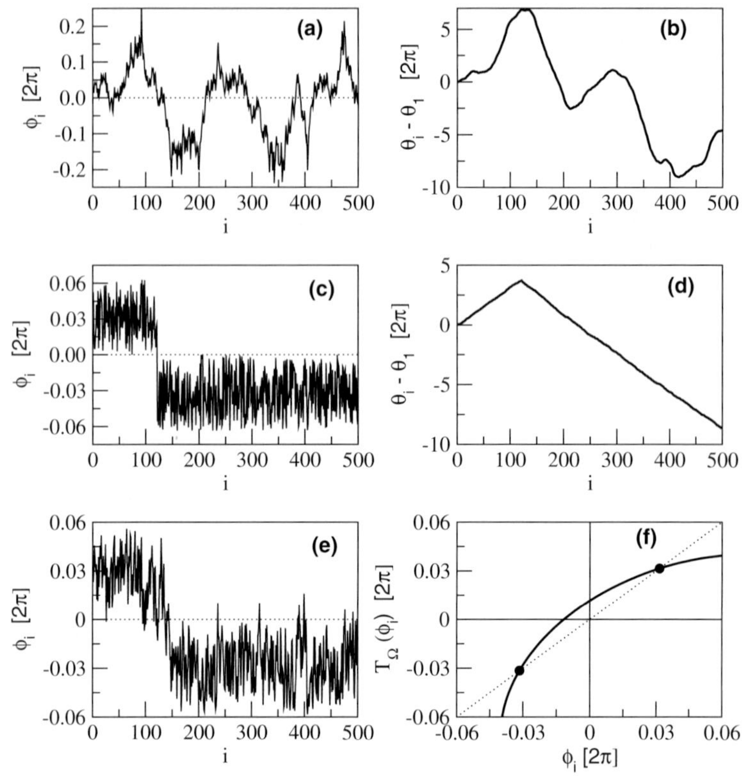

Fig. 15.15. Phase differences ϕ_i (left) and phase profile $\theta_i - \theta_1$ (b,d) in units of 2π for a chain of 500 phase oscillators (15.16) with uniformly distributed frequencies $\omega_i \in [-0.1, 0.1]$ and open boundaries. (a,b) Antisymmetric coupling (15.14) with $\gamma = 0$. (c,d) unidirectional coupling (15.17) with $\Gamma(\phi) = \phi\,\Theta(\phi)$. (e) System (15.14) with $\gamma = 2$. (f) Transfer map $T_\Omega(\phi_i)$ (15.20) for Eqs. (15.14) with $\gamma = 2$ (solid line) and fixed points ϕ^* (filled circles).

by autocorrelated fluctuations ζ_i (see Fig.15.15)

$$\phi_i = \pm\phi^* + \zeta_i. \tag{15.22}$$

After summation this leads to the quasi regular tent shape of the phase profile θ_i. In first approximation the ζ_i describe a linear autoregressive stochastic process

$$\zeta_i = a\zeta_{i-1} - b\omega_i. \tag{15.23}$$

We want to stress that the fluctuations ζ_i are an essential ingredient of the solution. Although the emerging concentric waves seem to be regular the underlying heterogeneity of the system does not permit analytical traveling wave solutions $\theta_i(t) = \Omega t - k|i - m|$.

The general solution (15.22) allows for very different phase profiles (see Fig. 15.15). The regularity of the wave pattern depends on the relative influence of the mean slope ϕ^* compared to the fluctuations ζ_i and can be measured by the quality factor $Q_\phi = {\phi^*}^2/\mathbf{var}(|\phi_i|)$ and the autocorrelation r of the ϕ_i. As demonstrated in Fig.15.16, for sufficiently large systems, both Q_ϕ and r only depend on the product $\gamma\sigma$ (see below). For $\gamma\sigma \to 0$ we find $r \to 1$, and the solution is essentially a random walk (see Fig.2a,b). With increasing values of $\gamma\sigma$ the correlations r are reduced and eventually become negative. Furthermore Q_ϕ increases with the product $\gamma\sigma$, and for $\gamma > 1$ can rise drastically (see Fig. 15.16c). Thus, with increasing disorder of the system we obtain more regular patterns until synchronization is lost.

A straightforward integration of system (15.13) can be problematic due to the long transients. Another approach, which also applies for two dimensional lattices, relies on the Cole-Hopf transformation of system (15.13). Assume that the ϕ_i are small so that it is possible to approximate the coupling function (15.14) around zero by $\Gamma(\phi) = \frac{1}{\gamma}\left(e^{\gamma\phi} - 1\right) + O(\phi^3)$. After the Cole-Hopf transformation $\theta_i = \frac{1}{\gamma}\ln q_i$ the synchronized lattice (15.13) is reduced to a linear system [37, 38, 40]

$$\dot{q_i} = Eq_i = \gamma\sigma\eta_i \cdot q_i + \sum_{j\in N_i}(q_j - q_i) \tag{15.24}$$

where the random frequencies $\eta_i = \omega_i/\sigma$ are of zero mean and variance one and $E = \gamma\Omega$ is some eigenvalue. System (15.24) is known as the tight binding model for a particle in a random potential on a lattice [45]. The eigenvector $\mathbf{q}^{\max}$ corresponding to the largest eigenvalue E_{max} will, in the re-transformed system of angles, outgrow the contribution of all other eigenvectors to the time dependent solution linearly in time. If the largest eigenvalue is non degenerate, the unique synchronized solution is

$$\theta_i(t) - \theta_1(t) = \frac{1}{\gamma}\log\left(\frac{q_i^{\max}}{q_1^{\max}}\right). \tag{15.25}$$

Eq. (15.25) is well defined since the components of $\mathbf{q}^{\max}$ do not change sign. Anderson localization theory [45] predicts exponentially decaying localized states for one and two dimensional lattices with some localization length l, which after applying the reverse Cole-Hopf transformation yields the observed tent-shape phase profile with wavelength $\lambda \sim \gamma l$. Concentric waves emerge when λ becomes smaller than the system size.

For extremal values of $\gamma\sigma$ the system (15.24) has well defined scaling properties $E_{max} \sim (\gamma\sigma)^{\alpha}$ and $l \sim (\gamma\sigma)^{-\beta}$ [45]. Perturbation theory yields $\alpha = 2$ for $\gamma\sigma \ll 1$ and $\alpha = 1$ for $\gamma\sigma \gg 1$. For the exponent β we find $\beta \lesssim 1$ in one dimensional systems, while $1 \leq \beta \leq 2$ in two dimensional lattices. This implies for the synchronization frequency Ω and the wavelength λ

$$\Omega \sim \gamma^{\alpha-1}\sigma^{\alpha}, \qquad \lambda \sim \gamma^{1-\beta}\sigma^{-\beta}. \tag{15.26}$$

Here, γ does not influence the wavelength as much as σ but while an increase of γ in one dimension leads also to an increasing wavelength the effect in two dimensions is the opposite.

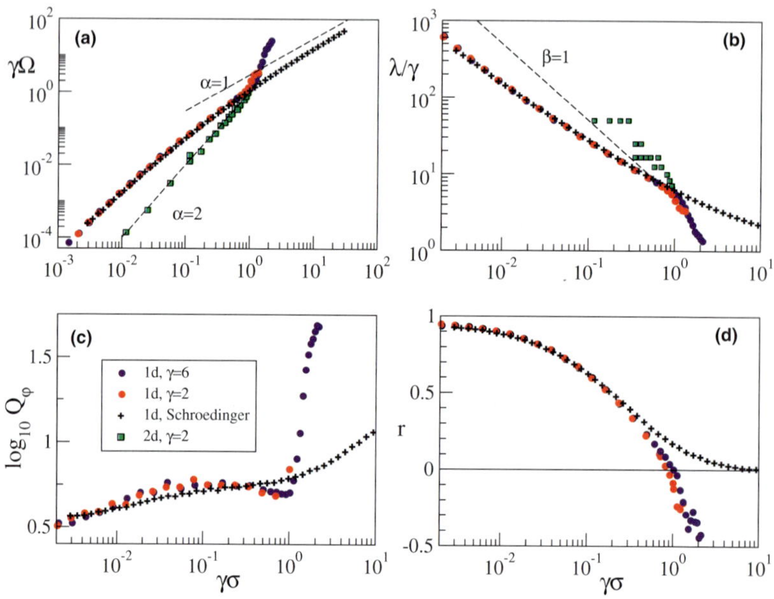

Fig. 15.16. Characterization of wave patterns by the locking frequency $\gamma\Omega$ (a), the wave length λ/γ (b), the quality factor Q_{ϕ} (c) and the cross correlation r between neighboring phase differences (d). Numerical solutions are obtained by integrating system (15.13,15.14) for one dimensional (circles) and two dimensional lattices (squares) in the synchronization regime. In one dimension the integrations were carried out with chains of length 500 and averaged over 50 simulation runs for $\gamma = 2$ (red circles) and $\gamma = 6$ (blue circles). In the two dimensional system each point (green squares) represents one single simulation in a 100x100 lattice with $\gamma = 2$. The results using the eigenvector method (15.24) are shown as (black +). Each point represents an average of 500 simulations with $N = 256$. Further, indicated in (a) (b) are straight lines with a given exponent α and β (dashed lines). The wavelength was obtained for one dimension as $\lambda = 2\pi/\phi^*$ and in the two dimensional system from a Fourier analysis of the phase profile. The plateaus in the wavelength plot are discretization effects.

In one dimension the scaling with $\gamma\sigma$ holds as long as Eq. (15.24) can approximate the transfer map reasonably. The approximation breaks down when the correlation r of the phase differences becomes negative. In this regime the quality factor Q_ϕ strongly depends on both $\gamma\sigma$ and γ. The noise term $b\omega_i$ in (15.23) can become small with increasing γ. This regime, which is not described by the Anderson approximation, can produce very regular concentric waves near the border of desynchronization. Our observables (Fig. 3) only depend on the system size for $N \lesssim 100$.

The constructive role of noise has often been studied [47]. It has been shown that in excitable systems spatial noise can enhance the pattern formation and for example is able to promote traveling waves [48]. In heterogeneous oscillatory media the emergence of target patterns was reported in [40]. Here, we have analysed the origin of these structures and, in particular, we have shown that fluctuations are essential for the observed dynamics. Whereas local coupling tends to synchronize the oscillators, the imposed disorder tends to desynchronize the array. The tension between these two opposing forces can give rise to quasi regular target patterns.

This work was supported by the German Volkswagen Stiftung and DFG (SFB 555).

References

[1] R. May, *Stability and Complexity in Model Ecosystems* (Princeton University Press, 1973).

[2] C. Elton & M. Nicholson. The ten-year cycle in numbers of the lynx in Canada. J. Anim. Ecol. **11**, 215-241 (1942).

[3] Keith, L. B. *Wildlife's ten year cycle.* (University of Wisconsin Press, Madison, 1963).

[4] Royama, T. *Analytical Population Dynamics.* (Chapman & Hall, London, 1992).

[5] Moran, P. A. P. The statistical analysis of the Canadian lynx cycle. Aust. J. Ecol. **1**, 291-298 (1953).

[6] E. Ranta, and V. Kaitala, Travelling waves in vole population dynamics. Nature **390**, 456 (1997).

[7] Johnson, D. M., Bjrnstad, O. N. & Liebhold, A. M. Landscape geometry and traveling waves in the larch budmoth. Ecology Letters **7** 967-974 (2004).

[8] O. N. Bjørnstad, R. A. Ims, and X. Lambin. Spatial population dynamics: analyzing patterns and processes of population synchrony. Trends in Ecology and Evolution **14** 427–432 (1999).

[9] Liebhold, A., Walter, D. K. & Bjørnstad, O. N. Spatial synchrony in population dynamics. Annual Review of Ecology and Systematics. **35** 467-490 (2004).

[10] Blasius, B., & Stone, L., Nonlinearity and the Moran effect. Nature, **406**, 846–847 (2000).

[11] Volterra V. Fluctuations in abundance of a species considered mathematically. Nature **118** 558-560 (1926).

[12] Lotka A. J. Undamped oscillations derived from the law of mass action. J. Amer. Chem. Soc. **42** 1595-1598 (1920).

[13] Armstrong, R.A. The effects of predator functional response and prey productivity on predator-prey stability: A graphical approach. Ecology **57**, 609-612 (1976).

[14] Rosenzweig, M.L. & R.H. MacArthur. Graphical Representation and Stability Conditions of Predator-prey interactions. Am. Nat. **97**, 209 (1963).

[15] Rosenzweig, M.L. Paradox of enrichment: destabilization of exploitation ecosystems in ecological time. Science **171**, 385-387 (1971).

[16] Gilpin, M.E. Enriched predator-prey systems: Theoretical stability. Science **177**, 902-904 (1972).

[17] May, R.M. (1972). Limit cycles in predator-prey communities. Science 177, 900-902.

[18] Persson, A., Hansson, L.A., Brönmark, C., Lundberg, P., Pettersson, L.B., Greenberg, L., Nilsson, P.A., Nyström, P., Romare, P., & Tranvik, L. Effects of enrichment on simple aquatic food webs. Am. Nat. **157** 654-669 (2001).

[19] Fußmann G.F. & Blasius B. Community response to enrichment is highly sensitive to model structure. Biol. Lett. **1** 9-12 (2005).

[20] Holling, C.S. The components of predation as revealed by a study of small-mammal predation of the European Pine Sawfly. The Canadian Entomologist **91**, 293-320 (1959).

[21] Ivlev, V.S. Experimental Ecology of the Feeding of Fishes. Yale University Press, New Haven (1961).

[22] Myerscough, M.R., Darwen, M.J. & Hogarth, W.L. 1996 Stability, persistence and structural stability in a classical predator-prey model. Ecol. Model. 89, 31-42.

[23] Wood, S.N. & Thomas, M.B. 1999 Super-sensitivity to structure in biological models. Proc. R. Soc. Lond. B. 266, 565-570.

[24] Gross, T., Ebenhöh, W. & Feudel, U. Enrichment and foodchain stability: the impact of different forms of predator-prey interaction. J. Theor. Biol. **227** 349-358 (2004).

[25] Ludwig, D., D.D. Jones & C.S. Holling Qualitative analysis of insect. outbreak systems: the spruce budworm and forest. J. Anim. Ecol. **47** 315-332 (1978).

[26] Lindner, B., García-Ojalvo, J., Neiman, A. & Schimansky-Geier, L. Effects of noise in excitable systems. Phys. Rep. **392** 321-424(2004).

[27] A.S. Pikovsky & J. Kurths. Coherence Resonance in a Noise-Driven Excitable System. Phys. Rev. Lett. **78** 775-778 (1997).

[28] B. Blasius, A. Huppert & L. Stone. Complex dynamics and phase synchronization in spatially extended ecological systems. Nature **399**, 354-359 (1999).

[29] B. Blasius & L. Stone. Chaos and phase synchronization in ecological sys-

tems. Int. J. Bif. and Chaos **10**, 2361-2380 (2000).
[30] O.E. Rössler, Phys. Lett. A **57**, 97 (1976).
[31] M.G. Rosenblum, A.S. Pikovsky, and J. Kurths, Phys. Rev. Lett. **76**, 1804 (1996).
[32] A.S. Pikovsky, M.G. Rosenblum, and J. Kurths, *Synchronization, a Universal Concept in Nonlinear Sciences* (Cambridge University Press, Cambridge, 2001).
[33] B. Blasius, E. Montbrió, & J. Kurths. Anomalous phase synchronization in populations of nonidentical oscillators. Phys. Rev. E **67**, 035204 (2003).
[34] E. Montbrio & B. Blasius. Using nonisochronicity to control synchronization in ensembles of non-identical oscillators. Chaos **13**, 291-308 (2003).
[35] B. Blasius. Anomalous phase synchronization in two asymmetrically coupled oscillators in the presence of noise. Phys. Rev. E **72** 066216 (2005).
[36] A. N. Zaikin and A. M. Zhabotinsky, Nature **255**, 535 (1970); P. S. Hagan, Adv. Appl. Math. **2**, 400 (1981).
[37] Y. Kuramoto *Chemical oscillations, waves and turbulence* (Springer, Berlin, 1985).
[38] A. S. Mikhailov, *Foundations of Synergetics I* (Springer, Berlin, 1994).
[39] A. S. Mikhailov and A. Engel, Physics Letters A **117**, 257 (1986).
[40] H. Sakaguchi, S. Shinomoto and Y. Kuramoto, Prog. of Theor. Phys. **79**, 1069 (1988).
[41] B. Blasius & R. Tönjes. Quasiregular concentric waves in heterogeneous lattices of coupled oscillators. Phys. Rev. Lett. **95** 084101 (2005).
[42] K. Koppel and G.B. Ermentrout, Comm. Pure and Appl. Math. **39**, 623 (1986); SIAM J. Appl. Math. **50**, 1014 (1990).
[43] S.C. Manrubia, A.S. Mikhailov, and D.H. Zanette, *Emergence of Dynamical Order: Synchronization Phenomena in Complex Systems* (World Scientific, Singapore, 2004).
[44] A. H. Cohen, P. J. Holmes and R. H. Rand, J. Math. Biology **13**, 345 (1982).
[45] P.W. Anderson, Phys. Rev. **109**, 1492 (1958) M. Janssen, Phys. Rep. **295**, 1 (1998).
[46] R. Kapral and K. Showalter (eds.) *Chemical Waves and Patterns* (Kluwer Academic Publishers, Dordrecht, 1995).
[47] T. Shinbrot and F. J. Muzzio, Nature **410**, 251 (2001).
[48] S. Kadar, J. Wang and K. Showalter, Nature **391**, 770 (1998).

Author Index

Subject Index